鸿蒙 HarmonyOS
手机应用开发实战

柳伟卫 / 著

清华大学出版社
北京

内 容 简 介

华为自主研发的 HarmonyOS（鸿蒙系统）是一款面向未来、面向全场景（移动办公、运动健康、社交通信、媒体娱乐等）的分布式操作系统。借助 HarmonyOS 全场景分布式系统和设备生态定义全新的硬件、交互和服务体验。本书采用新的 HarmonyOS 2 版本作为基石，详细介绍如何基于 HarmonyOS 进行手机应用的开发，内容涵盖 HarmonyOS 架构、DevEco Studio、应用结构、Ability、任务调度、公共事件、通知、剪切板、Java UI、JS UI、多模输入、线程管理、视频、图像、相机、音频、媒体会话管理、媒体数据管理、安全管理、二维码、通用文字识别、蓝牙、WLAN、网络管理、电话服务、设备管理、数据管理、原子化服务、流转等多个主题。本书列举了大量解决实际问题的案例，具有很强的前瞻性、应用性、趣味性。

本书主要面向的是对移动应用或 HarmonyOS 应用感兴趣的学生、开发人员、架构师。

本书封面贴有清华大学出版社防伪标签，无标签者不得销售。
版权所有，侵权必究。举报：010-62782989，beiqinquan@tup.tsinghua.edu.cn。

图书在版编目（CIP）数据

鸿蒙 HarmonyOS 手机应用开发实战 / 柳伟卫著. —北京：清华大学出版社，2021.12
ISBN 978-7-302-59642-4

Ⅰ.①鸿… Ⅱ.①柳… Ⅲ.①移动终端－操作系统－程序设计 Ⅳ.①TN929.53

中国版本图书馆 CIP 数据核字（2021）第 249669 号

责任编辑：王金柱
封面设计：王 翔
责任校对：闫秀华
责任印制：朱雨萌

出版发行：清华大学出版社
　　　网　　址：http://www.tup.com.cn，http://www.wqbook.com
　　　地　　址：北京清华大学学研大厦 A 座　　　邮　　编：100084
　　　社 总 机：010-62770175　　　邮　　购：010-62786544
　　　投稿与读者服务：010-62776969，c-service@tup.tsinghua.edu.cn
　　　质 量 反 馈：010-62772015，zhiliang@tup.tsinghua.edu.cn
印 装 者：三河市铭诚印务有限公司
经　　销：全国新华书店
开　　本：190mm×260mm　　　印　　张：46.25　　　字　　数：1247 千字
版　　次：2022 年 1 月第 1 版　　　印　　次：2022 年 1 月第 1 次印刷
定　　价：149.00 元

产品编号：094740-01

前　　言

写作背景

中国信息产业一直是"缺芯少魂",其中的"芯"指的是芯片,而"魂"则是指操作系统。而自2019年5月15日起,美国陆续把包括华为在内的中国高科技企业列入其所谓的"实体清单"(Entities List),标志着科技再次成为中美博弈的核心领域。

随着谷歌暂停与华为的部分合作,包括软件和技术服务的转让,华为在国外市场已经面临着升级Android版本、搭载谷歌服务等方面的困境。在这样的背景下,华为顺势推出HarmonyOS,以求在操作系统领域不受制于人。

HarmonyOS是一款面向未来、面向全场景(移动办公、运动健康、社交通信、媒体娱乐等)的全新的分布式操作系统。作为操作系统领域的新成员,HarmonyOS势必会面临bug多、学习资源缺乏等众多困难。为此,笔者在开源社区以开源方式推出了免费系列学习教程《跟老卫学HarmonyOS开发》[1],以帮助HarmonyOS爱好者入门。同时,为了让更多的人了解并使用HarmonyOS,笔者将自身工作、学习中遇到的问题、难题进行了总结,形成了本书,以补市场空白。

内容介绍

全书大致分为3部分:

- 入门(第1~4章):介绍HarmonyOS的背景、开发环境搭建,并创建一个简单的HarmonyOS应用。
- 进阶(第5~29章):介绍HarmonyOS的核心功能的开发,内容包括Ability、UI开发、线程管理、视频、图像、相机、音频、媒体会话管理、媒体数据管理、安全管理、二维码、通用文字识别、蓝牙、WLAN、网络管理、电话服务、设备管理、数据管理等。
- 实战(第30章):演示HarmonyOS在游戏领域的综合实战案例——俄罗斯方块游戏。

源代码下载

本书提供的素材和源代码可从以下网址下载:

https://github.com/waylau/harmonyos-tutorial

---

[1] 《跟老卫学 HarmonyOS 开发》主页见 https://github.com/waylau/harmonyos-tutorial

也可扫描右侧的二维码下载。

如果你在下载过程中遇到问题，可发送邮件至booksaga@126.com获得帮助，邮件标题为"鸿蒙HarmonyOS手机应用开发实战"。

本书所采用的技术及相关版本

技术的版本是非常重要的，因为不同版本之间存在兼容性问题，而且不同版本的软件所对应的功能也是不同的。本书所列出的技术在版本上相对较新，都是经过笔者大量测试的。这样读者在自行编写代码时，可以参考本书所列出的版本，从而避免版本兼容性所产生的问题。建议读者将相关开发环境设置得跟本书一致，或者不低于本书所列的配置。本书所涉及的技术及相关版本：

- 操作系统：Windows10 64位。
- 内存：8GB及以上。
- 硬盘：100GB及以上。
- 分辨率：1280×800像素及以上。
- DevEco Studio 2.2 Beta1。

读者对象

本书主要面向的是对HarmonyOS应用开发感兴趣的学生、开发人员、架构师。

勘误和交流

本书如有勘误，会在以下网址发布：

https://github.com/waylau/harmonyos-tutorial/issues

由于笔者能力有限、时间仓促，书中难免存在疏漏之处，欢迎读者通过GitHub：https://github.com/waylau与笔者联系。

致　　谢

感谢清华大学出版社的各位工作人员为本书的出版所做的努力。

感谢我的父母、妻子和两个女儿。由于撰写本书，我牺牲了很多陪伴家人的时间，谢谢他们对我的理解和支持。

感谢关心和支持我的朋友、读者、网友。

柳伟卫

2021年9月

目　　录

第 1 章　HarmonyOS 简介 1

1.1　HarmonyOS 产生的背景 1
- 1.1.1　为什么需要 HarmonyOS 1
- 1.1.2　什么是 HarmonyOS 3
- 1.1.3　鸿蒙生态、OpenHarmony、HarmonyOS 的区别与联系 4
- 1.1.4　HarmonyOS 应用开发 4

1.2　特性介绍 5
- 1.2.1　硬件互助，资源共享 5
- 1.2.2　一次开发，多端部署 8
- 1.2.3　统一 OS，弹性部署 8

1.3　架构介绍 8
- 1.3.1　内核层 9
- 1.3.2　系统服务层 9
- 1.3.3　框架层 10
- 1.3.4　应用层 11

1.4　获取开发支持 12

第 2 章　开发环境搭建 13

2.1　注册华为开发者联盟账号 13
- 2.1.1　开发者享受的权益 13
- 2.1.2　注册、认证准备的资料 14
- 2.1.3　注册账号 14
- 2.1.4　登录账号 15
- 2.1.5　实名认证 15

2.2　DevEco Studio 下载安装 19
- 2.2.1　运行环境要求 20
- 2.2.2　下载和安装 Node.js 20
- 2.2.3　下载和安装 DevEco Studio 20

2.3　设置 DevEco Studio（可选） 21
- 2.3.1　npm 设置 21
- 2.3.2　设置 Gradle 代理 22

2.3.3 设置 DevEco Studio 代理 23
2.3.4 下载 HarmonyOS SDK 24
2.4 DevEco Studio 功能介绍 26
2.4.1 创建新的工程 26
2.4.2 添加 Module 27
2.4.3 删除 Module 30
2.5 DevEco Studio 常见问题小结 30
2.5.1 问题 1：访问 Gradle 仓库慢 31
2.5.2 问题 2：模拟器端口被占用无法启动 32

第 3 章 开发第一个 HarmonyOS 应用 33

3.1 创建一个新工程 33
3.1.1 选择创建新工程 33
3.1.2 选择设备应用类型的模板 33
3.1.3 配置项目的信息 34
3.1.4 自动生成工程代码 35
3.2 运行工程 35
3.2.1 单击"运行"按钮 35
3.2.2 选择模拟器 36
3.2.3 启动模拟器 36
3.2.4 再次运行工程 38
3.3 在本地真机中运行应用 39
3.3.1 连接真实的设备 39
3.3.2 运行应用 40
3.4 使用远程真机运行应用 40
3.4.1 启动远程真机设备 40
3.4.2 运行应用 41
3.4.3 对应用进行签名 41
3.4.4 再次运行应用 44
3.5 使用 DevEco Studio 预览器 45
3.5.1 如何安装预览器 45
3.5.2 如何使用预览器 45

第 4 章 探索 HarmonyOS 应用 46

4.1 App 46
4.1.1 什么是 App 46
4.1.2 应用程序包结构 47
4.1.3 代码层次的应用 48
4.2 Ability 48

	4.2.1	Ability 类	48
	4.2.2	AbilitySlice 类	49
	4.2.3	UI 界面	49
4.3	库文件		50
4.4	资源文件		50
	4.4.1	限定词目录	51
	4.4.2	资源组目录	52
4.5	配置文件		53
	4.5.1	配置文件的组成	54
	4.5.2	app 对象的内部结构	56
	4.5.3	deviceConfig 对象的内部结构	56
	4.5.4	module 对象的内部结构	59
4.6	pack.info		63

第 5 章 Ability 基础知识 64

5.1	Ability 概述		64
	5.1.1	FA	65
	5.1.2	PA	65
	5.1.3	Ability 的配置	65
5.2	Ability 的三层架构		65
	5.2.1	应用的分层	66
	5.2.2	不分层的应用架构	66
	5.2.3	应用的三层架构	68
	5.2.4	Ability 的三层架构	69
5.3	Page Ability		69
	5.3.1	Page Ability 的基本概念	69
	5.3.2	多个 AbilitySlice 构成一个 Page	70
	5.3.3	AbilitySlice 路由配置	71
	5.3.4	不同 Page 间的导航	71
5.4	实战：多个 AbilitySlice 间的路由和导航		72
	5.4.1	创建应用	72
	5.4.2	创建多个 AbilitySlice	72
	5.4.3	修改 PayAbilitySlice 布局	74
	5.4.4	设置 PayAbilitySlice 的样式	74
	5.4.5	如何实现 AbilitySlice 之间的路由和导航	76
	5.4.6	运行	79
5.5	Page 与 AbilitySlice 的生命周期		79
	5.5.1	Page 的生命周期	80
	5.5.2	AbilitySlice 的生命周期	81

5.5.3 Page 与 AbilitySlice 生命周期的关联 82
5.6 实战：Page 与 AbilitySlice 生命周期的例子 82
　　5.6.1 修改 MainAbilitySlice 83
　　5.6.2 修改 PayAbilitySlice 84
　　5.6.3 修改 PayAbilitySlice 布局和文字 85
　　5.6.4 实现 AbilitySlice 之间的路由 87
　　5.6.5 运行 89
5.7 Service Ability 92
　　5.7.1 创建 Service 93
　　5.7.2 启动 Service 94
　　5.7.3 连接 Service 95
　　5.7.4 Service Ability 的生命周期 96
5.8 实战：Service Ability 生命周期的例子 97
　　5.8.1 创建 Service 97
　　5.8.2 创建远程对象 100
　　5.8.3 修改 MainAbilitySlice 100
　　5.8.4 修改 ability_main.xml 103
　　5.8.5 运行 104
5.9 Data Ability 105
　　5.9.1 URI 105
　　5.9.2 访问 Data 106
5.10 实战：使用 DataAbilityHelper 访问文件 107
　　5.10.1 创建 DataAbility 107
　　5.10.2 修改 UserDataAbility 110
　　5.10.3 创建文件 113
　　5.10.4 修改 MainAbilitySlice 114
　　5.10.5 创建 FileUtils 类 115
　　5.10.6 运行 116
5.11 实战：使用 DataAbilityHelper 访问数据库 117
　　5.11.1 创建 DataAbility 117
　　5.11.2 初始化数据库 120
　　5.11.3 重写 query 方法 121
　　5.11.4 重写 insert 方法 121
　　5.11.5 重写 update 方法 122
　　5.11.6 重写 delete 方法 122
　　5.11.7 修改 MainAbilitySlice 123
　　5.11.8 运行 126
5.12 Intent 127
　　5.12.1 Operation 与 Parameters 127

	5.12.2	根据 Ability 的全称启动应用	128
	5.12.3	实战：根据 Operation 的其他属性启动应用	129
	5.12.4	实战：启动系统应用	135

第6章 Ability 任务调度 ... 140

- 6.1 分布式任务调度概述 ... 140
  - 6.1.1 "超级虚拟终端"的能力互助 ... 140
  - 6.1.2 跨设备软件访问的系统服务 ... 141
  - 6.1.3 全场景下的任务调度 ... 142
- 6.2 分布式任务调度能力简介 ... 142
  - 6.2.1 全局查询 ... 143
  - 6.2.2 启动和关闭 ... 143
  - 6.2.3 连接和断开连接 ... 144
  - 6.2.4 轻量通信 ... 144
- 6.3 分布式任务调度实现原理 ... 145
  - 6.3.1 PRC ... 145
  - 6.3.2 HarmonyOS 设备之间的通信 ... 146
  - 6.3.3 HarmonyOS 设备与其他 OS 设备之间的通信 ... 146
- 6.4 实现分布式任务调度 ... 147
  - 6.4.1 如何实现分布式任务调度 ... 147
  - 6.4.2 分布式任务调度支持的场景 ... 148
- 6.5 实战：分布式任务调度启动远程 FA ... 148
  - 6.5.1 修改 RemoteFA 应用 ... 148
  - 6.5.2 修改 DistributedSchedulingStartRemoteFA 应用 ... 150
  - 6.5.3 运行 ... 156
- 6.6 实战：分布式任务调度启动和关闭远程 PA ... 157
  - 6.6.1 修改 RemotePA 应用 ... 157
  - 6.6.2 修改 DistributedSchedulingStartStopRemotePA 应用 ... 160
  - 6.6.3 运行 ... 167

第7章 Ability 公共事件与通知 ... 169

- 7.1 公共事件与通知概述 ... 169
  - 7.1.1 公共事件和通知 ... 169
  - 7.1.2 约束与限制 ... 170
- 7.2 公共事件服务 ... 170
  - 7.2.1 接口说明 ... 170
  - 7.2.2 发布公共事件 ... 172
  - 7.2.3 订阅公共事件 ... 174
  - 7.2.4 退订公共事件 ... 176

7.3 实战：公共事件服务发布事件 ... 176
7.3.1 修改 ability_main.xml ... 176
7.3.2 修改 MainAbilitySlice ... 177
7.3.3 自定义权限 ... 179
7.3.4 运行 ... 179
7.4 实战：公共事件服务订阅事件 ... 180
7.4.1 修改 ability_main.xml ... 180
7.4.2 创建 CommonEventSubscriber ... 180
7.4.3 修改 MainAbility ... 181
7.4.4 修改配置文件 ... 182
7.4.5 运行 ... 183
7.5 高级通知服务 ... 183
7.5.1 接口说明 ... 183
7.5.2 创建 NotificationSlot ... 186
7.5.3 发布通知 ... 186
7.5.4 取消通知 ... 187
7.6 实战：通知发布与取消 ... 188
7.6.1 修改 ability_main.xml ... 188
7.6.2 修改 MainAbilitySlice ... 189
7.6.3 运行 ... 191

第 8 章 剪贴板 ... 193

8.1 剪贴板概述 ... 193
8.2 场景介绍 ... 193
8.3 接口说明 ... 194
8.3.1 SystemPasteboard ... 194
8.3.2 PasteData ... 195
8.3.3 PasteData.Record ... 195
8.3.4 PasteData.DataProperty ... 196
8.3.5 IPasteDataChangedListener ... 196
8.4 实战：剪贴板数据的写入 ... 196
8.4.1 修改 ability_main.xml ... 196
8.4.2 修改 MainAbilitySlice ... 197
8.4.3 运行 ... 199
8.5 实战：剪切板数据的读取 ... 199
8.5.1 修改 ability_main.xml ... 199
8.5.2 修改 MainAbilitySlice ... 200
8.5.3 运行 ... 202

第 9 章 用 Java 开发 UI 204

9.1 用 Java 开发 UI 概述 204
9.1.1 组件和布局 204
9.1.2 Component 和 ComponentContainer 205
9.1.3 LayoutConfig 205
9.1.4 组件树 206

9.2 组件与布局 206
9.2.1 编写布局的方式 206
9.2.2 组件分类 207

9.3 实战：通过 XML 创建布局 207
9.3.1 理解 XML 布局文件 207
9.3.2 创建 XML 布局文件 209
9.3.3 加载 XML 布局 209
9.3.4 显示 XML 布局 210

9.4 实战：通过 Java 创建布局 211
9.4.1 新建 AbilitySlice 212
9.4.2 创建布局 212
9.4.3 在布局中添加组件 213
9.4.4 显示布局 214

9.5 实战：常用显示类组件——Text 215
9.5.1 设置背景 216
9.5.2 设置字体大小和颜色 217
9.5.3 设置字体风格和字重 218
9.5.4 设置文本对齐方式 219
9.5.5 设置文本换行和最大显示行数 221
9.5.6 设置自动调节字体大小 222
9.5.7 实现跑马灯效果 224
9.5.8 场景示例 225

9.6 实战：常用显示类组件——Image 227
9.6.1 创建 Image 227
9.6.2 设置透明度 228
9.6.3 设置缩放系数 229

9.7 实战：常用显示类组件——ProgressBar 230
9.7.1 创建 ProgressBar 230
9.7.2 设置方向 231
9.7.3 设置颜色 232
9.7.4 设置提示文字 233

9.8 实战：常用交互类组件——Button 234

	9.8.1 创建 Button	235
	9.8.2 设置点击事件	237
	9.8.3 设置椭圆按钮	239
	9.8.4 设置圆形按钮	241
	9.8.5 场景示例	242
9.9	实战：常用交互类组件——TextField	247
	9.9.1 创建 TextField	247
	9.9.2 设置多行显示	248
	9.9.3 场景示例	249
9.10	实战：常用交互类组件——Checkbox	251
	9.10.1 创建 Checkbox	251
	9.10.2 设置选中和取消选中时的颜色	252
9.11	实战：常用交互类组件——RadioButton/RadioContainer	253
	9.11.1 创建 RadioButton/RadioContainer	253
	9.11.2 设置显示单选结果	255
9.12	实战：常用交互类组件——Switch	259
	9.12.1 创建 Switch	259
	9.12.2 设置文本	260
9.13	实战：常用交互类组件——ScrollView	261
	9.13.1 创建 ScrollView	261
	9.13.2 配置 Text 显示的内容	262
9.14	实战：常用交互类组件——Tab/TabList	263
	9.14.1 创建 TabList	263
	9.14.2 响应焦点变化	266
9.15	实战：常用交互类组件——Picker	268
	9.15.1 创建 Picker	268
	9.15.2 格式化 Picker 的显示	268
	9.15.3 日期滑动选择器 DatePicker	270
	9.15.4 时间滑动选择器 TimePicker	271
9.16	实战：常用交互类组件——ListContainer	272
	9.16.1 创建 ListContainer	272
	9.16.2 创建 ListContainer 子布局	273
	9.16.3 创建 ListContainer 数据包装类	273
	9.16.4 创建 ListContainer 数据提供者	274
	9.16.5 修改 MainAbilitySlice	275
9.17	实战：常用交互类组件——RoundProgressBar	276
	9.17.1 创建 RoundProgressBar	276
	9.17.2 设置开始和结束角度	277
9.18	实战：常用交互类组件——DirectionalLayout	278

    9.18.1 创建 DirectionalLayout 279
    9.18.2 设置水平排列 280
    9.18.3 设置权重 281
  9.19 实战：常用交互类组件——DependentLayout 283
    9.19.1 创建 DependentLayout 283
    9.19.2 相对于同级组件 284
    9.19.3 相对于父组件 284
    9.19.4 场景示例 285
  9.20 实战：常用交互类组件——StackLayout 286
  9.21 实战：常用交互类组件——TableLayout 288

第 10 章 用 JS 开发 UI 290

  10.1 用 JS 开发 UI 概述 290
    10.1.1 基础能力 290
    10.1.2 整体架构 291
  10.2 实战：创建 JS FA 应用 291
    10.2.1 创建应用 292
    10.2.2 AceAbility 293
    10.2.3 如何加载 JS FA 294
    10.2.4 JS FA 开发目录 294
    10.2.5 运行 296
  10.3 组件与布局 297
    10.3.1 组件分类 297
    10.3.2 布局 297
  10.4 实战：点赞按钮 299
    10.4.1 应用概述 299
    10.4.2 修改应用 299
    10.4.3 运行应用 300
  10.5 实战：JS FA 调用 PA 301
    10.5.1 FA 调用 PA 接口 301
    10.5.2 创建应用及编写 FA 302
    10.5.3 编写 PA 303
    10.5.4 运行应用 306

第 11 章 多模输入 UI 开发 307

  11.1 多模输入概述 307
  11.2 接口说明 308
    11.2.1 MultimodalEvent 308
    11.2.2 CompositeEvent 308

11.2.3 RotationEvent 308
11.2.4 SpeechEvent 309
11.2.5 ManipulationEvent 309
11.2.6 KeyEvent 309
11.2.7 TouchEvent 310
11.2.8 KeyBoardEvent 310
11.2.9 MouseEvent 310
11.2.10 MmiPoint 311
11.2.11 EventCreator 311
11.2.12 InputDevice 311
11.3 实战：多模输入事件 311
11.3.1 修改 MainAbilitySlice 311
11.3.2 获取多模输入权限 313

第 12 章 线程管理 314
12.1 线程管理概述 314
12.2 场景介绍 314
12.2.1 传统 Java 多线程管理 315
12.2.2 HarmonyOS 多线程管理 317
12.3 接口说明 317
12.3.1 GlobalTaskDispatcher 317
12.3.2 ParallelTaskDispatcher 317
12.3.3 SerialTaskDispatcher 318
12.3.4 SpecTaskDispatcher 318
12.4 实战：线程管理示例 318
12.4.1 修改 ability_main.xml 319
12.4.2 自定义任务 320
12.4.3 执行任务派发 321
12.4.4 运行 322
12.5 线程间通信概述 323
12.5.1 基本概念 323
12.5.2 运作机制 324
12.5.3 约束限制 324
12.6 实战：线程间通信示例 324
12.6.1 修改 ability_main.xml 325
12.6.2 自定义事件处理器 325
12.6.3 执行事件发送 327
12.6.4 运行 328

第13章 视频 330

13.1 视频概述 330
13.2 实战：媒体编解码能力查询 330
13.2.1 接口说明 331
13.2.2 创建应用 331
13.2.3 修改 ability_main.xml 331
13.2.4 修改 MainAbilitySlice 332
13.2.5 运行 334
13.3 实战：视频编解码 334
13.3.1 接口说明 335
13.3.2 创建应用 335
13.3.3 修改 ability_main.xml 335
13.3.4 修改 MainAbilitySlice 336
13.3.5 运行 340
13.4 实战：视频播放 340
13.4.1 接口说明 340
13.4.2 创建应用 341
13.4.3 修改 ability_main.xml 341
13.4.4 修改 MainAbilitySlice 343
13.4.5 运行 347
13.5 实战：视频录制 347
13.5.1 接口说明 347
13.5.2 创建应用 348
13.5.3 修改 ability_main.xml 348
13.5.4 修改 MainAbilitySlice 350
13.5.5 运行 352

第14章 图像 353

14.1 图像概述 353
14.1.1 基本概念 353
14.1.2 约束与限制 354
14.2 实战：图像解码和编码 354
14.2.1 接口说明 354
14.2.2 创建应用 355
14.2.3 修改 ability_main.xml 355
14.2.4 修改 MainAbilitySlice 356
14.2.5 解码操作说明 360
14.2.6 编码操作说明 361

14.3 实战：位图操作 361
14.3.1 接口说明 361
14.3.2 创建应用 363
14.3.3 修改 ability_main.xml 363
14.3.4 修改 MainAbilitySlice 364
14.3.5 创建 PixelMap 操作说明 365
14.3.6 从位图对象中获取信息操作说明 366
14.3.7 读取和写入像素操作说明 366
14.3.8 释放资源 367
14.4 实战：图像属性解码 368
14.4.1 接口说明 368
14.4.2 创建应用 368
14.4.3 修改 ability_main.xml 368
14.4.4 修改 MainAbilitySlice 369
14.4.5 运行 372

第 15 章 相机 373
15.1 相机概述 373
15.1.1 基本概念 373
15.1.2 约束与限制 374
15.1.3 相机开发流程 374
15.1.4 核心接口 374
15.1.5 相机权限 374
15.2 实战：相机设备创建 375
15.2.1 接口说明 375
15.2.2 创建应用 375
15.2.3 声明相机权限 375
15.2.4 修改 ability_main.xml 376
15.2.5 修改 MainAbilitySlice 377
15.2.6 运行 381
15.3 实战：相机设备配置 382
15.3.1 添加类变量 382
15.3.2 新增 FrameStateCallbackImpl 383
15.3.3 修改 onStart 方法 384
15.3.4 修改 onCreated 方法 385
15.3.5 运行 386
15.4 实战：相机帧捕获 387
15.4.1 接口说明 387
15.4.2 启动预览（循环帧捕获） 387

		15.4.3	释放资源	388
		15.4.4	运行	388
		15.4.5	实现拍照（单帧捕获）	388
		15.4.6	声明存储权限	391
		15.4.7	运行	393

第16章 音频 394

16.1 音频概述 394
  16.1.1 基本概念 394
  16.1.2 约束与限制 395

16.2 实战：音频播放 395
  16.2.1 接口说明 395
  16.2.2 创建应用 396
  16.2.3 修改 ability_main.xml 396
  16.2.4 修改 MainAbilitySlice 397
  16.2.5 运行 401

16.3 实战：音频采集 402
  16.3.1 接口说明 402
  16.3.2 创建应用 403
  16.3.3 声明麦克风权限 403
  16.3.4 修改 ability_main.xml 404
  16.3.5 修改 MainAbilitySlice 405

16.4 实战：短音播放 408
  16.4.1 接口说明 408
  16.4.2 创建应用 409
  16.4.3 修改 ability_main.xml 409
  16.4.4 修改 MainAbilitySlice 410
  16.4.5 运行 413

第17章 媒体会话管理 414

17.1 媒体会话管理概述 414
  17.1.1 AVSession 框架主要的类 414
  17.1.2 约束与限制 415

17.2 接口说明 416
  17.2.1 AVBrowser 的主要接口 416
  17.2.2 AVBrowserService 的主要接口 416
  17.2.3 AVController 的主要接口 417
  17.2.4 AVSession 的主要接口 418
  17.2.5 AVElement 的主要接口 418

XVI 鸿蒙 HarmonyOS 手机应用开发实战

17.3 实战：AVSession 媒体框架客户端 419
   17.3.1 创建应用 419
   17.3.2 修改 ability_main.xml 419
   17.3.3 修改 MainAbilitySlice 420
17.4 实战：AVSession 媒体框架服务端 424
   17.4.1 创建 Service 424
   17.4.2 修改 AVService 425
   17.4.3 运行 428

第 18 章 媒体数据管理 429

18.1 媒体数据管理概述 429
   18.1.1 媒体数据管理基本概念 429
   18.1.2 约束与限制 429
18.2 实战：媒体元数据的获取 430
   18.2.1 接口说明 430
   18.2.2 创建应用 431
   18.2.3 修改 ability_main.xml 431
   18.2.4 修改 MainAbilitySlice 432
   18.2.5 运行 435
18.3 实战：媒体存储数据操作 436
   18.3.1 接口说明 436
   18.3.2 创建应用 437
   18.3.3 声明权限 437
   18.3.4 修改 ability_main.xml 438
   18.3.5 修改 MainAbilitySlice 439
   18.3.6 运行 443
18.4 实战：视频与图像缩略图获取 444
   18.4.1 接口说明 444
   18.4.2 创建应用 444
   18.4.3 修改 ability_main.xml 445
   18.4.4 修改 MainAbilitySlice 446
   18.4.5 运行 449

第 19 章 安全管理 451

19.1 权限的基本概念 451
19.2 权限运作机制 452
19.3 权限约束与限制 452
19.4 应用权限列表 453
   19.4.1 权限分类 453

	19.4.2	敏感权限	453
	19.4.3	非敏感权限	454
	19.4.4	受限开放的权限	455
19.5	应用权限开发流程		455
	19.5.1	权限申请	455
	19.5.2	自定义权限	456
	19.5.3	访问权限控制	458
	19.5.4	接口说明	459
	19.5.5	动态申请权限的开发步骤	460
19.6	生物特征识别认证概述		461
19.7	生物特征识别运作机制		461
19.8	生物特征识别的约束与限制		462
19.9	生物特征识别的开发流程		462
	19.9.1	接口说明	462
	19.9.2	开发准备	463
	19.9.3	开发过程	463

第20章 二维码 466

20.1	二维码概述		466
	20.1.1	什么是二维码	466
	20.1.2	二维码的发展历程	467
20.2	场景介绍		467
20.3	接口说明		468
20.4	实战：生成二维码		468
	20.4.1	创建应用	468
	20.4.2	修改 ability_main.xml	468
	20.4.3	修改 MainAbilitySlice	470
	20.4.4	运行	473

第21章 通用文字识别 474

21.1	通用文字识别概述		474
	21.1.1	什么是 OCR	474
	21.1.2	OCR 发展简史	474
21.2	场景介绍		475
21.3	接口说明		475
	21.3.1	setVisionConfiguration()方法	476
	21.3.2	detect()方法	476
	21.3.3	约束与限制	476
21.4	实战：通用文字识别示例		477

第 22 章 蓝牙

21.4.1 创建应用 477
21.4.2 修改 ability_main.xml 477
21.4.3 修改 MainAbilitySlice 479
21.4.4 运行 482

第 22 章 蓝牙 484

22.1 蓝牙概述 484
22.1.1 传统蓝牙 484
22.1.2 BLE 484
22.1.3 约束与限制 485
22.2 实战：传统蓝牙本机管理 485
22.2.1 接口说明 485
22.2.2 创建应用 486
22.2.3 声明权限 486
22.2.4 修改 ability_main.xml 487
22.2.5 修改 MainAbilitySlice 488
22.2.6 运行 491
22.3 实战：传统蓝牙远端设备操作 494
22.3.1 接口说明 494
22.3.2 创建应用 495
22.3.3 声明权限 495
22.3.4 修改 ability_main.xml 496
22.3.5 修改 MainAbilitySlice 497
22.3.6 运行 501
22.4 实战：BLE 扫描和广播 503
22.4.1 接口说明 503
22.4.2 创建应用 504
22.4.3 声明权限 504
22.4.4 修改 ability_main.xml 505
22.4.5 修改 MainAbilitySlice 507
22.4.6 运行 510

第 23 章 WLAN 512

23.1 WLAN 概述 512
23.1.1 WLAN 简介 512
23.1.2 约束与限制 513
23.2 实战：WLAN 的基础功能 513
23.2.1 接口说明 513
23.2.2 创建应用 514

目录 | XIX

- 23.2.3 声明权限 514
- 23.2.4 修改 ability_main.xml 515
- 23.2.5 修改 MainAbilitySlice 516
- 23.2.6 运行 520
- 23.3 实战：不信任热点配置 522
  - 23.3.1 接口说明 522
  - 23.3.2 创建应用 522
  - 23.3.3 声明权限 522
  - 23.3.4 修改 ability_main.xml 523
  - 23.3.5 修改 MainAbilitySlice 524
  - 23.3.6 运行 526
- 23.4 实战：WLAN 消息通知 527
  - 23.4.1 接口说明 527
  - 23.4.2 创建应用 528
  - 23.4.3 修改 ability_main.xml 528
  - 23.4.4 修改 MainAbilitySlice 529
  - 23.4.5 运行 531

第 24 章 网络管理 533

- 24.1 网络管理概述 533
  - 24.1.1 功能 533
  - 24.1.2 约束与限制 533
- 24.2 实战：使用当前网络打开一个 URL 链接 534
  - 24.2.1 接口说明 534
  - 24.2.2 创建应用 534
  - 24.2.3 声明权限 534
  - 24.2.4 修改 ability_main.xml 535
  - 24.2.5 修改 MainAbilitySlice 535
  - 24.2.6 运行 539
- 24.3 实战：使用当前网络进行 Socket 数据传输 540
  - 24.3.1 接口说明 540
  - 24.3.2 创建应用 540
  - 24.3.3 声明权限 540
  - 24.3.4 修改 ability_main.xml 541
  - 24.3.5 修改 MainAbilitySlice 541
  - 24.3.6 运行 545
- 24.4 实战：流量统计 546
  - 24.4.1 接口说明 546
  - 24.4.2 创建应用 546

	24.4.3	声明权限	546
	24.4.4	修改 ability_main.xml	547
	24.4.5	修改 MainAbilitySlice	548
	24.4.6	运行	551

第 25 章 电话服务 553

25.1 电话服务概述 553
- 25.1.1 主要的 API 553
- 25.1.2 约束与限制 553

25.2 实战：获取当前蜂窝网络信号信息 554
- 25.2.1 接口说明 554
- 25.2.2 创建应用 554
- 25.2.3 修改 ability_main.xml 554
- 25.2.4 修改 MainAbilitySlice 555
- 25.2.5 运行 557

25.3 实战：观察蜂窝网络的状态变化 559
- 25.3.1 接口说明 560
- 25.3.2 创建应用 560
- 25.3.3 声明权限 560
- 25.3.4 修改 ability_main.xml 560
- 25.3.5 修改 MainAbilitySlice 562
- 25.3.6 运行 564

第 26 章 设备管理 568

26.1 设备管理概述 568
- 26.1.1 传感器 569
- 26.1.2 控制类小器件 570
- 26.1.3 位置 571

26.2 实战：传感器示例 572
- 26.2.1 接口说明 572
- 26.2.2 创建应用 572
- 26.2.3 修改 ability_main.xml 573
- 26.2.4 修改 MainAbilitySlice 573
- 26.2.5 运行 576

26.3 实战：Light 示例 578
- 26.3.1 接口说明 578
- 26.3.2 创建应用 578
- 26.3.3 修改 ability_main.xml 579
- 26.3.4 修改 MainAbilitySlice 579

26.3.5 运行 ... 582
26.4 实战：获取设备的位置 ... 582
　26.4.1 接口说明 ... 583
　26.4.2 创建应用 ... 583
　26.4.3 声明权限 ... 583
　26.4.4 修改 ability_main.xml ... 584
　26.4.5 修改 MainAbilitySlice ... 585
　26.4.6 运行 ... 588
26.5 实战：（逆）地理编码转化 ... 589
　26.5.1 接口说明 ... 589
　26.5.2 创建应用 ... 590
　26.5.3 修改 ability_main.xml ... 590
　26.5.4 修改 MainAbilitySlice ... 591
　26.5.5 运行 ... 593

第 27 章　数据管理 ... 595

27.1 数据管理概述 ... 595
27.2 关系型数据库 ... 596
　27.2.1 基本概念 ... 596
　27.2.2 运作机制 ... 596
　27.2.3 默认配置 ... 597
　27.2.4 约束与限制 ... 597
　27.2.5 接口说明 ... 597
　27.2.6 开发过程 ... 600
27.3 对象关系映射数据库 ... 601
　27.3.1 基本概念 ... 601
　27.3.2 运作机制 ... 602
　27.3.3 实体对象属性支持的类型 ... 602
　27.3.4 接口说明 ... 603
27.4 实战：使用对象关系映射数据库 ... 605
　27.4.1 修改 build.gradle ... 605
　27.4.2 新增 User ... 606
　27.4.3 新增 UserStore ... 607
　27.4.4 创建 DataAbility ... 607
　27.4.5 初始化数据库 ... 609
　27.4.6 新增 queryAll 方法 ... 610
　27.4.7 新增 insert 方法 ... 611
　27.4.8 新增 update 方法 ... 611
　27.4.9 新增 deleteAll 方法 ... 611

		27.4.10	修改 ability_main.xml	612
		27.4.11	修改 MainAbilitySlice	614
		27.4.12	运行	616
	27.5	轻量级偏好数据库		619
		27.5.1	基本概念	619
		27.5.2	运作机制	620
		27.5.3	约束与限制	620
		27.5.4	接口说明	621
	27.6	实战：使用轻量级偏好数据库		622
		27.6.1	修改 ability_main.xml	622
		27.6.2	修改 MainAbilitySlice	624
		27.6.3	运行	627
	27.7	数据存储管理		630
		27.7.1	基本概念	630
		27.7.2	运作机制	630
		27.7.3	接口说明	631
	27.8	实战：使用数据存储管理		632
		27.8.1	修改 ability_main.xml	632
		27.8.2	修改 MainAbilitySlice	633
		27.8.3	运行	635

第 28 章 原子化服务 637

	28.1	原子化服务概述		637
		28.1.1	什么是原子化服务	637
		28.1.2	原子化服务特征	638
		28.1.3	原子化服务典型的使用场景	639
	28.2	服务中心		639
		28.2.1	服务中心入口	639
		28.2.2	常用服务	639
		28.2.3	我的收藏	639
		28.2.4	服务发现	640
	28.3	实战：原子化服务 HelloDog		641
		28.3.1	初始化原子化服务	641
		28.3.2	理解原子化服务的项目结构	641
		28.3.3	搜索原子化服务	655
		28.3.4	运行原子化服务	656

第 29 章 流转 657

	29.1	流转概述		657

29.2	流转架构		658
	29.2.1	跨端迁移流程	659
	29.2.2	多端协同流程	659
	29.2.3	兼容性设计	660
29.3	跨端迁移		661
	29.3.1	跨端迁移流程	661
	29.3.2	接口说明	663
	29.3.3	约束与限制	664
29.4	实战：实现跨端迁移与回迁		665
	29.4.1	MainAbility 实现 IAbilityContinuation 接口	665
	29.4.2	声明权限	666
	29.4.3	设计界面与布局	668
	29.4.4	修改 MainAbilitySlice	669
	29.4.5	新增 DeviceUtils	671
	29.4.6	运行	673
29.5	多端协同		674
	29.5.1	多端协同流程	674
	29.5.2	接口说明	675
	29.5.3	约束与限制	677
29.6	实战：实现多端协同		677
	29.6.1	MainAbility 实现 IAbilityContinuation 接口	677
	29.6.2	声明权限	678
	29.6.3	设计界面与布局	680
	29.6.4	修改 MainAbilitySlice	681
	29.6.5	新增 DeviceUtils	684
	29.6.6	运行	685

第 30 章 综合案例：俄罗斯方块游戏 ... 687

30.1	案例概述		687
	30.1.1	俄罗斯方块游戏概述	687
	30.1.2	俄罗斯方块游戏的规则	688
30.2	代码实现		688
	30.2.1	技术重点	688
	30.2.2	设置布局	689
	30.2.3	设置全屏	692
	30.2.4	应用的主体逻辑	693
	30.2.5	初始化游戏	696
	30.2.6	创建网格数据	697
	30.2.7	绘制网格	702

30.2.8 启动游戏 703
30.2.9 左移操作 705
30.2.10 右移操作 707
30.2.11 转换操作 708
30.2.12 重置操作 711
30.3 应用运行 711

参考文献 713

第 1 章

HarmonyOS 简介

鸿蒙是中国古代汉族传说中的一个时代，传说在开天辟地之前，世界是一团混沌的元气，这种自然的元气叫作鸿蒙。华为以此命名，寓意在国产操作系统领域开创一个新的时代。

本章介绍HarmonyOS（鸿蒙系统）产生的历史背景、特点及技术架构。

1.1 HarmonyOS 产生的背景

2020年9月10日，华为开发者大会（HDC.Together）正式在华为东莞松山湖基地拉开帷幕，华为如期为消费者带来了众多软件创新，其中最受期待的莫过于HarmonyOS 2.0的正式发布。

那么到底什么是HarmonyOS？为什么需要HarmonyOS？

1.1.1 为什么需要 HarmonyOS

早在1999年，时任中国科技部部长徐冠华曾说，"中国信息产业缺芯少魂"。其中的芯指的是芯片，而魂则是指操作系统。当时，中国曾大力扶持国产芯片和操作系统，也曾诞生过一些亮眼的产品，比如红旗Linux、龙芯等。然而，二十多年过去了，中国依然是缺芯少魂，这次美国对华为的封杀，第一个禁的是芯片，第二个禁的就是操作系统。

如今消费者身边的智能终端越来越多，但由于系统的碎片化，导致连接复杂、操控烦琐、体验割裂，并没有带来更好的体验。因此，业界急需这样一款智能终端操作系统，能够为不同设备的智能化、互联与协同提供统一的语言。华为看到了这个商机。华为消费者业务软件部总裁、HarmonyOS负责人王成录透露，其实早在2016年5月，自研操作系统就已经在华为内部立项（当然，当时还不叫HarmonyOS）。一年后，该操作系统的1.0版本研发完成。在2017年年中的华为消费者业务部务虚会上，王成录向参会成员分享了"开发一个多设备之间协同的分布式IoT操作系统"的议题，并展示了当时开发的操作系统1.0版本成果。

然而，华为自研的这款操作系统一直处于"备胎"的状态，并未正式商用。

2019年5月15日，美国商务部宣布，把华为及70家关联企业列入其所谓的"实体清单"（Entities List）[1]。这意味着，今后如果没有美国政府的批准，华为将无法向美国企业购买元器件和技术。"实体清单"是美国为维护其国家安全利益而设立的出口管制条例。在未得到许可证前，美国各出口商不得帮助这些名单上的企业获取受本条例管辖的任何物项。简单地说，"实体清单"就是一份"黑名单"，一旦进入此榜单，实际上就剥夺了相关企业在美国的贸易机会。

随着中国国力的崛起，自2019年5月15日起，美国的"实体清单"不断扩容，体现了美国对中国高科技企业的限制升级，科技再次成为中美博弈的核心领域。

作为中国科技领域的头部企业，华为首当其冲。华为虽然早就建立了自己的芯片企业——海思。但海思生产的芯片还不能完全覆盖自己的产品线，华为依然需要直接采购美国芯片厂商的产品。受到实体清单的影响，美国全面封锁华为在全球的芯片采购，直接导致华为忍痛出售旗下手机品牌——荣耀[2]。

除了芯片等硬件产品外，在"实体清单"的限制下，软件等技术同样受到限制。谷歌已暂停与华为的部分合作，包括软件和技术服务的转让。华为在国外市场面临着升级Android版本、搭载谷歌服务等方面的困境。

为了避免被人卡脖子，华为展开了自救和反击。2019年5月17日凌晨2点，华为海思总裁何庭波发表致员工的一封信[3]，信中称，"公司多年前做出了极限生存的假设，预计有一天，所有美国的先进芯片和技术将不可获得"，而华为"为了这个以为永远不会发生的假设，数千海思儿女，走上了科技史上最为悲壮的长征，为公司的生存打造'备胎'"。信中称，"今天，命运的年轮转到这个极限而黑暗的时刻，超级大国毫不留情地中断全球合作的技术与产业体系，做出了最疯狂的决定，在毫无依据的条件下，把华为公司放入了实体名单。"何庭波在信中说："今后的路，不会再有另一个10年来打造备胎，然后再换胎了，缓冲区已经消失，每一个新产品一出生，将必须同步'科技自立'的方案。"

因此，在这个背景下，除了加大海思的研发投入以外，华为也向世人公开了自己的"秘密武器"——HarmonyOS。正如其中文"鸿蒙"的寓意，意味着HarmonyOS将会开启一个开天辟地的时代，2020年12月16日，华为发布HarmonyOS 2.0手机开发者Beta版本，这意味着HarmonyOS能够覆盖手机应用场景。

2021年6月2日，华为发布HarmonyOS 2正式版，将支持超过百款手机、平板和智慧屏设备升级HarmonyOS 2，这也是华为史上规模最大的一次升级。

---

[1] 有关"实体清单"的介绍可见 https://zhuanlan.zhihu.com/p/66672579

[2] 该报道可见 http://www.yidianzixun.com/article/T_00b2I0Kf?COLLCC=2344939848&s=op398&appid=s3rd_op398

[3] 该报道可见 https://baijiahao.baidu.com/s?id=1633767680924123225&wfr=spider&for=pc

1.1.2 什么是HarmonyOS

HarmonyOS在2019年8月9日华为开发者大会上首次公开亮相[1]，华为消费者业务CEO余承东进行主题演讲。在演讲中，余承东正式发布了HarmonyOS，并确认HarmonyOS的核心能力将会以OpenHarmony项目的方式分阶段逐步开源[2]。

HarmonyOS也称为鸿蒙、鸿蒙系统，或者鸿蒙OS，是一个全新的面向全场景的分布式操作系统。HarmonyOS以人为中心，将人、设备、场景有机地联系在一起，尤其是面向IoT领域，将多种智能设备的体验实现系统级融合，使得人、设备、场景不再是孤立存在的，达到以人为中心，将IoT设备融为一体，以适应不同场景，进而带来最佳体验。

HarmonyOS是一款面向未来、面向全场景（移动办公、运动健康、社交通信、媒体娱乐等）的分布式操作系统。在传统的单设备系统能力的基础上，HarmonyOS提出了基于同一套系统能力、适配多种终端形态的分布式理念，能够支持手机、平板、智能穿戴、智慧屏、车机等多种终端设备。

- 对消费者而言，HarmonyOS用一个统一的软件系统从根本上解决了消费者使用大量终端体验割裂的问题。HarmonyOS能够将生活场景中的各类终端进行能力整合，可以实现不同的终端设备之间的快速连接、能力互助、资源共享，匹配合适的设备，为消费者提供统一、便利、安全、智慧化的全场景体验。
- 对应用开发者而言，HarmonyOS采用了多种分布式技术，整合各种终端硬件能力，形成一个虚拟的"超级终端"。开发者可以基于"超级终端"进行应用开发，使得应用程序的开发实现与不同终端设备的形态差异无关。这能够让开发者聚焦上层业务逻辑，无须关注硬件差异，更加便捷、高效地开发应用。
- 对设备开发者而言，HarmonyOS采用了组件化的设计方案，可以按需调用"超级终端"能力，可以带来"超级终端"的创新体验。根据设备的资源能力和业务特征进行灵活裁剪，满足不同形态的终端设备对于操作系统的要求。

举例来说，当用户走进厨房，用HarmonyOS手机一碰微波炉，就能实现设备极速联网；用HarmonyOS手机碰一下豆浆机，分分钟实现无屏变有屏。

自HarmonyOS诞生以来，经过一年多的发展，终于迎来了HarmonyOS 2.0。在2020年9月10日举办的华为开发者大会上[3]，发布的HarmonyOS 2.0带来了很多惊喜。

在大会上，王成录对HarmonyOS 2.0进行了详细的介绍。HarmonyOS 2.0在分布式能力上进行了全面的提升，升级后的分布式软总线、分布式数据管理和分布式安全为开发者和消费者带来了不少新鲜的感觉。

分布式软总线让多设备融合为"一个设备"，带来设备内和设备间高吞吐、低时延、高可靠的流畅连接体验。分布式数据管理让跨设备数据访问如同本地访问，大大提升了跨设备数据的远程

---

[1] 大会内容见 https://developer.huawei.com/consumer/cn/events/hdc2019/index.html
[2] 开源地址见 https://gitee.com/openharmony
[3] 大会内容见 https://developer.huawei.com/consumer/cn/events/hdc2020/index.html

读写和检索性能等。

分布式安全确保正确的人、用正确的设备、正确使用数据。当用户进行解锁、付款、登录等行为时，系统会主动拉出认证请求，并通过分布式技术可信互联能力，协同身份认证确保正确的人；HarmonyOS能够把手机的内核级安全能力扩展到其他终端，进而提升全场景设备的安全性，通过设备能力互助，共同抵御攻击，保障智能家居网络安全；HarmonyOS通过定义数据和设备的安全级别，对数据和设备都进行了分类分级保护，确保数据流通安全可信。

HarmonyOS不是手机系统的一个简单的替代，它是面向未来全场景融合的操作系统，这个系统的核心底座就是分布式技术。一方面，其分布式技术有了本质提升；另一方面，除了支持华为自身的设备之外，也开始支持第三方设备。目前，华为已经与美的、九阳、老板等设备厂商达成了合作，搭载了HarmonyOS 2.0的诸多设备也将陆续与广大消费者见面。

1.1.3 鸿蒙生态、OpenHarmony、HarmonyOS 的区别与联系

1. OpenHarmony

OpenHarmony是一个开源项目，由开放原子开源基金会（https://www.openatom.org/）进行管理。开放原子开源基金会由华为、阿里、腾讯、百度、小米、浪潮、招商银行、360、OPPO、VIVO等10家互联网企业共同发起组建。

OpenHarmony暂时还没有中文名字，名字还在申请中。项目地址为https://gitee.com/openharmony。

OpenHarmony开源项目主要包括两部分：一是华为捐献的"鸿蒙操作系统"的基础能力，二是其他参与者的贡献。

因此，目前OpenHarmony的核心贡献主力还是华为。OpenHarmony是HarmonyOS的底座。

2. HarmonyOS

HarmonyOS就是"鸿蒙操作系统"，或者简称为"鸿蒙OS"，是基于OpenHarmony、AOSP等开源项目的商用版本。

这里需要注意：

- HarmonyOS不是开源项目，而是商用版本。
- HarmonyOS手机和平板之所以能运行Android，是因为HarmonyOS实现了现有Android生态应用（即AOSP）的运行。

3. 鸿蒙生态

鸿蒙生态包括OpenHarmony和HarmonyOS，当然还包括开发工具以及周边的一些开发库。当我们在说"鸿蒙"的时候，也就是指鸿蒙生态。

1.1.4 HarmonyOS 应用开发

为了进一步扩大HarmonyOS的生态圈，面对广大的硬件设备厂商，HarmonyOS通过SDK、源代码、开发板/模组和HUAWEI DevEco Studio等装备共同构成了完备的开发平台与工具链，让HarmonyOS设备开发易如反掌。

应用创新是一个操作系统发展的关键，应用开发体验更是如此。一个完整的应用开发生态中，应用框架、编译器、IDE、API/SDK都是必不可少的。为了赋能开发者，HarmonyOS提供了一系列构建全场景应用的完整平台工具链与生态体系助力开发者，让应用能力可分、可合、可流转，轻松构筑全场景创新体验。

本书就是介绍如何针对HarmonyOS进行手机应用的开发。可以预见的是，HarmonyOS必将是近些年的热门话题。对于能在早期投身于HarmonyOS开发的技术人员而言，其意义不亚于当年早期Android的开发。HarmonyOS必将带给开发者广阔的前景。同时，基于HarmonyOS所提供的完善的平台工具链与生态体系，相信广大的读者一定能轻松入门HarmonyOS。

5G网络准备就绪，物联网产业链也已经渐趋成熟，在物联网即将爆发的前夜，亟需一套专为物联网准备的操作系统，华为的HarmonyOS正逢其时。Windows成就了微软，Android成就了谷歌，HarmonyOS是否能成就华为，让我们拭目以待。

1.2 特性介绍

本节介绍HarmonyOS的特性。

1.2.1 硬件互助，资源共享

HarmonyOS把各终端硬件的能力虚拟成可共享的能力资源池，让应用通过系统调用其所需的硬件能力。在这个架构下，硬件能力类似于活字印刷术中的一个个单字字模，可以被无限次重复使用。简单来说，各终端实现了硬件互助，资源共享；应用拥有了调用远程终端的能力，像调用本地终端一样方便；而用户则收获一个多设备组成的超级终端。

那么HarmonyOS是如何实现硬件互助、资源共享的呢？主要是基于以下几方面实现的。

1. 分布式软总线

分布式软总线是多种终端设备的统一基座，为设备之间的互联互通提供了统一的分布式通信能力，能够快速发现并连接设备，高效地分发任务和传输数据。分布式软总线示意图见图1-1。

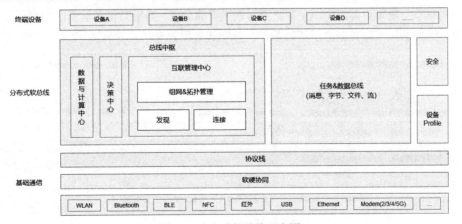

图1-1 分布式软总线示意图

简而言之，分布式软总线提供了多设备连接能力。

2. 分布式设备虚拟化

分布式设备虚拟化平台可以实现不同设备的资源融合、设备管理、数据处理，多种设备共同形成一个超级虚拟终端。针对不同类型的任务，为用户匹配并选择能力合适的执行硬件，让业务连续地在不同设备间流转，充分发挥不同设备的资源优势。分布式设备虚拟化示意图见图1-2。

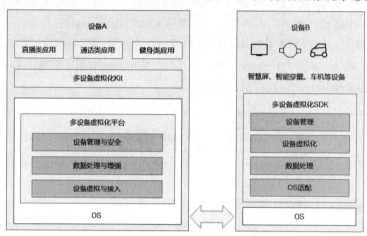

图1-2 分布式设备虚拟化示意图

以一个无人机为例，传统的无人机视频是按照如下步骤分享的：

（1）无人机拍摄画面。
（2）将无人机拍摄的视频保存下来。
（3）通过通信软件对视频进行分享。

而在分布式设备虚拟化后，无人机可以被当作手机的一个摄像头，在视频通话软件中，可以直接将无人机的摄像头做实时的分享。

3. 分布式数据管理

分布式数据管理基于分布式软总线的能力，实现应用程序数据和用户数据的分布式管理。用户数据不再与单一物理设备绑定，业务逻辑与数据存储分离，应用跨设备运行时数据无缝衔接，为打造一致、流畅的用户体验创造了基础条件。分布式数据管理示意图见图1-3。

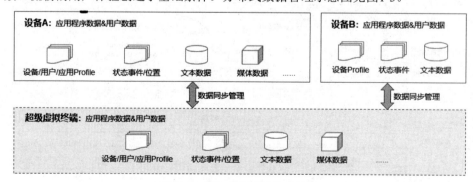

图1-3 分布式数据管理示意图

在全场景新时代，每个人拥有的设备越来越多，单一设备的数据往往无法满足用户的诉求，数据在设备间的流转变得越来越频繁。以一组照片数据在手机、平板、智慧屏和PC之间相互浏览和编辑为例，需要考虑到照片数据在多设备间是怎么存储、怎么共享和怎么访问的？HarmonyOS分布式数据管理的目标就是为开发者在系统层面解决这些问题，让应用开发变得简单。它能够保证多设备间的数据安全，解决多设备间数据同步、跨设备查找和访问的各种关键技术问题。

HarmonyOS分布式数据管理对开发者提供分布式数据库、分布式文件系统和分布式检索能力，开发者在多设备上开发应用时，对数据的操作、共享、检索可以跟使用本地数据一样处理，为开发者提供便捷、高效和安全的数据管理能力，大大降低了应用开发者实现数据分布式访问的门槛。同时，由于在系统层面实现了这样的功能，可以结合系统资源调度，大大提升跨设备数据远程访问和检索的性能，让更多的开发者可以快速上手实现流畅的分布式应用。

4. 分布式任务调度

分布式任务调度基于分布式软总线、分布式数据管理、分布式Profile等技术特性，构建统一的分布式服务管理（发现、同步、注册、调用）机制，支持对跨设备的应用进行远程启动、远程调用、远程连接以及迁移等操作，能够根据不同设备的能力、位置、业务运行状态、资源使用情况，以及用户的习惯和意图，选择合适的设备运行分布式任务。

图1-4以应用迁移为例，简要地展示了分布式任务的调度能力。

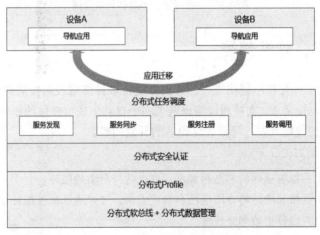

图1-4 分布式任务调度示意图

在传统的终端设备上做跨设备的应用访问时，需要应用自己完成服务发现、连接、命令监听、命令解析等一系列的工作，无论是应用开发者自己开发，还是使用第三方的库，都让应用开发过程变得沉重。分布式任务调度就是在系统层面为应用提供通用的分布式服务，让应用开发可以聚焦在业务实现上。HarmonyOS在分布式任务调度上充分考虑了应用开发者的使用便利性，提供了应用信息自动同步的能力，通过查询远程Ability接口，既可以指定Ability查询设备列表，又可以指定设备标识查询Ability列表，开发者可以根据实际场景灵活使用。在API形式上保持了和本地使用基本一致，仅仅增加了远程设备标识的参数，这让开发者使用起来完全没有障碍，开发者生态十分友好。举例来说，在手机和手表间进行应用间的协同，在游乐场游玩的场景用户可以全程不使用手机，解决了在游乐场游玩过程中手机容易丢失、损坏的痛点，非常好地提升了用户体验。

1.2.2 一次开发，多端部署

HarmonyOS提供了用户程序框架、Ability框架以及UI框架，支持应用开发过程中多终端的业务逻辑和界面逻辑进行复用，能够实现应用的一次开发、多端部署，提升了跨设备应用的开发效率。一次开发、多端部署示意图见图1-5。

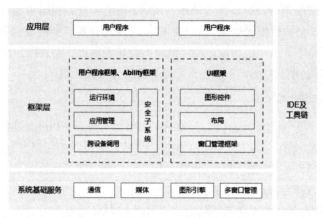

图1-5　一次开发、多端部署示意图

1.2.3 统一OS，弹性部署

HarmonyOS通过组件化和小型化等设计方法，支持多种终端设备按需弹性部署，能够适配不同类别的硬件资源和功能需求，支持通过编译链关系去自动生成组件化的依赖关系，形成组件树依赖图，支持产品系统的便捷开发，降低硬件设备的开发门槛。

组件化和小型化的好处在于：

- 组件可有可无：根据硬件的形态和需求，可以选择所需的组件。
- 组件可大可小：根据硬件的资源情况和功能需求，可以选择配置组件中的功能集。例如，选择配置图形框架组件中的部分控件。
- 平台可大可小：根据编译链关系，可以自动生成组件化的依赖关系。例如，选择图形框架组件，将会自动选择依赖的图形引擎组件等。

1.3 架构介绍

HarmonyOS整体遵从分层架构设计，从下向上依次为：内核层、系统服务层、框架层和应用层。系统功能按照"系统 > 子系统 > 功能/模块"逐级展开，在多设备部署场景下，支持根据实际需求裁剪某些非必要的子系统或功能/模块。HarmonyOS技术架构如图1-6所示。

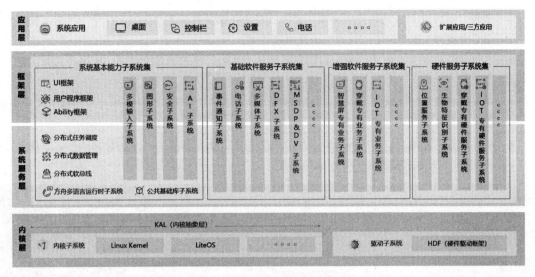

图 1-6 架构图

1.3.1 内核层

内核层主要分为两部分：

- 内核子系统：HarmonyOS采用多内核设计，支持针对不同资源受限设备选用适合的OS内核。内核抽象层（Kernel Abstract Layer，KAL）通过屏蔽多内核差异对上层提供基础的内核能力，包括进程/线程管理、内存管理、文件系统管理、网络管理和外设管理等，如图1-7所示。

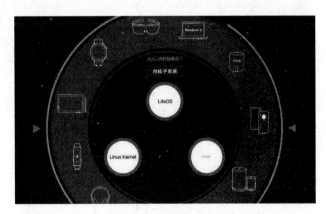

图 1-7 内核子系统

- 驱动子系统：鸿蒙驱动框架（HarmonyOS Driver Foundation，HDF）是HarmonyOS硬件生态开放的基础，提供统一外设访问能力和驱动开发、管理框架。

1.3.2 系统服务层

系统服务层是HarmonyOS的核心能力集合，通过框架层对应用程序提供服务。

图1-8展示了系统服务层的能力集合。该层包含以下几部分：

- 系统基本能力子系统集：为分布式应用在HarmonyOS多设备上的运行、调度、迁移等操作提供了基础能力，由分布式软总线、分布式数据管理、分布式任务调度、方舟多语言运行时、公共基础库、多模输入、图形、安全、AI等子系统组成。其中，方舟运行时提供了C/C++/JS

多语言运行时和基础的系统类库，也为使用方舟编译器静态化的Java程序（即应用程序或框架层中使用Java语言开发的部分）提供运行时。

- 基础软件服务子系统集：为HarmonyOS提供公共的、通用的软件服务，由事件通知、电话、多媒体、DFX（Design For X）、MSDP&DV等子系统组成。
- 增强软件服务子系统集：为HarmonyOS提供针对不同设备的、差异化的能力增强型软件服务，由智慧屏专有业务、穿戴专有业务、IoT专有业务等子系统组成。
- 硬件服务子系统集：为HarmonyOS提供硬件服务，由位置服务、生物特征识别、穿戴专有硬件服务、IoT专有硬件服务等子系统组成。

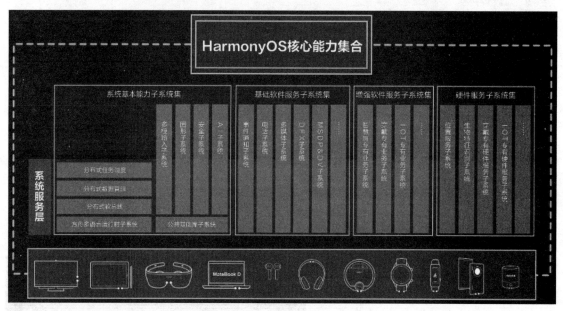

图1-8　系统服务层

根据不同设备形态的部署环境，基础软件服务子系统集、增强软件服务子系统集、硬件服务子系统集内部可以按子系统粒度裁剪，每个子系统内部又可以按功能粒度裁剪。

1.3.3　框架层

框架层为HarmonyOS应用开发提供了Java/C/C++/JS等多语言的用户程序框架和Ability框架、两种UI框架（包括适用于Java语言的Java UI框架、适用于JS语言的JS UI框架）以及各种软硬件服务对外开放的多语言框架API。根据系统的组件化裁剪程度，HarmonyOS设备支持的API也会有所不同。

图1-9展示了框架层所涵盖的功能。

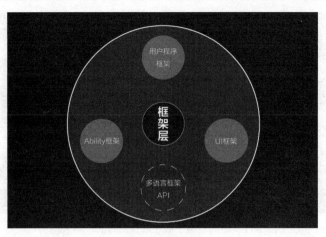

图 1-9　框架层

1.3.4　应用层

应用层包括系统应用和第三方非系统应用。HarmonyOS的应用由一个或多个FA（Feature Ability）或PA（Particle Ability）组成。其中，FA有UI界面，提供与用户交互的能力；而PA无UI界面，提供后台运行任务的能力以及统一的数据访问抽象。基于FA/PA开发的应用能够实现特定的业务功能，支持跨设备调度与分发，为用户提供一致、高效的应用体验。

图1-10展示的是一个视频通话应用的组成情况。

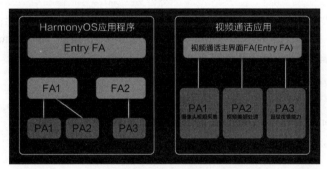

图 1-10　视频通话应用的组成

在一个视频通话应用中，往往会有一个FA作为视频通话的主界面，由若干个PA组成。FA提供UI界面用于与用户进行交互，PA1用于摄像头视频采集，PA2用于视频美颜处理，PA3用于超级夜景处理。这些FA、PA可以按需下载、加载和运行。

图1-11展示了不同设备下载相同应用时的不同表现。当手机下载该应用时，将同时拥有FA主界面、PA1摄像头视频采集、PA2视频美颜处理、PA3超级夜景处理；而当智慧屏下载该应用时，如果智慧屏不支持视频美颜处理、超级夜景处理功能，则只会下载FA主界面、PA1摄像头视频采集。

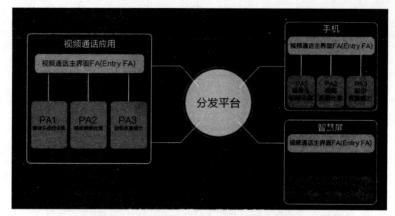

图 1-11　不同设备下载相同应用时的不同表现

1.4　获取开发支持

本书主要面向HarmonyOS应用开发，读者有关本书的任何问题都可以在本书的主页（https://github.com/waylau/harmonyos-tutorial/issues）进行提问。

除此之外，读者可以从以下网址获取新的HarmonyOS应用开发资讯：

- HarmonyOS官网：https://www.harmonyos.com/。
- OpenHarmony：https://gitee.com/openharmony。
- 华为开发者联盟：https://developer.huawei.com/consumer/cn/forum/blockdisplay?fid=0101303901040230869。
- 51CTO：https://harmonyos.51cto.com/。
- 电子发烧友：https://bbs.elecfans.com/harmonyos。
- 开源中国：https://www.oschina.net/group/harmonyos。
- CSDN：https://blog.csdn.net/harmonycommunity。

第 2 章

开发环境搭建

本章介绍HarmonyOS开发环境搭建。DevEco Studio是开发HarmonyOS应用常用的工具，本章会详细介绍DevEco Studio的安装和配置，以及一些常用的开发技巧。

2.1 注册华为开发者联盟账号

要进行HarmonyOS应用的开发，开发者首先要具备华为开发者联盟账号。华为开发者联盟开放诸多能力和服务，助力联盟成员打造优质应用。开发者需要注册华为开发者联盟账号，并且实名认证才能享受联盟开放的各类能力和服务。

账号注册完成后，可选择认证成为企业开发者或个人开发者。本书主要介绍如何认证成为个人开发者。

2.1.1 开发者享受的权益

表2-1所示是个人开发者和企业开发者的权益列表。从表中可以看到，企业开发者比个人开发者享受的服务更多。

表 2-1 个人开发者和企业开发者的权益

开发者类型	享受的服务/权益
个人开发者	应用市场、主题、商品管理、账号、PUSH、新游预约、互动评论、社交、HUAWEI HiAI、手表应用市场等
企业开发者	应用市场、主题、首发、支付、游戏礼包、应用市场推广、商品管理、游戏、账号、PUSH、新游预约、互动评论、社交、HUAWEI HiAI、手表应用市场、运动健康、云测、智能家居等

2.1.2 注册、认证准备的资料

对于个人开发者而言，注册、认证联盟账号需要准备以下资料：

- 注册：可接收验证码的手机号码或电子邮箱地址。
- 人工审核：身份证原件正反面扫描件或照片，以及手持身份证正面照片。
- 个人银行卡认证：个人银行卡号。

华为开发者联盟将在1至3个工作日内完成审核。审核完成后，将向提交的认证信息中的联系人邮箱发送审核结果。

2.1.3 注册账号

要注册华为开发者联盟账号，请按以下步骤注册：

1. 进入注册页面

打开华为开发者联盟官网（https://developer.huawei.com/consumer/cn/），单击"注册"按钮进入注册页面。

2. 进行注册

可以通过电子邮箱或手机号码注册华为开发者联盟账号。

如果使用电子邮箱注册，请输入正确的电子邮箱地址和验证码，设置密码后，单击"注册"按钮。电子邮箱注册页面如图2-1所示。

图 2-1 电子邮箱注册页面

如果使用手机号码注册，请输入正确的手机号码和验证码，设置密码后，单击"注册"按钮。手机号码注册页面如图2-2所示。

图 2-2　手机号码注册页面

2.1.4　登录账号

华为商城账号、华为云账号和花粉论坛账号均可登录联盟。

登录华为开发者联盟官网，单击右上角的"登录"，则会看到如图2-3所示的登录界面。输入账号和密码，单击"登录"按钮即可。或使用华为移动服务App扫一扫登录联盟。

图 2-3　登录页面

2.1.5　实名认证

华为开发者支持企业身份证验证和个人身份验证。本书主要介绍个人身份验证的过程。

打开华为开发者联盟官网，登录账号。单击"管理中心"跳转到开发者实名认证页面，如图2-4所示。

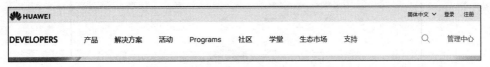

图 2-4　实名认证入口

1. 选择认证方式

在开发者实名认证页面，单击图2-5所示的"个人开发者"图标或"下一步"按钮，进入个人认证方式选择页面。

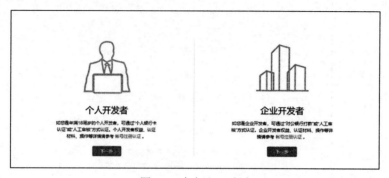

图 2-5　实名认证页面

个人实名认证有两种认证方式：个人银行卡认证和身份证人工审核认证，如图2-6所示。请根据上架应用的敏感性选择认证方式，如图2-7所示。

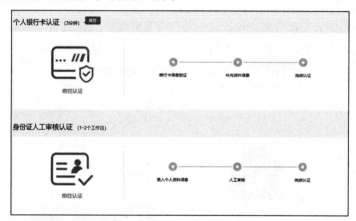

图 2-6　选择应用类型

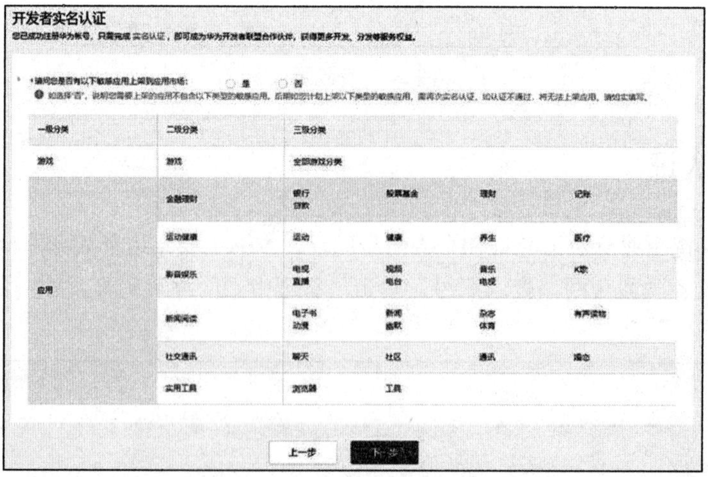

图 2-7　选择认证方式

如需上架的是敏感应用，请选择"是"，单击"下一步"按钮进入个人银行卡实名认证页面（如需上架的应用是敏感应用，只能通过个人银行卡认证方式进行实名认证）。

如需上架的是非敏感应用，请选择"否"，单击"下一步"按钮进入实名认证方式选择页面（如需上架的应用是非敏感应用，可选择个人银行卡认证或身份证人工审核认证中的一种认证方式进行认证）。

2. 个人银行卡认证

进入如图2-8所示的个人银行卡认证页面。

图2-8 个人银行卡认证页面

完善银行卡信息，若填写的信息正确，则单击"下一步"按钮跳转到完善信息页面，如图2-9所示。

图2-9 完善银行卡信息

完善个人信息，签署《华为开发者联盟与隐私的声明》和《华为开发者服务协议》，单击"提交"按钮，完成认证（标有红色星号（*）的项目是必填项），如图2-10所示。

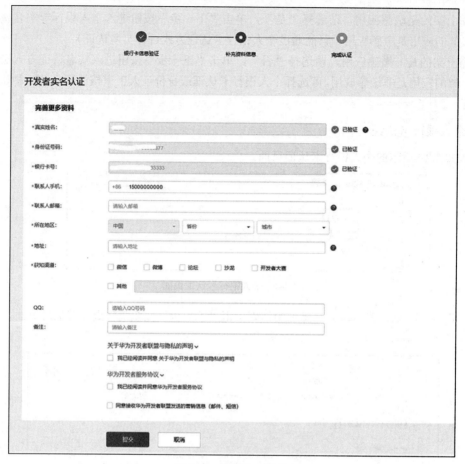

图 2-10　完善个人信息

3. 身份证人工审核认证

选择身份证人工审核认证，进入人工审核认证页面，如图2-11所示。

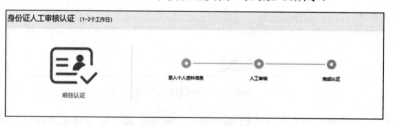

图 2-11　人工审核认证页面

完善个人信息，签署《华为开发者联盟与隐私的声明》和《华为开发者服务协议》，单击"下一步"按钮，等待审核，审核结果会在1~2个工作日发送至联系人邮箱（标有红色星号（*）的项目是必填项），如图2-12所示。

图 2-12 完善个人信息

2.2 DevEco Studio 下载安装

HUAWEI DevEco Studio（以下简称DevEco Studio）是基于IntelliJ IDEA Community开源版本打造、面向华为终端全场景多设备的一站式集成开发环境（Integrated Development Environment，IDE），为开发者提供工程模板创建、开发、编译、调试、发布等E2E的HarmonyOS应用开发服务。通过使用DevEco Studio，开发者可以更高效地开发具备HarmonyOS分布式能力的应用，进而提升创新效率。

作为一款开发工具，除了具有基本的代码开发、编译构建及调测等功能外，DevEco Studio还具有如下特点：

- 多设备统一开发环境：支持多种HarmonyOS设备的应用开发，包括手机（Phone）、平板电脑（Tablet）、车机（Car）、智慧屏（TV）、智能穿戴（Wearable）、轻量级智能穿戴（LiteWearable）和智慧视觉（Smart Vision）设备。

- 支持多语言的代码开发和调试：包括Java、XML（Extensible Markup Language）、C/C++、JS（JavaScript）、CSS（Cascading Style Sheets）和HML（HarmonyOS Markup Language）。
- 支持FA（Feature Ability）和PA（Particle Ability）快速开发：通过工程向导快速创建FA/PA工程模板，一键式打包成HAP（HarmonyOS Ability Package）。
- 支持分布式多端应用开发：开发者使用一个工程和一份代码就可以跨设备运行，支持不同设备界面的实时预览和差异化开发，实现代码的最大化重用。
- 支持多设备模拟器：提供多设备的模拟器资源，包括手机、平板电脑、车机、智慧屏、智能穿戴设备的模拟器，方便开发者高效调试。
- 支持多设备预览器：提供JS和Java预览器功能，可以实时查看应用的布局效果，支持实时预览和动态预览；还支持多设备同时预览，查看同一个布局文件在不同设备上的呈现效果。

DevEco Studio支持Windows和Mac版本，两个版本的安装步骤类似。因此，接下来将只针对Windows操作系统的软件安装方式进行介绍。

2.2.1 运行环境要求

为了保证DevEco Studio正常运行，建议计算机配置满足如下要求：

- 操作系统：Windows10 64位。
- 内存：8GB及以上。
- 硬盘：100GB及以上。
- 分辨率：1280×800像素及以上。

2.2.2 下载和安装 Node.js

Node.js软件仅在使用到JS语言开发HarmonyOS应用时才需要安装，使用其他语言开发不用安装Node.js，请跳过此章节。

登录Node.js官方网站（https://nodejs.org/en/download/），下载Node.js软件包。请选择LTS版本，Windows 64位对应的软件包。

单击下载后的软件包进行安装，全部按照默认设置，单击Next按钮，直至完成。安装过程中，Node.js会自动在系统的path环境变量中配置node.exe的目录路径。

有关Node.js的更多内容可以参考笔者所著的《Node.js企业级应用开发实战》。

2.2.3 下载和安装 DevEco Studio

DevEco Studio的编译构建依赖JDK，DevEco Studio预置了Open JDK，版本为1.8，安装过程中会自动安装JDK。

进入HUAWEI DevEco Studio产品页（https://developer.harmonyos.com/cn/develop/deveco-studio），下载DevEco Studio安装包。双击下载的deveco-studio-xxxx.exe，进入DevEco Studio安装向导，在安装选项界面勾选DevEco Studio launcher后，单击Next按钮，直至安装完成，如图2-13所示。

安装完成后，先不要勾选Run DevEco Studio选项（见图2-14），接下来根据配置开发环境检查和配置开发环境。

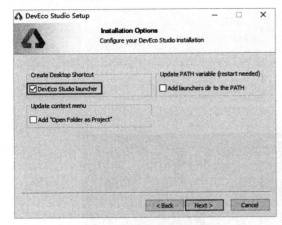

图 2-13　安装 DevEco Studio

图 2-14　安装 DevEco Studio 完成

2.3　设置 DevEco Studio（可选）

DevEco Studio开发环境需要依赖于网络环境，需要连接上网络才能确保工具的正常使用。可以先尝试是否能够正常使用DevEco Studio。如果可以正常使用，则可以跳过本节的设置；如果不能正常使用，再根据如下两种情况来配置开发环境：

- 如果可以直接访问互联网，只需进行设置npm仓库和下载HarmonyOS SDK操作。
- 如果不能直接访问互联网，需要通过代理服务器才可以访问，请根据本节内容逐条设置开发环境。

2.3.1　npm 设置

1. 设置 npm 代理

只有在同时满足以下两个条件时，需要配置npm代理，否则可跳过本章节。

- 需要使用JS语言开发HarmonyOS应用。
- 不能直接访问互联网，而是需要通过代理服务器才可以访问。这种情况下，配置npm代理，便于从npm服务器下载JS依赖。

打开命令行工具，按照如下方式进行npm代理设置和验证。

执行如下命令设置npm代理。

如果使用的代理服务器需要认证，请按照如下方式进行设置（对user、password、proxyserver和port按照实际代理服务器进行修改）。

```
npm config set proxy http://user:password@proxyserver:port
```

```
npm config set https-proxy http://user:password@proxyserver:port
```

如果使用的代理服务器不需要认证（不需要账号和密码），请按照如下方式进行设置：

```
npm config set proxy http://proxyserver:port
npm config set https-proxy http://proxyserver:port
```

代理设置完成后，执行如下命令进行验证：

```
npm info express
```

执行结果如图2-15所示，说明代理设置成功。

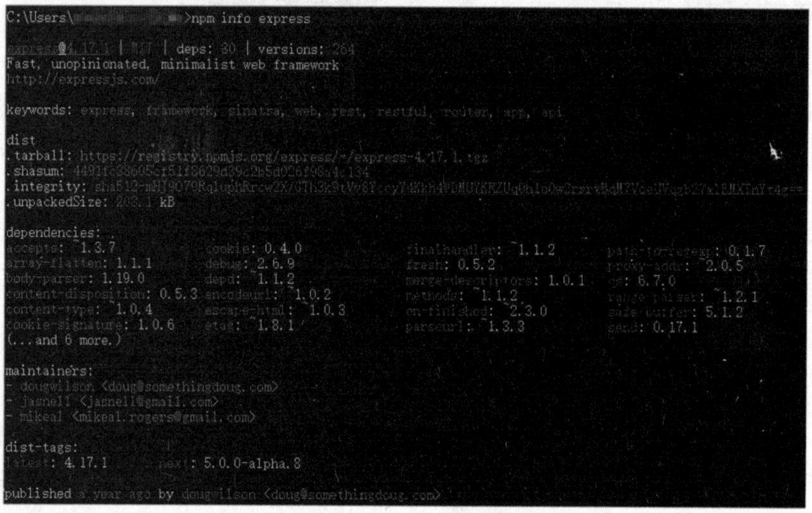

图 2-15　验证 npm 代理设置是否完成

2. 设置 npm 镜像仓库

国内或者企业内部环境访问npm官网仓库往往速度比较慢。为了提升下载JS SDK时使用npm安装JS依赖的速度，需要设置npm镜像仓库。

国内有非常多的网址都提供了npm镜像仓库，比如：

- 淘宝：http://registry.npm.taobao.org。
- 华为：https://mirrors.huaweicloud.com/repository/npm/。

重新设置npm仓库地址，用法如下：

```
npm config set registry https://mirrors.huaweicloud.com/repository/npm/
```

2.3.2　设置 Gradle 代理

如果不能直接访问互联网，而是需要通过代理服务器才可以访问，这种情况下，需要设置Gradle代理来访问和下载Gradle所需的依赖，否则可跳过本章节。

打开"此电脑"，在文件夹地址栏中输入%userprofile%，进入个人用户界面。

创建一个文件夹，命名为.gradle。如果已有.gradle文件夹，请跳过此操作。

进入.gradle文件夹，新建一个名为gradle.properties的文件，如图2-16所示。

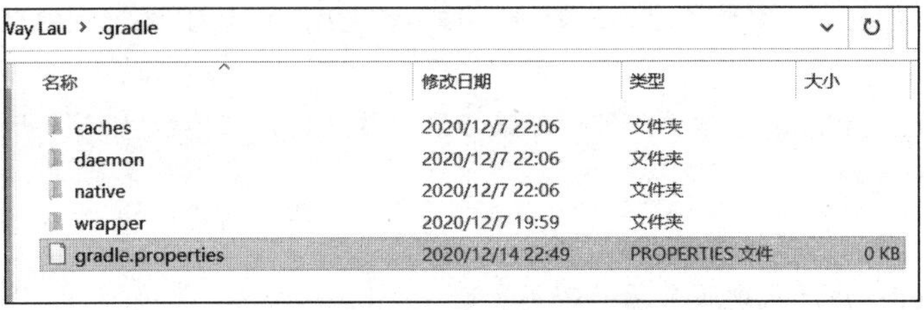

图 2-16 gradle.properties 文件

打开gradle.properties文件，添加如下脚本，然后保存。

```
systemProp.http.proxyHost=proxy.server.com
systemProp.http.proxyPort=8080
systemProp.http.nonProxyHosts=*.company.com|10.*|100.*
systemProp.http.proxyUser=userId
systemProp.http.proxyPassword=password
systemProp.https.proxyHost=proxy.server.com
systemProp.https.proxyPort=8080
systemProp.https.nonProxyHosts=*.company.com|10.*|100.*
systemProp.https.proxyUser=userId
systemProp.https.proxyPassword=password
```

其中代理服务器、端口、用户名、密码和不使用代理的域名，请根据实际代理情况进行修改。其中不使用代理的nonProxyHosts的配置间隔符是"|"。

2.3.3 设置 DevEco Studio 代理

如果不能直接访问互联网，而需要通过代理服务器才可以访问，这种情况下，需要设置DevEco Studio代理来访问和下载外部资源，否则可跳过本章节。

运行已安装的DevEco Studio，首次使用请选择Do not import settings，单击OK按钮。

根据DevEco Studio欢迎界面的提示，单击Setup Proxy，或者在欢迎页单击Configure→Settings→Appearance & Behavior→System Settings→HTTP Proxy，进入HTTP Proxy设置界面。

设置DevEco Studio的HTTP Proxy信息，如图2-17所示。

图 2-17 设置 DevEco Studio 代理

设置内容包括：

- HTTP配置项，设置代理服务器信息。
  - ➢ Host name：代理服务器主机名或IP地址。

- Port number：代理服务器对应的端口号。
- No proxy for：不需要通过代理服务器访问的 URL 或者 IP 地址（地址之间用英文逗号分隔）。
● Proxy authentication 配置项，如果代理服务器需要通过认证鉴权才能访问，则需要设置，否则可跳过该配置项。
- Login：访问代理服务器的用户名。
- Password：访问代理服务器的密码。
- Remember：勾选，记住密码。

配置完成后，单击 Check connection，输入网络地址（如：https://waylau.com），检查网络连通性。提示 Connection successful 表示代理设置成功。单击 OK 按钮完成 DevEco Studio 代理配置。

DevEco Studio 代理设置完成后，会提示安装 HarmonyOS SDK，可以单击 Next 按钮下载到默认目录中。如果想更改 SDK 的存储目录，可单击 Cancel 按钮，并根据下载 HarmonyOS SDK 进行操作。

2.3.4 下载 HarmonyOS SDK

Devco Studio 提供 SDK Manager 统一管理 SDK 及工具链，下载各种编程语言的 SDK 包时，SDK Manager 会自动下载该 SDK 包依赖的工具链。

SDK Manager 提供多种编程语言的 SDK 包和工具链，具体说明可参考表 2-2。

表 2-2 SDK 及工具链

包 名	说 明	默认是否下载
Native	C/C++ 语言 SDK 包	×
JS	JS 语言 SDK 包	×
Java	Java 语言 SDK 包	√
Toolchains	SDK 工具链，HarmonyOS 应用开发必备工具集，包括编译、打包、签名、数据库管理等工具的集合	√
Previewer	HarmonyOS 应用预览器，在开发过程中可以动态预览 Phone、TV、Wearable、LiteWearable 等设备的应用效果，支持 JS 和 Java 应用预览	×

由此可见，DevEco Studio 开箱即用支持 Java 语言来开发 HarmonyOS。如果是其他编程语言（比如 JS），则需要额外自行安装 JS SDK。

在菜单栏单击 Configure→Settings 或者按快捷键 Ctrl+Alt+S，打开 Settings 配置界面。

进入 Appearance & Behavior→System Settings→HarmonyOS SDK 菜单界面，单击 Edit 按钮，设置 HarmonyOS SDK 存储路径，如图 2-18 所示。

选择 HarmonyOS SDK 存储路径（不能包含中文），然后单击 Next 按钮。在弹出的 License Agreement 窗口单击 Accept 开始下载 SDK。如果本地已有 SDK 包，请选择本地已有 SDK 包的存储路径，DevEco Studio 会增量更新 SDK 及工具链。

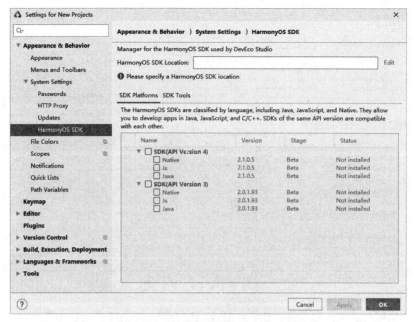

图 2-18　设置 HarmonyOS SDK 存储路径

等待HarmonyOS SDK及工具下载完成，单击Finish按钮，可以看到默认的SDK Platforms下的Java及SDK Tools下的Toolchains已完成下载，如图2-19所示。

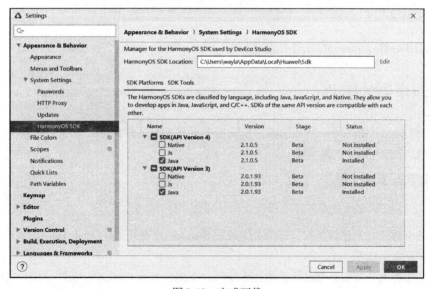

图 2-19　完成下载

如果工程还会用到JS或者C/C++语言，请在SDK Platform中勾选对应的SDK包，单击Apply按钮，SDK Manager会自动将SDK包和工具链下载到指定的SDK存储路径中，如图2-20所示。

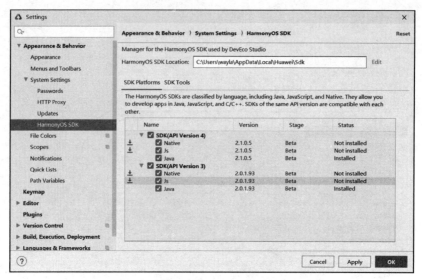

图 2-20　下载 JS 或者 C/C++ 语言 SDK 和工具链

开发环境配置完成后，可以通过运行HelloWorld工程来验证环境设置是否正确。

2.4　DevEco Studio 功能介绍

下面介绍DevEco Studio的常用功能。

2.4.1　创建新的工程

当开始开发一个HarmonyOS应用时，首先需要根据工程创建向导，创建一个新的工程，工具会自动生成对应的代码和资源模板。

1. 创建和配置新工程

首先，通过如下两种方式打开工程创建向导界面。

- 如果当前未打开任何工程，可以在DevEco Studio的欢迎页选择Create HarmonyOS Project开始创建一个新工程。
- 如果已经打开工程，可以在菜单栏选择File→New→New Project来创建一个新工程。

接着，根据工程创建向导选择需要开发的设备类型，然后选择对应的Ability模板，如图2-21所示。

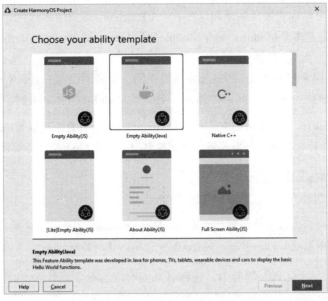

图 2-21　工程创建向导界面

单击Next按钮，进入工程配置阶段，需要根据向导配置工程的基本信息。

- Project Name：工程的名称，可以自定义。
- Project Type：工程的类型，可选Service或Application。
- Package Name：软件包名称，默认情况下，应用ID也会使用该名称，应用发布时，应用ID需要唯一。
- Save Location：工程文件本地存储路径，存储路径中不能包含中文字符。
- Compatible API Version：兼容的SDK最低版本。
- Device Type：设备类型。
- Show in Service Center：是否在服务中心露出，仅在Ability模板存在。

单击Finish按钮，工具会自动生成示例代码和相关资源，等待工程创建完成。

2. 打开现有工程

打开现有工程时请注意，待导入的工程文件存储路径不能包含中文字符。打开现有工程包括如下两种方式：

- 如果当前未打开任何工程，可以在DevEco Studio的欢迎页选择Open Project打开现有工程。
- 如果已经打开了工程，可以在菜单栏选择File→Open来打开现有工程。

打开现有工程时，DevEco Studio会提醒你可以选择在新的窗口打开工程，或者选择在当前窗口打开工程。

2.4.2　添加 Module

Module是HarmonyOS应用的基本功能单元，包含源代码、资源文件、第三方库及应用清单文件，每一个Module都可以独立进行编译和运行。一个HarmonyOS应用通常包含一个或多个Module，因此，可以在工程中创建多个Module，每个Module分为Ability和Library（HarmonyOS Library和Java Library）两种类型。

如HarmonyOS工程介绍的，在一个App中，对于同一类型的设备有且只有一个Entry Module，其余Module的类型均为Feature。因此，在创建一个类型为Ability的Module时，遵循如下原则：

- 若新增Module的设备类型为已有设备，则Module的类型将自动设置为Feature。
- 若新增Module的设备类型为当前还没有创建Module，则Module的类型将自动设置为Entry。

接下来介绍新增Module的步骤。

首先，通过如下两种方法在工程中添加新的Module。

- 方法1：鼠标移到工程目录顶部，右击选择New→Module，开始创建新的Module。
- 方法2：在菜单栏选择File→New→Module，开始创建新的Module。

接着，在New Project Module界面中，选择Module对应的设备类型和模板，如图2-22所示。

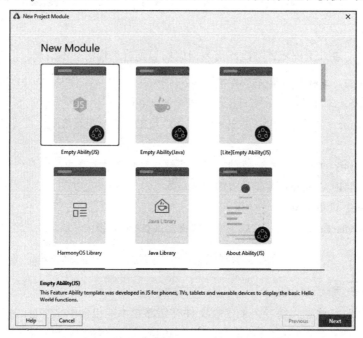

图2-22　选择 Module 对应的设备类型和模板

单击Next按钮，在Module配置页面设置新增Module的基本信息。

- Module类型为Ability或者HarmonyOS Library时，请根据如下内容进行设置，然后单击Next按钮。如图2-23所示。
  - Application/Library name：新增 Module 所属的类名称。
  - Module Name：新增模块的名称。
  - Module Type：仅在 Ability 模板存在，工具自动根据设备类型下的模块进行设置。
  - Package Name：软件包名称，可以单击 Edit 按钮修改默认包名称，需全局唯一。

- Compatible API Version：兼容的 SDK 最低版本。
- Device Type：选择模块的设备类型，如果新建模块的 Module Type 为 Feature，则只能选择该工程原有的设备类型；如果 Module Type 为 Entry，可以选择该 Module 支持的其他设备类型。
- Show in Service Center：是否在服务中心露出，仅在 Ability 模板存在。

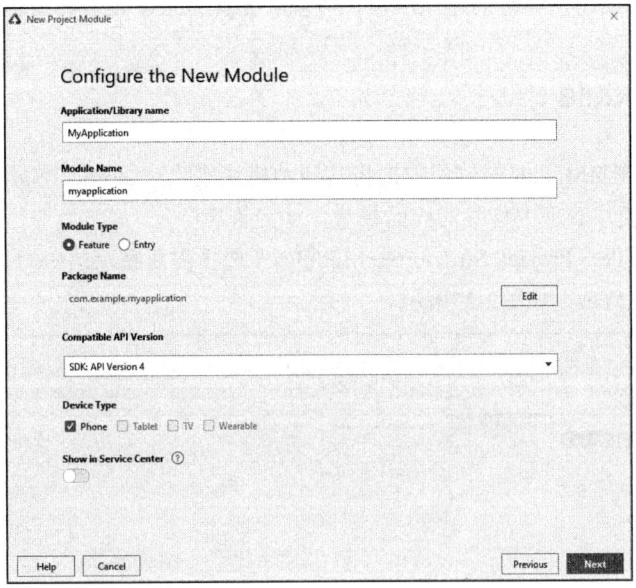

图 2-23　Module 类型为 Ability 或者 HarmonyOS Library

● Module类型为Java Library时，请根据如下内容进行设置，然后单击Finish按钮完成创建，如图2-24所示。
- Library name：Java Library 类名称。
- Java package name：软件包名称，可以单击 Edit 按钮修改默认包名称，需全局唯一。
- Java class name：class 文件名称。
- Create.gitignore file：是否自动创建.gitignore 文件，勾选表示创建。

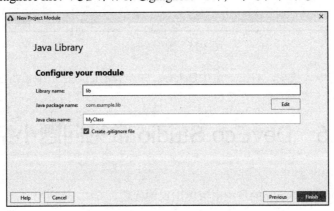

图 2-24　Module 类型为 Java Library

接着，设置新增Ability或HarmonyOS Library的Page Name。若该Module的模板类型为Ability，则还需要设置Visible参数，表示该Ability是否可以被其他应用调用。

- 勾选（true）：可以被其他应用调用。
- 不勾选（false）：不能被其他应用调用。

最后，单击Finish按钮，等待创建完成后，可以在工程目录中查看和编辑新增的Module。

2.4.3 删除Module

为防止开发者在删除Module的过程中误将其他的模块删除，DevEco Studio提供了统一的模块管理功能，需要先在模块管理中移除对应的模块，才允许删除。

在菜单栏中选择File→Project Structure→Modules，选择需要删除的Module，单击 按钮，并在弹出的对话框中单击Yes，如图2-25所示。

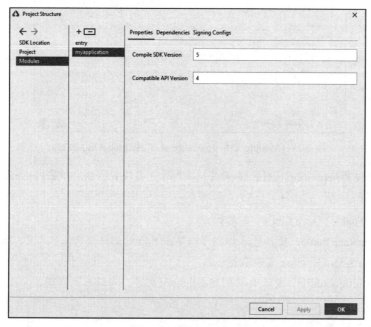

图2-25　删除Module

在工程目录中选中该模块，右击，选中Delete，并在弹出的对话框中单击Delete即可。

2.5　DevEco Studio常见问题小结

本节介绍使用DevEco Studio的过程中可能出现的问题。

2.5.1 问题 1：访问 Gradle 仓库慢

1. 原因

由于国内环境或者企业内网环境问题，往往访问 Maven 仓库比较困难，此时可以设置 Gradle 仓库为国内或者企业内部的镜像仓库。

2. 解决

设置步骤如下：

（1）在用户目录新建一个 .gradle 文件夹，比如我的机器登录账户是 waylau，具体路径为 C:.gradle。

（2）在这个目录下新建一个 init.gradle 文件放入以下内容（以下配置仅为示例，需根据实际的镜像仓库地址进行配置）：

```
allprojects {
    buildscript {
        repositories {
            mavenLocal()
            maven {
                url 'https://repo.huaweicloud.com/repository/maven/'
            }
            maven {
                url 'https://developer.huawei.com/repo/'
            }
        }

        dependencies {
        }
    }
    repositories {
        mavenLocal()
        maven {
            url 'https://repo.huaweicloud.com/repository/maven/'
        }
        maven {
            url 'https://developer.huawei.com/repo/'
        }
    }
}

settingsEvaluated { settings ->
    settings.pluginManagement {
        plugins {
        }
        resolutionStrategy {
```

```
        }
        repositories {
            maven {
                url 'https://repo.huaweicloud.com/repository/maven/'
            }
            maven {
                url 'https://developer.huawei.com/repo/'
            }
        }
    }
}
```

2.5.2　问题2：模拟器端口被占用无法启动

1. 问题

在内网环境下首次使用DevEco Studio创建应用时，可能会报如下问题：

```
server not running; starting it at tcp:5037
```

2. 原因

默认端口被占用了，需要重新指定一个。

3. 解决

如何指定？在系统变量中加一个HDC_SERVER_PORT，值就是你想要使用的端口，如图2-26所示。

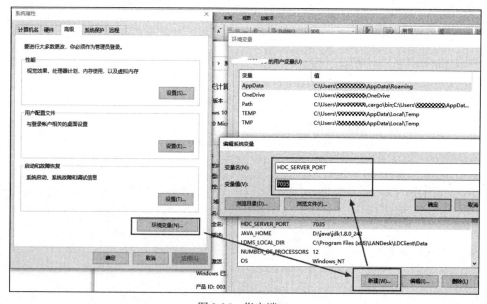

图 2-26　指定端口

上述例子指定使用的是7035端口，这样模拟器就能正常启动了。

第 3 章

开发第一个 HarmonyOS 应用

本章演示如何基于DevEco Studio来开发第一个HarmonyOS应用。在本章，我们不费吹灰之力（不用编写一行代码）就能开发出一个能够直接运行的HarmonyOS应用。

3.1 创建一个新工程

根据上一章的学习，我们已经安装好了DevEco Studio，终于可以进入激动人心的开发环节了。本节将演示如何基于DevEco Studio来开发第一个HarmonyOS应用。按照编程惯例，第一个应用称为Hello World应用。

3.1.1 选择创建新工程

在打开DevEco Studio后，单击Create HamonyOS Project来创建一个新工程。

3.1.2 选择设备应用类型的模板

看到如图3-1所示的界面时，选择支持不同设备应用类型的模板。不同的模板支持的设备是有差异的，比如本例子所选择的Empty Ability(Java)模板支持包括手机、平板、车机、智慧屏、智能穿戴设备等多种终端设备。有关Ability的概念，我们后续再介绍，这里就简单地认为Ability是应用的一个功能。换言之，我们将要创建的是一个没有功能的应用。单击Next按钮进行下一步。

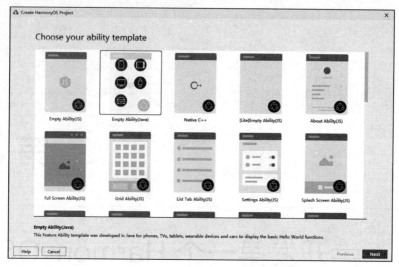

图 3-1　选择支持不同设备应用类型的模板

3.1.3　配置项目的信息

配置项目的信息，比如项目名称、包名、位置、SDK版本等，如图3-2所示。这个按照个人实际情况填写即可。

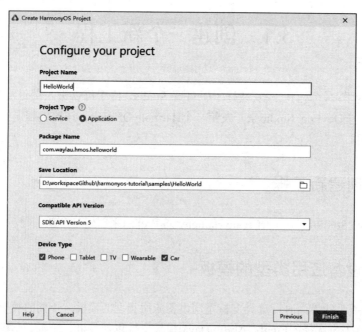

图 3-2　配置项目的信息

这里需要注意的是：

- Project Type这项选择Application，意味着这是一个独立的应用。
- Device Type用于配置目标安装的设备类型。因为本书是面向手机应用开发的，所以至少要选择Phone。当然，HarmonyOS是支持多设备应用开发的，如果同时选择了Phone和Car，则意味着这个项目支持在手机和车机上运行。

3.1.4　自动生成工程代码

单击Finish按钮之后，DevEco Studio就会为我们创建整个应用，并且自动生成工程代码。由于HarmonyOS应用是采用Gradle构建的，因此可以在控制台看到自动下载Gradle安装包。Gradle下载完成之后，会对工程进行配置，因此可以看到如图3-3所示的控制台配置成功的提示信息。

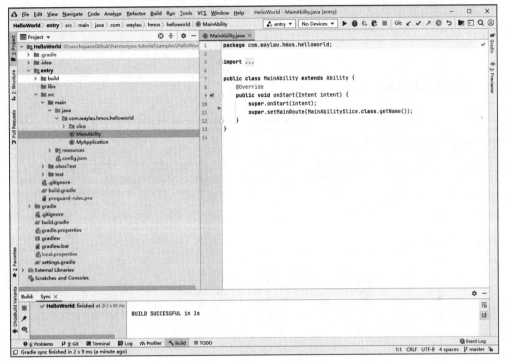

图 3-3　自动生成工程代码

3.2　运行工程

在构建好应用之后，如何来运行应用呢？

3.2.1　单击"运行"按钮

单击DevEco Studio工具栏中的"运行"按钮（三角形，见图3-4）运行工程，或使用默认快捷键Shift+F10运行工程。

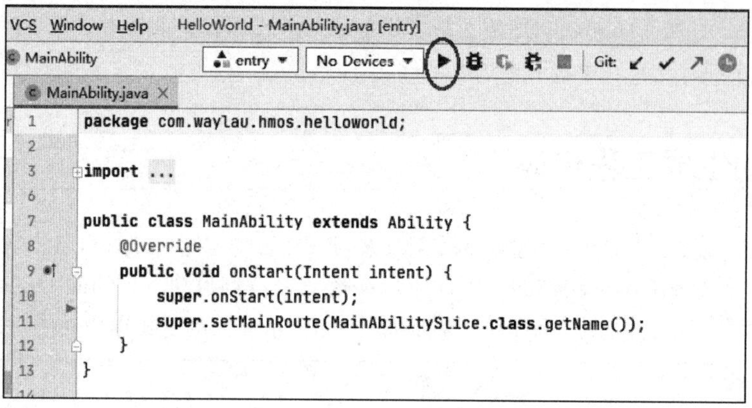

图 3-4　运行工程

3.2.2　选择模拟器

此时，在 DevEco Studio 右下角会弹出如图 3-5 所示的错误提示：Select a device first，意思是启动应用前先要选择相应设备的模拟器。

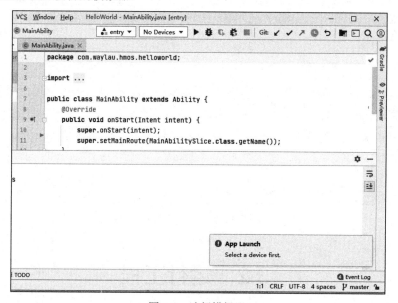

图 3-5　选择模拟器

3.2.3　启动模拟器

我们所创建的这个应用目标是可以在手机和车机上运行，因此需要相应的手机和车机设备模拟器来支持运行。在 DevEco Studio 菜单栏单击 Tools→Device Manager，启动设备模拟器，如图 3-6 所示。

访问 Device Manager 页面，此时需要使用华为开发者账号登录，并根据提示对设备进行授权，如图 3-7 所示。

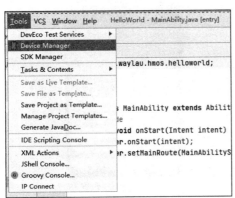

图 3-6　Device Manager

图 3-7　对设备进行授权

单击"允许"按钮进行下一步操作。看到如图3-8所示的界面，代表授权已经成功。

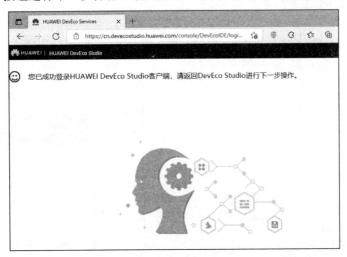

图 3-8　授权成功

授权完成之后，再次返回DevEco Studio，此时看到如图3-9所示的各种类型的设备模拟器列表，列表中就包含Phone和Car模拟器，单击启动Phone模拟器（以P40为例）。

这时，能看到Phone模拟器已经启动了，如图3-10所示。

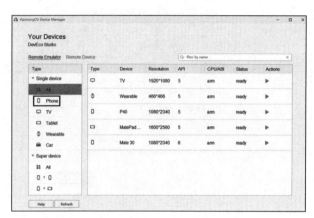

图 3-9　启动 Phone 模拟器

图 3-10　Phone 模拟器已经启动了

3.2.4　再次运行工程

再次运行工程，此时就能选中Phone模拟器了，即为图3-11的HUAWEI ANA-AN00。最终，工程运行效果如图3-12所示。

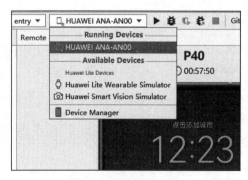

图 3-11　再次运行工程

图 3-12　应用运行效果

这样，我们就实现了一个应用的开发和运行了。

3.3 在本地真机中运行应用

上一节介绍了应用的开发，以及如何在Phone模拟器中运行。这种运行方式称为远程模拟器（Remote Emulator）运行。远程模拟器支持Phone、Tablet、Car、TV、Wearable等设备。本节将介绍如何在真实的本地设备中运行应用。

需要注意的是，无论是在本地真机还是在远程真机中运行应用，首先需要对应用进行签名。有关应用的签名可以参见3.4节。

3.3.1 连接真实的设备

如果你手上刚好有一台真实的设备，那么也可以将应用直接在设备中运行。

一般是可以用USB或IP方式连接的实体设备，支持USB方式连接的有Phone、Tablet、Car、Wearable等设备，支持IP方式连接的有TV等设备。

以下是连接不同设备的具体步骤。

1. Phone 或者 Tablet

（1）在Phone或者Tablet中，打开"开发者模式"，可在"设置"→"关于手机/关于平板"中连续多次单击"版本号"，直到提示"您正处于开发者模式"。

（2）使用USB方式将Phone或者Tablet与PC端进行连接。

（3）在Phone或者Tablet中，USB连接方式选择"传输文件"。

（4）在Phone或者Tablet中，打开"设置"→"系统和更新"→"开发人员选项"，打开"USB调试"开关。

2. TV

（1）将TV和PC连接到同一网络或设置为同一个网段。

（2）获取TV端的IP地址。

（3）TV上的5555端口为打开状态。

（4）在DevEco Studio菜单栏中，单击Tools→IP Connect，输入连接设备的IP地址，点击，连接正常后，设备状态为online，如图3-13所示。

3. Car 或者 Wearable

在Car或者Wearable中，直接使用USB方式连接Car或者Wearable和PC端即可。

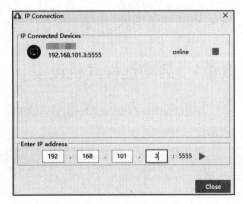

图 3-13 运行应用

3.3.2 运行应用

假设要在真实Phone设备中运行HelloWorld应用,那么步骤如下:

(1)使用USB方式连接Phone设备和你的开发PC。

(2)在DevEco Studio菜单栏中单击Run→Run 'entry'或单击放大,或使用快捷键Shift+F10运行应用,如图3-14所示。

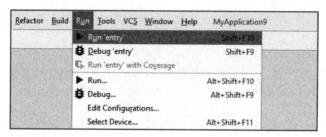

图3-14 运行应用

(3)在弹出的界面中选择已连接的Phone设备,单击OK按钮。

(4)DevEco Studio将会启动HAP的编译构建和安装。安装成功后,Phone会自动运行安装的HarmonyOS应用。

3.4 使用远程真机运行应用

如果开发者没有真机设备资源,则不能很方便地调试和验证HarmonyOS应用。为了方便开发者,DevEco Studio 2.2 Beta1及更高版本中提供了远程真机设备资源供开发者使用,以减少开发成本。目前,远程真机支持Phone和Wearable设备,开发者使用远程真机调试和运行应用时,同本地物理真机设备一样,需要对应用进行签名才能运行。

相比远程模拟器,远程真机是部署在云端的真机设备资源,远程真机的界面渲染和操作体验更加流畅,同时也可以更好地验证应用在真机设备上的运行效果,比如性能、手机网络环境等。

3.4.1 启动远程真机设备

单击Tools→Device Manager,在Remote Device页签中可以看到远程真机设备列表,如图3-15所示。

单击启动Phone远程真机设备(以P40为例),启动情况如图3-16所示。

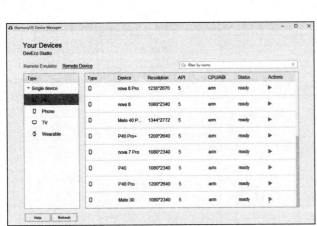
图 3-15 远程真机设备列表

图 3-16 Phone 远程真机设备

3.4.2 运行应用

此时就能选中Phone远程真机设备了，即为图3-17中的HUAWEI ANA-AN00。

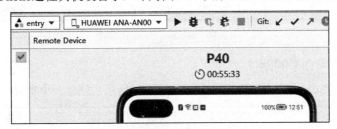

图 3-17 选中 Phone 远程真机设备

单击Run 'entry'运行应用。此时会报如图3-18所示的异常。

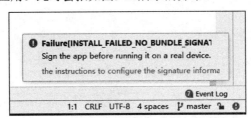

图 3-18 应用运行异常

该异常提示在真机设备中运行应用需要对应用进行签名。

3.4.3 对应用进行签名

接下来介绍对应用进行签名的步骤。

1. 进入签名配置界面

进入File→Project Structure→Project→Signing Configs界面，单击Sign In按钮进行登录，如图3-19所示。

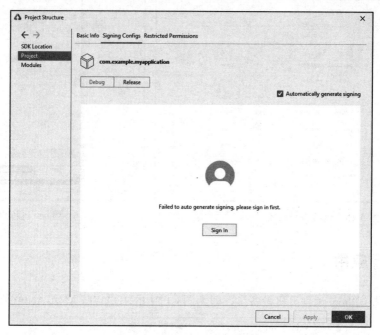

图3-19　Signing Configs 界面

2. 登录 AppGallery Connect

登录AppGallery Connect创建项目和应用。如图3-20所示，创建一个项目。

图3-20　创建项目

名称可以自定义，如图3-21所示。

第 3 章　开发第一个 HarmonyOS 应用　　43

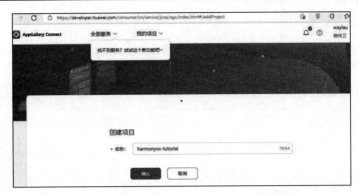

图 3-21　输入项目名称

接着单击"添加应用"按钮，如图3-22所示。

图 3-22　添加应用

完善应用信息，单击"确认"按钮即可，如图3-23所示。

图 3-23　完善应用信息

3. 再次进入签名配置界面

再次进入签名配置界面，此时可以看到已经能够正常获取用户的签名信息，单击OK按钮即可完成应用的签名，如图3-24所示。

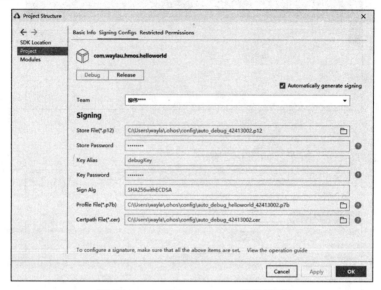

图 3-24　AppGallery Connect 登录界面

3.4.4　再次运行应用

完成签名之后，就能正常运行应用，如图3-25所示。

图 3-25　应用运行效果

3.5 使用 DevEco Studio 预览器

在前面的章节中，我们已经认识到了如何创建一个简单的HelloWorld应用，并且可以通过设备模拟器或者真机运行应用。

但是使用设备模拟器或者真机运行应用有一个缺点，那就是启动相对来说比较慢。如果只是调试一个简单的界面，却要等待非常久的时间，就太消磨人的耐心了。此时，推荐的方式是使用预览器。

3.5.1 如何安装预览器

在使用预览器查看应用界面的UI效果前，需要确保HarmonyOS SDK→SDK Tools中已下载Previewer资源，如图3-26所示。

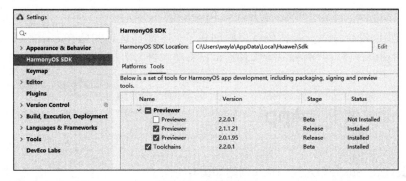

图 3-26　安装预览器

3.5.2 如何使用预览器

打开预览器有两种方式：

- 在菜单栏单击View→Tool Windows→Previewer，打开预览器。
- 在编辑窗口右上角的侧边工具栏单击Previewer，打开预览器。

显示效果如图3-27所示。

图 3-27　使用预览器

第 4 章

探索 HarmonyOS 应用

在第3章，我们没有输入一行代码就轻松创建了一个能够运行的HarmonyOS应用。那么这个HarmonyOS应用到底包含哪些内容，它为什么可以直接在设备上运行？本章将进行深入探索。

4.1 App

在第3章DevEco Studio会为我们自动创建如图4-1所示的目录结构的工程源码。那么这些目录结构到底有什么含义，每个文件的作用是什么？这些都是接下来我们要探索的话题。

本节将探索App这个话题。

4.1.1 什么是App

首先，我们来认识App。

App就是应用，是Application的简写。比如，在Android手机上安装一个软件，这个软件就称为应用。应用泛指运行在设备的操作系统之上，为用户提供特定服务的程序。

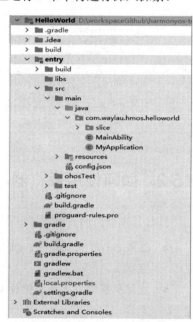

图 4-1 DevEco Studio 自动创建的工程

在HarmonyOS上运行的应用有两种形态：

- 传统方式需要安装的应用。
- 提供特定功能，免安装的应用，即原子化服务。

4.1.2 应用程序包结构

Android平台是以APK（Android Application Package，Android应用程序包）形式来发布的，当我们要在Android手机上安装一个App时，首先要找到这个App对应的APK安装包，执行该APK安装包就能安装App了。

同理，HarmonyOS的应用软件包以App Pack（Application Package）形式发布，它是由一个或多个HAP（HarmonyOS Ability Package）以及描述每个HAP属性的pack.info组成的。HAP是Ability的部署包，HarmonyOS应用代码围绕Ability组件展开。

一个HAP是由代码、资源、第三方库及应用配置文件组成的模块包，可分为entry和feature两种模块类型，如图4-2所示。

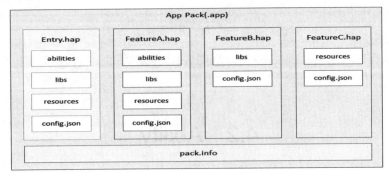

图 4-2　App 逻辑视图

其中：

- entry：应用的主模块。一个App中，对于同一设备类型必须有且只有一个entry类型的HAP，可独立安装运行。
- feature：应用的动态特性模块。一个App可以包含一个或多个feature类型的HAP，也可以不包含。只有包含Ability的HAP才能够独立运行。

我们在应用的build目录下可以找到名为entry-debug-rich-unsigned.hap或者entry-debug-rich-signed.hap的文件，如图4-3所示。这些.hap文件就是HarmonyOS的应用软件包。

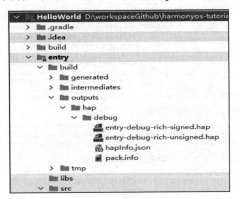

图 4-3　App 逻辑视图

4.1.3 代码层次的应用

在代码层次，我们可以看到如图4-4所示的MyApplication，这就是整个应用的入口。

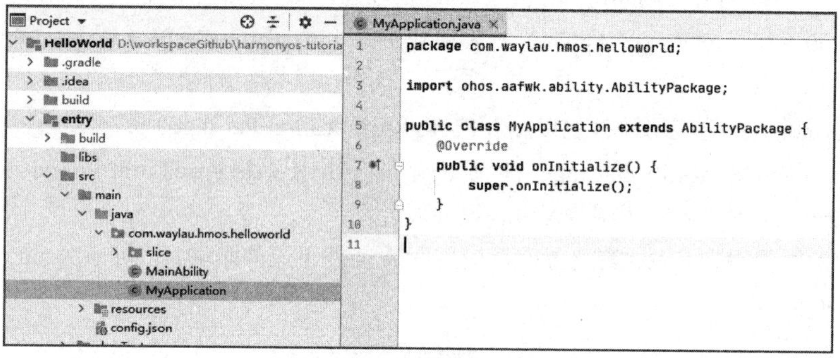

图4-4 App 逻辑视图

从代码可以看到，MyApplication继承自AbilityPackage。AbilityPackage用来初始化每个HAP的基类。

4.2 Ability

Ability是应用所具备的能力的抽象，一个应用可以包含一个或多个Ability。

Ability分为两种类型：FA（Feature Ability）和PA（Particle Ability）。

4.2.1 Ability 类

FA/PA是应用的基本组成单元，能够实现特定的业务功能。两者主要的区别是FA有UI界面，而PA无UI界面。

如图4-5所示的MainAbility就是一个FA。

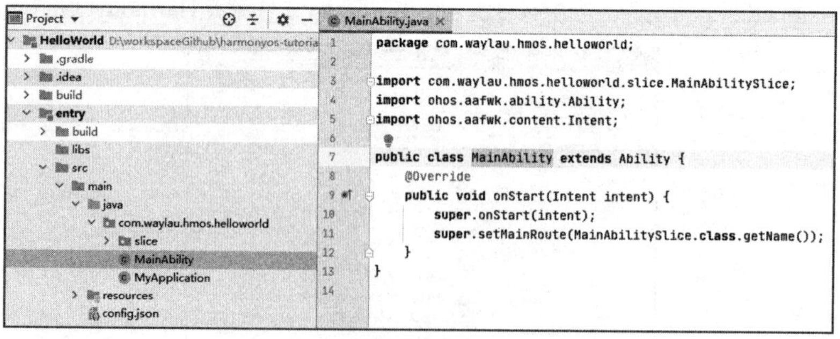

图4-5 MainAbility

4.2.2 AbilitySlice 类

MainAbility继承自Ability类。同时，从代码可以看出，MainAbility设置了一个路由，可以路由到MainAbilitySlice。

MainAbilitySlice继承自AbilitySlice类，如图4-6所示。而AbilitySlice就是用于呈现UI界面的。

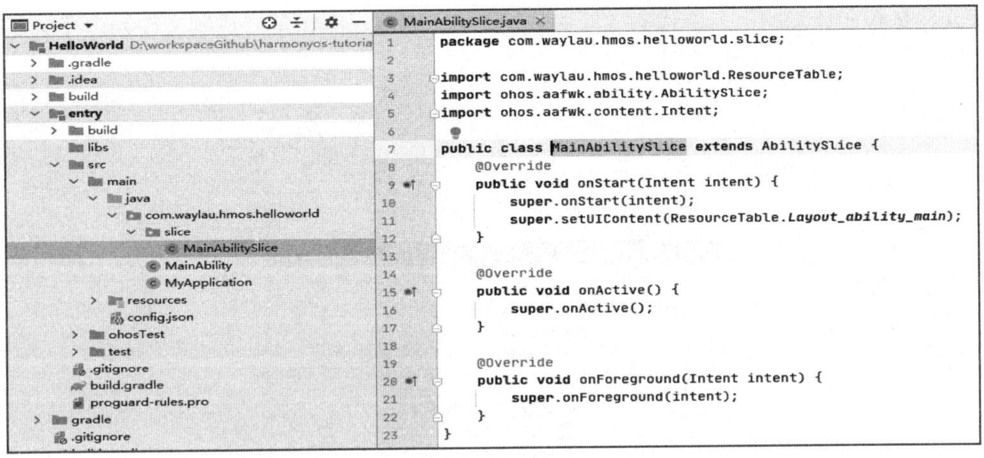

图 4-6 AbilitySlice 类

4.2.3 UI 界面

UI界面定义在哪里呢？我们可以打开resource目录，如图4-7所示，该目录就是整个应用所使用的UI界面元素。

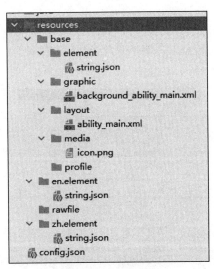

图 4-7 UI 界面元素

有关Ability的内容，将在后续章节深入探讨。

4.3 库 文 件

如果你有过Java开发或者Android开发的经验,那么对于库文件就不会陌生。

库文件是应用依赖的第三方代码(例如SO、JAR、BIN、HAR等二进制文件),存放在libs目录下。libs目录位置如图4-8所示。

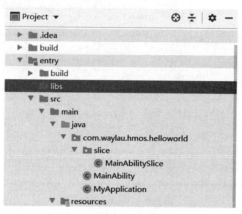

图4-8 libs 目录

4.4 资源文件

HarmonyOS应用采用Gradle进行项目管理,因此与Maven类似,应用的资源文件(字符串、图片、音频等)都存放在resources目录下,便于开发者使用和维护。

应用的resources目录如图4-9所示。

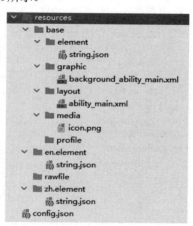

图4-9 resources 目录

resources目录包括两大类目录，一类为base目录与限定词目录，另一类为rawfile目录，详见表4-1。

表4-1 resources 的两大类目录

分 类	base 目录与限定词目录	rawfile 目录
组织形式	按照两级目录形式来组织，目录命名必须符合规范，以便根据设备状态去匹配相应目录下的资源文件。一级子目录为 base 目录和限定词目录。base 目录是默认存在的目录。当应用的 resources 资源目录中没有与设备状态匹配的限定词目录时，会自动引用该目录中的资源文件。限定词目录需要开发者自行创建。目录名称由一个或多个表征应用场景或设备特征的限定词组合而成。二级子目录为资源目录，用于存放字符串、颜色、布尔值等基础元素，以及媒体、动画、布局等资源文件	支持创建多层子目录，目录名称可以自定义，文件夹内可以自由放置各类资源文件。rawfile 目录的文件不会根据设备状态去匹配不同的资源
编译方式	目录中的资源文件会被编译成二进制文件，并赋予资源文件 ID	目录中的资源文件会被直接打包进应用，不经过编译，也不会被赋予资源文件 ID
引用方式	通过指定资源类型（type）和资源名称（name）来引用	通过指定文件路径和文件名来引用

4.4.1 限定词目录

限定词目录可以由一个或多个表征应用场景或设备特征的限定词组合而成，包括语言、文字、国家或地区、横竖屏、设备类型和屏幕密度6个维度，限定词之间通过下画线（_）或者中画线（-）连接。开发者在创建限定词目录时，需要掌握限定词目录的命名要求以及与限定词目录和设备状态的匹配规则。

1. 限定词目录的命名要求

限定词的组合顺序：语言_文字_国家或地区-横竖屏-设备类型-屏幕密度。开发者可以根据应用的使用场景和设备特征选择其中的一类或几类限定词组成目录名称。

限定词的连接方式：语言、文字、国家或地区之间采用下画线（_）连接，除此之外的其他限定词之间均采用中画线（-）连接，例如zh_Hant_CN、zh_CN-car-ldpi。

限定词的取值范围：每类限定词的取值必须符合表4-2中的条件，否则将无法匹配目录中的资源文件。

表4-2 限定词取值要求

限定词类型	含义与取值说明
移动国家码和移动网络码	移动国家码（MCC）和移动网络码（MNC）的值取自设备注册的网络。MCC 后面可以跟随 MNC，使用下画线（_）连接，也可以单独使用。例如，mcc460 表示中国，mcc460_mnc00 表示中国_中国移动。详细取值范围请查阅 ITU-T E.212（国际电联相关标准）
语言	表示设备使用的语言类型，由两个小写字母组成。例如，zh 表示中文，en 表示英语。详细取值范围参见 ISO 639（ISO 制定的语言编码标准）

（续表）

限定词类型	含义与取值说明
文字	表示设备使用的文字类型，由 1 个大写字母（首字母）和 3 个小写字母组成。例如，Hans 表示简体中文，Hant 表示繁体中文。详细取值范围参见 ISO 15924（ISO 制定的文字编码标准）
国家或地区	表示用户所在的国家或地区，由 2~3 个大写字母或者 3 个数字组成。例如，CN 表示中国，GB 表示英国。详细取值范围参见 ISO 3166-1（ISO 制定的国家和地区编码标准）
横竖屏	表示设备的屏幕方向，取值包括 vertical（竖屏）、horizontal（横屏）
设备类型	表示设备的类型，取值包括 phone（手机）、tablet（平台）、car（车机）、tv（智慧屏）、wearable（智能穿戴）等
颜色模式	表示设备的颜色模式，取值 dark（深色模式）或者 light（浅色模式）
屏幕密度	表示设备的屏幕密度（单位为 dpi）。sdpi 表示小规模的屏幕密度（Small-Scale Dots Per Inch），适用于 dpi 取值为(0, 120]的设备。mdpi 表示中规模的屏幕密度（Medium-Scale Dots Per Inch），适用于 dpi 取值为(120, 160]的设备。ldpi 表示大规模的屏幕密度（Large-Scale Dots Per Inch），适用于 dpi 取值为(160, 240]的设备。xldpi 表示特大规模的屏幕密度（Extra Large-Scale Dots Per Inch），适用于 dpi 取值为(240, 320]的设备。xxldpi 表示超大规模的屏幕密度（Extra Extra Large-Scale Dots Per Inch），适用于 dpi 取值为(320, 480]的设备。xxxldpi 表示超特大规模的屏幕密度（Extra Extra Extra Large-Scale Dots Per Inch），适用于 dpi 取值为(480, 640]的设备

2. 限定词目录与设备状态的匹配规则

在为设备匹配对应的资源文件时，限定词目录匹配的优先级从高到低依次为：

区域（语言_文字_国家或地区）> 横竖屏 > 设备类型 > 屏幕密度

如果限定词目录中包含语言、文字、横竖屏、设备类型限定词，则对应限定词的取值必须与当前的设备状态完全一致，该目录才能够参与设备的资源匹配。例如，限定词目录zh_CN-car-ldpi 不能参与en_US设备的资源匹配。

4.4.2 资源组目录

base目录与限定词目录下可以创建资源组目录（包括element、media、animation、layout、graphic、profile），用于存放特定类型的资源文件。

1. element

element表示元素资源，以下每一类数据都采用相应的JSON文件来表征。

- boolean：布尔型。
- color：颜色。
- float：浮点型。
- intarray：整型数组。
- integer：整型。
- pattern：样式。

- plural：复数形式。
- strarray：字符串数组。
- string：字符串。

element目录中的文件名称建议与下面的文件名保持一致。每个文件中只能包含同一类型的数据。

- boolean.json。
- color.json。
- float.json。
- intarray.json。
- integer.json。
- pattern.json。
- plural.json。
- strarray.json。
- string.json。

2. media

media表示媒体资源，包括图片、音频、视频等非文本格式的文件。

文件名可自定义，例如icon.png。

3. animation

animation表示动画资源，采用XML文件格式。

文件名可自定义，例如zoom_in.xml。

4. layout

layout表示布局资源，采用XML文件格式。

文件名可自定义，例如home_layout.xml。

5. graphic

graphic表示可绘制资源，采用XML文件格式。

文件名可自定义，例如notifications_dark.xml。

6. profile

profile表示其他类型文件，以原始文件形式保存。

文件名可自定义。

4.5 配置文件

HarmonyOS应用的每个HAP的根目录下都存在一个config.json配置文件，如图4-10所示。

图 4-10　配置文件

配置文件的内容主要涵盖以下3个方面：

- 应用的全局配置信息，包含应用的包名、生产厂商、版本号等基本信息。
- 应用在具体设备上的配置信息，包含应用的备份恢复、网络安全等能力。
- HAP包的配置信息，包含每个Ability必须定义的基本属性（如包名、类名、类型以及Ability提供的能力），以及应用访问系统或其他应用受保护部分所需的权限等。

4.5.1　配置文件的组成

配置文件config.json采用JSON文件格式，其中包含一系列配置项，每个配置项由属性和值两部分构成。

- 属性：属性出现顺序不分先后，且每个属性最多只允许出现一次。
- 值：每个属性的值为JSON的基本数据类型（数值、字符串、布尔值、数组、对象或者null类型）。

应用的配置文件config.json由app、deviceConfig和module三个部分组成，缺一不可。配置文件的内部结构说明参见表4-3。

表 4-3　配置文件的内部结构说明

属性名称	含　义	数据类型	是否可缺省
app	表示应用的全局配置信息。同一个应用的不同 HAP 包的 app 配置必须保持一致	对象	否
deviceConfig	表示应用在具体设备上的配置信息	对象	否
module	表示 HAP 包的配置信息。该标签下的配置只对当前 HAP 包生效	对象	否

以下是HelloWorld应用的配置文件：

```
{
    "app": {
```

```json
    "bundleName": "com.waylau.hmos.helloworld",
    "vendor": "waylau",
    "version": {
      "code": 1000000,
      "name": "1.0.0"
    }
  },
  "deviceConfig": {},
  "module": {
    "package": "com.waylau.hmos.helloworld",
    "name": ".MyApplication",
    "mainAbility": "com.waylau.hmos.helloworld.MainAbility",
    "deviceType": [
      "phone",
      "car"
    ],
    "distro": {
      "deliveryWithInstall": true,
      "moduleName": "entry",
      "moduleType": "entry",
      "installationFree": false
    },
    "abilities": [
      {
        "skills": [
          {
            "entities": [
              "entity.system.home"
            ],
            "actions": [
              "action.system.home"
            ]
          }
        ],
        "orientation": "unspecified",
        "name": "com.waylau.hmos.helloworld.MainAbility",
        "icon": "$media:icon",
        "description": "$string:mainability_description",
        "label": "$string:entry_MainAbility",
        "type": "page",
        "launchType": "standard"
      }
    ]
  }
}
```

接下来详细介绍上述配置的含义。

4.5.2　app 对象的内部结构

app对象包含应用的全局配置信息，内部结构说明如下。

- bundleName：表示应用的包名，用于标识应用的唯一性。包名是由字母、数字、下画线（_）和点号（.）组成的字符串，必须以字母开头。包名支持的字符串长度为7~127字节。包名通常采用业界常用的反域名形式表示（例如com.waylau.hmos）。建议第一级为域名后缀com，第二级为厂商/个人名，第三级为应用名，也可以采用多级。
- vendor：表示对应用开发厂商的描述。字符串长度不超过255字节。该值可缺省，缺省值为空。
- version：表示应用的版本信息。
  - code：表示应用的版本号，仅用于HarmonyOS管理该应用，不对应用的终端用户呈现。在 API 5 及更早的版本中，code 的取值规则为二进制 32 位以内的非负整数，需要从version.name 的值转换得到。从 API 6 版本起，code 的取值不与 version.name 字段的取值关联，开发者可以自定义 code 的取值，取值范围为小于 231 的非负整数，但是每次应用的版本更新均需更新 code 字段的值，新版本 code 的值必须大于旧版本 code 的值。
  - name：表示应用的版本号，用于向应用的终端用户呈现。取值可以自定义，长度不超过127字节。在 API 5 及更早版本中，推荐使用三段式数字版本号（也兼容两段式版本号），如 A.B.C（也兼容 A.B），其中 A、B、C 取值为 0~999 范围内的整数。从 API 6 版本起，推荐采用四段式数字版本号，如 A.B.C.D，其中 A、B、C 取值为 0~99 范围内的整数，D取值为 0~999 范围内的整数。
  - minCompatibleVersionCode：表示应用可兼容的最低版本号，用于在跨设备场景下判断其他设备上该应用的版本是否兼容。格式与 version.code 字段的格式要求相同。
- multiFrameworkBundle：表示应用是否为混合打包的HarmonyOS应用。混合打包场景配置为true，非混合打包场景配置为false。该标签值由IDE自动配置。
- smartWindowSize：该标签用于在悬浮窗场景下表示应用的模拟窗口的尺寸。配置格式为"正整数*正整数"，单位为vp。正整数取值范围为[200,2000]。
- smartWindowDeviceType：表示应用可以在哪些设备上使用模拟窗口打开，取值为phone（智能手机）、tablet（平板）、tv（智慧屏）。
- targetBundleList：表示允许以免安装方式拉起的其他HarmonyOS应用，列表取值为每个HarmonyOS应用的bundleName，多个bundleName之间用英文逗号","区分，最多配置10个bundleName。如果被拉起的应用不支持免安装方式，则拉起失败。

4.5.3　deviceConfig 对象的内部结构

deviceConfig包含在具体设备上的应用配置信息可以包含default、phone、tablet、tv、car、wearable、liteWearable和smartVision等属性。default标签内的配置是所有设备通用的，其他设备类型如果有特殊的需求，则需要在该设备类型的标签下进行配置。

- default：表示所有设备通用的应用配置信息。

- phone：表示手机类设备的应用信息配置。
- tablet：表示平板的应用配置信息。
- tv：表示智慧屏特有的应用配置信息。
- car：表示车机特有的应用配置信息。
- wearable：表示智能穿戴特有的应用配置信息。
- liteWearable：表示轻量级智能穿戴特有的应用配置信息。
- smartVision：表示智能摄像头特有的应用配置信息。

default、phone、tablet、tv、car、wearable、liteWearable和smartVision等对象的内部结构说明，可参见表4-4。

表4-4 不同设备的内部结构说明

属性名称	含义	数据类型	是否可缺省
process	表示应用或者Ability的进程名。如果在deviceConfig标签下配置了process标签，则该应用的所有Ability都运行在这个进程中。如果在abilities标签下也为某个Ability配置了process标签，则该Ability就运行在这个进程中。该标签仅适用于手机、平板、智慧屏、车机、智能穿戴设备	字符串	可缺省，缺省为应用的软件包名
jointUserId	表示应用的共享userid。通常情况下，不同的应用运行在不同的进程中，应用的资源是无法共享的。如果开发者的多个应用之间需要共享资源，则可以通过相同的jointUserId值实现，前提是这些应用的签名相同。该标签仅对系统应用生效，且仅适用于手机、平板、智慧屏、车机、智能穿戴设备。该字段在API Version 3及更高版本中不再支持配置	字符串	可缺省，缺省为空
supportBackup	表示应用是否支持备份和恢复。如果配置为false，则不支持为该应用执行备份或恢复操作。该标签仅适用于手机、平板、智慧屏、车机、智能穿戴设备	布尔类型	可缺省，缺省为false
compressNativeLibs	表示libs库是否以压缩存储的方式打包到HAP包。如果配置为false，则libs库以不压缩的方式存储，HAP包在安装时无须解压libs，运行时会直接从HAP内加载libs库	该标签仅适用于手机、平板、智慧屏、车机、智能穿戴设备	布尔类型
network	表示网络安全性配置。该标签允许应用通过配置文件的安全声明来自定义其网络安全，无须修改应用代码	对象	可缺省，缺省为空

表4-4中的network对象的内部结构说明见表4-5。

表 4-5 network 对象的内部结构说明

属性名称	含 义	数据类型	是否可缺省
usesCleartext	表示是否允许应用使用明文网络流量（例如明文HTTP）。true：允许应用使用明文流量的请求；false：拒绝应用使用明文流量的请求	布尔类型	可缺省，缺省为false
securityConfig	表示应用的网络安全配置信息	对象	可缺省，缺省为空

表4-5中的securityConfig对象的domainSettings属性的内部结构说明见表4-6。

表 4-6

属性名称	含 义	数据类型	是否可缺省
cleartextPermitted	表示自定义的网域范围内是否允许明文流量传输。当usesCleartext 和 securityConfig 同时存在时，自定义网域是否允许明文流量传输以 cleartextPermitted 的取值为准。true：允许明文流量传输；false：拒绝明文流量传输	布尔类型	否
domains	表示域名配置信息，包含两个参数：subDomains 和 name。subDomains（布尔类型）表示是否包含子域名。如果为true，此网域规则将与相应网域及所有子网域（包括子网域的子网域）匹配；否则，该规则仅适用于精确匹配项。name（字符串）表示域名名称	对象数组	否

deviceConfig示例如下：

```
"deviceConfig": {
  "default": {
    "process": "com.waylau.helloworld.example",
    "directLaunch": false,
    "supportBackup": false,
    "network": {
      "usesCleartext": true,
      "securityConfig": {
        "domainSettings": {
          "cleartextPermitted": true,
          "domains": [
            {
              "subDomains": true,
              "name": "example.ohos.com"
            }
          ]
        }
      }
    }
  }
}
```

4.5.4　module 对象的内部结构

module对象包含HAP包的配置信息，内部结构说明参见表4-7。

表 4-7　module 对象的内部结构说明

属性名称	含义	数据类型	是否可缺省
mainAbility	表示 HAP 包的入口 Ability 名称。该标签的值应配置为 module→abilities 中存在的 Page 类型 ability 的名称。该标签仅适用于手机、平板、智慧屏、车机、智能穿戴设备	字符串	如果存在 page 类型的 ability，则该字段不可缺省
package	表示 HAP 的包结构名称，在应用内应保证唯一性。采用反向域名格式（建议与 HAP 的工程目录保持一致）。字符串长度不超过 127 字节。该标签仅适用于手机、平板、智慧屏、车机、智能穿戴设备	字符串	否
name	表示 HAP 的类名。采用反向域名方式表示，前缀需要与同级的 package 标签指定的包名一致，也可采用"."开头的命名方式。字符串长度不超过 255 字节。该标签仅适用于手机、平板、智慧屏、车机、智能穿戴设备	字符串	否
description	表示 HAP 的描述信息。字符串长度不超过 255 字节。如果字符串超出长度或者需要支持多语言，可以采用资源索引的方式添加描述内容。该标签仅适用于手机、平板、智慧屏、车机、智能穿戴设备	字符串	可缺省，缺省值为空
supportedModes	表示应用支持的运行模式。当前只定义了驾驶模式（drive）。该标签仅适用于车机	字符串数组	可缺省，缺省值为空
deviceType	表示允许 Ability 运行的设备类型。系统预定义的设备类型包括：phone（手机）、tablet（平板）、tv（智慧屏）、car（车机）、wearable（智能穿戴设备）、liteWearable（轻量级智能穿戴设备）等	字符串数组	否
distro	表示 HAP 发布的具体描述。该标签仅适用于手机、平板、智慧屏、车机、智能穿戴设备	对象	否
metaData	表示 HAP 的元信息	对象	可缺省，缺省值为空
abilities	表示当前模块内的所有 Ability。采用对象数组格式，其中每个元素表示一个 Ability 对象	对象数组	可缺省，缺省值为空
js	表示基于 JS UI 框架开发的 JS 模块集合，其中的每个元素代表一个 JS 模块的信息	对象数组	可缺省，缺省值为空
shortcuts	表示应用的快捷方式信息。采用对象数组格式，其中的每个元素表示一个快捷方式对象	对象数组	可缺省，缺省值为空

(续表)

属性名称	含义	数据类型	是否可缺省
defPermissions	表示应用定义的权限。应用调用者必须申请这些权限，才能正常调用该应用	对象数组	可缺省，缺省值为空
reqPermissions	表示应用运行时向系统申请的权限	对象数组	可缺省，缺省值为空
colorMode	表示应用自身的颜色模式，可以取值 dark、light、auto。该标签仅适用于手机、平板、智慧屏、车机、智能穿戴设备	字符串	可缺省，缺省值为 auto
resizeable	表示应用是否支持多窗口特性。该标签仅适用于手机、平板、智慧屏、车机、智能穿戴设备	布尔类型	可缺省，缺省值为 true

在表4-7中，distro对象的内部结构说明参见表4-8。

表 4-8 distro 对象的内部结构说明

属性名称	含义	数据类型	是否可缺省
deliveryWithInstall	表示当前 HAP 是否支持随应用安装。true：支持随应用安装；false：不支持随应用安装	布尔类型	否
moduleName	表示当前 HAP 的名称	字符串	否
moduleType	表示当前 HAP 的类型，包括两种类型：entry 和 feature	字符串	否
installationFree	表示当前该 FA 是否支持免安装特性。true 表示支持免安装特性，且符合免安装约束；false 表示不支持免安装特性	布尔类型	entry.hap 可缺省，feature.hap 不可缺省

在表4-7中，distro对象的内部结构说明参见表4-9。

表 4-9 abilities 对象的内部结构说明

属性名称	含义	数据类型	是否可缺省
name	表示 Ability 名称。取值可采用反向域名方式表示，由包名和类名组成，如 com.example.myapplication.MainAbility；也可采用"."开头的类名方式表示，如.MainAbility。该标签仅适用于手机、平板、智慧屏、车机、智能穿戴设备	字符串	否
description	表示对 Ability 的描述。取值可以是描述性内容，也可以是对描述性内容的资源索引，以支持多语言	字符串	可缺省，缺省值为空
icon	表示 Ability 图标资源文件的索引。取值示例：$media:ability_icon。如果在该 Ability 的 skills 属性中，actions 的取值包含 action.system.home，entities 的取值包含 entity.system.home，则该 Ability 的 icon 将同时作为应用的 icon。如果存在多个符合条件的 Ability，则取位置靠前的 Ability 的 icon 作为应用的 icon	字符串	可缺省，缺省值为空

(续表)

属性名称	含 义	数据类型	是否可缺省
Label	表示Ability对用户显示的名称。取值可以是Ability名称，也可以是对该名称的资源索引，以支持多语言。如果在该Ability的skills属性中，actions的取值包含action.system.home，entities的取值包含entity.system.home，则该Ability的label将同时作为应用的label。如果存在多个符合条件的Ability，则取位置靠前的Ability的label作为应用的label	字符串	可缺省，缺省值为空
uri	表示Ability的统一资源标识符，格式为[scheme:][//authority][path][?query][#fragment]	字符串	可缺省，对于data类型的Ability不可缺省
launchType	表示Ability的启动模式，支持standard和singleton两种模式。standard：表示该Ability可以有多个实例。standard模式适用于大多数应用场景。singleton：表示该Ability只可以有一个实例。例如，具有全局唯一性的呼叫来电界面即采用singleton模式。该标签仅适用于手机、平板、智慧屏、车机、智能穿戴设备	字符串	可缺省，缺省值为standard
visible	表示Ability是否可以被其他应用调用。true：可以被其他应用调用；false：不能被其他应用调用	布尔类型	可缺省，缺省值为false
permissions	表示其他应用的Ability调用此Ability时需要申请的权限。通常采用反向域名格式，取值可以是系统预定义的权限，也可以是开发者自定义的权限。如果是自定义的权限，取值必须与defPermissions标签中定义的某个权限的name标签值一致	字符串数组	可缺省，缺省值为空
skills	表示Ability能够接收的Intent的特征	对象数组	可缺省，缺省值为空
deviceCapability	表示Ability运行时要求设备具有的能力，采用字符串数组的格式表示	字符串数组	可缺省，缺省值为空
metaData	表示Ability的元信息。调用Ability时调用参数的元信息，例如参数个数和类型。Ability执行完毕返回值的元信息，例如返回值个数和类型。该标签仅适用于智慧屏、智能穿戴设备、车机	对象	可缺省，缺省值为空
type	表示Ability的类型。page：表示基于Page模板开发的FA，用于提供与用户交互的能力；service：表示基于Service模板开发的PA，用于提供后台运行任务的能力；data：表示基于Data模板开发的PA，用于对外部提供统一的数据访问抽象	字符串	否

(续表)

属性名称	含义	数据类型	是否可缺省
Orientation	表示该 Ability 的显示模式。该标签仅适用于 page 类型的 Ability。unspecified：由系统自动判断显示方向；landscape：横屏模式；portrait：竖屏模式；followRecent：跟随栈中最近的应用	字符串	可缺省，缺省值为 unspecified
backgroundModes	表示后台服务的类型，可以为一个服务配置多个后台服务类型。该标签仅适用于 service 类型的 Ability。dataTransfer：通过网络/对端设备进行数据下载、备份、分享、传输等业务；audioPlayback：音频输出业务；audioRecording：音频输入业务；pictureInPicture：画中画、小窗口播放视频业务；voip：音视频电话、VOIP 业务；location：定位、导航业务；bluetoothInteraction：蓝牙扫描、连接、传输业务；wifiInteraction：WLAN 扫描、连接、传输业务；screenFetch：录屏、截屏业务	字符串数组	可缺省，缺省值为空
readPermission	表示读取 Ability 的数据所需的权限。该标签仅适用于 data 类型的 Ability，取值为长度不超过 255 字节的字符串。该标签仅适用于手机、平板、智慧屏、车机、智能穿戴设备	字符串	可缺省，缺省值为空
writePermission	表示向 Ability 写数据所需的权限。该标签仅适用于 data 类型的 Ability，取值为长度不超过 255 字节的字符串。该标签仅适用于手机、平板、智慧屏、车机、智能穿戴设备	字符串	可缺省，缺省值为空
configChanges	表示 Ability 关注的系统配置集合。当已关注的配置发生变更后，Ability 会收到 onConfigurationUpdated 回调。locale：表示语言区域发生变更；layout：表示屏幕布局发生变更；fontSize：表示字号发生变更；orientation：表示屏幕方向发生变更；density：表示显示密度发生变更	字符串数组	可缺省，缺省为空
mission	表示 Ability 指定的任务栈。该标签仅适用于 page 类型的 Ability。默认情况下，应用中所有 Ability 同属一个任务栈。该标签仅适用于手机、平板、智慧屏、车机、智能穿戴设备	字符串	可缺省，缺省值为应用的包名
targetAbility	表示当前 Ability 重用的目标 Ability。该标签仅适用于 page 类型的 Ability。如果配置了 targetAbility 属性，则当前 Ability（即别名 Ability）的属性中仅 name、icon、label、visible、permissions、skills 生效，其他属性均沿用 targetAbility 中的属性值。目标 Ability 必须与别名 Ability 在同一应用中，且在配置文件中目标 Ability 必须在别名之前声明。该标签仅适用于手机、平板、智慧屏、车机、智能穿戴设备	字符串	可缺省，缺省值为空。表示当前 Ability 不是一个别名 Ability
multiUserShared	表示 Ability 是否支持多用户状态进行共享，该标签仅适用于 data 类型的 Ability。配置为 true 时，表示在多用户下只有一份存储数据。需要注意的是，该属性会使 visible 属性失效。该标签仅适用于手机、平板、智慧屏、车机、智能穿戴设备	布尔类型	可缺省，缺省值为 false

（续表）

属性名称	含义	数据类型	是否可缺省
supportPipMode	表示 Ability 是否支持用户进入 PIP 模式（用于在页面最上层悬浮小窗口，俗称"画中画"，常见于视频播放等场景）。该标签仅适用于 page 类型的 Ability。该标签仅适用于手机、平板、智慧屏、车机、智能穿戴设备	布尔类型	可缺省，缺省值为 false
formsEnabled	表示 Ability 是否支持卡片（forms）功能。该标签仅适用于 page 类型的 Ability。true：支持卡片能力；false：不支持卡片能力	布尔类型	可缺省，缺省值为 false
forms	表示服务卡片的属性。该标签仅当 formsEnabled 为 true 时才能生效	对象数组	可缺省，缺省值为空
resizeable	表示 Ability 是否支持多窗口特性。该标签仅适用于手机、平板、智慧屏、车机、智能穿戴设备	布尔类型	可缺省，缺省值为 true

更多配置项的含义可参阅官方文档。

4.6　pack.info

pack.info用于描述应用软件包中每个HAP的属性，由IDE编译生成，应用市场根据该文件进行拆包和HAP的分类存储。HAP的具体属性包括：

- delivery-with-install：表示该HAP是否支持随应用安装，true表示支持随应用安装，false表示不支持随应用安装。
- name：HAP文件名。
- module-type：模块类型，为entry或feature。
- device-type：表示支持该HAP运行的设备类型。

第 5 章

Ability 基础知识

本章介绍 Ability 的基础知识。Ability 是 HarmonyOS 应用所具备的能力的抽象，也是 HarmonyOS 应用程序的核心组成部分。

5.1 Ability 概述

在 HarmonyOS 应用中，有一个非常核心的概念，那就是 Ability。正如其字面含义，Ability 可以理解为 HarmonyOS 应用所具备的能力的抽象。一个 HarmonyOS 应用可以具备多少种能力，就会包含多少个 Ability。HarmonyOS 支持应用以 Ability 为单位进行部署。

Ability 主要分为两种类型：FA（Feature Ability）和 PA（Particle Ability）。

每种类型为开发者提供了不同的模板，以便实现不同的业务功能。比如，我们在 DevEco Studio 中创建 HelloWorld 应用时，默认选择的是一个空的 FA，如图 5-1 所示。

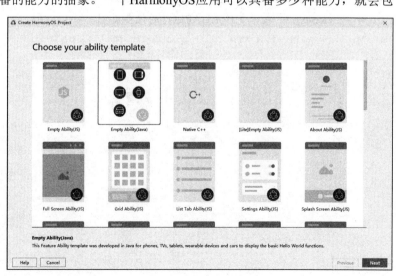

图 5-1 选择空的 FA 模板

5.1.1 FA

FA支持Page Ability。就目前而言，Page模板是FA唯一支持的模板，用于提供与用户交互的能力。一个Page实例可以包含一组相关页面，每个页面用一个AbilitySlice实例表示。简而言之，FA就是承担前端与用户交互的。

5.1.2 PA

PA支持Service Ability和Data Ability两种。其中：

- Service模板：用于提供后台运行任务的能力。
- Data模板：用于对外部提供统一的数据访问抽象。

5.1.3 Ability 的配置

在配置文件（config.json）中注册Ability时，可以通过配置Ability元素中的type属性来指定Ability模板类型。下面以HelloWorld应用为例进行介绍。

```
{
    "module": {
        ...
        "abilities": [
            {
                ...
                "type": "page"
                ...
            }
        ]
        ...
    }
    ...
}
```

在上述例子中，type的取值可以为page、service或data，分别代表Page模板、Service模板、Data模板。

基于Page模板、Service模板、Data模板实现的Ability分别可以简称为Page、Service、Data。

5.2 Ability 的三层架构

如果读者有大型企业级应用开发经验的话，应该对于分层架构并不陌生，Ability也采用分层架构。本节我们讨论以下话题：

- 为什么我们需要将应用程序进行分层?
- 如果不分层,系统将会出现哪些问题?
- 常见的分层方式有哪些?
- Ability是如何进行分层的?

5.2.1 应用的分层

随着面向对象程序设计和设计模式的出现,人们发现,现实生活中的建筑学有很多理论都可以用来指导软件工程(即程序的开发)。比如,在开发时,我们会先对要盖的楼房进行评估和核算(软件项目管理);根据需求设计楼房的图纸(软件设计),而后根据设计把楼房的地基、骨架搭建出来(搭建框架);再根据不同工种将人员进行分工,有些去砌墙,有些去贴砖(前端编码,后台编码);最后进行验收测试,交付给用户使用。

软件应用开发与建筑学的分层目的是一致的,都是旨在根据不同的业务、不同的技术、不同的组织,结合灵活性、可维护性、可扩展性等多种因素,将应用系统划分成不同的部分,并使这些部分彼此之间相互分工、相互协作,从而体现出最大化价值。对于一个良好分层的应用来说,一般具备如下特点:

1. 按业务功能进行分层

分层就是将相关的业务功能的类或组件放置在一起,而将不相关的业务功能的类或组件隔离开。比如我们会将与用户直接交互的部分分为"表示层"(Presentation Layer),将实现逻辑计算或者业务处理的部分分为"业务层"(Business Layer),将与数据库打交道的部分分为"数据访问层"(Data Access Layer)。

2. 良好的层次关系

设计良好的架构分层是上层依赖于下层,而下层支撑起上层,但却不能直接访问上层,层与层之间通过协作来共同完成特定的功能。

3. 每一层都能保持独立

层能够被单独构造,也能够被单独替换掉,最终不会影响整体功能。比如,我们将整个数据持久层的技术从Hibernate转成了EclipseLink,但不能对上层业务逻辑功能造成影响。

5.2.2 不分层的应用架构

为了更好地理解分层的好处,我们先来看一下不分层的应用架构是如何运作的。

下面举一个Web应用程序的例子。在Web应用程序开发的早期,所有的逻辑代码并没有明显的层次区分,因此代码之间的调用是相互交错的,整体代码看上去错综复杂。譬如,在早期使用诸如ASP、JSP以及PHP等动态网页技术时,常会将所有的页面逻辑、业务逻辑以及数据库访问逻辑放在一起,很多时候就在JSP页面里面写SQL语句,编码风格完全是过程化的。

以下代码就是一个JSP访问SQL Server数据库的例子。

```jsp
<%@ page language="java" contentType="text/html; charset=UTF-8"
pageEncoding="UTF-8"%>
<%@ page import = "java.sql.*"%>
<!DOCTYPE html PUBLIC "-//W3C//DTD XHTML 1.0 Transitional//EN"
 "http://www.w3.org/TR/xhtml1/DTD/xhtml1-transitional.dtd">
<html xmlns="http://www.w3.org/1999/xhtml">
<head>
</head>
<body>
<%

// 创建数据库连接
Class.forName("com.microsoft.sqlserver.jdbc.SQLServerDriver");
String url="jdbc:sqlserver://localhost:1433;databaseName=Book;user=sa;password=";
PreparedStatement pstmt;
String sql = "insert into students (UserName,WebSite) values(?,?)";
int returnValue = 0;
try{
    pstmt = (PreparedStatement) conn.prepareStatement(sql);
    pstmt.setString(1, student.getName());
    pstmt.setString(2, student.getSex());
    returnValue = pstmt.executeUpdate();

    // 判断是添加成功还是失败
    if(returnValue == 1){
        out.print("<li>添加成功!");
        out.print("<li>returnValue = " + returnValue);
    } else {
        out.print("<li>添加失败!");
    }
}catch(Exception ex){
    out.print(ex.getLocalizedMessage());
}finally{
    try{
        if(pstmt != null){
            pstmt.close();
            pstmt = null;
        }
        if(cn != null){
            conn.close();
            conn = null;
        }
    }catch(Exception e){
        e.printStackTrace();
    }
}
%>
</body>
</html>
```

先不论这段代码是否正确,从实现功能上来讲,这段代码既处理了数据库的访问操作,还做了页面的表示,其中又夹杂着业务逻辑判断。还好这段代码不长,读下来还能够理解,但如果是更加复杂的功能,这种代码肯定是非常不清晰的,维护起来也相当麻烦。

早期的这种不分层的架构主要存在如下弊端:

- 代码不够清晰,难以阅读。
- 代码职责不明,难以扩展。
- 代码错综复杂,难以维护。
- 代码没做分工,难以组织。

5.2.3 应用的三层架构

目前,比较常用的、典型的应用软件倾向于使用三层架构(Three-Tier Architecture),即:

- 表示层(Presentation Layer):提供与用户交互的界面。GUI(Graphical User Interface,图形用户界面)和Web页面是表示层的两个典型的例子。
- 业务层(Business Layer):也称为业务逻辑层,用于实现各种业务逻辑,比如处理数据验证,根据特定的业务规则和任务来响应特定的行为。
- 数据访问层(Data Access Layer):也称为数据持久层,负责存放和管理应用的持久性业务数据。

图5-2展示了三层架构的架构图。

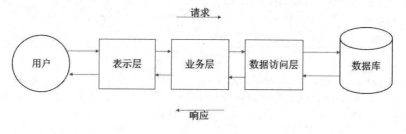

图 5-2 三层架构

如果你仔细看这些层,应该会看到每一层都需要不同的技能:

- 表示层需要诸如HTML、CSS、JavaScript等前端技能,以及具备UI设计能力。
- 业务层需要编程语言的技能,以便计算机可以处理业务规则。
- 数据访问层需要具有数据定义语言(Data Definition Language,DDL)、数据操作语言(Data Manipulation Language,DML)以及数据库设计形式的SQL技能。

虽然一个人有可能拥有上述所有技能,但这样的人是相当罕见的。在具有大型软件应用程序的大型组织中,一般将应用程序分割为单独的层,使得每个层都可以由具有相关专业技能的不同团队来开发和维护。

一个良好分层的架构系统应遵循如下的分层原则:

- 每个层的代码必须包含可以单独维护的单独文件。
- 每个层只能包含属于该层的代码。因此,业务逻辑只能驻留在业务层,表示逻辑只能在表示层,而数据访问逻辑只驻留在数据访问层中。
- 表示层只能接收来自外部代理的请求,并向外部代理返回响应。这通常是一个人,但也可能是另一个软件。
- 表示层只能向业务层发送请求,并从业务层接收响应。它不能直接访问数据库或数据访问层。
- 业务层只能接收来自表示层的请求,并返回对表示层的响应。
- 业务层只能向数据访问层发送请求,并从其接收响应。它不能直接访问数据库。
- 数据访问层只能从业务层接收请求并返回响应。它不能发出请求到除了它支持的数据库管理系统(Database Management System,DBMS)以外的地方。
- 每层应完全不知道其他层的内部工作原理。例如,业务层可以对数据库一无所知,并且可以不知道或不必关心数据访问对象的内部工作原理。业务层可以不知道或不关心表示层如何处理它的数据。表示层可以获取数据并构造HTML文档、PDF文档、CSV文档或以其他方式处理它,但是这应该与业务层完全无关。
- 每层应当可以用具有类似特征的替代组件来交换这个层,使得整体可以继续工作。

简而言之,应用在一开始设计的时候,就要考虑系统的架构设计,以及如何将系统有效地分层。由于系统架构设计属于比较高级别的话题,本书不会琢磨太多,对这方面感兴趣的读者朋友,可以参阅笔者所著的《分布式系统常用技术及案例分析》一书,该书的第2章详细介绍了软件系统的常见架构体系。

5.2.4　Ability 的三层架构

从三层架构中,我们很容易识别出,Ability同样遵循图5-2所示的三层架构。

- Page Ability:代表表示层。
- Service Ability:代表业务层。
- Data Ability:代表数据访问层。

因此,开发人员在设计Ability时,应先考虑这个Ability需要完成什么样的功能,代表了哪个层次的业务。

5.3　Page Ability

正如前文所讲,Page Ability代表应用的表示层的功能,用于提供与用户交互的能力。

5.3.1　Page Ability 的基本概念

Page模板(以下简称Page)是FA中目前来说唯一支持的模板。

一个Page可以由一个或多个AbilitySlice构成，AbilitySlice是指应用的单个页面及其控制逻辑的总和。

在HelloWorld应用中，MainAbility类就是一个Page。其代码如下：

```java
public class MainAbility extends Ability {
    @Override
    public void onStart(Intent intent) {
        super.onStart(intent);
        super.setMainRoute(MainAbilitySlice.class.getName());
    }
}
```

在HelloWorld应用中，MainAbility这个Page是由一个AbilitySlice构成的，这个AbilitySlice就是MainAbilitySlice类，代码如下：

```java
public class MainAbilitySlice extends AbilitySlice {
    @Override
    public void onStart(Intent intent) {
        super.onStart(intent);
        super.setUIContent(ResourceTable.Layout_ability_main);
    }

    @Override
    public void onActive() {
        super.onActive();
    }

    @Override
    public void onForeground(Intent intent) {
        super.onForeground(intent);
    }
}
```

5.3.2 多个 AbilitySlice 构成一个 Page

当一个Page由多个AbilitySlice共同构成时，这些AbilitySlice页面提供的业务能力应具有高度相关性。例如，新闻浏览功能可以通过一个Page来实现，其中包含两个AbilitySlice：一个AbilitySlice用于展示新闻列表，另一个AbilitySlice用于展示新闻详情。Page和AbilitySlice的关系如图5-3所示。

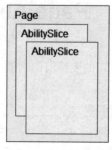

图 5-3 Page 和 AbilitySlice 的关系

相比于桌面场景，移动场景下应用之间的交互更为频繁。通常，单个应用专注于某个方面的能力开发，当它需要其他能力辅助时，会调用其他应用提供的能力。例如，快递应用提供了联系快递员的业务功能入口，当用户在使用该功能时，会跳转到通话应用的拨号页面。与此类似，HarmonyOS支持不同Page之间的跳转，并可以指定跳转到目标Page中某个具体的AbilitySlice。

5.3.3　AbilitySlice 路由配置

虽然一个Page可以包含多个AbilitySlice，但是Page进入前台时界面默认只能展示一个AbilitySlice。默认展示的AbilitySlice是通过setMainRoute()方法来指定的。

比如在HelloWorld应用中，MainAbility类就指定了MainAbilitySlice类作为默认展示的AbilitySlice。其代码如下：

```java
public class MainAbility extends Ability {
    @Override
    public void onStart(Intent intent) {
        super.onStart(intent);

        // 指定默认展示的AbilitySlice
        super.setMainRoute(MainAbilitySlice.class.getName());
    }
}
```

如果需要更改默认展示的AbilitySlice，可以通过addActionRoute()方法为此AbilitySlice配置一条路由规则。setMainRoute()方法与addActionRoute()方法的使用示例如下：

```java
public class MainAbility extends Ability {
    @Override
    public void onStart(Intent intent) {
        super.onStart(intent);

        // 指定默认展示的AbilitySlice
        super.setMainRoute(MainAbilitySlice.class.getName());

        // 配置路由规则
        addActionRoute("action.pay", PayAbilitySlice.class.getName());
    }
}
```

下一节将完整展示如何实现AbilitySlice的路由和导航。

5.3.4　不同 Page 间的导航

不同Page中的AbilitySlice相互不可见，因此无法通过present()或presentForResult()方法直接导航到其他Page的AbilitySlice。

AbilitySlice作为Page的内部单元，以Action的形式对外暴露，因此可以通过配置Intent的Action导航到目标AbilitySlice。Page间的导航可以使用startAbility()或startAbilityForResult()方法，获得返

回结果的回调为onAbilityResult()。在Ability中调用setResult()可以设置返回结果。详细用法可参考5.12节中的示例。

5.4 实战：多个AbilitySlice间的路由和导航

Page模板用于提供与用户交互的能力。一个Page可以由一个或多个AbilitySlice构成。当一个Page由多个AbilitySlice共同构成时，这些AbilitySlice页面提供的业务能力应具有高度相关性。本节主要演示在一个Page包含多个AbilitySlice时，这些AbilitySlice之间是如何路由和导航的。

5.4.1 创建应用

创建一个名为AbilitySliceNavigation的应用，如图5-4所示。该应用主要用于测试AbilitySlice之间的路由和导航。

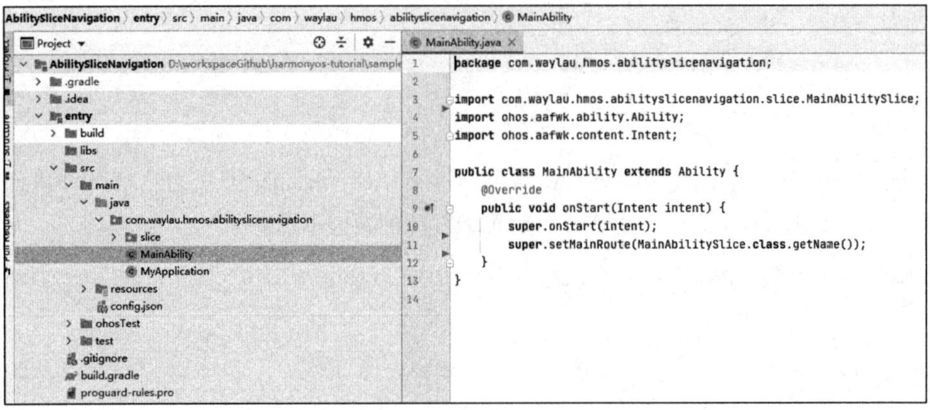

图 5-4　AbilitySliceNavigation 应用

5.4.2 创建多个 AbilitySlice

在初始化应用时，AbilitySliceNavigation应用已经包含一个主AbilitySlice，代码如下：

```
public class MainAbilitySlice extends AbilitySlice {
    @Override
    public void onStart(Intent intent) {
        super.onStart(intent);

        super.setUIContent(ResourceTable.Layout_ability_main);
    }

    @Override
    public void onActive() {
        super.onActive();
```

```
    }

    @Override
    public void onForeground(Intent intent) {
        super.onForeground(intent);
    }
}
```

因此，还需要再新增一个AbilitySlice。可以通过DevEco Studio的File→New→Ability→Empty Page Abilit(Java)来创建PayAbility，如图5-5所示。

创建PayAbility时，会同时创建一个PayAbilitySlice.java、background_ability_pay.xml和ability_pay.xml文件。

目前，MainAbilitySlice和PayAbilitySlice的代码基本是类似的，除了所引用的UT布局文件不同。

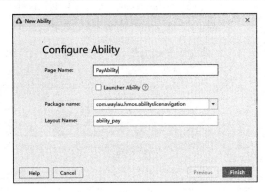

图 5-5　PayAbilitySlice

PayAbilitySlice代码如下：

```
package com.waylau.hmos.abilityslicenavigation.slice;

import com.waylau.hmos.abilityslicenavigation.ResourceTable;
import ohos.aafwk.ability.AbilitySlice;
import ohos.aafwk.content.Intent;

public class PayAbilitySlice extends AbilitySlice {
    @Override
    public void onStart(Intent intent) {
        super.onStart(intent);

        // UI界面内容引用了布局文件ability_pay.xml
        super.setUIContent(ResourceTable.Layout_ability_pay);
    }

    @Override
    public void onActive() {
        super.onActive();
    }

    @Override
    public void onForeground(Intent intent) {
        super.onForeground(intent);
    }
}
```

5.4.3 修改 PayAbilitySlice 布局

为了体现MainAbilitySlice和PayAbilitySlice的不同,我们需要在"面"上"整容"一下。从5.4.2节的代码可以得知,PayAbilitySlice通过setUIContent方法引用了ResourceTable.Layout_ability_pay的布局文件,而ResourceTable.Layout_ability_pay其实就是layout目录下的ability_pay.xml文件。

对ability_pay.xml的内容修改如下:

```xml
<?xml version="1.0" encoding="utf-8"?>
<DirectionalLayout
    xmlns:ohos="http://schemas.huawei.com/res/ohos"
    ohos:height="match_parent"
    ohos:width="match_parent"
    ohos:alignment="center"
    ohos:orientation="vertical">

    <Text
        ohos:id="$+id:text_pay"
        ohos:height="match_content"
        ohos:width="match_content"
        ohos:background_element="$graphic:background_ability_pay"
        ohos:layout_alignment="horizontal_center"
        ohos:text="$string:payability_money"
        ohos:text_size="40vp"
        />

</DirectionalLayout>
```

ability_pay.xml基本上与ability_main.xml类似,主要的差异点是:

- id设置为$+id:text_pay。
- background_element设置为$graphic:background_ability_pay。
- text设置为$string:payability_money。

5.4.4 设置 PayAbilitySlice 的样式

在5.4.3节,background_element设置为$graphic:background_ability_pay,其实这个background_ability_pay是一个独立的文件,即graphic目录下的background_ability_pay.xml文件,该文件定义了UI组件的背景样式,内容如下:

```xml
<?xml version="1.0" encoding="UTF-8" ?>
<shape
    xmlns:ohos="http://schemas.huawei.com/res/ohos"
    ohos:shape="rectangle">

    <solid
        ohos:color="#FFFFFF"/>
```

</shape>

上述样式的含义是组件的背景是一个白色的长方形。可以通过上述配置实现自定义的背景。

text设置为$string:payability_money，其实这个payability_money来自于en.json文件，用于定义在不同的语言环境下，所要显示的组件的文字内容。

以en.json文件为例，内容修改如下：

```
{
  "string": [
    {
      "name": "entry_MainAbility",
      "value": "entry_MainAbility"
    },
    {
      "name": "mainability_description",
      "value": "Java_Empty Ability"
    },
    {
      "name": "mainability_HelloWorld",
      "value": "Hello World"
    },
    {
      "name": "entry_PayAbility",
      "value": "entry_PayAbility"
    },
    {
      "name": "payability_description",
      "value": "Java_Empty Ability"
    },
    {
      "name": "payability_money",
      "value": "Pay me the money"
    }
  ]
}
```

上述代码定义了当组件引用$string:payability_money时，如果语言环境是en（英语），则会在组件上显示Pay me the money字样。

同理，我们修改zh.json内容如下：

```
{
  "string": [
    {
      "name": "entry_MainAbility",
      "value": "entry_MainAbility"
    },
    {
      "name": "mainability_description",
      "value": "Java_Empty Ability"
    },
```

```
    {
      "name": "mainability_HelloWorld",
      "value": "你好,世界"
    },
    {
      "name": "entry_PayAbility",
      "value": "entry_PayAbility"
    },
    {
      "name": "payability_description",
      "value": "Java_Empty Ability"
    },
    {
      "name": "payability_money",
      "value": "你好,支付"
    }
  ]
}
```

上述代码定义了当组件引用$string:payability_money时,如果语言环境是zh(中文),则会在组件上显示"你好,支付"字样。

5.4.5 如何实现 AbilitySlice 之间的路由和导航

实现AbilitySlice之间的路由和导航的步骤如下:

1. 设置路由

在MainAbility中,通过addActionRoute方法添加到PayAbilitySlice的路由。

```
package com.waylau.hmos.abilityslicenavigation;

import com.waylau.hmos.abilityslicenavigation.slice.MainAbilitySlice;
import com.waylau.hmos.abilityslicenavigation.slice.PayAbilitySlice;
import ohos.aafwk.ability.Ability;
import ohos.aafwk.content.Intent;

public class MainAbility extends Ability {
    @Override
    public void onStart(Intent intent) {
        super.onStart(intent);

        // 指定默认显示的AbilitySlice
        super.setMainRoute(MainAbilitySlice.class.getName());

        // 使用addActionRounte方法添加路由
        addActionRoute("action.pay", PayAbilitySlice.class.getName());
    }
}
```

第 5 章 Ability 基础知识 | 77

其中，上述action.pay用于指定路由动作的名称。这个名称还需要在config.json的actions数组中添加，配置如下：

```
...
"abilities": [
    {
      "skills": [
        {
          "entities": [
            "entity.system.home"
          ],
          "actions": [
            "action.system.home",
            "action.pay"   // 指定路由动作的名称
          ]
        }
      ],
      "orientation": "unspecified",
      "name": "com.waylau.hmos.abilityslicenavigation.MainAbility",
      "icon": "$media:icon",
      "description": "$string:mainability_description",
      "label": "$string:entry_MainAbility",
      "type": "page",
      "launchType": "standard"
    },
    {
      "orientation": "unspecified",
      "name": "com.waylau.hmos.abilityslicenavigation.PayAbility",
      "icon": "$media:icon",
      "description": "$string:payability_description",
      "label": "$string:entry_PayAbility",
      "type": "page",
      "launchType": "standard"
    }
  ]
...
```

2. 设置点击事件触发导航

在MainAbilitySlice中，为文本设置了点击事件，以便能够触发导航到PayAbilitySlice，代码如下：

```
package com.waylau.hmos.abilityslicenavigation.slice;

import com.waylau.hmos.abilityslicenavigation.ResourceTable;
import ohos.aafwk.ability.AbilitySlice;
import ohos.aafwk.content.Intent;
import ohos.agp.components.Text;

public class MainAbilitySlice extends AbilitySlice {
    @Override
```

```java
    public void onStart(Intent intent) {
        super.onStart(intent);

        // UI界面内容引用了布局文件ability_main.xml
        super.setUIContent(ResourceTable.Layout_ability_main);

        // 添加点击事件来触发导航到PayAbilitySlice
        Text text = (Text) findComponentById(ResourceTable.Id_text_helloworld);
        text.setClickedListener(listener ->
                present(new PayAbilitySlice(), new Intent()));
    }

    @Override
    public void onActive() {
        super.onActive();
    }

    @Override
    public void onForeground(Intent intent) {
        super.onForeground(intent);
    }
}
```

上述代码中,当发起导航的AbilitySlice和导航目标的AbilitySlice处于同一个Page时,可以通过present()方法实现导航。

同理,在PayAbilitySlice中,为文本设置了点击事件,已触发导航到MainAbilitySlice,代码如下:

```java
package com.waylau.hmos.abilityslicenavigation.slice;

import com.waylau.hmos.abilityslicenavigation.ResourceTable;
import ohos.aafwk.ability.AbilitySlice;
import ohos.aafwk.content.Intent;
import ohos.agp.components.Text;

public class PayAbilitySlice extends AbilitySlice {
    @Override
    public void onStart(Intent intent) {
        super.onStart(intent);

        // UI界面内容引用了布局文件ability_pay.xml
        super.setUIContent(ResourceTable.Layout_ability_pay);

        // 添加点击事件来触发导航到MainAbilitySlice
        Text text = (Text) findComponentById(ResourceTable.Id_text_pay);
        text.setClickedListener(listener ->
                present(new MainAbilitySlice(), new Intent()));
    }
```

```
    @Override
    public void onActive() {
        super.onActive();
    }

    @Override
    public void onForeground(Intent intent) {
        super.onForeground(intent);
    }
}
```

5.4.6 运行

在远程模拟器中运行应用后，可以看到如图5-6所示的界面效果。

点击文本"你好，世界"后，可以切换到"你好，支付"界面，如图5-7所示。

图 5-6　显示界面

图 5-7　切换到"你好，支付"界面

再点击文本"你好，支付"，可以切换到"你好，世界"。至此，实现了同个Page下多个AbilitySlice之间的路由和导航。

5.5　Page 与 AbilitySlice 的生命周期

本节介绍Page与AbilitySlice的生命周期。系统管理或用户操作等行为均会引起Page实例在其生命周期的不同状态之间进行转换。Ability类提供的回调机制能够让Page及时感知外界变化，从而正确地应对状态变化（比如释放资源），这有助于提升应用的性能和稳健性。

5.5.1 Page 的生命周期

Page的生命周期的不同状态转换及其对应的回调如图5-8所示。

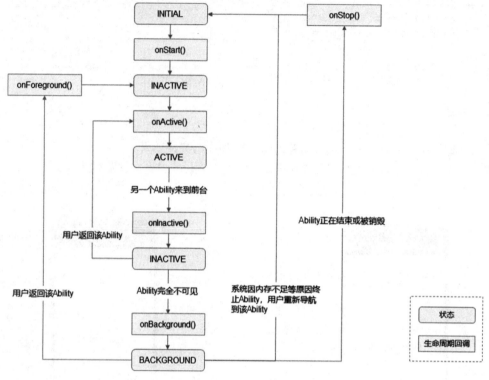

图 5-8　Page 生命周期

Page的生命周期包含以下回调方法：

1. onStart()

当系统首次创建Page实例时，触发该回调。对于一个Page实例，该回调在其生命周期过程中仅触发一次，Page在该逻辑后将进入INACTIVE状态。开发者必须重写该方法，并在此配置默认展示的AbilitySlice。

比如，在HelloWorld应用中，经常会在onStart()回调方法中设置默认显示的AbilitySlice，代码如下：

```
@Override
public void onStart(Intent intent) {
    super.onStart(intent);

    // 指定默认显示的AbilitySlice
    super.setMainRoute(MainAbilitySlice.class.getName());
}
```

2. onActive()

Page会在进入INACTIVE状态后来到前台，然后系统调用此回调。Page在此之后进入ACTIVE状态，该状态是应用与用户交互的状态。Page将保持在此状态，除非某类事件发生导致Page失去焦点，比如用户点击返回键或导航到其他Page。当此类事件发生时，会触发Page回到INACTIVE状态，系统将调用onInactive()回调。此后，Page可能重新回到ACTIVE状态，系统将再次调用onActive()回调。因此，开发者通常需要成对实现onActive()和onInactive()，并在onActive()中获取在onInactive()中被释放的资源。

3. onInactive()

当Page失去焦点时，系统将调用此回调，此后Page进入INACTIVE状态。开发者可以在此回调中实现Page失去焦点时应表现的恰当行为。

4. onBackground()

如果Page不再对用户可见，系统将调用此回调通知开发者用户进行相应的资源释放，此后Page进入BACKGROUND状态。开发者应该在此回调中释放Page不可见时无用的资源，或在此回调中执行较为耗时的状态保存操作。

5. onForeground()

处于BACKGROUND状态的Page仍然驻留在内存中，当重新回到前台时（比如用户重新导航到此Page），系统将先调用onForeground()回调通知开发者，而后Page的生命周期状态回到INACTIVE状态。开发者应当在此回调中重新申请在onBackground()中释放的资源，最后Page的生命周期状态进一步回到ACTIVE状态，系统将通过onActive()回调通知开发者用户。

6. onStop()

系统将要销毁Page时，将会触发此回调函数，通知用户进行系统资源的释放。销毁Page的可能原因包括以下几个方面：

- 用户通过系统管理能力关闭指定Page，例如使用任务管理器关闭Page。
- 用户行为触发Page的terminateAbility()方法调用，例如使用应用的退出功能。
- 配置变更导致系统暂时销毁Page并重建。
- 系统出于资源管理的目的，自动触发对处于BACKGROUND状态Page的销毁。

5.5.2 AbilitySlice 的生命周期

AbilitySlice作为Page的组成单元，其生命周期是依托于其所属Page的生命周期的。AbilitySlice和Page具有相同的生命周期状态和同名的回调（详见图5-8），当Page生命周期发生变化时，它的AbilitySlice也会发生相同的生命周期变化。此外，AbilitySlice还具有独立于Page的生命周期变化，这发生在同一Page中的AbilitySlice之间导航时，此时Page的生命周期状态不会改变。

AbilitySlice的生命周期回调与Page的相应回调类似，因此不再赘述。由于AbilitySlice承载具体的页面，开发者必须重写AbilitySlice的onStart()回调，并在此方法中通过setUIContent()方法设置页面。

比如，在AbilitySliceNavigation应用中，会在onStart()回调方法中通过super.setUIContent()来指定布局文件：

```
@Override
public void onStart(Intent intent) {
    super.onStart(intent);

    // UI界面内容引用了布局文件ability_pay.xml
    super.setUIContent(ResourceTable.Layout_ability_pay);
}
```

AbilitySlice实例创建和管理通常由应用负责，系统仅在特定情况下会创建AbilitySlice实例。例如，通过导航启动某个AbilitySlice时，是由系统负责实例化的，但是在同一个Page中，不同的AbilitySlice间导航时则由应用负责实例化。

5.5.3　Page 与 AbilitySlice 生命周期的关联

当AbilitySlice处于前台且具有焦点时，其生命周期状态随着所属Page的生命周期状态的变化而变化。当一个Page拥有多个AbilitySlice时，例如在AbilitySliceNavigation应用中，MainAbility下有MainAbilitySlice和PayAbilitySlice，当前MainAbilitySlice处于前台并获得焦点，并即将导航到PayAbilitySlice，在此期间的生命周期状态变化顺序为：

- MainAbilitySlice从ACTIVE状态变为INACTIVE状态。
- PayAbilitySlice则从INITIAL状态先变为INACTIVE状态，再变为ACTIVE状态（假定此前PayAbilitySlice未曾启动）。
- MainAbilitySlice从INACTIVE状态变为BACKGROUND状态。

对应两个slice的生命周期方法回调顺序为：

MainAbilitySlice.onInactive() → PayAbilitySlice.onStart() → PayAbilitySlice.onActive() → MainAbilitySlice.onBackground()

在整个流程中，MainAbility始终处于ACTIVE状态。但是，当Page被系统销毁时，其所有已实例化的AbilitySlice将联动销毁，而不仅是处于前台的AbilitySlice。

5.6　实战：Page 与 AbilitySlice 生命周期的例子

为了更好地理解Page与AbilitySlice的生命周期，我们将创建一个名为PageAndAbilitySliceLifeCycle的应用作为演示。同时，参考前文5.4节的步骤创建PayAbility。

5.6.1 修改 MainAbilitySlice

在初始化应用时，PageAndAbilitySliceLifeCycle 应用已经包含一个主 AbilitySlice，即 MainAbilitySlice。对 MainAbilitySlice 进行修改，代码如下：

```java
package com.waylau.hmos.pageandabilityslicelifecycle.slice;

import com.waylau.hmos.pageandabilityslicelifecycle.ResourceTable;
import ohos.aafwk.ability.AbilitySlice;
import ohos.aafwk.content.Intent;
import ohos.agp.components.Text;
import ohos.hiviewdfx.HiLog;
import ohos.hiviewdfx.HiLogLabel;

public class MainAbilitySlice extends AbilitySlice {
    private static final String TAG = MainAbilitySlice.class.getSimpleName();
    private static final HiLogLabel LABEL_LOG =
            new HiLogLabel(HiLog.LOG_APP, 0x00001, TAG);

    @Override
    public void onStart(Intent intent) {
        super.onStart(intent);

        // UI界面内容引用了布局文件ability_main.xml
        super.setUIContent(ResourceTable.Layout_ability_main);

        // 添加点击事件来触发导航到PayAbilitySlice
        Text text = (Text) findComponentById(ResourceTable.Id_text_helloworld);
        text.setClickedListener(listener ->
                present(new PayAbilitySlice(), new Intent()));

        HiLog.info(LABEL_LOG, "onStart");
    }

    @Override
    public void onActive() {
        super.onActive();
        HiLog.info(LABEL_LOG, "onActive");
    }

    @Override
    public void onForeground(Intent intent) {
        super.onForeground(intent);
        HiLog.info(LABEL_LOG, "onForeground");
    }

    @Override
```

```java
    public void onInactive() {
        HiLog.info(LABEL_LOG, "onInactive");
    }

    @Override
    public void onBackground() {
        HiLog.info(LABEL_LOG, "onBackground");
    }

    @Override
    public void onStop() {
        HiLog.info(LABEL_LOG, "onStop");
    }
}
```

在MainAbilitySlice类中:

- MainAbilitySlice重写了AbilitySlice的生命周期回调方法。
- 在每个生命周期回调方法中,通过HiLog日志工具将状态进行打印,以便能够在控制台看到实时的状态。
- 在Text中增加了点击事件,以便导航到其他AbilitySlice。

5.6.2 修改 PayAbilitySlice

创建PayAbility时会自动创建一个PayAbilitySlice类。修改PayAbilitySlice代码如下:

```java
package com.waylau.hmos.pageandabilityslicelifecycle.slice;

import com.waylau.hmos.pageandabilityslicelifecycle.ResourceTable;
import ohos.aafwk.ability.AbilitySlice;
import ohos.aafwk.content.Intent;
import ohos.agp.components.Text;
import ohos.hiviewdfx.HiLog;
import ohos.hiviewdfx.HiLogLabel;

public class PayAbilitySlice extends AbilitySlice {
    private static final String TAG = PayAbilitySlice.class.getSimpleName();
    private static final HiLogLabel LABEL_LOG =
            new HiLogLabel(HiLog.LOG_APP, 0x00001, TAG);

    @Override
    public void onStart(Intent intent) {
        super.onStart(intent);

        // UI界面内容引用了布局文件ability_pay.xml
        super.setUIContent(ResourceTable.Layout_ability_pay);

        // 添加点击事件来触发导航到MainAbilitySlice
```

```java
        Text text = (Text) findComponentById(ResourceTable.Id_text_pay);
        text.setClickedListener(listener ->
                present(new MainAbilitySlice(), new Intent()));

        HiLog.info(LABEL_LOG, "onStart");
    }

    @Override
    public void onActive() {
        super.onActive();
        HiLog.info(LABEL_LOG, "onActive");
    }

    @Override
    public void onForeground(Intent intent) {
        super.onForeground(intent);
        HiLog.info(LABEL_LOG, "onForeground");
    }

    @Override
    public void onInactive() {
        HiLog.info(LABEL_LOG, "onInactive");
    }

    @Override
    public void onBackground() {
        HiLog.info(LABEL_LOG, "onBackground");
    }

    @Override
    public void onStop() {
        HiLog.info(LABEL_LOG, "onStop");
    }
}
```

在PayAbilitySlice类中：

- PayAbilitySlice重写了AbilitySlice的生命周期回调方法。
- 在每个生命周期回调方法中，通过HiLog日志工具将状态进行打印，以便能够在控制台看到实时的状态。
- 在Text中增加了点击事件，以便导航到其他AbilitySlice。

5.6.3 修改 PayAbilitySlice 布局和文字

为了体现MainAbilitySlice和PayAbilitySlice的不同，我们需要在"面"上"整容"一下。从5.6.2节的代码可以得知，PayAbilitySlice通过setUIContent方法引用了ResourceTable.Layout_ability_pay的布局文件，而ResourceTable.Layout_ability_pay其实就是layout目录下的ability_pay.xml文件。

对ability_pay.xml的内容修改如下：

```xml
<?xml version="1.0" encoding="utf-8"?>
<DirectionalLayout
    xmlns:ohos="http://schemas.huawei.com/res/ohos"
    ohos:height="match_parent"
    ohos:width="match_parent"
    ohos:alignment="center"
    ohos:orientation="vertical">

    <Text
        ohos:id="$+id:text_pay"
        ohos:height="match_content"
        ohos:width="match_content"
        ohos:background_element="$graphic:background_ability_pay"
        ohos:layout_alignment="horizontal_center"
        ohos:text="$string:payability_money"
        ohos:text_size="40vp"
        />

</DirectionalLayout>
```

ability_pay.xml基本上与ability_main.xml类似,主要的差异点是:

- id设置为$+id:text_pay。
- background_element设置为$graphic:background_ability_pay。
- text设置为$string:payability_money。

上面这个payability_money来自于en.json文件,用于定义在不同的语言环境下所要显示的组件的文字内容。

以en.json文件为例,内容修改如下:

```
{
  "string": [
    {
      "name": "entry_MainAbility",
      "value": "entry_MainAbility"
    },
    {
      "name": "mainability_description",
      "value": "Java_Empty Ability"
    },
    {
      "name": "mainability_HelloWorld",
      "value": "Hello World"
    },
    {
      "name": "entry_PayAbility",
      "value": "entry_PayAbility"
    },
    {
      "name": "payability_description",
```

```
      "value": "Java_Empty Ability"
    },
    {
      "name": "payability_money",
      "value": "Pay me the money"
    }
  ]
}
```

上述代码定义了当组件引用$string:payability_money时，如果语言环境是en（英语），则会在组件上显示Pay me the money字样。

同理，我们修改zh.json的内容如下：

```
{
  "string": [
    {
      "name": "entry_MainAbility",
      "value": "entry_MainAbility"
    },
    {
      "name": "mainability_description",
      "value": "Java_Empty Ability"
    },
    {
      "name": "mainability_HelloWorld",
      "value": "你好，世界"
    },
    {
      "name": "entry_PayAbility",
      "value": "entry_PayAbility"
    },
    {
      "name": "payability_description",
      "value": "Java_Empty Ability"
    },
    {
      "name": "payability_money",
      "value": "你好，支付"
    }
  ]
}
```

上述代码定义了当组件引用$string:payability_money时，如果语言环境是zh（中文），则会在组件上显示"你好，支付"字样。

5.6.4 实现 AbilitySlice 之间的路由

为了实现AbilitySlice之间的路由和导航，在MainAbility中通过addActionRoute方法来添加到PayAbilitySlice的路由，代码如下：

```java
package com.waylau.hmos.abilityslicenavigation;

import com.waylau.hmos.abilityslicenavigation.slice.MainAbilitySlice;
import com.waylau.hmos.abilityslicenavigation.slice.PayAbilitySlice;
import ohos.aafwk.ability.Ability;
import ohos.aafwk.content.Intent;

public class MainAbility extends Ability {
    @Override
    public void onStart(Intent intent) {
        super.onStart(intent);

        // 指定默认显示的AbilitySlice
        super.setMainRoute(MainAbilitySlice.class.getName());

        // 使用addActionRounte方法添加路由
        addActionRoute("action.pay", PayAbilitySlice.class.getName());
    }
}
```

其中,上述action.pay是指定路由动作的名称。这个名称还需要在config.json的actions数组中添加,配置如下:

```json
...
"abilities": [
    {
        "skills": [
            {
                "entities": [
                    "entity.system.home"
                ],
                "actions": [
                    "action.system.home",
                    "action.pay"  // 指定路由动作的名称
                ]
            }
        ],
        "orientation": "unspecified",
        "name": "com.waylau.hmos.pageandabilityslicelifecycle.MainAbility",
        "icon": "$media:icon",
        "description": "$string:mainability_description",
        "label": "$string:entry_MainAbility",
        "type": "page",
        "launchType": "standard"
    },
    {
        "orientation": "unspecified",
        "name": "com.waylau.hmos.pageandabilityslicelifecycle.PayAbility",
        "icon": "$media:icon",
        "description": "$string:payability_description",
        "label": "$string:entry_PayAbility",
```

```
        "type": "page",
        "launchType": "standard"
      }
   ]
...
```

5.6.5 运行

将应用在模拟器中运行,如图5-9所示。

此时,能看到控制台输出如下内容:

```
08-29 11:40:34.932 13949-13949/? I 00001/MainAbilitySlice: onStart
08-29 11:40:34.952 13949-13949/? I 00001/MainAbilitySlice: onActive
```

从上述日志可以看出,MainAbilitySlice已经启动并处于ACTIVE状态。

点击文本"你好,世界"后,可以切换到"你好,支付"界面,如图5-10所示。

图5-9　在模拟器中运行应用　　　　图5-10　切换到"你好,支付"界面

此时,能看到控制台输出如下内容:

```
    08-29 11:43:34.820 13949-13949/com.waylau.hmos.pageandabilityslicelifecycle
I 00001/MainAbilitySlice: onInactive
    08-29 11:43:34.847 13949-13949/com.waylau.hmos.pageandabilityslicelifecycle
I 00001/PayAbilitySlice: onStart
    08-29 11:43:34.848 13949-13949/com.waylau.hmos.pageandabilityslicelifecycle
I 00001/PayAbilitySlice: onActive
    08-29 11:43:34.848 13949-13949/com.waylau.hmos.pageandabilityslicelifecycle
I 00001/MainAbilitySlice: onBackground
```

从上述日志可以看出,MainAbilitySlice失去了焦点并处于INACTIVE状态,而PayAbilitySlice启动并处于ACTIVE状态,最终MainAbilitySlice进入BACKGROUND状态。

再点击文本"你好,支付",可以切换到"你好,世界"界面,如图5-11所示。

图 5-11 切换到"你好,世界"界面

此时,能看到控制台输出如下内容:

```
    08-29 11:44:46.231 13949-13949/com.waylau.hmos.pageandabilityslicelifecycle
I 00001/PayAbilitySlice: onInactive
    08-29 11:44:46.234 13949-13949/com.waylau.hmos.pageandabilityslicelifecycle
I 00001/MainAbilitySlice: onStart
    08-29 11:44:46.234 13949-13949/com.waylau.hmos.pageandabilityslicelifecycle
I 00001/MainAbilitySlice: onActive
    08-29 11:44:46.234 13949-13949/com.waylau.hmos.pageandabilityslicelifecycle
I 00001/PayAbilitySlice: onBackground
```

从上述日志可以看出,PayAbilitySlice失去了焦点并处于INACTIVE状态,而MainAbilitySlice启动并处于ACTIVE状态。最终PayAbilitySlice进入BACKGROUND状态。

点击模拟器的返回按钮,返回"你好,支付"界面,如图5-12所示。

图 5-12 返回"你好,支付"界面

此时,能看到控制台输出如下内容:

```
    08-29 11:45:58.747 13949-13949/com.waylau.hmos.pageandabilityslicelifecycle
I 00001/MainAbilitySlice: onInactive
    08-29 11:45:58.747 13949-13949/com.waylau.hmos.pageandabilityslicelifecycle
I 00001/PayAbilitySlice: onForeground
    08-29 11:45:58.751 13949-13949/com.waylau.hmos.pageandabilityslicelifecycle
I 00001/PayAbilitySlice: onActive
    08-29 11:45:58.751 13949-13949/com.waylau.hmos.pageandabilityslicelifecycle
I 00001/MainAbilitySlice: onBackground
    08-29 11:45:58.751 13949-13949/com.waylau.hmos.pageandabilityslicelifecycle
I 00001/MainAbilitySlice: onStop
```

从上述日志可以看出，MainAbilitySlice失去了焦点并处于INACTIVE状态，而PayAbilitySlice重新回到前台，并处于ACTIVE状态。最终MainAbilitySlice进入BACKGROUND状态，最后被销毁。

再次点击模拟器的返回按钮，返回"你好，世界"界面。此时，能看到控制台输出如下内容：

```
    08-29 11:48:37.999 13949-13949/com.waylau.hmos.pageandabilityslicelifecycle
I 00001/PayAbilitySlice: onInactive
    08-29 11:48:37.999 13949-13949/com.waylau.hmos.pageandabilityslicelifecycle
I 00001/MainAbilitySlice: onForeground
    08-29 11:48:38.002 13949-13949/com.waylau.hmos.pageandabilityslicelifecycle
I 00001/MainAbilitySlice: onActive
    08-29 11:48:38.003 13949-13949/com.waylau.hmos.pageandabilityslicelifecycle
I 00001/PayAbilitySlice: onBackground
    08-29 11:48:38.003 13949-13949/com.waylau.hmos.pageandabilityslicelifecycle
I 00001/PayAbilitySlice: onStop
```

从上述日志可以看出，PayAbilitySlice失去了焦点并处于INACTIVE状态，而MainAbilitySlice重新回到前台，并处于ACTIVE状态。最终PayAbilitySlice进入BACKGROUND状态，最后被销毁。

再次点击模拟器的返回按钮，返回系统主界面，如图5-13所示。

图5-13 返回系统主界面

此时，能看到控制台输出如下内容：

```
    08-29 11:49:52.104 13949-13949/com.waylau.hmos.pageandabilityslicelifecycle
I 00001/MainAbilitySlice: onInactive
    08-29 11:49:53.827 13949-13949/com.waylau.hmos.pageandabilityslicelifecycle
I 00001/MainAbilitySlice: onBackground
    08-29 11:49:53.895 13949-13949/com.waylau.hmos.pageandabilityslicelifecycle
I 00001/MainAbilitySlice: onStop
```

从上述日志可以看出，MainAbilitySlice处于INACTIVE状态，接着进入BACKGROUND状态，最后被销毁。

当点击系统主界面的PageAndAbilitySliceLifeCycle应用图标时，将进入PageAndAbilitySliceLifeCycle应用，如图5-14所示。

图5-14　进入PageAndAbilitySliceLifeCycle应用

此时，能看到控制台输出如下内容：

```
    08-29 11:53:21.660 13949-13949/com.waylau.hmos.pageandabilityslicelifecycle
I 00001/MainAbilitySlice: onStart
    08-29 11:53:21.689 13949-13949/com.waylau.hmos.pageandabilityslicelifecycle
I 00001/MainAbilitySlice: onActive
```

从上述日志可以看出，MainAbilitySlice启用并处于ACTIVE状态。

5.7　Service Ability

基于Service模板的Ability简称为Service，其主要用于后台运行任务，比如执行音乐播放、文件下载等，但不提供用户交互界面。Service可由其他应用或Ability启动，即使用户切换到其他应用，Service仍将在后台继续运行。

Service是单实例的。在一个设备上，相同的Service只会存在一个实例。如果多个Ability共用这个实例，只有当与Service绑定的所有Ability都退出后，Service才能够退出。由于Service是在主线程里执行的，因此，如果在Service里面的操作时间过长，开发者必须在Service中创建新的线程来处理，防止造成主线程阻塞，应用程序无响应。

有关线程方面的内容详见第12章。

5.7.1 创建 Service

接下来介绍如何创建一个Service。

1. 继承 Ability

每个Service都是Ability的子类，需要实现Service相关的生命周期方法。

Ability为Service提供了以下生命周期方法，用户可以重写这些方法来添加自己的处理。

- onStart()：该方法在创建Service的时候调用，用于Service的初始化。在Service的整个生命周期只会调用一次，调用时传入的Intent应为空。该方法在前面章节的Page中已经做了介绍。
- onCommand()：在Service创建完成之后调用，该方法在客户端每次启动该Service时都会调用，用户可以在该方法中做一些调用统计、初始化类的操作。
- onConnect()：在Ability和Service连接时调用，该方法返回IRemoteObject对象，用户可以在该回调函数中生成对应Service的IPC通信通道，以便Ability与Service交互。Ability可以多次连接同一个Service，系统会缓存该Service的IPC通信对象，只有第一个客户端连接Service时，系统才会调用Service的onConnect方法来生成IRemoteObject对象，而后系统会将同一个RemoteObject对象传递至其他连接同一个Service的所有客户端，而无须再次调用onConnect方法。
- onDisconnect()：在Ability与绑定的Service断开连接时调用。
- onStop()：在Service销毁时调用。Service应通过实现此方法来清理任何资源，如关闭线程、注册的侦听器等。该方法在前面章节的Page中已经做了介绍。

创建Service的示例代码如下：

```
public class ServiceAbility extends Ability {
    @Override
    public void onStart(Intent intent) {
        super.onStart(intent);
    }

    @Override
    public void onCommand(Intent intent, boolean restart, int startId) {
        super.onCommand(intent, restart, startId);
    }

    @Override
    public IRemoteObject onConnect(Intent intent) {
        super.onConnect(intent);
        return null;
    }

    @Override
    public void onDisconnect(Intent intent) {
```

```
        super.onDisconnect(intent);
    }

    @Override
    public void onStop() {
        super.onStop();
    }
}
```

2. 注册 Service

Service也需要在应用配置文件中进行注册，注册类型type需要设置为service。配置内容如下：

```
{
    "module": {
        "abilities": [
            {
                "name": ".ServiceAbility",
                "type": "service",
                "visible": true
                ...
            }
        ]
        ...
    }
    ...
}
```

5.7.2 启动 Service

接下来介绍通过startAbility()启动Service以及对应的停止方法。

1. 启动 Service

Ability为开发者提供了startAbility()方法来启动另一个Ability。因为Service也是Ability的一种，开发者同样可以通过将Intent传递给该方法来启动Service，不仅支持启动本地Service，还支持启动远程Service。

开发者可以通过构造包含DeviceId、BundleName与AbilityName的Operation对象来设置目标Service信息。这3个参数的含义如下：

- DeviceId：表示设备ID。如果是本地设备，则可以直接留空；如果是远程设备，则可以通过ohos.distributedschedule.interwork.DeviceManager提供的getDeviceList获取设备列表。
- BundleName：表示包名称。
- AbilityName：表示待启动的Ability名称。

启动本地设备Service的示例代码如下：

```
Intent intent = new Intent();
Operation operation = new Intent.OperationBuilder()
    .withDeviceId("")
```

```
            .withBundleName("com.waylau.hmos.serviceabilitylifecycle")
            .withAbilityName("com.waylau.hmos.serviceabilitylifecycle.ServiceAbi
lity")
            .build();
    intent.setOperation(operation);
    startAbility(intent);
启动远程设备Service的代码示例如下:
Operation operation = new Intent.OperationBuilder()
            .withDeviceId("deviceId")
            .withBundleName("com.waylau.hmos.serviceabilitylifecycle")
            .withAbilityName("com.waylau.hmos.serviceabilitylifecycle.ServiceAbi
lity")
            .withFlags(Intent.FLAG_ABILITYSLICE_MULTI_DEVICE)  // 设置支持分布式调度
系统多设备启动的标识
            .build();
    Intent intent = new Intent();
    intent.setOperation(operation);
    startAbility(intent);
```

执行上述代码后，Ability将通过startAbility()方法来启动Service。

- 如果Service尚未运行，则系统会先调用onStart()来初始化Service，再回调Service的onCommand()方法来启动Service。
- 如果Service正在运行，则系统会直接回调Service的onCommand()方法来启动Service。

2. 停止 Service

Service一旦创建就会一直保持在后台运行，除非必须回收内存资源，否则系统不会停止或销毁Service。开发者可以在Service中通过terminateAbility()来停止本Service，或在其他Ability调用stopAbility()来停止Service。

停止Service同样支持停止本地设备Service和停止远程设备Service，使用方法与启动Service一样。一旦调用停止Service的方法，系统便会尽快销毁Service。

5.7.3 连接 Service

如果Service需要与Page Ability或其他应用的Service Ability进行交互，则应创建用于连接的Connection。Service支持其他Ability通过connectAbility()方法与其进行连接。

在使用connectAbility()处理回调时，需要传入目标Service的Intent与IAbilityConnection的实例。IAbilityConnection提供了两个方法供开发者实现：

- onAbilityConnectDone()用来处理连接的回调。
- onAbilityDisconnectDone()用来处理断开连接的回调。

连接Service的代码示例如下：

```
// 创建连接回调实例
private IAbilityConnection connection = new IAbilityConnection() {
    // 连接到Service的回调
```

```java
    @Override
    public void onAbilityConnectDone(ElementName elementName,
        IRemoteObject iRemoteObject, int resultCode) {
        // Client侧需要定义与Service侧相同的IRemoteObject实现类
        // 开发者获取服务端传过来IRemoteObject对象,并从中解析出服务端传过来的信息
    }

    // 断开与连接的回调
    @Override
    public void onAbilityDisconnectDone(ElementName elementName, int resultCode) {
    }
};

// 连接Service
connectAbility(intent, connection);
```

同时,Service侧也需要在onConnect()时返回IRemoteObject,从而定义与Service进行通信的接口。onConnect()需要返回一个IRemoteObject对象,HarmonyOS提供了IRemoteObject的默认实现,用户可以通过继承LocalRemoteObject来创建自定义的实现类。Service侧把自身的实例返回给调用侧的示例代码如下:

```java
// 创建自定义IRemoteObject实现类
private class MyRemoteObject extends LocalRemoteObject {
    public MyRemoteObject() {
        super("MyRemoteObject");
    }
}

// 把IRemoteObject返回给客户端
@Override
protected IRemoteObject onConnect(Intent intent) {
    return new MyRemoteObject();
}
```

5.7.4　Service Ability 的生命周期

与Page类似,Service也拥有生命周期,如图5-15所示。

根据调用方法的不同,其生命周期有以下两种路径:

- 启动Service:该Service在其他Ability调用startAbility()时创建,然后保持运行。其他Ability通过调用stopAbility()来停止Service,Service停止后,系统会将其销毁。
- 连接Service:该Service在其他Ability调用connectAbility()时创建,客户端可通过调用disconnectAbility()断开连接。多个客户端可以绑定到相同的Service,而且当所有绑定全部取消后,系统就会销毁该Service。connectAbility()也可以连接通过startAbility()创建的Service。

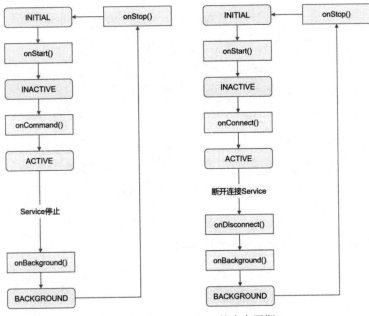

图 5-15　Service Ability 的生命周期

5.8　实战：Service Ability 生命周期的例子

为了更好地理解Service Ability的生命周期，我们将创建一个名为ServiceAbilityLifeCycle的应用作为演示。

5.8.1　创建 Service

在DevEco Studio中，可以如图5-16所示的方式创建一个Empty Service Ability。

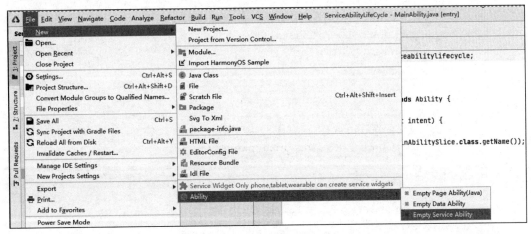

图 5-16　创建一个 Empty Service Ability

根据如图5-17所示的引导,创建一个名为TimeServiceAbility的Service。

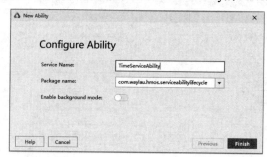

图5-17　创建一个名为 TimeServiceAbility 的 Service

注意,上述步骤中的Enable backgroud mode(后台模式)先不要启用。

在自动创建的TimeServiceAbility的基础上修改代码如下:

```java
package com.waylau.hmos.serviceabilitylifecycle;

import ohos.aafwk.ability.Ability;
import ohos.aafwk.content.Intent;
import ohos.rpc.IRemoteObject;
import ohos.hiviewdfx.HiLog;
import ohos.hiviewdfx.HiLogLabel;

import java.time.LocalDateTime;

public class TimeServiceAbility extends Ability {
    private static final String TAG =
TimeServiceAbility.class.getSimpleName();
    private static final HiLogLabel LABEL_LOG =
        new HiLogLabel(HiLog.LOG_APP, 0x00001, TAG);

    private TimeRemoteObject timeRemoteObject = new TimeRemoteObject();

    @Override
    public void onStart(Intent intent) {
        HiLog.info(LABEL_LOG, "onStart");
        super.onStart(intent);
    }

    @Override
    public void onBackground() {
        super.onBackground();
        HiLog.info(LABEL_LOG, "onBackground");
    }

    @Override
    public void onStop() {
        super.onStop();
        HiLog.info(LABEL_LOG, "onStop");
```

```java
    }

    @Override
    public void onCommand(Intent intent, boolean restart, int startId) {
        super.onCommand(intent, restart, startId);
        HiLog.info(LABEL_LOG, "onCommand");
    }

    @Override
    public IRemoteObject onConnect(Intent intent) {
        super.onConnect(intent);
        HiLog.info(LABEL_LOG, "onConnect");

        LocalDateTime now = LocalDateTime.now();
        timeRemoteObject.setTime(now);

        return timeRemoteObject;
    }

    @Override
    public void onDisconnect(Intent intent) {
        super.onDisconnect(intent);
        HiLog.info(LABEL_LOG, "onDisconnect");
    }
}
```

其中，timeRemoteObject是一个IRemoteObject子类的实例，可以通过timeRemoteObject将当前时间返回给调用侧。稍后，我们将创建TimeRemoteObject类。

同时，在配置文件中会自动新增TimeServiceAbility相关的配置信息。

```
...
"abilities": [
    {
        "skills": [
            {
                "entities": [
                    "entity.system.home"
                ],
                "actions": [
                    "action.system.home"
                ]
            }
        ],
        "orientation": "unspecified",
        "name": "com.waylau.hmos.serviceabilitylifecycle.MainAbility",
        "icon": "$media:icon",
        "description": "$string:mainability_description",
        "label": "$string:entry_MainAbility",
        "type": "page",
        "launchType": "standard"
```

```
        },
        // 新增的TimeServiceAbility
        {
          "name": 
"com.waylau.hmos.serviceabilitylifecycle.TimeServiceAbility",
          "icon": "$media:icon",
          "description": "$string:timeserviceability_description",
          "type": "service"
        }
     ]
     ...
```

5.8.2 创建远程对象

timeRemoteObject是TimeRemoteObject类的实例。TimeRemoteObject类是远程对象，继承自LocalRemoteObject，代码如下：

```
package com.waylau.hmos.serviceabilitylifecycle;

import ohos.aafwk.ability.LocalRemoteObject;

import java.time.LocalDateTime;

public class TimeRemoteObject extends LocalRemoteObject {
    private LocalDateTime time;

    public TimeRemoteObject() {
    }

    public void setTime(LocalDateTime time) {
        this.time = time;
    }

    public LocalDateTime getTime() {
        return time;
    }
}
```

其中，time属性用于返回给调用方服务器的当前时间。

5.8.3 修改 MainAbilitySlice

修改MainAbilitySlice的代码如下：

```
package com.waylau.hmos.serviceabilitylifecycle.slice;

import com.waylau.hmos.serviceabilitylifecycle.ResourceTable;
import com.waylau.hmos.serviceabilitylifecycle.TimeRemoteObject;
import com.waylau.hmos.serviceabilitylifecycle.TimeServiceAbility;
```

```java
import ohos.aafwk.ability.AbilitySlice;
import ohos.aafwk.ability.IAbilityConnection;
import ohos.aafwk.content.Intent;
import ohos.aafwk.content.Operation;
import ohos.agp.components.Text;
import ohos.bundle.ElementName;
import ohos.hiviewdfx.HiLog;
import ohos.hiviewdfx.HiLogLabel;
import ohos.rpc.IRemoteObject;

public class MainAbilitySlice extends AbilitySlice {
    private static final String TAG = TimeServiceAbility.class.getSimpleName();
    private static final HiLogLabel LABEL_LOG =
            new HiLogLabel(HiLog.LOG_APP, 0x00001, TAG);

    private TimeRemoteObject timeRemoteObject;

    @Override
    public void onStart(Intent intent) {
        super.onStart(intent);
        super.setUIContent(ResourceTable.Layout_ability_main);

        // 添加点击事件
        Text textStart = (Text) findComponentById(ResourceTable.Id_text_start);
        textStart.setClickedListener(listener -> {
            // 启动本地服务
            startupLocalService(intent);

            // 连接本地服务
            connectLocalService(intent);
        });

        // 添加点击事件
        Text textStop = (Text) findComponentById(ResourceTable.Id_text_stop);
        textStop.setClickedListener(listener -> {
            // 断开本地服务
            disconnectLocalService(intent);

            // 关闭本地服务
            stopLocalService(intent);
        });

        HiLog.info(LABEL_LOG, "onStart");
    }

    @Override
    public void onActive() {
```

```java
        super.onActive();
    }

    @Override
    public void onForeground(Intent intent) {
        super.onForeground(intent);
    }

    /**
     * 启动本地服务
     */
    private void startupLocalService(Intent intent) {
        //Intent intent = new Intent();
        //构建操作方式
        Operation operation = new Intent.OperationBuilder()
                // 设备ID
                .withDeviceId("")
                // 包名称
                .withBundleName("com.waylau.hmos.serviceabilitylifecycle")
                // 待启动的Ability名称
                .withAbilityName("com.waylau.hmos.serviceabilitylifecycle.TimeServiceAbility")
                .build();
        //设置操作
        intent.setOperation(operation);
        startAbility(intent);

        HiLog.info(LABEL_LOG, "startupLocalService");
    }

    /**
     * 关闭本地服务
     */
    private void stopLocalService(Intent intent) {
        stopAbility(intent);

        HiLog.info(LABEL_LOG, "stopLocalService");
    }

    // 创建连接回调实例
    private IAbilityConnection connection = new IAbilityConnection() {
        // 连接到Service的回调
        @Override
        public void onAbilityConnectDone(ElementName elementName,
                                IRemoteObject iRemoteObject, int resultCode) {
            // Client侧需要定义与Service侧相同的IRemoteObject实现类
            // 开发者获取服务端传过来IRemoteObject对象,并从中解析出服务端传过来的信息
            timeRemoteObject = (TimeRemoteObject) iRemoteObject;
```

```java
            HiLog.info(LABEL_LOG, "onAbilityConnectDone, time: %{public}s",
                timeRemoteObject.getTime());
        }

        // 断开与连接的回调
        @Override
        public void onAbilityDisconnectDone(ElementName elementName, int resultCode) {
            HiLog.info(LABEL_LOG, "onAbilityDisconnectDone");
        }
    };

    /**
     * 连接本地服务
     */
    private void connectLocalService(Intent intent) {
        // 连接Service
        connectAbility(intent, connection);

        HiLog.info(LABEL_LOG, "connectLocalService");
    }

    /**
     * 断开连接本地服务
     */
    private void disconnectLocalService(Intent intent) {
        // 断开连接Service
        disconnectAbility(connection);

        HiLog.info(LABEL_LOG, "disconnectLocalService");
    }
}
```

上述代码在onStart()方法中增加了对Text的事件监听。当点击textStart时，会启动本地服务并连接本地服务；当点击textEnd时，会断开本地服务并关闭本地服务。

5.8.4 修改 ability_main.xml

修改ability_main.xml的内容如下：

```xml
<?xml version="1.0" encoding="utf-8"?>
<DirectionalLayout
    xmlns:ohos="http://schemas.huawei.com/res/ohos"
    ohos:height="match_parent"
    ohos:width="match_parent"
    ohos:orientation="vertical">

    <Text
        ohos:id="$+id:text_start"
```

```
            ohos:height="match_content"
            ohos:width="match_content"
            ohos:background_element="$graphic:background_ability_main"
            ohos:layout_alignment="horizontal_center"
            ohos:text="Start"
            ohos:text_size="40vp"
            />

    <Text
            ohos:id="$+id:text_stop"
            ohos:height="match_parent"
            ohos:width="match_content"
            ohos:background_element="$graphic:background_ability_main"
            ohos:layout_alignment="horizontal_center"
            ohos:text="End"
            ohos:text_size="40vp"
            />

</DirectionalLayout>
```

上述代码主要定义了两个Text，一个用于触发Start点击事件，一个用于触发End点击事件。

5.8.5 运行

将应用在模拟器中运行，如图5-18所示。

图 5-18　在模拟器中运行应用

此时，能看到控制台输出如下内容：

```
08-29 14:30:38.115 25562-25562/com.waylau.hmos.serviceabilitylifecycle I
00001/TimeServiceAbility: onStart
```

从上述日志可以看出，MainAbilitySlice已经启动。

点击文本Start后，触发点击事件，此时能看到控制台输出如下内容：

```
08-29 14:32:58.807 25562-25562/com.waylau.hmos.serviceabilitylifecycle I
00001/TimeServiceAbility: startupLocalService
08-29 14:32:58.813 25562-25562/com.waylau.hmos.serviceabilitylifecycle I
00001/TimeServiceAbility: connectLocalService
08-29 14:32:58.817 25562-25562/com.waylau.hmos.serviceabilitylifecycle I
00001/TimeServiceAbility: onStart
08-29 14:32:58.819 25562-25562/com.waylau.hmos.serviceabilitylifecycle I
00001/TimeServiceAbility: onCommand
08-29 14:32:58.822 25562-25562/com.waylau.hmos.serviceabilitylifecycle I
00001/TimeServiceAbility: [2532d12a1ac57b7, f8a3c0, d1978e] onConnect
08-29 14:32:58.838 25562-25562/com.waylau.hmos.serviceabilitylifecycle I
00001/TimeServiceAbility: [2532d12a1ac57b7, 21de4fa, 2fd6c0a]
onAbilityConnectDone, time: 2021-08-29T14:32:58.828
```

当点击文本Start时，会启动本地服务并连接本地服务。而TimeServiceAbility分别执行了onStart、onCommand以及onConnect等生命周期，并将当前时间返回给了MainAbilitySlice。

点击文本End后，触发点击事件，此时能看到控制台输出如下内容：

```
08-29 14:33:43.856 25562-25562/com.waylau.hmos.serviceabilitylifecycle I
00001/TimeServiceAbility: disconnectLocalService
08-29 14:33:43.859 25562-25562/com.waylau.hmos.serviceabilitylifecycle I
00001/TimeServiceAbility: stopLocalService
08-29 14:33:43.862 25562-25562/com.waylau.hmos.serviceabilitylifecycle I
00001/TimeServiceAbility: [2532d22a47d0292, 27497f3, 105ac6e] onDisconnect
08-29 14:33:43.864 25562-25562/com.waylau.hmos.serviceabilitylifecycle I
00001/TimeServiceAbility: onBackground
08-29 14:33:43.864 25562-25562/com.waylau.hmos.serviceabilitylifecycle I
00001/TimeServiceAbility: onStop
```

当点击文本End时，会断开本地服务并关闭本地服务。而TimeServiceAbility分别执行了onDisconnect、onBackground以及onStop等生命周期。

5.9 Data Ability

使用Data模板的Ability也简称Data，主要职责是管理其自身应用和其他应用存储数据的访问，并提供与其他应用共享数据的方法。Data既可用于同一设备不同应用的数据共享，也支持跨设备不同应用的数据共享。

数据的存储方式多种多样，可以是传统意义上的数据库系统，也可以是本地磁盘上的文件。Data对外提供对数据的增、删、改、查，以及打开文件等接口，这些接口的具体实现由开发者提供。

5.9.1 URI

Data的提供方和使用方都通过URI（Uniform Resource Identifier，统一资源定位符）来标识一个具体的数据,例如数据库中的某个表或磁盘上的某个文件。HarmonyOS的URI是基于URI通用标准的,具体格式如图5-19所示。

图 5-19 URI 格式

其中:

- scheme:协议方案名,固定为dataability,代表Data Ability所使用的协议类型。
- authority:设备ID。如果为跨设备场景,则为目标设备的ID;如果为本地设备场景,则不需要填写。
- path:资源的路径信息,代表特定资源的位置信息。
- query:查询参数。
- fragment:用于指示要访问的子资源。

以下是具体的URI示例:

跨设备场景:

```
dataability://device_id/com.waylau.hmos.dataabilityhelperaccessfile.dataability.persondata/person/10
```

本地设备:

```
dataability:///com.waylau.hmos.dataabilityhelperaccessfile.dataability.persondata/person/10
```

注:本地设备的device_id字段为空,因此在dataability:后面有3个"/"。

5.9.2 访问 Data

可以通过DataAbilityHelper类来访问当前应用或其他应用提供的共享数据。DataAbilityHelper作为客户端,与提供方的Data进行通信。Data接收到请求后,执行相应的处理,并返回结果。DataAbilityHelper提供了一系列与Data Ability对应的方法。

下面介绍DataAbilityHelper具体的使用步骤。

1. 声明使用权限

如果待访问的Data声明了访问需要的权限,则访问此Data需要在配置文件中声明需要此权限,示例如下:

```
"reqPermissions": [
    {
        "name": "com.waylau.hmos.dataabilityhelperaccessfile.DataAbility.DATA"
    }
]
```

如果访问的数据是文件，则还需要添加访问存储读、写的权限：ohos.permission.READ_USER_STORAGE和ohos.permission.WRITE_USER_STORAGE。

2. 创建DataAbilityHelper

DataAbilityHelper为开发者提供了creator()方法来创建DataAbilityHelper实例。该方法为静态方法，有多个重载。常见的方法是通过传入一个context对象来创建DataAbilityHelper对象。

以下为获取helper对象的示例：

```
DataAbilityHelper helper = DataAbilityHelper.creator(this);
```

3. 访问Data Ability

DataAbilityHelper为开发者提供了一系列的接口来访问不同类型的数据，比如文件、数据库等。

- 访问文件：DataAbilityHelper为开发者提供了FileDescriptor openFile（Uri uri, String mode）方法来操作文件。此方法需要传入两个参数，其中uri用来确定目标资源路径，mode用来指定打开文件的方式，可选方式包含r（读）、w（写）、rw（读写）、wt（覆盖写）、wa（追加写）、rwt（覆盖写且可读）。该方法返回一个目标文件的FD（文件描述符），把文件描述符封装成流，开发者就可以对文件流进行自定义处理。
- 访问数据库：DataAbilityHelper为开发者提供了增、删、改、查以及批量处理等方法来操作数据库。

在接下来的章节将会详细介绍如何使用DataAbilityHelper来访问文件和数据库。

5.10 实战：使用DataAbilityHelper访问文件

本节演示如何通过DataAbilityHelper类来访问当前应用的文件数据。创建一个名为DataAbilityHelperAccessFile的应用作为演示。

下面介绍DataAbilityHelper具体的使用步骤。

5.10.1 创建DataAbility

在DevEco Studio中，可以通过如图5-20所示的方式创建一个Empty Data Ability。

根据如图5-21所示的引导创建一个名为UserDataAbility的Data。

鸿蒙 HarmonyOS 手机应用开发实战

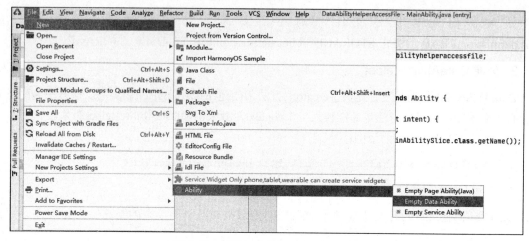

图 5-20　创建一个 Empty Data Ability

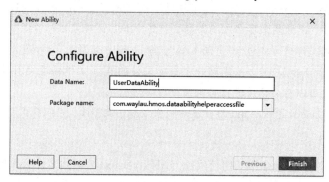

图 5-21　创建一个名为 UserDataAbility 的 Data

自定义生成的UserDataAbility代码如下：

```java
package com.waylau.hmos.dataabilityhelperaccessfile;

import ohos.aafwk.ability.Ability;
import ohos.aafwk.content.Intent;
import ohos.data.resultset.ResultSet;
import ohos.data.rdb.ValuesBucket;
import ohos.data.dataability.DataAbilityPredicates;
import ohos.hiviewdfx.HiLog;
import ohos.hiviewdfx.HiLogLabel;
import ohos.utils.net.Uri;
import ohos.utils.PacMap;

import java.io.FileDescriptor;

public class UserDataAbility extends Ability {
    private static final HiLogLabel LABEL_LOG = new HiLogLabel(3, 0xD001100, "Demo");

    @Override
    public void onStart(Intent intent) {
```

```java
        super.onStart(intent);
        HiLog.info(LABEL_LOG, "UserDataAbility onStart");
    }

    @Override
    public ResultSet query(Uri uri, String[] columns, DataAbilityPredicates predicates) {
        return null;
    }

    @Override
    public int insert(Uri uri, ValuesBucket value) {
        HiLog.info(LABEL_LOG, "UserDataAbility insert");
        return 999;
    }

    @Override
    public int delete(Uri uri, DataAbilityPredicates predicates) {
        return 0;
    }

    @Override
    public int update(Uri uri, ValuesBucket value, DataAbilityPredicates predicates) {
        return 0;
    }

    @Override
    public FileDescriptor openFile(Uri uri, String mode) {
        return null;
    }

    @Override
    public String[] getFileTypes(Uri uri, String mimeTypeFilter) {
        return new String[0];
    }

    @Override
    public PacMap call(String method, String arg, PacMap extras) {
        return null;
    }

    @Override
    public String getType(Uri uri) {
        return null;
    }
}
```

上述代码除了基本的增删改查、打开文件、获取URI类型、获取文件类型外，还有一个回调，再加上一个onStart方法，总共是9个方法。

UserDataAbility自动在配置文件中添加了相应的配置，内容如下：

```
...
"abilities": [
    {
    "skills": [
        {
        "entities": [
            "entity.system.home"
        ],
        "actions": [
            "action.system.home"
        ]
        }
    ],
    "orientation": "unspecified",
    "name": "com.waylau.hmos.dataabilityhelperaccessfile.MainAbility",
    "icon": "$media:icon",
    "description": "$string:mainability_description",
    "label": "$string:entry_MainAbility",
    "type": "page",
    "launchType": "standard"
    },
    // 新增UserDataAbility配置
    {
    "permissions": [
"com.waylau.hmos.dataabilityhelperaccessfile.DataAbilityShellProvider.PROVIDER"
    ],
    "name": "com.waylau.hmos.dataabilityhelperaccessfile.UserDataAbility",
    "icon": "$media:icon",
    "description": "$string:userdataability_description",
    "type": "data",
    "uri": "dataability://com.waylau.hmos.dataabilityhelperaccessfile.UserDataAbility"
    }
]
...
```

从上述配置可以看出，

- type：类型设置为data。
- uri：对外提供的访问路径，全局唯一。
- permissions：访问该Data Ability时需要申请的访问权限。

5.10.2 修改 UserDataAbility

由于本示例只涉及文件，因此修改UserDataAbility时只需重写onStart和openFile方法：

```java
package com.waylau.hmos.dataabilityhelperaccessfile;

import ohos.aafwk.ability.Ability;
import ohos.aafwk.content.Intent;
import ohos.data.resultset.ResultSet;
import ohos.data.rdb.ValuesBucket;
import ohos.data.dataability.DataAbilityPredicates;
import ohos.global.resource.RawFileEntry;
import ohos.global.resource.Resource;
import ohos.hiviewdfx.HiLog;
import ohos.hiviewdfx.HiLogLabel;
import ohos.rpc.MessageParcel;
import ohos.utils.net.Uri;
import ohos.utils.PacMap;

import java.io.*;
import java.nio.file.Paths;

public class UserDataAbility extends Ability {
    private static final String TAG = UserDataAbility.class.getSimpleName();
    private static final HiLogLabel LABEL_LOG = new HiLogLabel(HiLog.LOG_APP, 0x00001, TAG);

    private File targetFile;

    @Override
    public void onStart(Intent intent) {
        super.onStart(intent);
        HiLog.info(LABEL_LOG, "UserDataAbility onStart");

        try {
            // 初始化目标文件数据
            initFile();
        } catch (IOException e) {
            e.printStackTrace();
        }
    }

    private void initFile() throws IOException {
        // 获取数据目录
        File dataDir = new File(this.getDataDir().toString());
        if(!dataDir.exists()){
            dataDir.mkdirs();
        }

        // 构建目标文件
        targetFile = new File(Paths.get(dataDir.toString(),"users.txt").toString());

        // 获取源文件
```

```java
            RawFileEntry rawFileEntry =
this.getResourceManager().getRawFileEntry("resources/rawfile/users.txt");
            Resource resource = rawFileEntry.openRawFile();

            // 新建目标文件
            FileOutputStream fos = new FileOutputStream(targetFile);

            byte[] buffer = new byte[4096];
            int count = 0;

            // 源文件内容写入目标文件
            while((count = resource.read(buffer)) >= 0){
                fos.write(buffer,0,count);
            }

            resource.close();
            fos.close();
        }

        @Override
        public FileDescriptor openFile(Uri uri, String mode) {
            FileDescriptor fd = null;

            try {
                // 获取目标文件FileDescriptor
                FileInputStream fileIs = new FileInputStream(targetFile);
                fd = fileIs.getFD();

                HiLog.info(LABEL_LOG, "fd: %{public}s", fd);
            } catch (IOException e) {
                e.printStackTrace();
            }

            // 创建MessageParcel
            MessageParcel messageParcel = MessageParcel.obtain();

            // 复制FileDescriptor
            fd = messageParcel.dupFileDescriptor(fd);
            return fd;
        }

        @Override
        public ResultSet query(Uri uri, String[] columns, DataAbilityPredicates predicates) {
            return null;
        }

        @Override
        public int insert(Uri uri, ValuesBucket value) {
            HiLog.info(LABEL_LOG, "UserDataAbility insert");
```

```
            return 999;
    }

    @Override
    public int delete(Uri uri, DataAbilityPredicates predicates) {
        return 0;
    }

    @Override
    public int update(Uri uri, ValuesBucket value, DataAbilityPredicates
predicates) {
        return 0;
    }

    @Override
    public String[] getFileTypes(Uri uri, String mimeTypeFilter) {
        return new String[0];
    }

    @Override
    public PacMap call(String method, String arg, PacMap extras) {
        return null;
    }

    @Override
    public String getType(Uri uri) {
        return null;
    }
}
```

initFile方法用于将源文件写入目标文件。this.getDataDir()方法可以获取到数据目录，目标文件最终写入该目录下。

HarmonyOS提供了一个ResourceManager资源管理器，通过该资源管理器可以方便读取到resouece目录下的资源文件。其中，RawFileEntry代表rawfile目录下的文件。可以通过rawFileEntry.openRawFile()方法方便地获取到指定文件。

最后在方法返回前，需要通过MessageParcel对FileDescriptor进行复制。

5.10.3 创建文件

在resouece目录的rawfile目录下创建如图5-22所示的测试用文件users.txt。

该文件的测试内容比较简单，就是一个用户的名字：

```
Way Lau
```

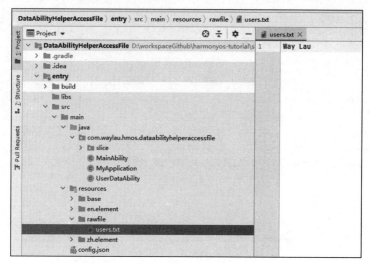

图 5-22　创建一个文件

5.10.4　修改 MainAbilitySlice

修改MainAbilitySlice的onStart方法，代码如下：

```
package com.waylau.hmos.dataabilityhelperaccessfile.slice;

import com.waylau.hmos.dataabilityhelperaccessfile.FileUtils;
import com.waylau.hmos.dataabilityhelperaccessfile.ResourceTable;
import ohos.aafwk.ability.AbilitySlice;
import ohos.aafwk.ability.DataAbilityHelper;
import ohos.aafwk.ability.DataAbilityRemoteException;
import ohos.aafwk.content.Intent;
import ohos.agp.components.Text;
import ohos.hiviewdfx.HiLog;
import ohos.hiviewdfx.HiLogLabel;
import ohos.utils.net.Uri;

import java.io.*;

public class MainAbilitySlice extends AbilitySlice {
    private static final String TAG = MainAbilitySlice.class.getSimpleName();
    private static final HiLogLabel LABEL_LOG = new HiLogLabel(HiLog.LOG_APP, 0x00001, TAG);

    @Override
    public void onStart(Intent intent) {
        super.onStart(intent);
        super.setUIContent(ResourceTable.Layout_ability_main);

        // 添加点击事件来触发访问数据
        Text text = (Text) findComponentById(ResourceTable.Id_text_helloworld);
```

```
            text.setClickedListener(listener -> this.getFile());
        }

        private void getFile() {
            DataAbilityHelper helper = DataAbilityHelper.creator(this);

            // 访问数据用的URI，注意用3个斜杠
            Uri uri =
                    Uri.parse("dataability:///com.waylau.hmos.dataabilityhelperaccessfile.UserDataAbility");

            // 使用DataAbilityHelper的openFile方法来访问文件
            try {
                FileDescriptor fd = helper.openFile(uri, "r");

                HiLog.info(LABEL_LOG, "fd: %{public}s", fd);
                HiLog.info(LABEL_LOG, "file content: %{public}s",
FileUtils.getFileContent(fd));
            } catch (DataAbilityRemoteException | IOException e) {
                e.printStackTrace();
            }
        }

        @Override
        public void onActive() {
            super.onActive();
        }

        @Override
        public void onForeground(Intent intent) {
            super.onForeground(intent);
        }
    }
```

在上述方法中，在Text中添加点击事件来触发访问文件的操作。

getFile方法用于访问文件。借助DataAbilityHelper类的openFile方法来访问当前UserDataAbility提供的文件数据。

FileUtils.getFileContent方法用于将文件内容转为字符串，这样可以在日志中方便查看文件的具体内容。FileUtils工具类稍后介绍。

注：上述代码中访问的URI与UserDataAbility在配置文件中添加的URI基本是一致的，唯一的区别是上述代码中访问的URI用3个斜杠。

5.10.5　创建 FileUtils 类

FileUtils类是一个工具类，其中getFileContent方法用于将文件内容转为字符串，代码如下：

```
package com.waylau.hmos.dataabilityhelperaccessfile;
```

```java
import java.io.FileDescriptor;
import java.io.FileInputStream;
import java.io.IOException;

public class FileUtils {

    /**
     * 输出文件内容
     * @param fd
     * @return 文件内容
     * @throws IOException
     */
    public static String getFileContent(FileDescriptor fd) throws IOException {
        // 根据FileDescriptor创建FileInputStream对象
        FileInputStream fis = new FileInputStream(fd);

        int b = 0;
        StringBuilder sb = new StringBuilder();
        while((b = fis.read()) != -1){
            sb.append((char)b);
        }
        fis.close();

        return sb.toString();
    }
}
```

5.10.6 运行

运行应用后，效果如图5-23所示。

图 5-23　运行应用的效果

点击"你好，世界"触发了访问文件的操作，可以看到控制台HiLog输出内容如下：

```
08-29 23:46:36.416 18923-18923/com.waylau.hmos.dataabilityhelperaccessfile I
00001/UserDataAbility: fd: java.io.FileDescriptor@900efa4
08-29 23:46:36.417 18923-18923/com.waylau.hmos.dataabilityhelperaccessfile I
00001/MainAbilitySlice: fd: java.io.FileDescriptor@32f08c2
08-29 23:46:36.417 18923-18923/com.waylau.hmos.dataabilityhelperaccessfile I
00001/MainAbilitySlice: file content: Way Lau
```

至此，DataAbilityHelper访问文件的示例演示完毕。

5.11 实战：使用 DataAbilityHelper 访问数据库

上一节演示了如何通过DataAbilityHelper类来访问当前应用的文件数据。本节将演示如何通过DataAbilityHelper类来访问当前应用的数据库数据。创建一个名为DataAbilityHelperAccessDatabase的应用作为演示。

下面介绍DataAbilityHelper具体的使用步骤。

5.11.1 创建 DataAbility

在DevEco Studio中创建一个名为UserDataAbility的Data，如图5-24所示。

图 5-24　创建一个名为 UserDataAbility 的 Data

UserDataAbility初始化时代码如下：

```
package com.waylau.hmos.dataabilityhelperaccessdatabase;

import ohos.aafwk.ability.Ability;
import ohos.aafwk.content.Intent;
import ohos.data.resultset.ResultSet;
import ohos.data.rdb.ValuesBucket;
import ohos.data.dataability.DataAbilityPredicates;
import ohos.hiviewdfx.HiLog;
import ohos.hiviewdfx.HiLogLabel;
import ohos.utils.net.Uri;
```

```java
    import ohos.utils.PacMap;

    import java.io.FileDescriptor;

    public class UserDataAbility extends Ability {
        private static final HiLogLabel LABEL_LOG = new HiLogLabel(3, 0xD001100, "Demo");

        @Override
        public void onStart(Intent intent) {
            super.onStart(intent);
            HiLog.info(LABEL_LOG, "UserDataAbility onStart");
        }

        @Override
        public ResultSet query(Uri uri, String[] columns, DataAbilityPredicates predicates) {
            return null;
        }

        @Override
        public int insert(Uri uri, ValuesBucket value) {
            HiLog.info(LABEL_LOG, "UserDataAbility insert");
            return 999;
        }

        @Override
        public int delete(Uri uri, DataAbilityPredicates predicates) {
            return 0;
        }

        @Override
        public int update(Uri uri, ValuesBucket value, DataAbilityPredicates predicates) {
            return 0;
        }

        @Override
        public FileDescriptor openFile(Uri uri, String mode) {
            return null;
        }

        @Override
        public String[] getFileTypes(Uri uri, String mimeTypeFilter) {
            return new String[0];
        }

        @Override
        public PacMap call(String method, String arg, PacMap extras) {
            return null;
```

```java
    }

    @Override
    public String getType(Uri uri) {
        return null;
    }
}
```

UserDataAbility自动在配置文件中添加了相应的配置,内容如下:

```json
...
"abilities": [
    {
    "skills": [
        {
        "entities": [
            "entity.system.home"
        ],
        "actions": [
            "action.system.home"
        ]
        }
    ],
    "orientation": "unspecified",
    "name": "com.waylau.hmos.dataabilityhelperaccessdatabase.MainAbility",
    "icon": "$media:icon",
    "description": "$string:mainability_description",
    "label": "$string:entry_MainAbility",
    "type": "page",
    "launchType": "standard"
    },
    // 新增UserDataAbility配置
    {
    "permissions": [
"com.waylau.hmos.dataabilityhelperaccessdatabase.DataAbilityShellProvider.PROVIDER"
    ],
    "name":
"com.waylau.hmos.dataabilityhelperaccessdatabase.UserDataAbility",
    "icon": "$media:icon",
    "description": "$string:userdataability_description",
    "type": "data",
    "uri":
"dataability://com.waylau.hmos.dataabilityhelperaccessdatabase.UserDataAbility"
    }
]
...
```

从上述配置可以看出:

- type：类型设置为data。
- uri：对外提供的访问路径，全局唯一。
- permissions：访问该Data Ability时需要申请的访问权限。

5.11.2 初始化数据库

HarmonyOS关系型数据库对外提供通用的操作接口，底层使用SQLite作为持久化存储引擎，支持SQLite具有的所有数据库特性，包括但不限于事务、索引、视图、触发器、外键、参数化查询和预编译SQL语句。

```java
public class UserDataAbility extends Ability {
    private static final String TAG = UserDataAbility.class.getSimpleName();
    private static final HiLogLabel LABEL_LOG =
            new HiLogLabel(HiLog.LOG_APP, 0x00001, TAG);
    private static final String DATABASE_NAME = "RdbStoreTest.db";
    private static final String TABLE_NAME = "user_t";
    private RdbStore store = null;

    @Override
    public void onStart(Intent intent) {
        super.onStart(intent);
        HiLog.info(LABEL_LOG, "UserDataAbility onStart");

        // 创建数据库连接，并获取连接对象
        DatabaseHelper helper = new DatabaseHelper(this);
        StoreConfig config = StoreConfig.newDefaultConfig(DATABASE_NAME);
        RdbOpenCallback callback = new RdbOpenCallback() {

            // 初始化数据库
            public void onCreate(RdbStore store) {
                store.executeSql("CREATE TABLE IF NOT EXISTS " + TABLE_NAME +
                        "(user_id INTEGER PRIMARY KEY, user_name TEXT NOT NULL, user_age INTEGER)");
            }

            @Override
            public void onUpgrade(RdbStore store, int oldVersion, int newVersion) {
            }
        };

        store = helper.getRdbStore(config, 1, callback, null);
    }
    ...
```

在上述代码中：

- 初始化了一个名为RdbStoreTest.db的数据库。
- 创建了一个名为user_t的表结构。该表包含user_id、user_name和user_age三个字段，其中user_id为主键。
- 初始化了RdbStore，用于方便处理后续关系型数据库（RDB）的管理。

5.11.3 重写 query 方法

重写UserDataAbility的query方法，代码如下：

```
@Override
public ResultSet query(Uri uri, String[] columns, DataAbilityPredicates predicates) {

    if (store == null) {
        HiLog.error(LABEL_LOG, "failed to query, ormContext is null");
        return null;
    }

    // 查询数据库
    RdbPredicates rdbPredicates = DataAbilityUtils.createRdbPredicates(predicates, TABLE_NAME);
    ResultSet resultSet = store.query(rdbPredicates, columns);
    if (resultSet == null) {
        HiLog.info(LABEL_LOG, "resultSet is null");
    }

    // 返回结果
    return resultSet;
}
```

上述代码中：

- DataAbilityUtils工具类将入参DataAbilityPredicates转为了RdbStore所能处理的RdbPredicates。
- 通过RdbStore查询所需要的字段。

5.11.4 重写 insert 方法

重写UserDataAbility的insert方法，代码如下：

```
@Override
public int insert(Uri uri, ValuesBucket value) {
    HiLog.info(LABEL_LOG, "UserDataAbility insert");

    // 参数校验
    if (store == null) {
        HiLog.error(LABEL_LOG, "failed to insert, ormContext is null");
        return -1;
    }
```

```
    // 插入数据库
    long userId = store.insert(TABLE_NAME, value);

    return Integer.valueOf(userId + "");
}
```

上述代码中：

- 通过RdbStore插入表对应的数据。
- RdbStore插入数据后的返回值是该插入数据的主键，即user_id字段值。

5.11.5 重写 update 方法

重写UserDataAbility的update方法，代码如下：

```
@Override
public int update(Uri uri, ValuesBucket value, DataAbilityPredicates
predicates) {
    if (store == null) {
        HiLog.error(LABEL_LOG, "failed to update, ormContext is null");
        return -1;
    }

    RdbPredicates rdbPredicates =
            DataAbilityUtils.createRdbPredicates(predicates, TABLE_NAME);
    int index = store.update(value, rdbPredicates);
    return index;
}
```

上述代码中：

- DataAbilityUtils工具类将入参DataAbilityPredicates转为了RdbStore所能处理的RdbPredicates。
- 通过RdbStore更新了对应条件的数据。
- 更新后返回的value值代表更新的数量。

5.11.6 重写 delete 方法

重写UserDataAbility的delete方法，代码如下：

```
@Override
public int delete(Uri uri, DataAbilityPredicates predicates) {
    if (store == null) {
        HiLog.error(LABEL_LOG, "failed to delete, ormContext is null");
        return -1;
    }

    RdbPredicates rdbPredicates =
        DataAbilityUtils.createRdbPredicates(predicates, TABLE_NAME);
```

```
        int value = store.delete(rdbPredicates);
        return value;
}
```

上述代码中：

- DataAbilityUtils工具类将入参DataAbilityPredicates转为了RdbStore所能处理的RdbPredicates。
- 通过RdbStore删除了对应条件的数据。
- 删除后返回的value值代表删除的数量。

5.11.7 修改MainAbilitySlice

修改MainAbilitySlice的onStart方法，代码如下：

```
package com.waylau.hmos.dataabilityhelperaccessdatabase.slice;

import com.waylau.hmos.dataabilityhelperaccessdatabase.ResourceTable;
import ohos.aafwk.ability.AbilitySlice;
import ohos.aafwk.ability.DataAbilityHelper;
import ohos.aafwk.ability.DataAbilityRemoteException;
import ohos.aafwk.content.Intent;
import ohos.agp.components.Text;
import ohos.data.dataability.DataAbilityPredicates;
import ohos.data.rdb.ValuesBucket;
import ohos.data.resultset.ResultSet;
import ohos.hiviewdfx.HiLog;
import ohos.hiviewdfx.HiLogLabel;
import ohos.utils.net.Uri;

public class MainAbilitySlice extends AbilitySlice {
    private static final String TAG = MainAbilitySlice.class.getSimpleName();
    private static final HiLogLabel LABEL_LOG =
            new HiLogLabel(HiLog.LOG_APP, 0x00001, TAG);

    @Override
    public void onStart(Intent intent) {
        super.onStart(intent);
        super.setUIContent(ResourceTable.Layout_ability_main);

        // 添加点击事件来触发访问数据
        Text text = (Text) findComponentById(ResourceTable.Id_text_helloworld);
        text.setClickedListener(listener -> this.doDatabaseAction());
    }

    private void doDatabaseAction() {

        DataAbilityHelper helper = DataAbilityHelper.creator(this);
        // 访问数据用的URI，注意用3个斜杠
```

```java
            Uri uri = Uri.parse("dataability:///com.waylau.hmos.dataabilityhelperaccessdatabase.UserDataAbility");
            String[] columns = {"user_id", "user_name", "user_age"};

            // 查询
            doQuery(helper, uri, columns);

            // 插入
            doInsert(helper, uri, columns);

            // 查询
            doQuery(helper, uri, columns);

            // 更新
            doUpdate(helper, uri, columns);

            // 查询
            doQuery(helper, uri, columns);

            // 删除
            doDelete(helper, uri, columns);

            // 查询
            doQuery(helper, uri, columns);
        }

        private void doQuery(DataAbilityHelper helper, Uri uri, String[] columns) {
            // 构造查询条件
            DataAbilityPredicates predicates = new DataAbilityPredicates();
            predicates.between("user_id", 101, 200);

            // 进行查询
            ResultSet resultSet = null;
            try {
                resultSet = helper.query(uri, columns, predicates);
            } catch (DataAbilityRemoteException e) {
                e.printStackTrace();
            }

            if (resultSet != null && resultSet.getRowCount() > 0) {
                // 在此处理ResultSet中的记录
                while (resultSet.goToNextRow()) {
                    // 处理结果
                    HiLog.info(LABEL_LOG, "resultSet user_id: %{public}s, user_name: %{public}s, user_age: %{public}s",
```

第 5 章 Ability 基础知识 | 125

```
                        resultSet.getInt(0), resultSet.getString(1),
resultSet.getInt(2));
            }
        } else {
            HiLog.info(LABEL_LOG, "resultSet is null or row count is 0");
        }
    }

    private void doInsert(DataAbilityHelper helper, Uri uri, String[] columns)
{
        // 构造查询条件
        DataAbilityPredicates predicates = new DataAbilityPredicates();
        predicates.between("user_Id", 101, 103);

        // 构造插入数据
        ValuesBucket valuesBucket = new ValuesBucket();
        valuesBucket.putInteger(columns[0], 101);
        valuesBucket.putString(columns[1], "Way Lau");
        valuesBucket.putInteger(columns[2], 33);

        try {
            int result = helper.insert(uri, valuesBucket);
            HiLog.info(LABEL_LOG, "insert result: %{public}s", result);
        } catch (DataAbilityRemoteException e) {
            e.printStackTrace();
        }
    }

    private void doUpdate(DataAbilityHelper helper, Uri uri, String[] columns)
{
        // 构造查询条件
        DataAbilityPredicates predicates = new DataAbilityPredicates();
        predicates.equalTo("user_id", 101);

        // 构造更新数据
        ValuesBucket valuesBucket = new ValuesBucket();
        valuesBucket.putInteger(columns[0], 101);
        valuesBucket.putString(columns[1], "Way Lau");
        valuesBucket.putInteger(columns[2], 35);

        try {
            int result = helper.update(uri, valuesBucket, predicates);

            HiLog.info(LABEL_LOG, "update result: %{public}s", result);
        } catch (DataAbilityRemoteException e) {
            e.printStackTrace();
        }
    }
```

```
    private void doDelete(DataAbilityHelper helper, Uri uri, String[] columns)
{
        // 构造查询条件
        DataAbilityPredicates predicates = new DataAbilityPredicates();
        predicates.equalTo("user_id", 101);

        try {
            int result = helper.delete(uri, predicates);

            HiLog.info(LABEL_LOG, "delete result: %{public}s", result);
        } catch (DataAbilityRemoteException e) {
            e.printStackTrace();
        }
    }

    @Override
    public void onActive() {
        super.onActive();
    }

    @Override
    public void onForeground(Intent intent) {
        super.onForeground(intent);
    }
}
```

在上述方法中，在Text中添加点击事件来触发访问数据库的操作。

doDatabaseAction方法用于执行数据库操作。借助DataAbilityHelper类的方法来访问当前UserDataAbility提供的数据库数据，分别执行了查询、插入、查询、更新、查询、删除和查询动作。

注：上述代码中访问的URI与UserDataAbility在配置数据库中添加的URI基本是一致的，唯一的区别是上述代码中访问的URI用3个斜杠。

5.11.8 运行

运行应用后，效果如图5-25所示。

点击"你好，世界"触发了访问数据库的操作，可以看到控制台HiLog输出内容如下：

图5-25 运行应用效果

```
  08-29 23:57:33.017 
19282-19282/com.waylau.hmos.dataabilityhelperaccessdatabase I
00001/MainAbilitySlice:  resultSet is null or row count is 0
  08-29 23:57:33.019 
19282-19282/com.waylau.hmos.dataabilityhelperaccessdatabase I
```

```
00001/UserDataAbility:  UserDataAbility insert
    08-29 23:57:33.023
19282-19282/com.waylau.hmos.dataabilityhelperaccessdatabase I
00001/MainAbilitySlice:  insert result: 101
    08-29 23:57:33.027
19282-19282/com.waylau.hmos.dataabilityhelperaccessdatabase I
00001/MainAbilitySlice:  resultSet user_id: 101, user_name: Way Lau, user_age: 33
    08-29 23:57:33.032
19282-19282/com.waylau.hmos.dataabilityhelperaccessdatabase I
00001/MainAbilitySlice:  update result: 1
    08-29 23:57:33.036
19282-19282/com.waylau.hmos.dataabilityhelperaccessdatabase I
00001/MainAbilitySlice:  resultSet user_id: 101, user_name: Way Lau, user_age: 35
    08-29 23:57:33.040
19282-19282/com.waylau.hmos.dataabilityhelperaccessdatabase I
00001/MainAbilitySlice:  delete result: 1
    08-29 23:57:33.043
19282-19282/com.waylau.hmos.dataabilityhelperaccessdatabase I
00001/MainAbilitySlice:  resultSet is null or row count is 0
```

从上述运行日志可以验证程序的运行。

- 首先执行查询，数据库的数据是空的。
- 接着插入一条用户数据，该数据的user_id是101。
- 再次查询，返回了user_id是101的用户数据，user_name是Way Lau，user_age是33。
- 执行了更新操作。
- 再次查询，可以看到user_id是101的用户数据，user_age被改为35。
- 执行了删除操作。
- 再次执行查询，数据库的数据是空的。

5.12　Intent

在HarmonyOS中，Intent是对象之间传递信息的载体。例如，当一个Ability需要启动另一个Ability时，或者一个AbilitySlice需要导航到另一个AbilitySlice时，可以通过Intent指定启动的目标同时携带相关数据。

Intent的构成元素包括Operation与Parameters。

5.12.1　Operation 与 Parameters

Operation由表5-1所示的属性组成。

表 5-1　Operation 属性

属 性	描 述
Action	表示动作，通常使用系统预置 Action，应用也可以自定义 Action。例如 IntentConstants.ACTION_HOME 表示返回桌面动作
Entity	表示类别，通常使用系统预置 Entity，应用也可以自定义 Entity。例如 Intent.ENTITY_HOME 表示在桌面显示图标
Uri	表示 Uri 描述。如果在 Intent 中指定了 Uri，则 Intent 将匹配指定的 Uri 信息，包括 scheme、schemeSpecificPart、authority 和 path 信息
Flags	表示处理 Intent 的方式。例如 Intent.FLAG_ABILITY_CONTINUATION 表示标记在本地的一个 Ability 是否可以迁移到远端设备继续运行
BundleName	表示包描述。如果在 Intent 中同时指定了 BundleName 和 AbilityName，则 Intent 可以直接匹配到指定的 Ability
AbilityName	表示待启动的 Ability 名称。如果在 Intent 中同时指定了 BundleName 和 AbilityName，则 Intent 可以直接匹配到指定的 Ability
DeviceId	表示运行指定 Ability 的设备 ID

除了上述属性之外，开发者也可以通过Parameters传递某些请求所需的额外信息。Parameters是一种支持自定义的数据结构。

当Intent用于发起请求时，根据指定元素的不同，分为两种类型：

- 如果同时指定了 BundleName 与 AbilityName，则根据 Ability 的全称（例如 com.waylau.hmos.PayAbility）来直接启动应用。
- 如果未同时指定BundleName和AbilityName，则根据Operation中的其他属性来启动应用。

注：Intent设置属性时，必须先使用Operation来设置属性。如果需要新增或修改属性，则必须在设置Operation后再执行操作。

5.12.2　根据 Ability 的全称启动应用

通过构造包含BundleName与AbilityName的Operation对象可以启动一个Ability，并导航到该Ability。

在5.8节的ServiceAbilityLifeCycle应用中，启动本地服务的方式就是根据Ability的全称启动应用的一个例子。其代码如下：

```
/**
 * 启动本地服务
 */
private void startupLocalService(Intent intent) {
    //构建操作方式
    Operation operation = new Intent.OperationBuilder()
            // 设备ID
            .withDeviceId("")
            // 包名称
            .withBundleName("com.waylau.hmos.serviceabilitylifecycle")
```

```
            // 待启动的Ability名称
            .withAbilityName("com.waylau.hmos.serviceabilitylifecycle.TimeSe
rviceAbility")
            .build();
    //设置操作
    intent.setOperation(operation);
    startAbility(intent);
    
    HiLog.info(LOG_LABEL, "startupLocalService");
}
```

在上面的代码中，通过Intent中的OperationBuilder类构造operation对象，指定设备标识（空串表示当前设备）、应用包名、Ability名称。

5.12.3 实战：根据 Operation 的其他属性启动应用

在有些场景下，开发者需要在应用中使用其他应用提供的某种能力，而不感知提供该能力的具体是哪一个应用。例如，开发者需要通过浏览器打开一个链接，而不关心用户最终选择哪一个浏览器应用，则可以通过Operation的其他属性（除BundleName与AbilityName之外的属性）描述需要的能力。如果设备上存在多个应用提供同种能力，系统则弹出候选列表，由用户选择由哪个应用处理请求。以下示例展示使用Intent跨Ability查询天气信息。

创建一个名为IntentOperationWithAction的应用作为演示。

1. 创建 Page

通过DevEco Studio创建一个关于天气信息的Page——WeatherAbility，如图5-26所示。

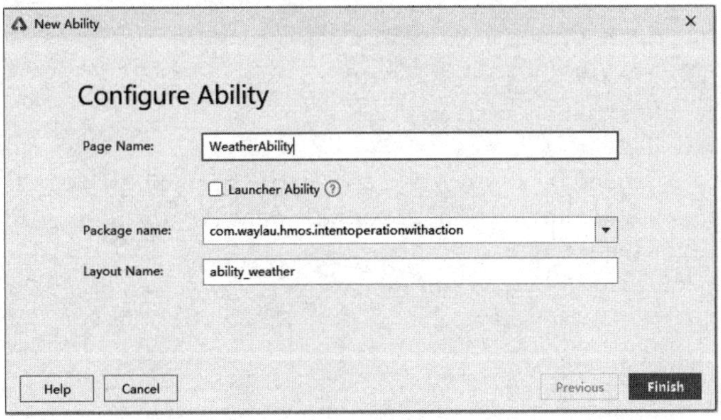

图 5-26　创建 Page

自动创建了如下的WeatherAbilitySlice类：

```
package com.waylau.hmos.intentoperationwithaction.slice;

import com.waylau.hmos.intentoperationwithaction.ResourceTable;
import ohos.aafwk.ability.AbilitySlice;
import ohos.aafwk.content.Intent;
import ohos.hiviewdfx.HiLog;
```

```java
import ohos.hiviewdfx.HiLogLabel;

public class WeatherAbilitySlice extends AbilitySlice {
    @Override
    public void onStart(Intent intent) {
        super.onStart(intent);
        super.setUIContent(ResourceTable.Layout_ability_weather);
    }

    @Override
    public void onActive() {
        super.onActive();
    }

    @Override
    public void onForeground(Intent intent) {
        super.onForeground(intent);
    }
}
```

WeatherAbility使用的布局名称是ability_weather。修改ability_weather.xml如下:

```xml
<?xml version="1.0" encoding="utf-8"?>
<DirectionalLayout
    xmlns:ohos="http://schemas.huawei.com/res/ohos"
    ohos:height="match_parent"
    ohos:width="match_parent"
    ohos:alignment="center"
    ohos:orientation="vertical">

    <Text
        ohos:id="$+id:text_helloworld"
        ohos:height="match_content"
        ohos:width="match_content"
        ohos:background_element="$graphic:background_ability_weather"
        ohos:layout_alignment="horizontal_center"
        ohos:text="Weather"
        ohos:text_size="40vp"
        />

</DirectionalLayout>
```

上述布局会在页面正中显示Weather字样。

2. 配置路由

修改配置文件,增加action.weather路由信息内容如下:

```
...
"abilities": [
  {
    "skills": [
```

```json
    {
      "entities": [
        "entity.system.home"
      ],
      "actions": [
        "action.system.home"
      ]
    }
  ],
  "orientation": "unspecified",
  "name": "com.waylau.hmos.intentoperationwithaction.MainAbility",
  "icon": "$media:icon",
  "description": "$string:mainability_description",
  "label": "$string:entry_MainAbility",
  "type": "page",
  "launchType": "standard"
},
{
  "skills": [
    {
      "actions": [
        "action.weather"// 指定路由的动作
      ]
    }
  ],
  "orientation": "unspecified",
  "name": "com.waylau.hmos.intentoperationwithaction.WeatherAbility",
  "icon": "$media:icon",
  "description": "$string:weatherability_description",
  "label": "$string:entry_WeatherAbility",
  "type": "page",
  "launchType": "standard"
}
]
...
```

上述配置是为了配置路由，以便支持以此action导航到对应的AbilitySlice。

3. 修改 WeatherAbility

重写WeatherAbility类的onActive的方法，代码如下：

```
package com.waylau.hmos.intentoperationwithaction;

import com.waylau.hmos.intentoperationwithaction.slice.WeatherAbilitySlice;
import ohos.aafwk.ability.Ability;
import ohos.aafwk.content.Intent;
import ohos.hiviewdfx.HiLog;
import ohos.hiviewdfx.HiLogLabel;

public class WeatherAbility extends Ability {
    private static final String TAG = WeatherAbility.class.getSimpleName();
```

```java
    private static final HiLogLabel LABEL_LOG =
            new HiLogLabel(HiLog.LOG_APP, 0x00001, TAG);
    private final static int CODE = 1;
    private static final String TEMP_KEY = "temperature";

    @Override
    public void onStart(Intent intent) {
        super.onStart(intent);
        super.setMainRoute(WeatherAbilitySlice.class.getName());
    }

    @Override
    protected void onActive() {
        HiLog.info(LABEL_LOG, "before onActive");

        super.onActive();

        Intent resultIntent = new Intent();
        resultIntent.setParam(TEMP_KEY, "17");

        setResult(CODE, resultIntent); // 暂存返回结果

        HiLog.info(LABEL_LOG, "after onActive");
    }
}
```

上述代码中：

- resultIntent是Intent的示例，用于返回结果。
- 通过Parameters的方式来传递天气信息。Parameters是一种支持自定义的数据结构，通过setParam方法将温度17放到resultIntent中。
- 调用setResult方法暂存返回结果。

4. 修改 MainAbilitySlice

MainAbilitySlice作为请求方，修改代码如下：

```java
package com.waylau.hmos.intentoperationwithaction.slice;

import com.waylau.hmos.intentoperationwithaction.ResourceTable;
import ohos.aafwk.ability.AbilitySlice;
import ohos.aafwk.content.Intent;
import ohos.aafwk.content.Operation;
import ohos.agp.components.Text;
import ohos.hiviewdfx.HiLog;
import ohos.hiviewdfx.HiLogLabel;

public class MainAbilitySlice extends AbilitySlice {
    private static final String TAG = MainAbilitySlice.class.getSimpleName();
    private static final HiLogLabel LABEL_LOG =
```

```java
            new HiLogLabel(HiLog.LOG_APP, 0x00001, TAG);
    private final static int CODE = 1;
    private static final String TEMP_KEY = "temperature";

    @Override
    public void onStart(Intent intent) {
        super.onStart(intent);
        super.setUIContent(ResourceTable.Layout_ability_main);

        // 添加点击事件来触发请求
        Text text = (Text) findComponentById(ResourceTable.Id_text_helloworld);
        text.setClickedListener(listener -> this.queryWeather());
    }
    private void queryWeather() {
        HiLog.info(LABEL_LOG, "before queryWeather");

        Intent intent = new Intent();

        Operation operation = new Intent.OperationBuilder()
                .withAction("action.weather")
                .build();
        intent.setOperation(operation);

        // 上述方式等同于intent.setAction("action.weather");

        startAbilityForResult(intent, CODE);
    }

    @Override
    protected void onAbilityResult(int requestCode, int resultCode, Intent resultData) {
        HiLog.info(LABEL_LOG, "onAbilityResult");

        switch (requestCode) {
            case CODE:
                HiLog.info(LABEL_LOG, "code 1 result: %{public}s",
                        resultData.getStringParam(TEMP_KEY));
                break;
            default:
                HiLog.info(LABEL_LOG, "defualt result: %{public}s", resultData.getAction());
                break;
        }
    }

    ...
}
```

上述代码中：

- Text增加了点击事件以触发请求，并路由到action.weather。
- 重写了onAbilityResult方法，用来接收请求的返回值。注意resultData获取的返回参数的值是字符串，以此需要用getStringParam方法；如果是整型，则可以使用getIntParam方法。

5. 运行应用

运行应用，界面效果如图5-27所示。

点击"你好，世界"字样，触发路由到action.weather的请求，界面效果如图5-28所示。

图5-27 运行效果

图5-28 运行效果

此时，控制台HiLog输出内容如下：

```
08-30 23:12:16.912 12981-12981/com.waylau.hmos.intentoperationwithaction I
00001/MainAbilitySlice: before queryWeather
08-30 23:12:17.103 12981-12981/com.waylau.hmos.intentoperationwithaction I
00001/WeatherAbility: [a269a1f550dfbaf, 370df30, 22b4094] before onActive
08-30 23:12:17.103 12981-12981/com.waylau.hmos.intentoperationwithaction I
00001/WeatherAbility: [a269a1f550dfbaf, 370df30, 22b4094] after onActive
```

点击模拟器返回按钮，返回"你好，世界"界面，效果如图5-29所示。

此时，控制台HiLog输出内容如下：

```
08-30 23:13:57.338 12981-12981/com.waylau.hmos.intentoperationwithaction I
00001/MainAbilitySlice: onAbilityResult
08-30 23:13:57.338 12981-12981/com.waylau.hmos.intentoperationwithaction I
00001/MainAbilitySlice: code 1 result: 17
```

从上述日志文件可以看出，MainAbilitySlice已经能够拿到WeatherAbility的返回值，温度是17。

图 5-29　运行效果

5.12.4　实战：启动系统应用

5.12.3节演示了通过指定Action来打开指定的应用，其实系统自带的应用也可以用类似的方式打开。

本节演示如何来打开系统应用拨号盘。创建名为IntentOperationWithActionDial的应用作为演示。

1. 修改 MainAbilitySlice

修改MainAbilitySlice代码如下：

```
package com.waylau.hmos.intentoperationwithactiondial.slice;

import com.waylau.hmos.intentoperationwithactiondial.ResourceTable;
import ohos.aafwk.ability.AbilitySlice;
import ohos.aafwk.content.Intent;
import ohos.aafwk.content.Operation;
import ohos.agp.components.Text;
import ohos.hiviewdfx.HiLog;
import ohos.hiviewdfx.HiLogLabel;
import ohos.utils.IntentConstants;

public class MainAbilitySlice extends AbilitySlice {
    private static final String TAG = MainAbilitySlice.class.getSimpleName();
    private static final HiLogLabel LABEL_LOG =
            new HiLogLabel(HiLog.LOG_APP, 0x00001, TAG);

    @Override
    public void onStart(Intent intent) {
        super.onStart(intent);
```

```java
        super.setUIContent(ResourceTable.Layout_ability_main);

        // 添加点击事件来触发请求
        Text text = (Text) findComponentById(ResourceTable.Id_text_helloworld);
        text.setClickedListener(listener -> this.doCall());

    }

    private void doCall() {
        HiLog.info(LABEL_LOG, "before doCall");

        Operation operation = new Intent.OperationBuilder()
                .withAction(IntentConstants.ACTION_DIAL) // 系统应用拨号盘
                .build();

        Intent intent = new Intent();
        intent.setOperation(operation);

        // 启动Ability
        startAbility(intent);

        HiLog.info(LABEL_LOG, "after doCall");
    }

    @Override
    public void onActive() {
        super.onActive();
    }

    @Override
    public void onForeground(Intent intent) {
        super.onForeground(intent);
    }
}
```

上述代码中：

- Text增加了点击事件以触发请求，点击就会执行doCall方法。
- doCall构建了Intent，该对象指明了withAction为系统应用拨号盘。拨号盘的Action常量值定义在IntentConstants.ACTION_DIAL中。
- 使用startAbility方法来启动Ability。

IntentConstants定义了众多的系统应用常量值，比如搜索、发短信等，这些系统应用都可以参照拨号盘的方式来启动。IntentConstants源码如下：

```java
public final class IntentConstants {
    public static final String ACTION_APPLICATION_DETAILS_SETTINGS = "ohos.settings.application.details";
    public static final String ACTION_CHOOSE = "ohos.intent.action.choose";
```

```
        public static final String ACTION_DIAL = "ohos.intent.action.dial";
        public static final String ACTION_DISMISS_ALARM =
"ohos.intent.action.dismissAlarm";
        public static final String ACTION_DISMISS_TIMER =
"ohos.intent.action.dismissTimer";
        public static final String ACTION_HIDISK_CHOOSE =
"ohos.hidisk.intent.action.choose";
        public static final String ACTION_HIDISK_CHOOSE_PATH =
"ohos.hidisk.intent.action.choosePath";
        public static final String ACTION_HOME = "action.system.home";
        public static final String ACTION_MANAGE_APPLICATIONS_SETTINGS =
"ohos.settings.manage.applications";
        public static final String ACTION_SEARCH = "ohos.intent.action.search";
        public static final String ACTION_SEND_SMS = "ohos.intent.action.sendSms";
        public static final String ACTION_SET_ALARM =
"ohos.intent.action.setAlarm";
        public static final String ACTION_SHOW_ALARMS =
"ohos.intent.action.showAlarms";
        public static final String ACTION_SNOOZE_ALARM =
"ohos.intent.action.snoozeAlarm";
        public static final String ACTION_WIRELESS_SETTINGS =
"ohos.settings.wireless";
        public static final String ENTITY_DEFAULT = "entity.system.default";
        public static final String ENTITY_HOME = "entity.system.home";
        public static final String ENTITY_VOICE = "entity.system.voice";

        IntentConstants() {
            throw new RuntimeException("Stub!");
        }
    }
```

2. 运行

运行应用，界面显示如图5-30所示。

图5-30　运行效果

点击"你好,世界"本文,触发点击事件。此时,界面切换到了系统应用拨号盘,如图5-31所示。

图 5-31　运行效果

3. 设置参数

那么,能否在启动Ability的时候来设置参数呢?比如,启动拨号盘的时候设置拨号的号码。可以将代码修改如下:

```
@Override
public void onStart(Intent intent) {
    super.onStart(intent);
    super.setUIContent(ResourceTable.Layout_ability_main);

    // 添加点击事件来触发请求
    Text text = (Text) findComponentById(ResourceTable.Id_text_helloworld);
    // text.setClickedListener(listener -> this.doCall());
    text.setClickedListener(listener -> this.doCall("18088888888"));
}

private void doCall() {
    this.doCall(null);
}

private void doCall(String destinationNum) {
    HiLog.info(LABEL_LOG, "before doCall");
    Operation operation;

    if (destinationNum == null) {
        operation = new Intent.OperationBuilder()
                .withAction(IntentConstants.ACTION_DIAL) // 系统应用拨号盘
                .build();
    } else {
        operation = new Intent.OperationBuilder()
                .withAction(IntentConstants.ACTION_DIAL) // 系统应用拨号盘
```

```
                .withUri(Uri.parse("tel:" + destinationNum)) // 设置号码
                .build();
}

Intent intent = new Intent();
intent.setOperation(operation);

// 启动Ability
startAbility(intent);

HiLog.info(LABEL_LOG, "after doCall");
}
```

上述代码在Operation中通过withUri来传递电话号码。再次运行应用,可以看到界面效果如图5-32所示。

图 5-32　运行效果

第 6 章

Ability 任务调度

在HarmonyOS中,分布式任务调度平台对搭载HarmonyOS的多设备构筑的"超级虚拟终端"提供统一的组件管理能力,为应用定义统一的能力基线、接口形式、数据结构、服务描述语言,屏蔽硬件差异,支持远程启动、远程调用、业务无缝迁移等分布式任务。

分布式任务调度平台在底层实现Ability跨设备的启动、关闭、连接、断开连接以及迁移等能力,实现跨设备的组件管理。

6.1 分布式任务调度概述

本节介绍分布式任务调度的核心价值。

6.1.1 "超级虚拟终端"的能力互助

在HarmonyOS中,分布式任务调度平台是支持"超级虚拟终端"的关键技术和能力,提供针对多设备场景下的统一的组件管理能力。

图6-1展示了分布式任务调度平台在HarmonyOS中的地位。分布式任务调度平台助力"超级虚拟终端"实现两方面的能力的互助:

- 硬件能力:比如,在手机中玩游戏时,玩家将游戏画面切换到大屏电视上。
- 软件能力:比如,在骑行过程中,骑手通过手表获取到了手机所提供的地图定位服务。

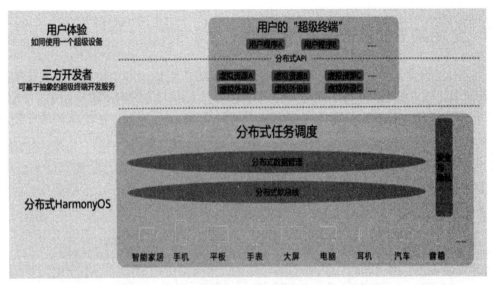

图 6-1　分布式任务调度平台在 HarmonyOS 中的地位

6.1.2　跨设备软件访问的系统服务

传统的跨设备软件开发是极其烦琐的，主要体现在以下几个方面：

- 平台烦琐。目前，针对硬件而言，CPU处理器技术架构主要分为X86和ARM，而针对操作系统而言，主流的平台有Linux、macOS、Windows、Android等。开发者不得不针对不同的平台做兼容性开发，这带来了极大的开发难度和工作量。因此，有些软件厂商只针对部分平台推出软件，比如微信聊天软件在Windows、macOS、Android平台下就能提供丰富的功能，而在Linux平台下却没有对应的安装包。
- 开发复杂。开发一个分布式系统本身是复杂的，需要考虑设备通信、序列化、事务、安全性、可用性等方面的内容，如果开发者需要自己从零开始构建一套分布式系统，则开发工作相当复杂。有兴趣的读者可以参阅笔者所著的《分布式系统常用技术及案例分析》。

为了降低开发者开发跨设备应用的难度，分布式任务调度平台提供了跨设备软件访问的系统服务。借助分布式任务调度平台，开发者在调用跨设备的服务时，实际上跟调用本地服务基本上没有差别。

图6-2展示了跨设备之间的软件访问示意图。

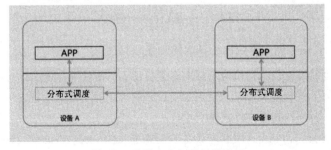

图 6-2　跨设备之间的软件访问

6.1.3 全场景下的任务调度

分布式任务调度无论是在HarmonyOS的富设备上，还是在HarmonyOS的轻设备上，都是支持的。除此之外，通过HarmonyOS分布式中间件还能够支持其他OS的任务调度。

图6-3展示了HarmonyOS与HarmonyOS设备之间以及HarmonyOS与其他OS之间的任务调度示意图。

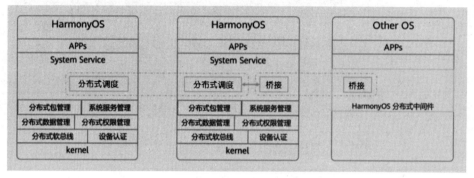

图 6-3　HarmonyOS 以及其他 OS 之间的任务调度

6.2　分布式任务调度能力简介

分布式任务调度平台在HarmonyOS底层实现Ability跨设备的组件管理、控制和访问，如图6-4所示。

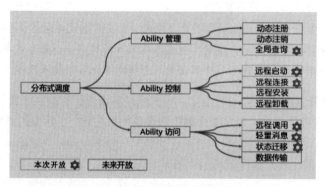

图 6-4　分布式任务调度能力

截至目前，分布式任务调度平台已经开放的功能包括：

- 全局查询：支持查询在相同组网下到底有哪些设备，这些设备是在线的还是离线的，等等。
- 启动和关闭：向开发者提供管理远程Ability的能力，即支持启动Page模板的Ability，以及启动、关闭Service和Data模板的Ability。
- 连接和断开连接：向开发者提供跨设备控制服务（Service和Data模板的Ability）的能力，开发者可以通过与远程服务连接及断开连接实现获取或注销跨设备管理服务的对象，达到和本地一致的服务调度。
- 迁移能力：向开发者提供跨设备业务的无缝迁移能力，开发者可以通过调用Page模板Ability的迁移接口将本地业务无缝迁移到指定设备中，打通设备间的壁垒。
- 轻量通信：可以通过远程对象的方式来实现设备之间的轻量通信。

6.2.1 全局查询

图6-5所示的是一个全局查询的示意图。

图6-5　全局查询

全局查询可以分为两个维度：针对设备的查询以及针对Ability的查询。

针对设备的查询是指在相同组网下支持查询这个网络下到底有哪些设备，这些设备是在线的还是离线的，这个设备具备哪些Ability，等等。

针对Ability的查询是指查询到底哪些设备支持具体的特定功能。比如，当我们需要进行投屏的时候，可以查看周边哪些设备支持投屏功能。

6.2.2 启动和关闭

图6-6所示的是一个启动和关闭的示意图。

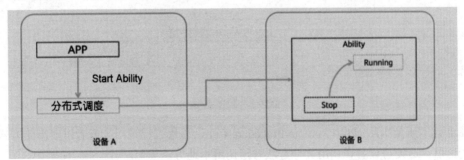

图6-6　启动和关闭

跟PC不同，移动终端的一个短板在于其硬件资源和电池存在一定瓶颈，这决定了在为移动终端设计Ability时，这些Ability需要按需启动或者关闭。分布式任务调度平台提供了管理远程Ability的能力，即支持启动Page模板的Ability，以及启动、关闭Service和Data模板的Ability。

6.2.3 连接和断开连接

图6-7所示的是一个连接和断开连接的示意图。

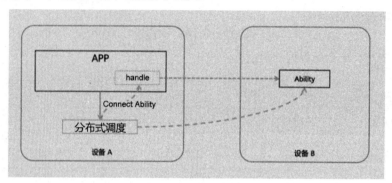

图 6-7　连接和断开连接

在连接到远程设备之后，就可以对设备进行一些列的操作了。操作完成之后，也可以断开连接。

6.2.4 轻量通信

图6-8所示的是一个轻量通信的示意图。

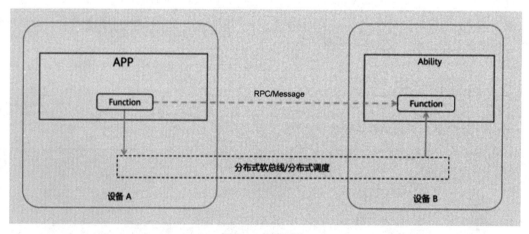

图 6-8　轻量通信

轻量通信本质是指从RPC（Remote Procedure Call，远程过程调用）或者以消息的方式实现设备之间的通信。这使得设备在调用其他方法时跟调用本地方法类似。

6.3 分布式任务调度实现原理

分布式任务调度实现原理最为核心的问题是设备之间的通信问题。

6.3.1 PRC

PRC主要涉及三方面，即接口定义、序列化和反序列化。要实现PRC，必须要实现IRemoteBroker接口。同时，需要在本地及对端分别实现对外接口一致的代理。一个具备加法能力的代理示例如下：

```java
public class MyRemoteProxy implements IRemoteBroker{
    private static final int ERR_OK = 0;
    private static final int COMMAND_PLUS = IRemoteObject.MIN_TRANSACTION_ID;
    private final IRemoteObject remote;

    public MyRemoteProxy(IRemoteObject remote) {
        this.remote = remote;
    }

    @Override
    public IRemoteObject asObject() {
        return remote;
    }

    public int plus(int a, int b) throws RemoteException {
        MessageParcel data = MessageParcel.obtain();
        MessageParcel reply = MessageParcel.obtain();

        // option不同的取值决定采用同步或异步方式跨设备控制PA
        // 本例需要同步获取对端PA执行加法的结果，因此采用同步的方式，即 MessageOption.TF_SYNC
        // 具体MessageOption的设置可参考相关API文档
        MessageOption option = new MessageOption(MessageOption.TF_SYNC);
        data.writeInt(a);
        data.writeInt(b);

        try {
            remote.sendRequest(COMMAND_PLUS, data, reply, option);
            int ec = reply.readInt();
            if (ec != ERR_OK) {
                throw new RemoteException();
            }
            int result = reply.readInt();
            return result;
        } catch (RemoteException e) {
            throw new RemoteException();
```

```
        } finally {
          data.reclaim();
          reply.reclaim();
        }
    }
}
```

6.3.2 HarmonyOS 设备之间的通信

HarmonyOS设备之间的通信原理如图6-9所示。

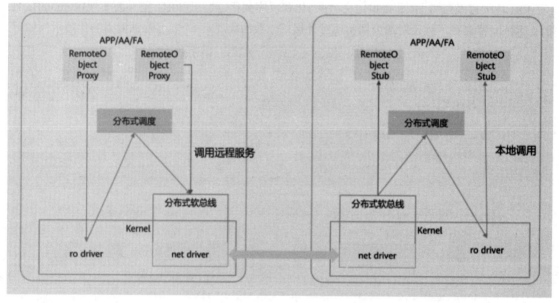

图 6-9 HarmonyOS 设备之间的通信

无论是调用本地设备还是远程设备的Ability，HarmonyOS都是通过RemoteObject来实现的。当初次调用远程设备时，会先通过分布式调度平台获取到远程设备的一个句柄。在后续的通信过程中，本地设备就可以不必再依赖分布式调度平台而直接通过句柄去跟远程设备进行通信，从而提升通信效率。

6.3.3 HarmonyOS 设备与其他 OS 设备之间的通信

HarmonyOS设备与其他OS设备之间的通信原理如图6-10所示。

与HarmonyOS设备之间的通信不同，HarmonyOS设备与其他OS设备之间无法直接通过句柄去调用，因此分布式调度平台充当了HarmonyOS设备与其他OS设备之间的代理。所有的通信必须经过分布式调度平台，分布式调度平台会做调用过程中的序列化和反序列化。因此，从通信效率而言，HarmonyOS设备与其他OS设备之间的通信效率肯定要低于HarmonyOS设备之间的通信。

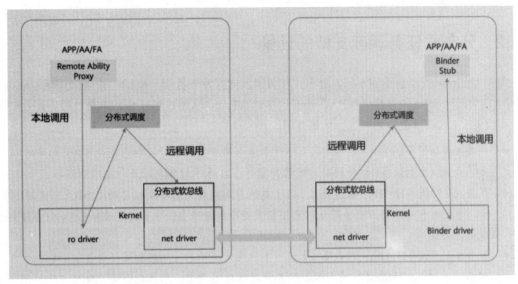

图 6-10　HarmonyOS 设备与其他 OS 设备之间的通信

6.4　实现分布式任务调度

在了解了分布式任务调度的实现原理之后，接下来将介绍如何实现分布式任务调度。

6.4.1　如何实现分布式任务调度

要实现分布式任务调度，开发者需要在应用中做如下操作：

- 在Intent中设置支持分布式的标记（例如Intent.FLAG_ABILITYSLICE_MULTI_DEVICE表示该应用支持分布式调度），否则将无法获得分布式能力。
- 在config.json中的reqPermissions字段中添加多设备协同访问的权限申请：三方应用使用{"name": "ohos.permission.DISTRIBUTED_DATASYNC"}。
- PA的调用支持连接及断开连接、启动及关闭这4类行为，在进行调度时：
  - 必须在 Intent 中指定 PA 对应的 bundleName 和 abilityName。
  - 当需要跨设备启动、关闭或连接 PA 时，需要在 Intent 中指定对端设备的 deviceId。可通过如设备管理类 DeviceManager 提供的 getDeviceList 获取指定条件下匿名化处理的设备列表，实现对指定设备 PA 的启动/关闭以及连接管理。
- FA的调用支持启动和迁移行为，在进行调度时：
  - 当启动 FA 时，需要开发者在 Intent 中指定对端设备的 deviceId、bundleName 和 abilityName。
  - FA 的迁移实现相同 bundleName 和 abilityName 的 FA 跨设备迁移，因此需要指定迁移设备的 deviceId。

6.4.2 分布式任务调度支持的场景

根据Ability模板及意图的不同，分布式任务调度向开发者提供了6种能力：启动远程FA、启动远程PA、关闭远程PA、连接远程PA、断开连接远程PA和FA跨设备迁移。下面以设备A（本地设备）和设备B（远端设备）为例进行场景介绍：

- 设备A启动设备B的FA：在设备A上通过本地应用提供的启动按钮启动设备B上对应的FA。例如，设备A控制设备B打开相册，只需开发者在启动FA时指定打开相册的意图即可。
- 设备A启动设备B的PA：在设备A上通过本地应用提供的启动按钮启动设备B上指定的PA。例如，开发者在启动远程服务时通过意图指定音乐播放服务，即可实现设备A启动设备B音乐播放的能力。
- 设备A关闭设备B的PA：在设备A上通过本地应用提供的关闭按钮关闭设备B上指定的PA。类似启动的过程，开发者在关闭远程服务时通过意图指定音乐播放服务，即可实现关闭设备B上该服务的能力。
- 设备A连接设备B的PA：在设备A上通过本地应用提供的连接按钮连接设备B上指定的PA。连接后，通过其他功能相关按钮实现控制对端PA的能力。通过连接关系，开发者可以实现跨设备的同步服务调度，实现如大型计算任务互助等价值场景。
- 设备A与设备B的PA断开连接：在设备A上通过本地应用提供的断开连接的按钮将之前已连接的PA断开连接。
- 设备A的FA迁移至设备B：设备A上通过本地应用提供的迁移按钮将设备A的业务无缝迁移到设备B中。通过业务迁移能力打通设备A和设备B间的壁垒，实现如文档跨设备编辑、视频从客厅到房间跨设备接续播放等场景。

6.5 实战：分布式任务调度启动远程FA

分别创建名为DistributedSchedulingStartRemoteFA和RemoteFA的应用，用于演示分布式任务调度时如何启动远程FA。

在"设备A启动设备B的FA"场景中，DistributedSchedulingStartRemoteFA作为"设备A"角色，而RemoteFA应用作为"设备B"角色。

6.5.1 修改RemoteFA应用

RemoteFA应用作为被调用方，修改内容比较简单。

1. 修改ability_main.xml

修改ability_main.xml文件如下：

```
<?xml version="1.0" encoding="utf-8"?>
```

```xml
<DirectionalLayout
    xmlns:ohos="http://schemas.huawei.com/res/ohos"
    ohos:height="match_parent"
    ohos:width="match_parent"
    ohos:alignment="center"
    ohos:orientation="vertical">

    <Text
        ohos:id="$+id:text_helloworld"
        ohos:height="match_content"
        ohos:width="match_content"
        ohos:background_element="$graphic:background_ability_main"
        ohos:layout_alignment="horizontal_center"
        ohos:text="I am Remote FA"
        ohos:text_size="40vp"
        />

</DirectionalLayout>
```

上述修改只是将显示的文本内容改为"I am Remote FA",在预览器中的预览效果如图6-11所示。

图6-11 修改显示的文本内容

2. 修改配置文件

修改配置文件如下:

```
...
"abilities": [
    {
    "skills": [
```

```
        {
    "entities": [
        "entity.system.home"
    ],
    "actions": [
        "action.system.home"
    ]
    }
],
"orientation": "unspecified",
"name": "com.waylau.hmos.remotefa.MainAbility",
"icon": "$media:icon",
"description": "$string:mainability_description",
"label": "$string:entry_MainAbility",
"type": "page",
"launchType": "standard",
"visible":true  // 被其他应用可见
}
]
...
```

上述主要改动点为设置visible为true，这样就能被其他应用发现。

6.5.2 修改 DistributedSchedulingStartRemoteFA 应用

DistributedSchedulingStartRemoteFA应用作为调用方，修改内容相对来说比较多。

1. 修改 ability_main.xml

修改ability_main.xml文件如下：

```
<?xml version="1.0" encoding="utf-8"?>
<DirectionalLayout
    xmlns:ohos="http://schemas.huawei.com/res/ohos"
    ohos:height="match_parent"
    ohos:width="match_parent"
    ohos:alignment="center"
    ohos:orientation="vertical">

    <Text
        ohos:id="$+id:text_helloworld"
        ohos:height="match_content"
        ohos:width="match_content"
        ohos:background_element="$graphic:background_ability_main"
        ohos:layout_alignment="horizontal_center"
        ohos:text="Start Remote FA"
        ohos:text_size="40vp"
        />

</DirectionalLayout>
```

上述修改只是将显示的文本内容改为"Start Remote FA",在预览器中的预览效果如图6-12所示。

图 6-12 修改显示的文本内容

2. 修改配置文件

修改配置文件如下:

```
...
"abilities": [
    {
    "skills": [
        {
        "entities": [
            "entity.system.home"
        ],
        "actions": [
            "action.system.home"
        ]
        }
    ],
    "orientation": "unspecified",
    "name":
"com.waylau.hmos.distributedschedulingstartremotefa.MainAbility",
    "icon": "$media:icon",
    "description": "$string:mainability_description",
    "label": "$string:entry_MainAbility",
    "type": "page",
    "launchType": "standard"
```

```
        }
    ],
    // 声明权限
    "reqPermissions": [
        {
            "name": "ohos.permission.DISTRIBUTED_DATASYNC"
        },
        {
            "name": "ohos.permission.DISTRIBUTED_DEVICE_STATE_CHANGE"
        },
        {
            "name": "ohos.permission.GET_DISTRIBUTED_DEVICE_INFO"
        },
        {
            "name": "ohos.permission.GET_BUNDLE_INFO"
        }
    ]
    ...
```

上述主要改动点是新增了reqPermissions，用于声明多设备协同访问的权限、查询设备列表权限及查询设备信息的权限。

3. 修改 MainAbilitySlice

修改MainAbilitySlice如下：

```
package com.waylau.hmos.distributedschedulingstartremotefa.slice;

import com.waylau.hmos.distributedschedulingstartremotefa.DeviceUtils;
import com.waylau.hmos.distributedschedulingstartremotefa.ResourceTable;
import ohos.aafwk.ability.AbilitySlice;
import ohos.aafwk.content.Intent;
import ohos.aafwk.content.Operation;
import ohos.agp.components.Text;
import ohos.hiviewdfx.HiLog;
import ohos.hiviewdfx.HiLogLabel;


public class MainAbilitySlice extends AbilitySlice {
    private static final String TAG = MainAbilitySlice.class.getSimpleName();
    private static final HiLogLabel LABEL_LOG =
            new HiLogLabel(HiLog.LOG_APP, 0x00001, TAG);

    private static final String BUNDLE_NAME = "com.waylau.hmos.remotefa";
    private static final String ABILITY_NAME = BUNDLE_NAME + ".MainAbility";

    @Override
    public void onStart(Intent intent) {
        super.onStart(intent);
        super.setUIContent(ResourceTable.Layout_ability_main);
```

```java
        // 添加点击事件来触发
        Text text = (Text) findComponentById(ResourceTable.Id_text_helloworld);
        text.setClickedListener(listener -> startRemoteFA());
    }

    // 启动远程FA
    private void startRemoteFA() {
        HiLog.info(LABEL_LOG, "before startRemoteFA");

        String deviceId = DeviceUtils.getDeviceId();

        HiLog.info(LABEL_LOG, "get deviceId: %{public}s", deviceId);

        // 指定待启动FA的deviceId、bundleName和abilityName
        Operation operation = new Intent.OperationBuilder()
                .withDeviceId(deviceId)
                .withBundleName(BUNDLE_NAME)
                .withAbilityName(ABILITY_NAME)
                .build();

        Intent intent = new Intent();
        intent.setOperation(operation);

        // 如果deviceId不为空，则为分布式调用
        if (!deviceId.isEmpty()) {
            intent.addFlags(Intent.FLAG_ABILITYSLICE_MULTI_DEVICE); // 设置分布式标记
        }

        // 通过AbilitySlice包含的startAbility接口实现跨设备启动FA
        startAbility(intent);

        HiLog.info(LABEL_LOG, "after startRemoteFA");
    }

    @Override
    public void onActive() {
        super.onActive();
    }

    @Override
    public void onForeground(Intent intent) {
        super.onForeground(intent);
    }

}
```

上述代码中：

- DeviceUtils.getDeviceId()方法主要用于返回设备ID。该方法内部是调用DeviceManager的getDeviceList接口，通过FLAG_GET_ONLINE_DEVICE标记获得在线设备列表，并从中选择任意一个（这里选择第一个）作为返回的设备ID。
- 在onStart中的Text上增加了点击事件，用于触发startRemotePA和stopRemotePA方法。
- 执行startRemotePA和stopRemotePA方法时：
  - 指定待启动FA的bundleName和abilityName。
  - 这里需要判断deviceId是否为空字符。如果不是，则设置分布式标记，表明当前涉及分布式能力；如果是，则说明没有远程设备，启动本地设备的FA。
- 通过AbilitySlice包含的startAbility接口实现跨设备启动FA。

4. 新增DeviceUtils

DeviceUtils类是一个查询设备列表的工具类，核心内容如下：

```java
package com.waylau.hmos.distributedschedulingstartremotefa;

import ohos.distributedschedule.interwork.DeviceInfo;
import ohos.distributedschedule.interwork.DeviceManager;
import ohos.hiviewdfx.HiLog;
import ohos.hiviewdfx.HiLogLabel;

import java.util.ArrayList;
import java.util.List;

public class DeviceUtils {
    private static final String TAG = DeviceUtils.class.getSimpleName();
    private static final HiLogLabel LABEL_LOG =
            new HiLogLabel(HiLog.LOG_APP, 0x00001, TAG);

    private DeviceUtils() {
    }

    // 获取当前组网下可迁移的设备ID列表
    public static List<String> getAvailableDeviceId() {
        List<String> deviceIds = new ArrayList<>();

        List<DeviceInfo> deviceInfoList =
DeviceManager.getDeviceList(DeviceInfo.FLAG_GET_ALL_DEVICE);
        if (deviceInfoList == null) {
            return deviceIds;
        }

        if (deviceInfoList.size() == 0) {
            HiLog.warn(LABEL_LOG, "did not find other device");
            return deviceIds;
        }

        for (DeviceInfo deviceInfo : deviceInfoList) {
            deviceIds.add(deviceInfo.getDeviceId());
```

```
        }
            return deviceIds;
        }

        // 获取当前组网下可迁移的设备ID
        // 如果有多个,则取第一个
        public static String getDeviceId() {
            String deviceId = "";
            List<String> outerDevices = DeviceUtils.getAvailableDeviceId();

            if (outerDevices == null || outerDevices.size() == 0) {
                HiLog.warn(LABEL_LOG, "did not find other device");
            } else {
                for (String item : outerDevices) {
                    HiLog.info(LABEL_LOG, "outerDevices:%{public}s", item);
                }
                deviceId = outerDevices.get(0);
            }
            HiLog.info(LABEL_LOG, "getDeviceId:%{public}s", deviceId);
            return deviceId;
        }

    ;
}
```

DeviceManager提供了查询设备列表的管理器,其中DeviceInfo主要分为3种:

- FLAG_GET_ALL_DEVICE:获取所有设备。
- FLAG_GET_OFFLINE_DEVICE:获取离线设备。
- FLAG_GET_ONLINE_DEVICE:获取在线设备。

5. 显式声明需要使用的权限

此外,对于三方应用还要求在实现Ability的代码中显式声明需要使用的权限,代码如下:

```
package com.waylau.hmos.distributedschedulingstartremotefa;

import
com.waylau.hmos.distributedschedulingstartremotefa.slice.MainAbilitySlice;
import ohos.aafwk.ability.Ability;
import ohos.aafwk.content.Intent;

import java.util.ArrayList;
import java.util.List;

public class MainAbility extends Ability {
    @Override
    public void onStart(Intent intent) {
        super.onStart(intent);
        super.setMainRoute(MainAbilitySlice.class.getName());
```

```
        requestPermission();
    }

    // 显式声明需要使用的权限
    private void requestPermission() {
        String[] permission = {
                "ohos.permission.DISTRIBUTED_DATASYNC",
                "ohos.permission.DISTRIBUTED_DEVICE_STATE_CHANGE",
                "ohos.permission.GET_DISTRIBUTED_DEVICE_INFO",
                "ohos.permission.GET_BUNDLE_INFO"};
        List<String> applyPermissions = new ArrayList<>();
        for (String element : permission) {
            if (verifySelfPermission(element) != 0) {
                if (canRequestPermission(element)) {
                    applyPermissions.add(element);
                }
            }
        }
        requestPermissionsFromUser(applyPermissions.toArray(new String[0]), 0);
    }
}
```

6.5.3 运行

先在设备管理器的Super device中同时启动两台Phone模拟器实现跨设备的环境，如图6-13所示。

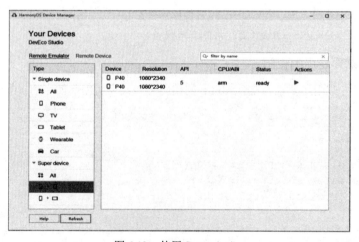

图 6-13 使用 Super device

分别在设备A（左侧）和设备B（右侧）中安装了DistributedSchedulingStartRemoteFA和RemoteFA应用，而后运行DistributedSchedulingStartRemoteFA应用，效果如图6-14所示。

点击设备A（左侧）界面的Start Remote FA文本，此时会启动设备B（右侧）中的RemoteFA应用，效果如图6-15所示。

图 6-14　两台模拟器的界面

图 6-15　RemoteFA 应用被启动的界面

上述操作验证了在设备 A（DistributedSchedulingStartRemoteFA 应用）中启动设备 B（RemoteFA 应用）的 FA 功能。

6.6　实战：分布式任务调度启动和关闭远程 PA

分别创建名为 DistributedSchedulingStartStopRemotePA 和 RemotePA 的应用，用于演示分布式任务调度时如何启动和关闭远程 PA。

在"设备 A 启动和关闭设备 B 的 PA"场景中，DistributedSchedulingStartStopRemotePA 作为"设备 A"角色，而 RemotePA 应用作为"设备 B"角色。

6.6.1　修改 RemotePA 应用

RemotePA 应用作为被调用方，修改内容比较简单。

1. 修改 ability_main.xml

修改 ability_main.xml 文件如下：

```xml
<?xml version="1.0" encoding="utf-8"?>
<DirectionalLayout
    xmlns:ohos="http://schemas.huawei.com/res/ohos"
    ohos:height="match_parent"
    ohos:width="match_parent"
    ohos:alignment="center"
    ohos:orientation="vertical">
```

```xml
    <Text
        ohos:id="$+id:text_helloworld"
        ohos:height="match_content"
        ohos:width="match_content"
        ohos:background_element="$graphic:background_ability_main"
        ohos:layout_alignment="horizontal_center"
        ohos:text="I am Remote PA"
        ohos:text_size="40vp"
        />

</DirectionalLayout>
```

上述修改只是将显示的文本内容改为"I am Remote PA",在预览器中的预览效果如图6-16所示。

图 6-16　修改显示的文本内容

2. 创建 Service

我们创建了一个名为TimeServiceAbility的Service。

在自动创建的TimeServiceAbility的基础上修改代码如下:

```java
package com.waylau.hmos.remotepa;

import ohos.aafwk.ability.Ability;
import ohos.aafwk.content.Intent;
import ohos.rpc.IRemoteObject;
import ohos.hiviewdfx.HiLog;
import ohos.hiviewdfx.HiLogLabel;

public class TimeServiceAbility extends Ability {
    private static final String LOG_TAG =
TimeServiceAbility.class.getSimpleName();
    private static final HiLogLabel LABEL_LOG =
```

```
                new HiLogLabel(HiLog.LOG_APP, 0x00001, LOG_TAG);

    @Override
    public void onStart(Intent intent) {
        HiLog.info(LABEL_LOG, "onStart");
        super.onStart(intent);
    }

    @Override
    public void onBackground() {
        super.onBackground();
        HiLog.info(LABEL_LOG, "onBackground");
    }

    @Override
    public void onStop() {
        super.onStop();
        HiLog.info(LABEL_LOG, "onStop");
    }

    @Override
    public void onCommand(Intent intent, boolean restart, int startId) {
        super.onCommand(intent, restart, startId);
        HiLog.info(LABEL_LOG, "onCommand");
    }

    @Override
    public IRemoteObject onConnect(Intent intent) {
        super.onConnect(intent);
        HiLog.info(LABEL_LOG, "onConnect");

        return null;
    }

    @Override
    public void onDisconnect(Intent intent) {
        super.onDisconnect(intent);
        HiLog.info(LABEL_LOG, "onDisconnect");
    }
}
```

代码比较简单，只是在各个生命周期内加上日志，这样便于观察。

3. 修改配置文件

创建Service时，在配置文件中会自动新增TimeServiceAbility相关的配置信息。修改配置文件如下：

```
...
"abilities": [
    {
```

```
      "skills": [
         {
            "entities": [
               "entity.system.home"
            ],
            "actions": [
               "action.system.home"
            ]
         }
      ],
      "orientation": "unspecified",
      "name": "com.waylau.hmos.remotepa.MainAbility",
      "icon": "$media:icon",
      "description": "$string:mainability_description",
      "label": "$string:entry_MainAbility",
      "type": "page",
      "launchType": "standard"
   },
   // 新增的TimeServiceAbility
   {
      "name": "com.waylau.hmos.remotepa.TimeServiceAbility",
      "icon": "$media:icon",
      "description": "$string:timeserviceability_description",
      "type": "service",
      "visible": true  // 被其他应用可见
   }
]
...
```

上述主要改动点为设置visible为true，这样就能被其他应用发现。

6.6.2 修改 DistributedSchedulingStartStopRemotePA 应用

DistributedSchedulingStartStopRemotePA应用作为调用方，修改内容相对来说比较多。

1. 修改 ability_main.xml

修改ability_main.xml文件如下：

```xml
<?xml version="1.0" encoding="utf-8"?>
<DirectionalLayout
   xmlns:ohos="http://schemas.huawei.com/res/ohos"
   ohos:height="match_parent"
   ohos:width="match_parent"
   ohos:alignment="center"
   ohos:orientation="vertical">

   <Text
      ohos:id="$+id:text_startpa"
      ohos:height="match_content"
```

```
        ohos:width="match_content"
        ohos:background_element="$graphic:background_ability_main"
        ohos:layout_alignment="horizontal_center"
        ohos:text="Start Remote PA"
        ohos:text_size="40vp"
        />

    <Text
        ohos:id="$+id:text_stoppa"
        ohos:height="match_parent"
        ohos:width="match_content"
        ohos:background_element="$graphic:background_ability_main"
        ohos:layout_alignment="horizontal_center"
        ohos:text="Stop Remote PA"
        ohos:text_size="40vp"
        />

</DirectionalLayout>
```

上述修改将显示两端文本内容Start Remote PA和Stop Remote PA，在预览器中的预览效果如图6-17所示。

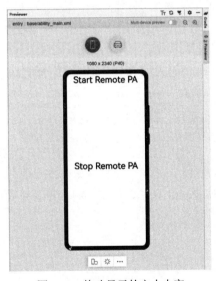

图6-17　修改显示的文本内容

2. 修改配置文件

修改配置文件如下：

```
...
"abilities": [
    {
    "skills": [
        {
        "entities": [
            "entity.system.home"
```

```json
            ],
            "actions": [
                "action.system.home"
            ]
            }
        ],
        "orientation": "unspecified",
        "name": 
"com.waylau.hmos.distributedschedulingstartstopremotepa.MainAbility",
        "icon": "$media:icon",
        "description": "$string:mainability_description",
        "label": "$string:entry_MainAbility",
        "type": "page",
        "launchType": "standard"
        }
    ],
    // 声明权限
    "reqPermissions": [
        {
        "name": "ohos.permission.DISTRIBUTED_DATASYNC"
        },
        {
        "name": "ohos.permission.DISTRIBUTED_DEVICE_STATE_CHANGE"
        },
        {
        "name": "ohos.permission.GET_DISTRIBUTED_DEVICE_INFO"
        },
        {
        "name": "ohos.permission.GET_BUNDLE_INFO"
        }
    ]
    ...
```

上述主要改动点是新增了reqPermissions，用于声明多设备协同访问的权限、查询设备列表的权限及查询设备信息的权限。

3. 修改 MainAbilitySlice

修改MainAbilitySlice如下：

```
package com.waylau.hmos.distributedschedulingstartstopremotepa.slice;

import com.waylau.hmos.distributedschedulingstartstopremotepa.DeviceUtils;
import com.waylau.hmos.distributedschedulingstartstopremotepa.ResourceTable;
import ohos.aafwk.ability.AbilitySlice;
import ohos.aafwk.content.Intent;
import ohos.aafwk.content.Operation;
import ohos.agp.components.Text;
import ohos.hiviewdfx.HiLog;
import ohos.hiviewdfx.HiLogLabel;
```

```java
public class MainAbilitySlice extends AbilitySlice {
    private static final String TAG = MainAbilitySlice.class.getSimpleName();
    private static final HiLogLabel LABEL_LOG =
            new HiLogLabel(HiLog.LOG_APP, 0x00001, TAG);

    private static final String BUNDLE_NAME = "com.waylau.hmos.remotepa";
    private static final String ABILITY_NAME = BUNDLE_NAME +
".TimeServiceAbility";


    @Override
    public void onStart(Intent intent) {
        super.onStart(intent);
        super.setUIContent(ResourceTable.Layout_ability_main);

        // 添加点击事件来触发启动PA
        Text textStartPA = (Text) findComponentById(ResourceTable.Id_text_startpa);
        textStartPA.setClickedListener(listener -> startRemotePA());

        // 添加点击事件来触发关闭PA
        Text textStopPA = (Text) findComponentById(ResourceTable.Id_text_stoppa);
        textStopPA.setClickedListener(listener -> stopRemotePA());
    }

    // 启动远程PA
    private void startRemotePA() {
        HiLog.info(LABEL_LOG, "before startRemotePA");

        String deviceId = DeviceUtils.getDeviceId();

        HiLog.info(LABEL_LOG, "get deviceId: %{public}s", deviceId);

        // 指定待启动FA的deviceId、bundleName和abilityName
        Operation operation = new Intent.OperationBuilder()
                .withDeviceId(deviceId)
                .withBundleName(BUNDLE_NAME)
                .withAbilityName(ABILITY_NAME)
                .build();

        Intent intent = new Intent();
        intent.setOperation(operation);

        // 如果deviceId不为空，则为分布式调用
        if (!deviceId.isEmpty()) {
            intent.addFlags(Intent.FLAG_ABILITYSLICE_MULTI_DEVICE); // 设置分
```
布式标记

```java
        }

        // 通过AbilitySlice包含的startAbility接口实现跨设备启动PA
        startAbility(intent);

        HiLog.info(LABEL_LOG, "after startRemotePA");
    }


    // 关闭远程PA
    private void stopRemotePA() {
        HiLog.info(LABEL_LOG, "before stopRemotePA");

        String deviceId = DeviceUtils.getDeviceId();

        HiLog.info(LABEL_LOG, "get deviceId: %{public}s", deviceId);

        // 指定待启动FA的deviceId、bundleName和abilityName
        Operation operation = new Intent.OperationBuilder()
                .withDeviceId(deviceId)
                .withBundleName(BUNDLE_NAME)
                .withAbilityName(ABILITY_NAME)
                .build();

        Intent intent = new Intent();
        intent.setOperation(operation);

        // 如果deviceId不为空，则为分布式调用
        if (!deviceId.isEmpty()) {
            intent.addFlags(Intent.FLAG_ABILITYSLICE_MULTI_DEVICE); // 设置分布式标记
        }

        // 通过AbilitySlice包含的stopAbility接口实现跨设备关闭PA
        stopAbility(intent);

        HiLog.info(LABEL_LOG, "after stopRemotePA");
    }


    @Override
    public void onActive() {
        super.onActive();
    }

    @Override
    public void onForeground(Intent intent) {
        super.onForeground(intent);
    }
```

}

上述代码中：

- DeviceUtils.getDeviceId()方法主要用于返回设备ID。该方法内部是调用DeviceManager的getDeviceList接口，通过FLAG_GET_ONLINE_DEVICE标记获得在线设备列表，并从中选择任意一个（这里选择第一个）作为返回的设备ID。
- 在onStart中的Text上增加了点击事件，用于触发startRemotePA和stopRemotePA方法。
- 执行startRemotePA和stopRemotePA方法时：
  - 指定待启动FA的bundleName和abilityName。
  - 这里需要判断deviceId是否为空字符。如果不是，则设置分布式标记，表明当前涉及分布式能力；如果是，则说明没有远程设备，启动本地设备的FA。
- 通过AbilitySlice包含的startAbility接口实现跨设备启动PA。
- 通过AbilitySlice包含的stopAbility接口实现跨设备关闭PA。

4. 新增DeviceUtils

DeviceUtils类是一个或者设备列表的工具类，核心内容如下：

```
package com.waylau.hmos.distributedschedulingstartstopremotepa;

import ohos.distributedschedule.interwork.DeviceInfo;
import ohos.distributedschedule.interwork.DeviceManager;
import ohos.hiviewdfx.HiLog;
import ohos.hiviewdfx.HiLogLabel;

import java.util.ArrayList;
import java.util.List;

public class DeviceUtils {
    private static final String TAG = DeviceUtils.class.getSimpleName();
    private static final HiLogLabel LABEL_LOG =
            new HiLogLabel(HiLog.LOG_APP, 0x00001, TAG);

    private DeviceUtils() {
    }

    // 获取当前组网下可迁移的设备ID列表
    public static List<String> getAvailableDeviceId() {
        List<String> deviceIds = new ArrayList<>();

        List<DeviceInfo> deviceInfoList =
                DeviceManager.getDeviceList(DeviceInfo.FLAG_GET_ALL_DEVICE);
        if (deviceInfoList == null) {
            return deviceIds;
        }

        if (deviceInfoList.size() == 0) {
            HiLog.warn(LABEL_LOG, "did not find other device");
```

```java
            return deviceIds;
        }

        for (DeviceInfo deviceInfo : deviceInfoList) {
            deviceIds.add(deviceInfo.getDeviceId());
        }

        return deviceIds;
    }

    // 获取当前组网下可迁移的设备ID
    // 如果有多个,则取第一个
    public static String getDeviceId() {
        String deviceId = "";
        List<String> outerDevices = DeviceUtils.getAvailableDeviceId();

        if (outerDevices == null || outerDevices.size() == 0) {
            HiLog.warn(LABEL_LOG, "did not find other device");
        } else {
            for (String item : outerDevices) {
                HiLog.info(LABEL_LOG, "outerDevices:%{public}s", item);
            }
            deviceId = outerDevices.get(0);
        }
        HiLog.info(LABEL_LOG, "getDeviceId:%{public}s", deviceId);
        return deviceId;
    }
}
```

该类在前文已经做过介绍,这里不再赘述。

5. 显式声明需要使用的权限

此外,对于三方应用还要求在实现Ability的代码中显式声明需要使用的权限,代码如下:

```java
package com.waylau.hmos.distributedschedulingstartstopremotepa;

import com.waylau.hmos.distributedschedulingstartstopremotepa.slice.MainAbilitySlice;
import ohos.aafwk.ability.Ability;
import ohos.aafwk.content.Intent;

import java.util.ArrayList;
import java.util.List;

public class MainAbility extends Ability {
    @Override
    public void onStart(Intent intent) {
        super.onStart(intent);
        super.setMainRoute(MainAbilitySlice.class.getName());
        requestPermission();
    }
```

```
// 显式声明需要使用的权限
private void requestPermission() {
    String[] permission = {
            "ohos.permission.DISTRIBUTED_DATASYNC",
            "ohos.permission.DISTRIBUTED_DEVICE_STATE_CHANGE",
            "ohos.permission.GET_DISTRIBUTED_DEVICE_INFO",
            "ohos.permission.GET_BUNDLE_INFO"};
    List<String> applyPermissions = new ArrayList<>();
    for (String element : permission) {
        if (verifySelfPermission(element) != 0) {
            if (canRequestPermission(element)) {
                applyPermissions.add(element);
            }
        }
    }
    requestPermissionsFromUser(applyPermissions.toArray(new String[0]),
0);
}
```

该类在前文已经做过介绍，这里不再赘述。

6.6.3 运行

我们先在设备管理器的Super device中同时启动两台Phone模拟器，在其中一台设备A（左侧）上安装DistributedSchedulingStartStopRemotePA应用，而在另一台设备B（右侧）上安装RemotePA应用。两个设备同时启动相应的应用，如图6-18所示。

图6-18　两个设备同时启动相应的应用

当点击设备A（左侧）界面的Start Remote PA文本时，会启动设备B（右侧）的Service。这点可以从控制台日志看出。

```
    09-01 23:52:43.909 22193-22193/com.waylau.hmos.remotepa I
00001/TimeServiceAbility: [b17d23a1cbc80b6, 31d65d1, 3a2a3ae] onStart
    09-01 23:52:43.910 22193-22193/com.waylau.hmos.remotepa I
00001/TimeServiceAbility: [b17d23a1cbc80b6, 2e8336f, 1962a4e] onCommand
```

当点击设备A（左侧）界面的Stop Remote PA文本时，会关闭设备B（右侧）的Service。这点也可以从控制台日志看出。

```
    09-01 23:53:21.629 22193-22193/com.waylau.hmos.remotepa I
00001/TimeServiceAbility: onBackground
    09-01 23:53:21.629 22193-22193/com.waylau.hmos.remotepa I
00001/TimeServiceAbility: onStop
```

上述操作验证了在设备A（DistributedSchedulingStartStopRemotePA应用）中启动和关闭设备B（RemotePA应用）的PA功能。

第 7 章

Ability 公共事件与通知

HarmonyOS通过CES（Common Event Service，公共事件服务）为应用程序提供订阅、发布、退订公共事件的能力，通过ANS（Advanced Notification Service，即高级通知服务）系统服务来为应用程序提供发布通知的能力。

7.1 公共事件与通知概述

在应用里面往往会有事件和通知。比如，朋友给我手机发了一条信息，未读信息会在手机的通知栏给出提示。

在HarmonyOS里面，系统给应用发送提示一般分为两种方式，即公共事件和通知。

7.1.1 公共事件和通知

公共事件可分为系统公共事件和自定义公共事件。

- 系统公共事件：系统将收集到的事件信息根据系统策略发送给订阅该事件的用户程序，例如用户可感知亮灭屏事件、系统关键服务发送的系统事件（例如USB插拔、网络连接、系统升级等）。
- 自定义公共事件：应用自定义公共事件来处理业务逻辑。

IntentAgent封装了一个指定行为的Intent，可以通过IntentAgent启动Ability和发送公共事件。应用如果需要接收公共事件，则需要订阅相应的事件。

通知提供应用的即时消息或通信消息，用户可以直接删除或点击通知触发进一步的操作。比如，收到一条未读信息的通知，则可以选择删除该通知，或者点击该通知进入短信应用查看该信息。

7.1.2 约束与限制

1. 公共事件的约束与限制

- 目前公共事件仅支持动态订阅。部分系统事件需要具有指定的权限。
- 目前公共事件订阅不支持多用户。
- ThreadMode表示线程模型，目前仅支持HANDLER模式，即在当前UI线程上执行回调函数。
- deviceId用来指定订阅本地公共事件还是远端公共事件。deviceId为null、空字符串或本地设备deviceId时，表示订阅本地公共事件，否则表示订阅远端公共事件。

2. 通知的约束与限制

- 通知目前支持6种样式：普通文本、长文本、图片、社交、多行文本和媒体样式。创建通知时必须包含其中一种样式。
- 通知支持快捷回复。

3. IntentAgent的限制

使用IntentAgent启动Ability时，Intent必须指定Ability的包名和类名。

7.2 公共事件服务

每个应用都可以订阅自己感兴趣的公共事件，订阅成功且公共事件发布后，系统会把其发送给应用。这些公共事件可能来自系统、其他应用和应用自身。HarmonyOS提供了一套完整的API，支持用户订阅、发送和接收公共事件。发送公共事件需要借助CommonEventData对象，接收公共事件需要继承CommonEventSubscriber类并实现onReceiveEvent回调函数。

7.2.1 接口说明

公共事件相关基础类包含：

- CommonEventData。
- CommonEventPublishInfo。
- CommonEventSubscribeInfo。
- CommonEventSubscriber。
- CommonEventManager。

上述基础类之间的关系如图7-1所示。

图 7-1 基础类之间的关系

1. CommonEventData

CommonEventData 封装公共事件相关信息，用于在发布、分发和接收时处理数据。在构造 CommonEventData 对象时，相关参数需要注意以下事项：

- code 为有序公共事件的结果码，data 为有序公共事件的结果数据，仅用于有序公共事件场景。
- intent 不允许为空，否则发布公共事件失败。

2. CommonEventPublishInfo

CommonEventPublishInfo 封装公共事件发布相关属性、限制等信息，包括公共事件类型（有序或粘性）、接收者权限等。

- 有序公共事件：主要场景是多个订阅者有依赖关系或者对处理顺序有要求，例如，高优先级订阅者可修改公共事件内容或处理结果，包括终止公共事件处理；低优先级订阅者依赖高优先级的处理结果等。有序公共事件的订阅者可以通过 CommonEventSubscribeInfo.setPriority() 方法指定优先级，默认为0，优先级范围为[-1000, 1000]，值越大优先级越高。
- 粘性公共事件：指公共事件的订阅动作是在公共事件发布之后进行的，订阅者也能收到的公共事件类型。主要场景是由公共事件服务记录某些系统状态，如蓝牙、WLAN、充电等事件和状态。不使用粘性公共事件机制时，应用可以通过直接访问系统服务获取该状态；在状态变化时，系统服务、硬件需要提供类似 observer 等方式通知应用。发布粘性公共事件可以通过 setSticky() 方法设置，发布粘性公共事件需要申请 ohos.permission.COMMONEVENT_STICKY 权限。

3. CommonEventSubscribeInfo

CommonEventSubscribeInfo 封装公共事件订阅相关信息，比如优先级、线程模式、事件范围等。

- 线程模式：设置订阅者的回调方法执行的线程模式，主要有 HANDLER、POST、ASYNC、BACKGROUND 四种模式。
  - HANDLER：在 Ability 的主线程执行。
  - POST：在事件分发线程执行。
  - ASYNC：在一个新创建的异步线程执行。
  - BACKGROUND：在后台线程执行。

截至目前只支持HANDLER模式。

4. CommonEventSubscriber

CommonEventSubscriber封装公共事件订阅者及相关参数。

CommonEventSubscriber.AsyncCommonEventResult类处理有序公共事件异步执行。目前只能通过调用CommonEventManager的subscribeCommonEvent()进行订阅。

5. CommonEventManager

CommonEventManager是为应用提供订阅、退订和发布公共事件的静态接口类。主要接口如下：

- 发布公共事件：publishCommonEvent(CommonEventData event)。
- 发布公共事件指定发布信息：publishCommonEvent(CommonEventData event, CommonEventPublishInfo publishinfo)。
- 发布有序公共事件，指定发布信息和最后一个接收者：publishCommonEvent(CommonEventData event, CommonEventPublishInfo publishinfo, CommonEventSubscriber resultSubscriber)。
- 订阅公共事件：subscribeCommonEvent(CommonEventSubscriber subscriber)。
- 退订公共事件：unsubscribeCommonEvent(CommonEventSubscriber subscriber)。

7.2.2 发布公共事件

开发者可以发布4种公共事件：无序的公共事件、带权限的公共事件、有序的公共事件、粘性的公共事件。

1. 发布无序的公共事件

发布无序的公共事件时，构造CommonEventData对象，设置Intent，通过构造operation对象把需要发布的公共事件信息传入intent对象。然后调用 CommonEventManager.publishCommonEvent(CommonEventData) 接口发布公共事件。

示例代码如下：

```
try {
    Intent intent = new Intent();
    Operation operation = new Intent.OperationBuilder()
            .withAction("com.my.test")
            .build();
    intent.setOperation(operation);
    CommonEventData eventData = new CommonEventData(intent);
    CommonEventManager.publishCommonEvent(eventData);
} catch (RemoteException e) {
    HiLog.info(LABEL, "publishCommonEvent occur exception.");
}
```

2. 发布携带权限的公共事件

发布携带权限的公共事件时，构造CommonEventPublishInfo对象，设置订阅者的权限。订阅者在config.json中申请所需的权限，各字段含义详见权限定义字段说明。

示例如下：

```
{
    "reqPermissions": [{
       "name": "com.example.MyApplication.permission",
       "reason": "get right",
       "usedScene": {
         "ability": [
          ".MainAbility"
         ],
         "when": "inuse"
       }
    }, {
       ...
    }]
}
```

发布带权限的公共事件示例代码如下：

```
Intent intent = new Intent();
Operation operation = new Intent.OperationBuilder()
       .withAction("com.my.test")
       .build();
intent.setOperation(operation);
CommonEventData eventData = new CommonEventData(intent);
CommonEventPublishInfo publishInfo = new CommonEventPublishInfo();
String[] permissions = {"com.example.MyApplication.permission" };
publishInfo.setSubscriberPermissions(permissions); // 设置权限
try {
    CommonEventManager.publishCommonEvent(eventData, publishInfo);
} catch (RemoteException e) {
    HiLog.info(LABEL, "publishCommoneEvent occur exception.");
}
```

3. 发布有序的公共事件

发布有序的公共事件时，构造CommonEventPublishInfo对象，通过setOrdered(true)指定公共事件属性为有序公共事件，也可以指定一个最后的公共事件接收者。

示例代码如下：

```
CommonEventSubscriber resultSubscriber = new MyCommonEventSubscriber();
CommonEventPublishInfo publishInfo = new CommonEventPublishInfo();
publishInfo.setOrdered(true); // 设置属性为有序公共事件
try {
    // 指定resultSubscriber为有序公共事件最后一个接收者
    CommonEventManager.publishCommonEvent(eventData, publishInfo,
resultSubscriber);
} catch (RemoteException e) {
    HiLog.info(LABEL, "publishCommoneEvent occur exception.");
}
```

4. 发布粘性公共事件

发布粘性公共事件时，构造CommonEventPublishInfo对象，通过setSticky(true)指定公共事件属性为粘性公共事件。

发布者首先在config.json中申请发布粘性公共事件所需的权限，配置如下：

```
{
    "reqPermissions": [{
       "name": "ohos.permission.COMMONEVENT_STICKY",
       "reason": "get right",
       "usedScene": {
          "ability": [
           ".MainAbility"
          ],
          "when": "inuse"
       }
    }, {
       ...
    }]
}
```

发布粘性公共事件代码如下：

```
CommonEventPublishInfo publishInfo = new CommonEventPublishInfo();
publishInfo.setSticky(true); // 设置属性为粘性公共事件
try {
    CommonEventManager.publishCommonEvent(eventData, publishInfo);
} catch (RemoteException e) {
    HiLog.info(LABEL, "publishCommoneEvent occur exception.");
}
```

7.2.3 订阅公共事件

订阅公共事件时，首先创建CommonEventSubscriber派生类，在onReceiveEvent()回调函数中处理公共事件。示例代码如下：

```
class MyCommonEventSubscriber extends CommonEventSubscriber {
    MyCommonEventSubscriber(CommonEventSubscribeInfo info) {
        super(info);
    }
    @Override
    public void onReceiveEvent(CommonEventData commonEventData) {
    }
}
```

注意：此处不能执行耗时操作，否则会阻塞UI线程，产生用户点击没有反应等异常。

接着构造MyCommonEventSubscriber对象，调用CommonEventManager.subscribeCommonEvent()接口进行订阅。示例代码如下：

```
String event = "com.my.test";
```

```
    MatchingSkills matchingSkills = new MatchingSkills();
    matchingSkills.addEvent(event); // 自定义事件
    matchingSkills.addEvent(CommonEventSupport.COMMON_EVENT_SCREEN_ON);//亮屏事件
    CommonEventSubscribeInfo subscribeInfo = new
CommonEventSubscribeInfo(matchingSkills);
    MyCommonEventSubscriber subscriber = new
MyCommonEventSubscriber(subscribeInfo);
    try {
        CommonEventManager.subscribeCommonEvent(subscriber);
    } catch (RemoteException e) {
        HiLog.info(LABEL, "subscribeCommonEvent occur exception.");
    }
```

如果订阅拥有指定权限应用发布的公共事件，发布者需要在config.json中申请权限，各字段含义详见权限申请字段说明。

```
    "reqPermissions": [
      {
          "name": "ohos.abilitydemo.permission.PROVIDER",
          "reason": "get right",
          "usedScene": {
              "ability": ["com.huawei.hmi.ivi.systemsetting.MainAbility"],
              "when": "inuse"
          }
      }
    ]
```

如果订阅的公共事件是有序的，则可以调用setPriority()指定优先级。

```
    String event = "com.my.test";
    MatchingSkills matchingSkills = new MatchingSkills();
    matchingSkills.addEvent(event ); // 自定义事件

    CommonEventSubscribeInfo subscribeInfo = new
CommonEventSubscribeInfo(matchingSkills);
    subscribeInfo.setPriority(100); // 设置优先级，优先级取值范围为[-1000,1000]，值
默认为0
    MyCommonEventSubscriber subscriber = new
MyCommonEventSubscriber(subscribeInfo);
    try {
        CommonEventManager.subscribeCommonEvent(subscriber);
    } catch (RemoteException e) {
        HiLog.info(LABEL, "subscribeCommonEvent occur exception.");
    }
```

最后，针对在onReceiveEvent中不能执行耗时操作的限制，可以使用CommonEventSubscriber的goAsyncCommonEvent()来实现异步操作，函数返回后仍保持该公共事件活跃，且执行完成后必须调用AsyncCommonEventResult.finishCommonEvent()来结束。示例代码如下：

```
    // 创建新线程，将耗时的操作放到新的线程上执行
    EventRunner runner = EventRunner.create();
```

```
//MyEventHandler为EventHandler的派生类，在不同线程间分发和处理事件和Runnable任务
MyEventHandler myHandler = new MyEventHandler(runner);

@Override
public void onReceiveEvent(CommonEventData commonEventData){
    final AsyncCommonEventResult result = goAsyncCommonEvent();

    Runnable task = new Runnable() {
       @Override
       public void run() {
            ........            // 待执行的操作，由开发者定义
            result.finishCommonEvent(); // 调用finish结束异步操作
       }
    };
    myHandler.postTask(task);
}
```

7.2.4 退订公共事件

在Ability的onStop()中调用CommonEventManager.unsubscribeCommonEvent()方法来退订公共事件。调用后，之前订阅的所有公共事件均被退订。示例代码如下：

```
try {
    CommonEventManager.unsubscribeCommonEvent(subscriber);
} catch (RemoteException e) {
    HiLog.info(LABEL, "unsubscribeCommonEvent occur exception.");
}
```

7.3 实战：公共事件服务发布事件

本节演示如何使用公共事件服务发布携带权限的公共事件的例子。为了演示该功能，创建了一个名为CommonEventPublisher的应用，作为公共事件的发布者。

7.3.1 修改 ability_main.xml

修改ability_main.xml文件如下：

```
<?xml version="1.0" encoding="utf-8"?>
<DirectionalLayout
    xmlns:ohos="http://schemas.huawei.com/res/ohos"
    ohos:height="match_parent"
    ohos:width="match_parent"
    ohos:alignment="center"
    ohos:orientation="vertical">
```

```xml
<Text
    ohos:id="$+id:Id_text_publish_event"
    ohos:height="match_content"
    ohos:width="match_content"
    ohos:background_element="$graphic:background_ability_main"
    ohos:layout_alignment="horizontal_center"
    ohos:text="Publish Event"
    ohos:text_size="40vp"
    />

</DirectionalLayout>
```

上述修改只是将显示的文本内容改为"Publish Event",在预览器中的预览效果如图7-2所示。

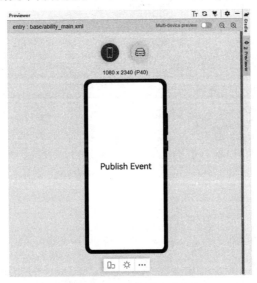

图 7-2　修改显示的文本内容

7.3.2　修改 MainAbilitySlice

在初始化应用时,应用已经包含一个主AbilitySlice,即MainAbilitySlice。对MainAbilitySlice进行修改,代码如下:

```java
package com.waylau.hmos.commoneventpublisher.slice;

import com.waylau.hmos.commoneventpublisher.ResourceTable;
import ohos.aafwk.ability.AbilitySlice;
import ohos.aafwk.content.Intent;
import ohos.aafwk.content.Operation;
import ohos.agp.components.Text;
import ohos.event.commonevent.CommonEventData;
import ohos.event.commonevent.CommonEventManager;
import ohos.event.commonevent.CommonEventPublishInfo;
import ohos.hiviewdfx.HiLog;
```

```java
import ohos.hiviewdfx.HiLogLabel;
import ohos.rpc.RemoteException;

public class MainAbilitySlice extends AbilitySlice {
    private static final String TAG = MainAbilitySlice.class.getSimpleName();
    private static final HiLogLabel LABEL_LOG =
            new HiLogLabel(HiLog.LOG_APP, 0x00001, TAG);

    private static final String EVENT_PERMISSION =
            "com.waylau.hmos.commoneventpublisher.PERMISSION";
    private static final String EVENT_NAME =
            "com.waylau.hmos.commoneventpublisher.EVENT";
    private static final int EVENT_CODE = 1;
    private static final String EVENT_DATA = "Welcome to waylau.com";
    private int index = 0;  // 递增的序列

    @Override
    public void onStart(Intent intent) {
        super.onStart(intent);
        super.setUIContent(ResourceTable.Layout_ability_main);

        // 添加点击事件来触发
        Text text = (Text) findComponentById(ResourceTable.Id_text_publish_event);
        text.setClickedListener(listener -> publishEvent());
    }

    private void publishEvent() {
        HiLog.info(LABEL_LOG, "before publishEvent");

        Operation operation = new Intent.OperationBuilder()
                .withAction(EVENT_NAME) // 设置事件名称
                .build();

        Intent intent = new Intent();
        intent.setOperation(operation);

        CommonEventData eventData = new CommonEventData(intent);
        CommonEventPublishInfo publishInfo = new CommonEventPublishInfo();
        String[] permissions = {EVENT_PERMISSION};
        publishInfo.setSubscriberPermissions(permissions); // 设置权限
        try {
            CommonEventManager.publishCommonEvent(eventData, publishInfo);
        } catch (RemoteException e) {
            HiLog.info(LABEL_LOG, "publishCommonEvent occur exception.");
        }

        HiLog.info(LABEL_LOG, "end publishEvent, event data %{public}s", eventData);
    }
```

```java
@Override
public void onActive() {
    super.onActive();
}

@Override
public void onForeground(Intent intent) {
    super.onForeground(intent);
}
}
```

在MainAbilitySlice类中：

- 在Text中增加了点击事件，以便触发发送事件的方法。
- 通过CommonEventManager.publishCommonEvent发送了事件。该事件是一个携带权限信息的自定义事件，事件名称为com.waylau.hmos.commoneventpublisher.PERMISSION。

7.3.3 自定义权限

需要在config.json文件的defPermissions字段中自定义所需的权限：

```json
...
// 自定义权限
"defPermissions": [
    {
        "name": "com.waylau.hmos.commoneventpublisher.PERMISSION"
    }
]
...
```

7.3.4 运行

运行应用之后，点击Publish Event文本内容，可以看到控制台日志输出内容如下：

```
09-03 00:43:21.621 14045-14045/com.waylau.hmos.commoneventpublisher I
00001/MainAbilitySlice: before publishEvent
09-03 00:43:21.623 14045-14045/com.waylau.hmos.commoneventpublisher I
00001/MainAbilitySlice: end publishEvent, event data CommonEventData[ code = 0
data = null]
09-03 00:43:21.837 14045-14045/com.waylau.hmos.commoneventpublisher I
00001/MainAbilitySlice: before publishEvent
09-03 00:43:21.839 14045-14045/com.waylau.hmos.commoneventpublisher I
00001/MainAbilitySlice: end publishEvent, event data CommonEventData[ code = 0
data = null]
09-03 00:43:22.079 14045-14045/com.waylau.hmos.commoneventpublisher I
00001/MainAbilitySlice: before publishEvent
09-03 00:43:22.083 14045-14045/com.waylau.hmos.commoneventpublisher I
```

```
00001/MainAbilitySlice: end publishEvent, event data CommonEventData[ code = 0
data = null]
```

可以看到每次点击都会发送一次事件。

7.4 实战：公共事件服务订阅事件

本节演示如何使用公共事件服务订阅携带权限的公共事件的例子。为了演示该功能，创建了一个名为CommonEventSubscriber的应用，作为公共事件的订阅者。

7.4.1 修改 ability_main.xml

修改ability_main.xml文件如下：

```xml
<?xml version="1.0" encoding="utf-8"?>
<DirectionalLayout
    xmlns:ohos="http://schemas.huawei.com/res/ohos"
    ohos:height="match_parent"
    ohos:width="match_parent"
    ohos:alignment="center"
    ohos:orientation="vertical">

    <Text
        ohos:id="$+id:text_eventsubscriber"
        ohos:height="match_content"
        ohos:width="match_content"
        ohos:background_element="$graphic:background_ability_main"
        ohos:layout_alignment="horizontal_center"
        ohos:text="EventSubscriber"
        ohos:text_size="40vp"
        />

</DirectionalLayout>
```

上述修改只是将显示的文本内容改为"EventSubscriber"，在预览器中的预览效果如图7-3所示。

7.4.2 创建 CommonEventSubscriber

创建一个名为WelcomeCommonEventSubscriber的事件订阅者，该类继承了CommonEventSubscriber。

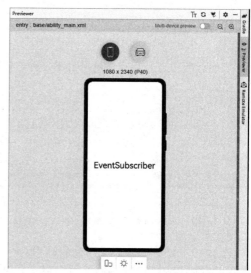

图7-3 修改显示的文本内容

代码如下:

```java
package com.waylau.hmos.commoneventsubscriber;

import ohos.event.commonevent.CommonEventData;
import ohos.event.commonevent.CommonEventSubscribeInfo;
import ohos.event.commonevent.CommonEventSubscriber;
import ohos.hiviewdfx.HiLog;
import ohos.hiviewdfx.HiLogLabel;

public class WelcomeCommonEventSubscriber extends CommonEventSubscriber {
    private static final String TAG =
        WelcomeCommonEventSubscriber.class.getSimpleName();
    private static final HiLogLabel LABEL_LOG =
        new HiLogLabel(HiLog.LOG_APP, 0x00001, TAG);

    public WelcomeCommonEventSubscriber(CommonEventSubscribeInfo info) {
        super(info);
    }

    @Override
    public void onReceiveEvent(CommonEventData commonEventData) {
        HiLog.info(LABEL_LOG, "receive event data %{public}s",
            commonEventData);
    }
}
```

在上述代码中,当接收到事件时,只是简单地将事件内容在日志中打印出来。

7.4.3 修改 MainAbility

在初始化应用时,应用已经包含一个MainAbility。对MainAbility进行修改,代码如下:

```java
package com.waylau.hmos.commoneventsubscriber.slice;

import com.waylau.hmos.commoneventsubscriber.ResourceTable;
import com.waylau.hmos.commoneventsubscriber.WelcomeCommonEventSubscriber;
import ohos.aafwk.ability.AbilitySlice;
import ohos.aafwk.content.Intent;
import ohos.event.commonevent.CommonEventManager;
import ohos.event.commonevent.CommonEventSubscribeInfo;
import ohos.event.commonevent.MatchingSkills;
import ohos.hiviewdfx.HiLog;
import ohos.hiviewdfx.HiLogLabel;
import ohos.rpc.RemoteException;

public class MainAbilitySlice extends AbilitySlice {
    private static final String TAG = MainAbilitySlice.class.getSimpleName();
```

```java
        private static final HiLogLabel LABEL_LOG =
                new HiLogLabel(HiLog.LOG_APP, 0x00001, TAG);
        private static final String EVENT_NAME =
                "com.waylau.hmos.commoneventpublisher.EVENT";
        @Override
        public void onStart(Intent intent) {
            super.onStart(intent);
            super.setUIContent(ResourceTable.Layout_ability_main);

            // 订阅事件
            subscribeEvent();
        }

        private void subscribeEvent() {
            MatchingSkills matchingSkills = new MatchingSkills();
            matchingSkills.addEvent(EVENT_NAME);
            CommonEventSubscribeInfo subscribeInfo = new
CommonEventSubscribeInfo(matchingSkills);
            subscribeInfo.setPriority(100);

            WelcomeCommonEventSubscriber subscriber = new
WelcomeCommonEventSubscriber(subscribeInfo);
            try {
                CommonEventManager.subscribeCommonEvent(subscriber);
            } catch (RemoteException e) {
                HiLog.error(LABEL_LOG, "%{public}s", "subscribeEvent
remoteException.");
            }
        }

        @Override
        public void onActive() {
            super.onActive();
        }

        @Override
        public void onForeground(Intent intent) {
            super.onForeground(intent);
        }
    }
```

在MainAbility类中：

- 在该类启动时，会执行subscribeEvent()方法。
- subscribeEvent()方法内容会通过CommonEventManager.subscribeCommonEvent订阅指定的事件。

7.4.4 修改配置文件

修改配置文件，增加权限的申请。

```
...
// 声明权限
"reqPermissions": [{
    "name": "com.waylau.hmos.commoneventpublisher.PERMISSION"
}]
...
```

上述com.waylau.hmos.commoneventpublisher.PERMISSION要与发布者定义的权限一致。

7.4.5 运行

先运行CommonEventSubscriber应用，而后运行7.3节介绍的CommonEventPublisher应用，并点击CommonEventPublisher应用的Publish Event文本内容，可以看到控制台日志输出内容如下：

```
09-03 00:36:50.704 28349-28349/com.waylau.hmos.commoneventsubscriber I
00001/WelcomeCommonEventSubscriber: receive event data CommonEventData[ code = 0
data = null]
09-03 00:36:55.279 28349-28349/com.waylau.hmos.commoneventsubscriber I
00001/WelcomeCommonEventSubscriber: receive event data CommonEventData[ code = 0
data = null]
09-03 00:37:03.504 28349-28349/com.waylau.hmos.commoneventsubscriber I
00001/WelcomeCommonEventSubscriber: receive event data CommonEventData[ code = 0
data = null]
```

可以看到每次点击CommonEventPublisher应用都会发送一次事件，而CommonEventPublisher应用都能收到相应的事件。

7.5　高级通知服务

读者对手机的通知功能应该不会陌生。HarmonyOS也提供了通知功能，即在一个应用的UI界面之外显示的消息主要用来提醒用户有来自某个应用的信息。当应用向系统发出通知时，它将先以图标的形式显示在通知栏中，用户可以下拉通知栏查看通知的详细信息。常见的使用场景如下：

- 显示接收到短消息、即时消息等。
- 显示应用的推送消息，如广告、版本更新等。
- 显示当前正在进行的事件，如播放音乐、导航、下载等。

7.5.1 接口说明

通知相关基础类包含：

- NotificationSlot。
- NotificationRequest。

- NotificationHelper。

上述基础类之间的关系如图7-4所示。

图7-4　通知基础类的关系

1. NotificationSlot

NotificationSlot可以对提示音、振动、锁屏显示和重要级别等进行设置。一个应用可以创建一个或多个NotificationSlot，在发送通知时，通过绑定不同的NotificationSlot实现不同的用途。

NotificationSlot需要先通过NotificationHelper的addNotificationSlot(NotificationSlot)方法发布后，通知才能绑定使用；所有绑定该NotificationSlot的通知在发布后都具备相应的特性，对象在创建后将无法更改这些设置，对于是否启动相应设置，用户有最终控制权。

不指定NotificationSlot时，当前通知会使用默认的NotificationSlot，默认的NotificationSlot优先级为LEVEL_DEFAULT。

NotificationSlot的级别目前支持如下几种，由低到高：

- LEVEL_NONE：表示通知不发布。
- LEVEL_MIN：表示通知可以发布，但是不显示在通知栏，不自动弹出，无提示音。该级别不适用于前台服务的场景。
- LEVEL_LOW：表示通知可以发布且显示在通知栏，不自动弹出，无提示音。
- LEVEL_DEFAULT：表示通知发布后可在通知栏显示，不自动弹出，触发提示音。
- LEVEL_HIGH：表示通知发布后可在通知栏显示，自动弹出，触发提示。

2. NotificationRequest

NotificationRequest用于设置具体的通知对象，包括设置通知的属性，如通知的分发时间、小图标、大图标、自动删除等参数，以及设置具体的通知类型，如普通文本、长文本等。

通知的常用属性包括：

- 通知分组：对于同一类型的通知，比如电子邮件，可以放在一个群组内展示。
- 小图标、大图标：分别通过NotificationRequest的setLittleIcon(PixelMap)、setBigIcon(PixelMap)设置的小图标、大图标。
- 显示时间戳：通知除了显示时间戳外，还可以显示计时器功能，包含正计时和倒计时。通知通过NotificationRequest的setCreateTime(Long)、setShowCreateTime(boolean)设置并显示时间戳。通知通过NotificationRequest的setShowStopwatch(boolean)显示计时器功能。通知通过NotificationRequest的setShowStopwatch(boolean)、setCountdownTimer(boolean)显示倒计时功

能。
- 进度条：主要用于播放音乐、下载等场景。通知通过NotificationRequest的setProgressBar(int, int, boolean)显示进度条。
- 从通知启动Ability：点击通知栏的通知，可以通过启动Ability触发新的事件。通知通过NotificationRequest的setIntentAgent(IntentAgent)设置IntentAgent后，点击通知栏上发布的通知，将触发通知中的IntentAgent承载的事件。IntentAgent的设置请参考IntentAgent开发指导。
- 通知设置ActionButton：通过点击通知按钮，可以触发按钮承载的事件。通过NotificationRequest的addActionButton(NotificationActionButton)附加按钮，点击按钮后可以触发相关的事件，具体事件内容如何设置需要参考NotificationActionButton。
- 通知设置ComponentProvider：通过ComponentProvider设置自定义的布局。通过NotificationRequest的setCustomView(ComponentProvider)配置自定义布局，替代系统布局，具体布局信息如何设置需要参考ComponentProvider。

目前支持6种通知类型，包括：

- 普通文本（NotificationNormalContent）：若为通知的标题，则通过NotificationRequest的setTitle(String)方法设置。若为通知的内容，则通过NotificationRequest的setText(String)方法设置。
- 长文本（NotificationLongTextContent）：长文本的内容通过setLongText(String)设置，文本长度最大支持1024个字符。
- 图片（NotificationPictureContent）：具有图片的通知。
- 多行（NotificationMultiLineContent）：若为折叠状态下的多行通知样式的标题，则通过NotificationMultiLineContent的setTitle(String)方法设置。若为折叠状态下的多行通知样式的内容，则通过NotificationMultiLineContent的setText(String)方法设置。若为展开状态下的多行通知样式的标题，则通过NotificationMultiLineContent的setExpandedTitle(String)方法设置。若为展开状态下的多行通知样式的内容，则通过NotificationMultiLineContent的addSingleLine(String)方法设置。
- 社交（NotificationConversationalContent）：若为社交通知样式的标题，则通过NotificationConversationalContent的setConversationTitle(String)方法设置。若为社交通知样式中的消息内容，则通过NotificationConversationalContent的addConversationalMessage(ConversationalMessage)方法设置。
- 媒体（NotificationMediaContent）：若为媒体通知样式的标题，则通过NotificationMediaContent的setTitle(String)方法设置。若为媒体通知样式中的消息内容，则通过NotificationMediaContent的setText(String)方法设置。若为媒体通知样式对应的多媒体按钮，具备控制音频媒体的用途，则通过NotificationMediaContent的setAVToken(AVToken)、NotificationMediaContent的setShownActions(int[])方法设置。

3. NotificationHelper

NotificationHelper封装了发布、更新、删除通知等静态方法。主要接口如下：

- 发布一条通知：publishNotification(NotificationRequest request)。

- 发布一条带TAG的通知：publishNotification(String tag, NotificationRequest)。
- 取消指定的通知：cancelNotification(int notificationId)。
- 取消指定的带TAG的通知：cancelNotification(String tag, int notificationId)。
- 取消之前发布的所有通知：cancelAllNotifications()。
- 创建一个NotificationSlot：addNotificationSlot(NotificationSlot slot)。
- 获取NotificationSlo：getNotificationSlot(String slotId)。
- 删除一个NotificationSlot：removeNotificationSlot(String slotId)。
- 获取当前应用发布的活跃通知：getActiveNotifications()。
- 获取系统中当前应用发布的活跃通知的数量：getActiveNotificationNums()。
- 设置通知的角标：setNotificationBadgeNum(int num)。
- 设置当前应用中活跃状态通知的数量在角标显示：setNotificationBadgeNum()。

7.5.2 创建 NotificationSlot

NotificationSlot可以设置公共通知的震动、锁屏模式、重要级别等，并通过调用NotificationHelper.addNotificationSlot()发布NotificationSlot对象。

示例代码如下：

```
 // 创建notificationSlot对象
NotificationSlot slot =
    new NotificationSlot("slot_001","slot_default",
NotificationSlot.LEVEL_MIN);

slot.setDescription("NotificationSlotDescription");
slot.setEnableVibration(true);  // 设置振动提醒
slot.setLockscreenVisibleness(NotificationRequest.VISIBLENESS_TYPE_PUBLIC)
;//设置锁屏模式
slot.setEnableLight(true);  // 设置开启呼吸灯提醒
slot.setLedLightColor(Color.RED.getValue());// 设置呼吸灯的提醒颜色

try {
    NotificationHelper.addNotificationSlot(slot);
} catch (RemoteException ex) {
    HiLog.warn(LABEL, "addNotificationSlot occur exception.");
}
```

7.5.3 发布通知

发布通知分为以下几个步骤。

1. 构建 NotificationRequest 对象

应用发布通知前，通过NotificationRequest的setSlotId()方法与NotificationSlot绑定，使该通知在发布后具备该对象的特征。示例如下：

```
int notificationId = 1;
NotificationRequest request = new NotificationRequest(notificationId);
request.setSlotId(slot.getId());
```

2. 设置通知内容

调用setContent()设置通知的内容,代码如下:

```
String title = "Welcome";
String text = "Welcome to waylau.com!";
NotificationNormalContent content = new NotificationNormalContent();
content.setTitle(title)
    .setText(text);
NotificationContent notificationContent = new NotificationContent(content);
request.setContent(notificationContent); // 设置通知的内容
```

3. 发送通知

调用publishNotification()发送通知。示例代码如下:

```
try {
    NotificationHelper.publishNotification(request);
} catch (RemoteException ex) {
    HiLog.warn(LABEL, "publishNotification occur exception.");
}
```

7.5.4 取消通知

取消通知分为取消指定单条通知和取消所有通知,应用只能取消自己发布的通知。

1. 取消指定的单条通知

调用cancelNotification()取消指定的单条通知,示例代码如下:

```
int notificationId = 1;
try {
    NotificationHelper.cancelNotification(notificationId);
} catch (RemoteException ex) {
    HiLog.warn(LABEL, "cancelNotification occur exception.");
}
```

2. 取消所有通知

调用cancelAllNotifications()取消所有通知,示例代码如下:

```
try {
    NotificationHelper.cancelAllNotifications();
} catch (RemoteException ex) {
    HiLog.warn(LABEL, "cancelAllNotifications occur exception.");
}
```

7.6 实战：通知发布与取消

本节演示如何使用通知发布与取消的例子。为了演示该功能，创建一个名为Notification的应用，用于通知的发布与取消。

7.6.1 修改 ability_main.xml

修改ability_main.xml文件如下：

```xml
<?xml version="1.0" encoding="utf-8"?>
<DirectionalLayout
    xmlns:ohos="http://schemas.huawei.com/res/ohos"
    ohos:height="match_parent"
    ohos:width="match_parent"
    ohos:alignment="center"
    ohos:orientation="vertical">

    <Text
        ohos:id="$+id:text_publish_notification"
        ohos:height="match_content"
        ohos:width="match_content"
        ohos:background_element="$graphic:background_ability_main"
        ohos:layout_alignment="horizontal_center"
        ohos:text="Publish Notification"
        ohos:text_size="40vp"
        />

    <Text
        ohos:id="$+id:text_cancel_notification"
        ohos:height="match_parent"
        ohos:width="match_content"
        ohos:background_element="$graphic:background_ability_main"
        ohos:layout_alignment="horizontal_center"
        ohos:text="Cancel Notification"
        ohos:text_size="40vp"
        />
</DirectionalLayout>
```

上述修改将显示两端文本内容Publish Notification与Cancel Notification，在预览器中的预览效果如图7-5所示。

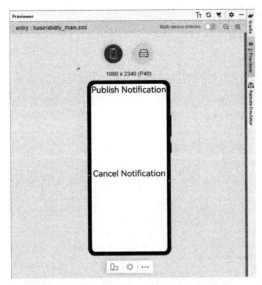

图 7-5　修改显示的文本内容

7.6.2　修改 MainAbilitySlice

在初始化应用时，应用已经包含一个主 AbilitySlice，即 MainAbilitySlice。对 MainAbilitySlice 进行修改，代码如下：

```
package com.waylau.hmos.notification.slice;

import com.waylau.hmos.notification.ResourceTable;
import ohos.aafwk.ability.AbilitySlice;
import ohos.aafwk.content.Intent;
import ohos.agp.components.Text;
import ohos.agp.utils.Color;
import ohos.event.notification.NotificationHelper;
import ohos.event.notification.NotificationRequest;
import ohos.event.notification.NotificationSlot;
import ohos.hiviewdfx.HiLog;
import ohos.hiviewdfx.HiLogLabel;
import ohos.rpc.RemoteException;

public class MainAbilitySlice extends AbilitySlice {
    private static final String TAG = MainAbilitySlice.class.getSimpleName();
    private static final HiLogLabel LABEL_LOG =
        new HiLogLabel(HiLog.LOG_APP, 0x00001, TAG);

    private int notificationId = 0; // 递增的序列

    @Override
    public void onStart(Intent intent) {
        super.onStart(intent);
        super.setUIContent(ResourceTable.Layout_ability_main);
```

```java
            // 添加点击事件来触发
        Text textPublishNotification = (Text) findComponentById(ResourceTable.Id_text_publish_notification);
        textPublishNotification.setClickedListener(listener -> publishNotification());


            // 添加点击事件来触发
        Text textCancelNotification = (Text) findComponentById(ResourceTable.Id_text_cancel_notification);
        textCancelNotification.setClickedListener(listener -> cancelNotification());
    }

    private void publishNotification() {
        HiLog.info(LABEL_LOG, "before publishNotification");

        // 创建notificationSlot对象
        NotificationSlot slot =
                new NotificationSlot("slot_001", "slot_default", NotificationSlot.LEVEL_HIGH);

        slot.setLevel(NotificationSlot.LEVEL_HIGH); // 设置提醒级别
        slot.setDescription("NotificationSlotDescription"); // 设置提示内容
        slot.setEnableVibration(true); // 设置振动提醒
        slot.setLockscreenVisibleness(NotificationRequest.VISIBLENESS_TYPE_PUBLIC);// 设置锁屏模式
        slot.setEnableLight(true); // 设置开启呼吸灯提醒
        slot.setLedLightColor(Color.RED.getValue());// 设置呼吸灯的提醒颜色

        try {
            NotificationHelper.addNotificationSlot(slot);

            String title = "title";
            String text = "There is a normal notification content.";
            NotificationRequest.NotificationNormalContent content = new NotificationRequest.NotificationNormalContent();
            content.setTitle(title)
                    .setText(text);
            notificationId++;
            NotificationRequest request = new NotificationRequest(notificationId);

            NotificationRequest.NotificationContent notificationContent = new NotificationRequest.NotificationContent(content);
            request.setContent(notificationContent); // 设置通知的内容
            request.setSlotId(slot.getId());

            NotificationHelper.publishNotification(request);
```

```java
        } catch (RemoteException ex) {
            HiLog.warn(LABEL_LOG, "publishNotification occur exception.");
        }
        HiLog.info(LABEL_LOG, "end publishNotification");
    }

    private void cancelNotification() {
        HiLog.info(LABEL_LOG, "before cancelNotification");
        try {
            NotificationHelper.cancelNotification(notificationId);
        } catch (RemoteException ex) {
            HiLog.warn(LABEL_LOG, "cancelNotification occur exception.");
        }

        HiLog.info(LABEL_LOG, "end cancelNotification");
    }


    @Override
    public void onActive() {
        super.onActive();
    }

    @Override
    public void onForeground(Intent intent) {
        super.onForeground(intent);
    }
}
```

在MainAbilitySlice类中：

- 在Text中增加了点击事件，以便触发发送事件的方法。
- publishNotification方法用于触发发布通知，而cancelNotification方法用于触发取消通知。
- 通过NotificationHelper.publishNotification发送通知。
- 通过NotificationHelper.cancelNotification取消通知。

7.6.3 运行

运行应用后，点击Publish Notification，可以在通知栏看到所定义的通知，效果如图7-6所示。点击Cancel Notification后，通知会被删除，效果如图7-7所示。

图 7-6　收到的通知　　　　　　　　图 7-7　通知被删除

第 8 章

剪 贴 板

用户通过系统剪贴板服务可以实现应用之间的简单数据传递。

8.1 剪贴板概述

读者对于剪切板的功能应该不会陌生,大多数软件都提供了剪贴板的功能。剪贴板的功能是这样的,用户通过系统剪贴板服务可以将内容数据从一个应用传递到另一个应用。

例如,在应用A中复制了一段数据,那么可以在应用B中粘贴,反之亦可。

针对剪贴板,HarmonyOS提供以下支持:

- 提供系统剪贴板服务的操作接口,支持用户程序从系统剪贴板中读取、写入和查询剪贴板数据,以及添加、移除系统剪贴板数据变化的回调。
- 提供剪贴板数据的对象定义,包含内容对象和属性对象。

8.2 场景介绍

同一设备的应用程序A、B之间可以借助系统剪贴板服务完成简单数据的传递,即从应用程序A向剪贴板服务写入数据后,可以在应用程序B中读取出数据。

图8-1展示了剪贴板服务的工作示意图。

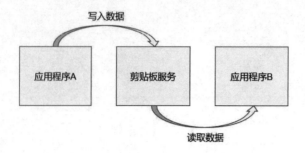

图 8-1　剪贴板服务的工作示意图

在使用剪贴板服务时，需要注意以下几点：

- 只有在前台获取到焦点的应用才有读取系统剪贴板的权限（系统默认输入法应用除外）。
- 写入剪贴板服务中的剪贴板数据不会随应用程序的结束而销毁。
- 对同一用户而言，写入剪贴板服务的数据会被下一次写入剪贴板的数据所覆盖。
- 在同一设备内，剪贴板单次传递的内容不应超过500KB。

8.3　接口说明

在HarmonyOS中，SystemPasteboard提供系统剪贴板操作的相关接口，比如复制、粘贴、配置回调等。其中：

- PasteData是剪贴板服务操作的数据对象，一个PasteData由若干个内容节点（PasteData.Record）和一个属性集合对象（PasteData.DataProperty）组成。
- Record是存放剪贴板数据内容信息的最小单位，每个Record都有其特定的MIME类型，如纯文本、HTML、URI、Intent。
- 剪贴板数据的属性信息存放在DataProperty中，包括标签、时间戳等。

8.3.1　SystemPasteboard

SystemPasteboard提供系统剪贴板服务的操作接口，比如复制、粘贴、配置回调等。
SystemPasteboard的主要接口如下：

- 获取系统剪贴板服务的对象实例：getSystemPasteboard(Context context)。
- 读取当前系统剪贴板中的数据：getPasteData()。
- 判断当前系统剪贴板中是否有内容：hasPasteData()。
- 将剪贴板数据写入系统剪贴板：setPasteData(PasteData data)。
- 清空系统剪贴板数据：clear()。
- 用户程序添加系统剪贴板数据变化的回调：addPasteDataChangedListener (IPasteDataChangedListener listener)。

- 用户程序移除系统剪贴板数据变化的回调：removePasteDataChangedListener(IPasteDataChangedListener listener)。

8.3.2 PasteData

PasteData是剪贴板服务操作的数据对象，其中内容节点定义为PasteData.Record，属性集合定义为PasteData.DataProperty。

PasteData的主要接口如下：

- 构造器：PasteData()。
- 构建一个包含纯文本内容节点的数据对象：createPlainTextData(CharSequence text)。
- 构建一个包含HTML内容节点的数据对象：creatHtmlData(String htmlText)。
- 构建一个包含URI内容节点的数据对象：creatUriData(Uri uri)。
- 构建一个包含Intent内容节点的数据对象：creatIntentData(Intent intent)。
- 获取数据对象中首个内容节点的MIME类型：getPrimaryMimeType()。
- 获取数据对象中首个内容节点的纯文本内容：getPrimaryText()。
- 向数据对象中添加一个纯文本内容节点：addTextRecord(CharSequence text)。
- 向数据对象中添加一个内容节点：addRecord(Record record)。
- 获取数据对象中内容节点的数量：getRecordCount()。
- 获取数据对象在指定下标处的内容节点：getRecordAt(int index)。
- 移除数据对象在指定下标处的内容节点：removeRecordAt(int index)。
- 获取数据对象中所有内容节点的MIME类型列表：getMimeTypes()。
- 获取该数据对象的属性集合成员：getProperty()。

8.3.3 PasteData.Record

一个PasteData中包含若干个特定MIME类型的PasteData.Record，每个Record是存放剪贴板数据内容信息的最小单位。

PasteData.Record的主要接口如下：

- 构造一个MIME类型为纯文本的内容节点：createPlainTextRecord(CharSequence text)。
- 构造一个MIME类型为HTML的内容节点：createHtmlTextRecord(String htmlText)。
- 构造一个MIME类型为URI的内容节点：createUriRecord(Uri uri)。
- 构造一个MIME类型为Intent的内容节点：createIntentRecord(Intent intent)。
- 获取该内容节点中的文本内容：getPlainText()。
- 获取该内容节点中的HTML内容：getHtmlText()。
- 获取该内容节点中的URI内容：getUri()。
- 获取该内容节点中的Intent内容：getIntent()。
- 获取该内容节点的MIME类型：getMimeType()。
- 将该内容节点的内容转为文本形式：convertToText(Context context)。

8.3.4　PasteData.DataProperty

每个PasteData中都有一个PasteData.DataProperty成员，其中存放着该数据对象的属性集合，例如自定义标签、MIME类型集合列表等。

PasteData.DataProperty的主要接口如下：

- 获取所属数据对象的MIME类型集合列表：getMimeTypes()。
- 判断所属数据对象中是否包含特定MIME类型的内容：hasMimeType(String mimeType)。
- 获取所属数据对象被写入系统剪贴板时的时间戳：getTimestamp()。
- 设置自定义标签：setTag(CharSequence tag)。
- 获取自定义标签：getTag()。
- 设置一些附加键值对信息：setAdditions(PacMap extraProps)。
- 获取附加键值对信息：getAdditions()。

8.3.5　IPasteDataChangedListener

IPasteDataChangedListener是定义剪贴板数据变化回调的接口类，开发者需要实现此接口来编码触发回调时的处理逻辑。

IPasteDataChangedListener的主要接口是onChanged()，这个是当系统剪贴板数据发生变化时的回调接口。

8.4　实战：剪贴板数据的写入

本节演示使用剪切板的例子。为了演示该功能，创建了一个名为SystemPasteboardSetter的应用，作为剪贴板数据的写入者。

8.4.1　修改 ability_main.xml

修改ability_main.xml文件如下：

```xml
<?xml version="1.0" encoding="utf-8"?>
<DirectionalLayout
    xmlns:ohos="http://schemas.huawei.com/res/ohos"
    ohos:height="match_parent"
    ohos:width="match_parent"
    ohos:alignment="center"
    ohos:orientation="vertical">

    <Text
```

```xml
    ohos:id="$+id:text_set_paste_data"
    ohos:height="match_content"
    ohos:width="match_content"
    ohos:background_element="$graphic:background_ability_main"
    ohos:layout_alignment="horizontal_center"
    ohos:text="Set Paste Data"
    ohos:text_size="40vp"
    />

</DirectionalLayout>
```

上述修改只是将显示的文本内容改为Set Paste Data，在预览器中的预览效果如图8-2所示。

图8-2　修改显示的文本内容

8.4.2　修改 MainAbilitySlice

在初始化应用时，应用已经包含一个主AbilitySlice，即MainAbilitySlice。对MainAbilitySlice进行修改，代码如下：

```
package com.waylau.hmos.systempasteboardsetter.slice;

import com.waylau.hmos.systempasteboardsetter.ResourceTable;
import ohos.aafwk.ability.AbilitySlice;
import ohos.aafwk.content.Intent;
import ohos.agp.components.Text;
import ohos.hiviewdfx.HiLog;
import ohos.hiviewdfx.HiLogLabel;
import ohos.miscservices.pasteboard.PasteData;
import ohos.miscservices.pasteboard.SystemPasteboard;

public class MainAbilitySlice extends AbilitySlice {
    private static final String TAG = MainAbilitySlice.class.getSimpleName();
    private static final HiLogLabel LABEL_LOG =
```

```java
                new HiLogLabel(HiLog.LOG_APP, 0x00001, TAG);

    @Override
    public void onStart(Intent intent) {
        super.onStart(intent);
        super.setUIContent(ResourceTable.Layout_ability_main);

        // 添加点击事件来触发
        Text textSetPasteData = (Text) findComponentById(ResourceTable.Id_text_set_paste_data);
        textSetPasteData.setClickedListener(listener -> setPasteData());
    }

    private void setPasteData() {
        HiLog.info(LABEL_LOG, "before setPasteData");

        // 获取系统剪贴板服务
        SystemPasteboard pasteboard = SystemPasteboard.getSystemPasteboard(this.getContext());

        // 向系统剪贴板中写入一条纯文本数据
        if (pasteboard != null) {
            pasteboard.setPasteData(PasteData.creatPlainTextData("Welcome to waylau.com!"));
        }

        HiLog.info(LABEL_LOG, "end setPasteData");
    }

    @Override
    public void onActive() {
        super.onActive();
    }

    @Override
    public void onForeground(Intent intent) {
        super.onForeground(intent);
    }
}
```

在MainAbilitySlice类中：

- 在Text中增加了点击事件，以便触发设置剪贴板数据的方法。
- 通过SystemPasteboard.getSystemPasteboard方法获取系统剪贴板服务。
- 通过pasteboard.setPasteData方法写入了PasteData，这个PasteData就是一条文本内容"Welcome to waylau.com!"。

8.4.3 运行

运行应用之后,点击Set Paste Data文本内容,可以看到控制台日志输出内容如下:

```
09-03 22:40:20.005 13545-13545/com.waylau.hmos.systempasteboardsetter I
00001/MainAbilitySlice:  before setPasteData
09-03 22:40:20.019 13545-13545/com.waylau.hmos.systempasteboardsetter I
00001/MainAbilitySlice:  end setPasteData
```

从上述日志可以看出,setPasteData方法已经能正常触发了。

接下来将演示如何读取剪贴板的数据。

8.5 实战:剪切板数据的读取

本节演示读取剪贴板数据的例子。为了演示该功能,创建了一个名为SystemPasteboardGetter的应用,作为剪贴板数据的读取者。

8.5.1 修改 ability_main.xml

修改ability_main.xml文件如下:

```xml
<?xml version="1.0" encoding="utf-8"?>
<DirectionalLayout
    xmlns:ohos="http://schemas.huawei.com/res/ohos"
    ohos:height="match_parent"
    ohos:width="match_parent"
    ohos:alignment="center"
    ohos:orientation="vertical">

    <Text
        ohos:id="$+id:text_get_paste_data"
        ohos:height="match_content"
        ohos:width="match_content"
        ohos:background_element="$graphic:background_ability_main"
        ohos:layout_alignment="horizontal_center"
        ohos:text="Get Paste Data"
        ohos:text_size="40vp"
        />
    <Text
        ohos:id="$+id:text_paste_data"
        ohos:height="match_parent"
        ohos:width="match_content"
        ohos:background_element="$graphic:background_ability_main"
        ohos:layout_alignment="horizontal_center"
```

```
            ohos:text="Ready To Get"
            ohos:text_size="40vp"
            ohos:multiple_lines="true"
            ohos:max_text_lines="2"
            />

</DirectionalLayout>
```

上述修改将显示的文本内容改为Get Paste Data和Ready To Get，在预览器中的预览效果如图8-3所示。

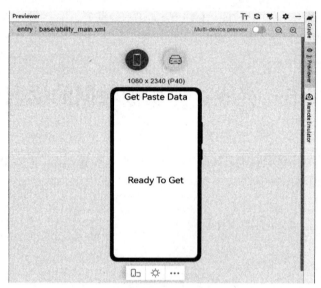

图8-3　修改显示的文本内容

8.5.2　修改 MainAbilitySlice

在初始化应用时，应用已经包含一个主AbilitySlice，即MainAbilitySlice。对MainAbilitySlice进行修改，代码如下：

```
package com.waylau.hmos.systempasteboardgetter.slice;

import com.waylau.hmos.systempasteboardgetter.ResourceTable;
import ohos.aafwk.ability.AbilitySlice;
import ohos.aafwk.content.Intent;
import ohos.agp.components.Text;
import ohos.hiviewdfx.HiLog;
import ohos.hiviewdfx.HiLogLabel;
import ohos.miscservices.pasteboard.PasteData;
import ohos.miscservices.pasteboard.SystemPasteboard;

public class MainAbilitySlice extends AbilitySlice {
    private static final String TAG = MainAbilitySlice.class.getSimpleName();
    private static final HiLogLabel LABEL_LOG =
```

```java
            new HiLogLabel(HiLog.LOG_APP, 0x00001, TAG);

    @Override
    public void onStart(Intent intent) {
        super.onStart(intent);
        super.setUIContent(ResourceTable.Layout_ability_main);

        // 添加点击事件来触发
        Text textGetPasteData =
            (Text) findComponentById(ResourceTable.Id_text_get_paste_data);
        textGetPasteData.setClickedListener(listener -> getPasteData());
    }

    private void getPasteData() {
        HiLog.info(LABEL_LOG, "before getPasteData");

        // 获取系统剪贴板服务
        SystemPasteboard pasteboard =
SystemPasteboard.getSystemPasteboard(this.getContext());

        // 从系统剪贴板中读取纯文本数据
        if (pasteboard != null) {
            PasteData pasteData = pasteboard.getPasteData();

            if (pasteData != null) {
                PasteData.DataProperty dataProperty = pasteData.getProperty();
                boolean hasHtml =
dataProperty.hasMimeType(PasteData.MIMETYPE_TEXT_HTML);
                boolean hasText =
dataProperty.hasMimeType(PasteData.MIMETYPE_TEXT_PLAIN);

                if (hasHtml || hasText) {
                    String textString = "";

                    // 遍历剪贴板中所有的记录
                    for (int i = 0; i < pasteData.getRecordCount(); i++) {
                        PasteData.Record record = pasteData.getRecordAt(i);
                        String mimeType = record.getMimeType();
                        if (mimeType.equals(PasteData.MIMETYPE_TEXT_HTML)) {
                            textString = record.getHtmlText();
                            break;
                        } else if
(mimeType.equals(PasteData.MIMETYPE_TEXT_PLAIN)) {//纯文本数据
                            textString = record.getPlainText().toString();
                            break;
                        }
                    }

                    // 将内容输出到界面
                    Text text = (Text)
```

```
findComponentById(ResourceTable.Id_text_paste_data);
                text.setText(textString);
            }
        } else {
            HiLog.info(LABEL_LOG, "PasteData is null");
        }


    }

    HiLog.info(LABEL_LOG, "end getPasteData");
}

@Override
public void onActive() {
    super.onActive();
}

@Override
public void onForeground(Intent intent) {
    super.onForeground(intent);
}
```

在MainAbilitySlice类中：

- 在Text中增加了点击事件，以便触发获取剪贴板数据的方法。
- 通过SystemPasteboard.getSystemPasteboard方法获取系统剪贴板服务。
- 通过pasteData.getProperty方法获取剪贴板数据的属性。
- 通过pasteData.getRecordAt方法获取具体剪贴板的数据。

8.5.3 运行

首先运行8.4节创建的SystemPasteboardSetter应用，而后点击Set Paste Data文本内容以写入剪贴板数据。可以看到控制台日志输出内容如下：

```
09-03 22:22:47.747 20369-20369/com.waylau.hmos.systempasteboardsetter I
00001/MainAbilitySlice: before setPasteData
09-03 22:22:47.753 20369-20369/com.waylau.hmos.systempasteboardsetter I
00001/MainAbilitySlice: end setPasteData
```

而后运行SystemPasteboardGetter应用，点击Get Paste Data文本内容，可以看到控制台日志输出内容如下：

```
09-03 22:46:47.158 23928-23928/com.waylau.hmos.systempasteboardgetter I
00001/MainAbilitySlice: before getPasteData
09-03 22:46:47.164 23928-23928/com.waylau.hmos.systempasteboardgetter I
00001/MainAbilitySlice: end getPasteData
```

此时，界面显示如图8-4所示。

图8-4　界面显示内容

可以看到，已经获取到了剪贴板中的数据"Welcome to waylau.com!"，并回写到了界面上。

第 9 章

用 Java 开发 UI

Java UI框架提供了用于创建用户界面（User Interface，UI）的各类组件，包括一些常用的组件和常用的布局。用户可通过组件进行交互操作，并获得响应。

9.1 用 Java 开发 UI 概述

在前面的应用开发过程中，我们已经初步接触了UI编程。以SystemPasteboardGetter应用为例，在开发一个Page Ability时，往往需要涉及AbilitySlice和ability_main.xml文件的修改，这其实就是Java UI编程的一部分。

应用的Ability在屏幕上将显示一个用户界面，该界面用来显示所有可被用户查看和交互的内容。应用中所有的用户界面元素都是由Component和ComponentContainer对象构成的。Component是绘制在屏幕上的一个对象，用户能与之交互。ComponentContainer是一个用于容纳其他Component和ComponentContainer对象的容器。

Java UI框架提供了一部分Component和ComponentContainer的具体子类，即创建用户界面的各类组件，包括一些常用的组件（比如文本、按钮、图片、列表等）和常用的布局（比如DirectionalLayout和DependentLayout）。用户可通过组件进行交互操作，并获得响应。所有的UI操作都应该在主线程进行设置。

9.1.1 组件和布局

用户界面元素统称为组件，组件根据一定的层级结构进行组合形成布局。组件在未被添加到布局中时，既无法显示又无法交互，因此一个用户界面至少包含一个布局。在UI框架中，具体的布局类通常以XXLayout命名，完整的用户界面是一个布局，用户界面中的一部分也可以是一个布局。布局中容纳Component与ComponentContainer对象。

9.1.2 Component 和 ComponentContainer

Component用于提供内容显示，是界面中所有组件的基类，开发者可以给Component设置事件处理回调来创建一个可交互的组件。Java UI框架提供了一些常用的界面元素，也可称之为组件，组件一般直接继承Component或它的子类，如Text、Image等。

ComponentContainer作为容器容纳Component或ComponentContainer对象，并对它们进行布局。Java UI框架提供了一些标准布局功能的容器，它们继承自ComponentContainer，一般以Layout结尾，如DirectionalLayout、DependentLayout等。

图9-1展示了Component和ComponentContainer的结构组成。

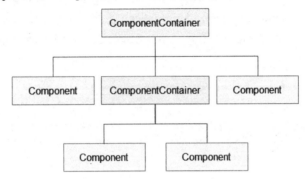

图 9-1　Component 和 ComponentContainer 的结构组成

9.1.3 LayoutConfig

每种布局都根据自身的特点提供LayoutConfig供子Component设定布局属性和参数，通过指定布局属性可以对子Component在布局中的显示效果进行约束。例如width、height是基本的布局属性，它们指定了组件的大小。

图9-2展示了LayoutConfig在布局中的作用。

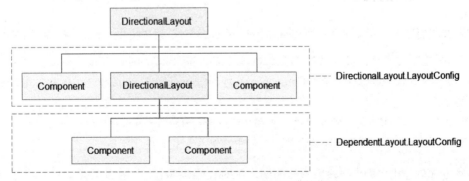

图 9-2　LayoutConfig 在布局中的作用

9.1.4 组件树

正如图9-1所展示的那样，布局把Component和ComponentContainer以树状的层级结构进行组织，这样的一个布局就称为组件树。组件树的特点是仅有一个根组件，其他组件有且仅有一个父节点，组件之间的关系受到父节点的规则约束。

9.2 组件与布局

正如前面章节所介绍的那样，HarmonyOS提供了Ability和AbilitySlice两个基础类。一个Page Ability可以由一个或多个AbilitySlice构成，AbilitySlice主要用于承载单个页面的具体逻辑实现和界面UI，是应用显示、运行和跳转的最小单元。AbilitySlice通过setUIContent为界面设置布局，示例如下：

```java
public class MainAbilitySlice extends AbilitySlice {

    @Override
    public void onStart(Intent intent) {
        super.onStart(intent);

        // 设置布局
        super.setUIContent(ResourceTable.Layout_ability_main);
    }
    ...
}
```

上述代码中，ResourceTable.Layout_ability_main即为界面组件的树根节点。

9.2.1 编写布局的方式

组件需要进行组合，并添加到界面的布局中。在Java UI框架中提供了两种编写布局的方式：

- 在代码中创建布局：用代码创建Component和ComponentContainer对象，为这些对象设置合适的布局参数和属性值，并将Component添加到ComponentContainer中，从而创建完整的界面。
- 在XML中声明UI布局：按层级结构来描述Component和ComponentContainer的关系，给组件节点设定合适的布局参数和属性值，代码中可直接加载生成此布局。

这两种方式创建出来的布局没有本质差别，在XML中声明布局，在加载后同样可以在代码中对该布局进行修改。在前面的章节中所介绍的实战案例多是在XML中声明布局。这也是本书所推崇的布局方式。

9.2.2 组件分类

根据组件的功能，可以将组件分为布局类、显示类、交互类3类。表9-1总结了各个组件的分类。

表9-1 组件的分类

组件类别	组件名称	功能描述
布局类	PositionLayout、DirectionalLayout、StackLayout、DependentLayout、TableLayout、AdaptiveBoxLayout	提供了不同布局规范的组件容器，例如以单一方向排列的 DirectionalLayout、以相对位置排列的 DependentLayout、以确切位置排列的 PositionLayout 等
显示类	Text、Image、Clock、TickTimer、ProgressBar	提供了单纯的内容显示，例如用于文本显示的 Text、用于图像显示的 Image 等
交互类	TextField、Button、Checkbox、RadioButton/RadioContainer、Switch、ToggleButton、Slider、Rating、ScrollView、TabList、ListContainer、PageSlider、PageFlipper、PageSliderIndicator、Picker、TimePicker、DatePicker、SurfaceProvider、ComponentProvider	提供了具体场景下与用户交互响应的功能，例如 Button 提供了点击响应功能，Slider 提供了进度选择功能，等等

框架提供的组件使应用界面开发更加便利。

9.3 实战：通过 XML 创建布局

本节演示如何通过XML创建DirectionalLayout布局。为了演示该功能，创建了一个名为DirectionalLayoutWithXml的应用。

XML声明布局的方式是非常简便直观的。在初始化DirectionalLayoutWithXml应用时，已经为应用创建了默认的界面布局，即MainAbilitySlice和ability_main.xml。

9.3.1 理解 XML 布局文件

每一个Component和ComponentContainer对象大部分属性都支持在XML中进行设置，它们都有各自的XML属性列表。某些属性仅适用于特定的组件，例如只有Text支持text_color属性，但不支持该属性的组件如果添加了该属性，该属性就会被忽略。具有继承关系的组件子类将继承父类的属性列表，Component作为组件的基类，拥有各个组件常用的属性，比如ID、布局参数等。

下面是初始化DirectionalLayoutWithXml应用时ability_main.xml文件的内容：

```
<?xml version="1.0" encoding="utf-8"?>
<DirectionalLayout
    xmlns:ohos="http://schemas.huawei.com/res/ohos"
```

```xml
    ohos:height="match_parent"
    ohos:width="match_parent"
    ohos:alignment="center"
    ohos:orientation="vertical">

    <Text
        ohos:id="$+id:text_helloworld"
        ohos:height="match_content"
        ohos:width="match_content"
        ohos:background_element="$graphic:background_ability_main"
        ohos:layout_alignment="horizontal_center"
        ohos:text="$string:mainability_HelloWorld"
        ohos:text_size="40vp"
        />

</DirectionalLayout>
```

接下来详细介绍ID和布局参数。

1. ID

在上述配置中，ohos:id="$+id:text_helloworld"就是在XML中使用此格式声明一个对开发者友好的ID，它会在编译过程中转换成一个常量。尤其在DependentLayout布局中，组件之间需要描述相对位置关系，描述时要通过ID来指定对应的组件。

布局中的组件通常要设置独立的ID，以便在程序中查找该组件。如果布局中有不同组件设置了相同的ID，在通过ID查找组件时会返回查找到的第一个组件，因此尽量保证在所要查找的布局中为组件设置独立的ID值，避免出现与预期不符的问题。

比如在SystemPasteboardGetter应用中，就为不同的Text设置了不同的ID，代码如下：

```xml
<Text
    ohos:id="$+id:text_get_paste_data"
    ohos:height="match_content"
    ohos:width="match_content"
    ohos:background_element="$graphic:background_ability_main"
    ohos:layout_alignment="horizontal_center"
    ohos:text="Get Paste Data"
    ohos:text_size="40vp"
    />
<Text
    ohos:id="$+id:text_paste_data"
    ohos:height="match_parent"
    ohos:width="match_content"
    ohos:background_element="$graphic:background_ability_main"
    ohos:layout_alignment="horizontal_center"
    ohos:text="Ready To Get"
    ohos:text_size="40vp"
    ohos:multiple_lines="true"
    ohos:max_text_lines="2"
    />
```

2. 布局参数

在上述配置中，ohos:width和ohos:height都是布局参数，在XML中它们的取值可以是：

- 具体的数值：10（以像素为单位）、10vp（以屏幕相对像素为单位）。
- match_parent：表示组件大小将扩展为父组件允许的最大值，它将占据父组件方向上的剩余大小。
- match_content：表示组件大小与它的内容占据的大小范围相适应。

9.3.2 创建 XML 布局文件

如果要新建XML布局文件，则可以在DevEco Studio的Project窗口打开entry→src→main→resources→base，右击layout文件夹，选择New→Layout Resource File来创建布局文件。比如，本例将布局文件命名为ability_pay.xml。

当然，另一种快捷的创建布局文件的方式是直接复制ability_main.xml的内容来修改。

打开新创建的ability_pay.xml布局文件，修改其中的内容，对布局和组件的属性和层级进行描述。

```
<?xml version="1.0" encoding="utf-8"?>
<DirectionalLayout
    xmlns:ohos="http://schemas.huawei.com/res/ohos"
    ohos:height="match_parent"
    ohos:width="match_parent"
    ohos:orientation="vertical">

    <Text
        ohos:id="$+id:text_pay"
        ohos:width="match_content"
        ohos:height="match_content"
        ohos:layout_alignment="horizontal_center"
        ohos:text="Show me the money"
        ohos:text_size="25vp"/>

</DirectionalLayout>
```

在预览器中可以对上述布局进行实时预览，如图9-3所示。

9.3.3 加载 XML 布局

在代码中需要加载XML布局，并添加为根布局或作为其他布局的子Component，代码如下：

图 9-3　对 ability_pay.xml 布局文件的预览

```java
package com.waylau.hmos.directionallayoutwithxml.slice;

import com.waylau.hmos.directionallayoutwithxml.ResourceTable;
import ohos.aafwk.ability.AbilitySlice;
import ohos.aafwk.content.Intent;
import ohos.agp.colors.RgbColor;
import ohos.agp.components.Text;
import ohos.agp.components.element.ShapeElement;

public class PayAbilitySlice extends AbilitySlice {
    @Override
    public void onStart(Intent intent) {
        super.onStart(intent);

        // 加载XML布局作为根布局
        super.setUIContent(ResourceTable.Layout_ability_pay);

        // 获取组件
        Text textPay = (Text) findComponentById(ResourceTable.Id_text_pay);

        // 设置组件的属性
        ShapeElement background = new ShapeElement();
        background.setRgbColor(new RgbColor(0, 125, 255));
        background.setCornerRadius(25);
        textPay.setBackground(background);
    }

    @Override
    public void onActive() {
        super.onActive();
    }

    @Override
    public void onForeground(Intent intent) {
        super.onForeground(intent);
    }
}
```

上述代码中：

- 通过setUIContent的方式来加载XML布局。
- 通过findComponentById方法来获取组件。
- 组件可以重新设置属性。上述例子中，我们设置了文本的背景。

9.3.4 显示 XML 布局

那么如何来显示PayAbilitySlice所设置的布局呢？一种方式是在5.6.4节介绍的，通过导航的方式从MainAbilitySlice导航到PayAbilitySlice。另一种简单的方式是直接将PayAbilitySlice设置为主AbilitySlice，代码如下：

```
package com.waylau.hmos.directionallayoutwithxml;

import com.waylau.hmos.directionallayoutwithxml.slice.MainAbilitySlice;
import com.waylau.hmos.directionallayoutwithxml.slice.PayAbilitySlice;
import ohos.aafwk.ability.Ability;
import ohos.aafwk.content.Intent;

public class MainAbility extends Ability {
    @Override
    public void onStart(Intent intent) {
        super.onStart(intent);
        //super.setMainRoute(MainAbilitySlice.class.getName());

        super.setMainRoute(PayAbilitySlice.class.getName());
    }
}
```

这样，运行应用后，可以看到界面显示如图9-4所示。

图 9-4　应用主界面显示效果

9.4　实战：通过 Java 创建布局

如果有Java Swing或者Java AWT编程经验的话，那么对于用Java语言通过代码方式创建布局就不会陌生。本节演示如何通过Java来创建布局。为了演示该功能，创建了一个名为DirectionalLayoutWithJava的应用。

在初始化DirectionalLayoutWithJava应用时，已经为应用创建了默认的界面布局，即MainAbilitySlice和ability_main.xml。我们需要再创建一个新的DirectionalLayout布局。

9.4.1 新建 AbilitySlice

创建一个新的AbilitySlice，将其命名为PayAbilitySlice。PayAbilitySlice需要继承AbilitySlice，代码如下：

```java
package com.waylau.hmos.directionallayoutwithjava.slice;

import ohos.aafwk.ability.AbilitySlice;

public class PayAbilitySlice extends AbilitySlice {
    package com.waylau.hmos.directionallayoutwithjava.slice;

import ohos.aafwk.ability.AbilitySlice;
import ohos.aafwk.content.Intent;

public class PayAbilitySlice extends AbilitySlice {
    @Override
    public void onStart(Intent intent) {
        super.onStart(intent);
    }

    @Override
    public void onActive() {
        super.onActive();
    }

    @Override
    public void onForeground(Intent intent) {
        super.onForeground(intent);
    }
}
```

上述代码重写了onStart、onActive、onForeground方法。

9.4.2 创建布局

在PayAbilitySlice的onStart方法中创建布局并使用，代码如下：

```java
@Override
public void onStart(Intent intent) {
    super.onStart(intent);

    // 声明布局
    DirectionalLayout directionalLayout = new DirectionalLayout(getContext());

    // 设置布局大小
```

```
directionalLayout.setWidth(ComponentContainer.LayoutConfig.MATCH_PARENT);

directionalLayout.setHeight(ComponentContainer.LayoutConfig.MATCH_PARENT);

    // 设置布局属性
    directionalLayout.setOrientation(Component.VERTICAL);

    // 将布局添加到组件树中
    setUIContent(directionalLayout);
}
```

上述代码中：

- 声明了布局。
- 设置了布局的大小和属性。
- 将布局添加到了组件树中。

这样，一个布局就创建完成了。

点击预览器，可以对布局进行预览，如图9-5所示。

图9-5　预览布局

9.4.3　在布局中添加组件

只有布局，界面只会显示一片空白，此时需要在布局中添加组件，代码如下：

```
// 声明Text组件
Text textPay = new Text(getContext());
textPay.setText("Show me the money");
textPay.setTextSize(25, Text.TextSizeType.VP);
textPay.setId(1);

// 设置组件的属性
```

```
ShapeElement background = new ShapeElement();
background.setRgbColor(new RgbColor(0, 125, 255));
background.setCornerRadius(25);
textPay.setBackground(background);

// 为组件添加对应布局的布局属性
DirectionalLayout.LayoutConfig layoutConfig =
        new
DirectionalLayout.LayoutConfig(ComponentContainer.LayoutConfig.MATCH_CONTENT,
            ComponentContainer.LayoutConfig.MATCH_CONTENT);
layoutConfig.alignment = LayoutAlignment.HORIZONTAL_CENTER;
textPay.setLayoutConfig(layoutConfig);

// 将组件添加到布局中（视布局需要对组件设置布局属性进行约束）
directionalLayout.addComponent(textPay);
```

上述代码中：

- 声明了Text组件。
- 设置了组件的属性。上述例子中，我们设置了文本的背景。
- 为组件添加了对应布局的布局属性。
- 将组件添加到了布局中。

点击预览器，可以对布局进行预览，如图9-6所示。

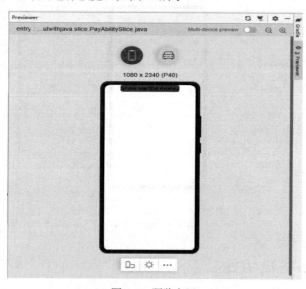

图 9-6　预览布局

9.4.4　显示布局

那么如何来显示PayAbilitySlice所设置的布局呢？一种方式是在5.6.4节介绍的，通过导航的方式从MainAbilitySlice导航到PayAbilitySlice。另一种简单的方式是直接将PayAbilitySlice设置为主AbilitySlice，代码如下：

```
package com.waylau.hmos.directionallayoutwithjava;

import com.waylau.hmos.directionallayoutwithjava.slice.MainAbilitySlice;
import com.waylau.hmos.directionallayoutwithjava.slice.PayAbilitySlice;
import ohos.aafwk.ability.Ability;
import ohos.aafwk.content.Intent;

public class MainAbility extends Ability {
    @Override
    public void onStart(Intent intent) {
        super.onStart(intent);
        //super.setMainRoute(MainAbilitySlice.class.getName());

        super.setMainRoute(PayAbilitySlice.class.getName());
    }
}
```

这样，运行应用后，可以看到界面显示如图9-7所示。

图9-7 应用主界面显示效果

9.5 实战：常用显示类组件——Text

常用显示类组件包括Text、Image、ProgressBar等，这些组件一般提供单纯的内容显示，例如用于文本显示的Text，用于图像显示的Image等。本节介绍Text组件的用法。

Text是前面章节中介绍最多的组件。Text是用来显示字符串的组件，在界面上显示为一块文本区域。Text作为一个基本组件，有很多扩展，常见的有按钮组件Button，文本编辑组件TextField。

创建一个名为Text的应用来作为演示。

9.5.1 设置背景

以下是创建Text应用时产生的ability_main.xml文件,内容如下:

```xml
<?xml version="1.0" encoding="utf-8"?>
<DirectionalLayout
    xmlns:ohos="http://schemas.huawei.com/res/ohos"
    ohos:height="match_parent"
    ohos:width="match_parent"
    ohos:alignment="center"
    ohos:orientation="vertical">

    <Text
        ohos:id="$+id:text_helloworld"
        ohos:height="match_content"
        ohos:width="match_content"
        ohos:background_element="$graphic:background_ability_main"
        ohos:layout_alignment="horizontal_center"
        ohos:text="$string:mainability_HelloWorld"
        ohos:text_size="40vp"
        />

</DirectionalLayout>
```

上述文件中:

- ohos:text可以用来配置组件显示的文字内容。上述配置引用了element、zh、en.json文件里面的内容,主要是为了处理本组件显示文字的国际化问题。也可以不用引用,直接把文字内容赋值上去即可,比如ohos:text="Hello World"。
- ohos:background_element可以用来配置常用的背景,如常见的文本背景、按钮背景。上述配置引用了background_ability_main.xml文件里面的内容,该文件放置在graphic目录下。

background_ability_main.xml文件内容如下:

```xml
<?xml version="1.0" encoding="UTF-8" ?>
<shape xmlns:ohos="http://schemas.huawei.com/res/ohos"
       ohos:shape="rectangle">
    <solid
        ohos:color="#FFFFFF"/>
</shape>
```

修改上述配置,可以设置Text背景的效果。修改后的background_ability_main.xml文件内容如下:

```xml
<?xml version="1.0" encoding="UTF-8" ?>
<shape xmlns:ohos="http://schemas.huawei.com/res/ohos"
       ohos:shape="rectangle">
    <corners
        ohos:radius="20"/>
    <solid
```

```
        ohos:color="#878787"/>
</shape>
```

最终效果如图9-8所示。

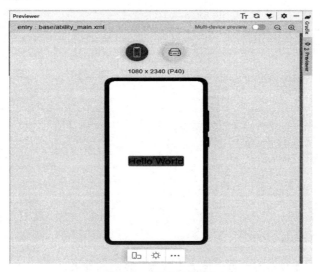

图9-8　界面显示效果

9.5.2　设置字体大小和颜色

为了演示字体大小和颜色的设置过程，通过DevEco Studio创建了名为ColorSizeAbility的Page，同时会自动创建如图9-9所示的4个文件。

图9-9　创建一个Page

这4个文件分别是ColorSizeAbility、ColorSizeAbilitySlice、ability_color_size.xml和background_ability_color_size.xml。

修改ability_color_size.xml内容如下：

```
<?xml version="1.0" encoding="utf-8"?>
<DirectionalLayout
    xmlns:ohos="http://schemas.huawei.com/res/ohos"
    ohos:height="match_parent"
    ohos:width="match_parent"
```

```
        ohos:alignment="center"
        ohos:orientation="vertical">

        <Text
            ohos:id="$+id:text_helloworld"
            ohos:height="match_content"
            ohos:width="match_content"
            ohos:background_element="$graphic:background_ability_color_size"
            ohos:layout_alignment="horizontal_center"
            ohos:text="Hello World"
            ohos:text_size="28fp"
            ohos:text_color="#0000FF"
            ohos:left_margin="15vp"
            ohos:bottom_margin="15vp"
            ohos:right_padding="15vp"
            ohos:left_padding="15vp"
            />

</DirectionalLayout>
```

图9-10展示了预览器显示的设置字体大小和颜色之后的效果。

图 9-10　设置字体大小和颜色之后的效果

9.5.3　设置字体风格和字重

为了演示设置字体风格和字重的过程，通过DevEco Studio创建了名为ItalicWeightAbility的Page，同时会自动创建4个文件。

这4个文件分别是ItalicWeightAbility、ItalicWeightAbilitySlice、ability_italic_weight.xml和background_ability_italic_weight.xml。

修改ability_italic_weight.xml内容如下：

```
<?xml version="1.0" encoding="utf-8"?>
<DirectionalLayout
```

```xml
    xmlns:ohos="http://schemas.huawei.com/res/ohos"
    ohos:height="match_parent"
    ohos:width="match_parent"
    ohos:alignment="center"
    ohos:orientation="vertical">

    <Text
        ohos:id="$+id:text_helloworld"
        ohos:height="match_content"
        ohos:width="match_content"
        ohos:background_element="$graphic:background_ability_italic_weight"
        ohos:layout_alignment="horizontal_center"
        ohos:text="Hello World"
        ohos:text_size="28fp"
        ohos:text_color="#0000FF"
        ohos:italic="true"
        ohos:text_weight="700"
        ohos:text_font="serif"
        ohos:left_margin="15vp"
        ohos:bottom_margin="15vp"
        ohos:right_padding="15vp"
        ohos:left_padding="15vp"
        />

</DirectionalLayout>
```

图9-11展示了预览器显示的设置italic字体风格和字重之后的效果。

图9-11 设置字体风格和字重之后的效果

9.5.4 设置文本对齐方式

为了演示设置文本对齐方式的过程，通过DevEco Studio创建了名为AlignmentAbility的Page，同时会自动创建4个文件。

这4个文件分别是AlignmentAbility、AlignmentAbilitySlice、ability_alignment.xml和background_ability_alignment.xml。

修改ability_alignment.xml内容如下：

```xml
<?xml version="1.0" encoding="utf-8"?>
<DirectionalLayout
    xmlns:ohos="http://schemas.huawei.com/res/ohos"
    ohos:height="match_parent"
    ohos:width="match_parent"
    ohos:alignment="center"
    ohos:orientation="vertical">

    <Text
        ohos:id="$+id:text_helloworld"
        ohos:background_element="$graphic:background_ability_alignment"
        ohos:text="Hello World"
        ohos:width="300vp"
        ohos:height="100vp"
        ohos:text_size="28fp"
        ohos:italic="true"
        ohos:text_weight="700"
        ohos:text_font="serif"
        ohos:left_margin="15vp"
        ohos:bottom_margin="15vp"
        ohos:right_padding="15vp"
        ohos:left_padding="15vp"
        ohos:text_alignment="horizontal_center|bottom"
        />

</DirectionalLayout>
```

图9-12展示了预览器显示的设置文本对齐方式之后的效果。

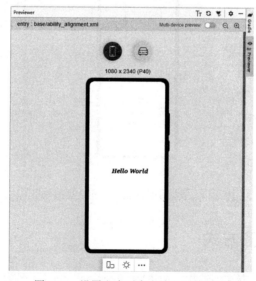

图9-12　设置文本对齐方式之后的效果

9.5.5 设置文本换行和最大显示行数

为了演示设置文本换行和最大显示行数的过程，通过DevEco Studio创建了名为LinesAbility的Page，同时会自动创建4个文件。

这4个文件分别是LinesAbility、LinesAbilitySlice、ability_lines.xml和background_ability_lines.xml。

修改ability_lines.xml内容如下：

```xml
<?xml version="1.0" encoding="utf-8"?>
<DirectionalLayout
    xmlns:ohos="http://schemas.huawei.com/res/ohos"
    ohos:height="match_parent"
    ohos:width="match_parent"
    ohos:alignment="center"
    ohos:orientation="vertical">

    <Text
        ohos:id="$+id:text_helloworld"
        ohos:background_element="$graphic:background_ability_lines"
        ohos:layout_alignment="horizontal_center"
        ohos:text="Hello World"
        ohos:width="75vp"
        ohos:height="match_content"
        ohos:text_size="28fp"
        ohos:multiple_lines="true"
        ohos:max_text_lines="2"
        />

</DirectionalLayout>
```

图9-13展示了预览器显示的设置文本换行和最大显示行数之后的效果。由于文本长度超过了限制，因此文本内容无法显示完整。

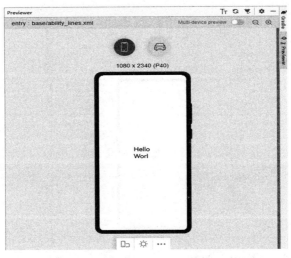

图9-13 设置文本换行和最大显示行数之后的效果

9.5.6 设置自动调节字体大小

为了演示设置自动调节字体大小的过程，通过DevEco Studio创建了名为AutoFontSizeAbility的Page，同时会自动创建4个文件。

这4个文件分别是AutoFontSizeAbility、AutoFontSizeAbilitySlice、ability_auto_font_size.xml和background_ability_auto_font_size.xml。

修改ability_auto_font_size.xml内容如下：

```xml
<?xml version="1.0" encoding="utf-8"?>
<DirectionalLayout
    xmlns:ohos="http://schemas.huawei.com/res/ohos"
    ohos:height="match_parent"
    ohos:width="match_parent"
    ohos:alignment="center"
    ohos:orientation="vertical">

    <Text
        ohos:id="$+id:text_auto_font_size"
        ohos:background_element="$graphic:background_ability_auto_font_size"
        ohos:layout_alignment="horizontal_center"
        ohos:text="Hello World"
        ohos:width="75vp"
        ohos:height="match_content"
        ohos:text_size="28fp"
        ohos:multiple_lines="true"
        ohos:max_text_lines="2"
        ohos:auto_font_size="true"
        />

</DirectionalLayout>
```

图9-14展示了预览器显示的设置自动调节字体大小之后的效果。

图9-14　设置自动调节字体大小之后的效果

修改AutoFontSizeAbilitySlice，代码如下：

```java
package com.waylau.hmos.text.slice;

import com.waylau.hmos.text.ResourceTable;
import ohos.aafwk.ability.AbilitySlice;
import ohos.aafwk.content.Intent;
import ohos.agp.components.Component;
import ohos.agp.components.Text;

public class AutoFontSizeAbilitySlice extends AbilitySlice {
    @Override
    public void onStart(Intent intent) {
        super.onStart(intent);
        super.setUIContent(ResourceTable.Layout_ability_auto_font_size);

        Text textAutoFontSize = (Text) findComponentById(ResourceTable.Id_text_auto_font_size);

        // 设置自动调整规则
        textAutoFontSize.setAutoFontSizeRule(30, 100, 1);

        // 设置点击一次增多一个"!"
        textAutoFontSize.setClickedListener(listener ->
            textAutoFontSize.setText(textAutoFontSize.getText() + "!"));
    }

    @Override
    public void onActive() {
        super.onActive();
    }

    @Override
    public void onForeground(Intent intent) {
        super.onForeground(intent);
    }
}
```

上述代码设置了点击事件。当点击Text内容之后，每点击一次，文本内容就增加一个"i"，同时可以看到字体也跟着缩小了。

图9-15显示了点击多次之后，字体缩小之后的效果。

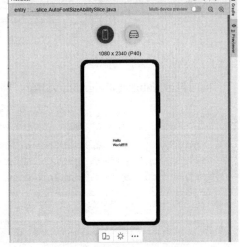

图9-15　字体缩小之后的效果

9.5.7 实现跑马灯效果

当文本过长时，可以设置跑马灯效果，实现文本滚动显示。前提是文本换行关闭且最大显示行数为1，默认情况下即可满足前提要求。

为了演示跑马灯效果的设置过程，通过DevEco Studio创建了名为AutoScrollingAbility的Page，同时会自动创建4个文件。这4个文件分别是AutoScrollingAbility、AutoScrollingAbilitySlice、ability_auto_scrolling.xml和background_ability_auto_scrolling.xml。

修改ability_auto_scrolling.xml内容如下：

```xml
<?xml version="1.0" encoding="utf-8"?>
<DirectionalLayout
    xmlns:ohos="http://schemas.huawei.com/res/ohos"
    ohos:height="match_parent"
    ohos:width="match_parent"
    ohos:alignment="center"
    ohos:orientation="vertical">

    <Text
        ohos:id="$+id:text_auto_scrolling"
        ohos:background_element="$graphic:background_ability_auto_scrolling"
        ohos:text="Hello World"
        ohos:height="match_content"
        ohos:width="75vp"
        ohos:text_size="28fp"
        ohos:text_color="#0000FF"
        ohos:italic="true"
        ohos:text_weight="700"
        ohos:text_font="serif"
        />

</DirectionalLayout>
```

同时修改AutoScrollingAbilitySlice，代码如下：

```java
package com.waylau.hmos.text.slice;

import com.waylau.hmos.text.ResourceTable;
import ohos.aafwk.ability.AbilitySlice;
import ohos.aafwk.content.Intent;
import ohos.agp.components.Text;

public class AutoScrollingAbilitySlice extends AbilitySlice {
    @Override
    public void onStart(Intent intent) {
        super.onStart(intent);
        super.setUIContent(ResourceTable.Layout_ability_auto_scrolling);

        Text textAutoScrolling =
```

```
                (Text) findComponentById(ResourceTable.Id_text_auto_scrolling);
        // 跑马灯效果
textAutoScrolling.setTruncationMode(Text.TruncationMode.AUTO_SCROLLING);
        // 始终处于自动滚动状态
textAutoScrolling.setAutoScrollingCount(Text.AUTO_SCROLLING_FOREVER);
        // 启动跑马灯效果
        textAutoScrolling.startAutoScrolling();
    }

    @Override
    public void onActive() {
        super.onActive();
    }

    @Override
    public void onForeground(Intent intent) {
        super.onForeground(intent);
    }
}
```

上述代码启动了跑马灯效果，如图9-16所示。

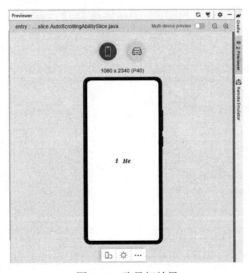

图9-16　跑马灯效果

9.5.8　场景示例

接下来是一个场景示例，利用文本组件实现一个包含标题栏、详细内容以及提交按钮的界面。

为了演示该示例，通过DevEco Studio创建了名为TitleDetailAbility的Page，同时会自动创建4个文件。这4个文件分别是TitleDetailAbility、TitleDetailAbilitySlice、ability_title_detail.xml和

background_ability_title_detail.xml。

修改ability_title_detail.xml内容如下：

```xml
<?xml version="1.0" encoding="utf-8"?>
<DirectionalLayout
    xmlns:ohos="http://schemas.huawei.com/res/ohos"
    ohos:height="match_parent"
    ohos:width="match_parent"
    ohos:alignment="center"
    ohos:orientation="vertical">

    <Text
        ohos:id="$+id:text_title"
        ohos:height="match_content"
        ohos:width="match_parent"
        ohos:background_element="$graphic:background_ability_title_detail"
        ohos:left_margin="15vp"
        ohos:right_margin="15vp"
        ohos:text="Title"
        ohos:text_alignment="horizontal_center"
        ohos:text_size="25fp"
        ohos:text_weight="1000"
        ohos:top_margin="15vp"/>

    <Text
        ohos:id="$+id:text_content"
        ohos:height="100vp"
        ohos:width="match_parent"
        ohos:background_element="$graphic:background_ability_title_detail"
        ohos:below="$id:text_title"
        ohos:bottom_margin="15vp"
        ohos:left_margin="15vp"
        ohos:right_margin="15vp"
        ohos:text="Content"
        ohos:text_alignment="center"
        ohos:text_font="serif"
        ohos:text_size="25fp"
        ohos:top_margin="15vp"/>

    <Text
        ohos:id="$+id:text_submit"
        ohos:height="match_content"
        ohos:width="75vp"
        ohos:align_parent_end="true"
        ohos:background_element="$graphic:background_ability_title_detail"
        ohos:below="$id:text_content"
        ohos:bottom_margin="15vp"
        ohos:left_padding="5vp"
        ohos:right_margin="15vp"
        ohos:right_padding="5vp"
        ohos:text="Submit"
```

```
            ohos:text_font="serif"
            ohos:text_size="15fp"/>
</DirectionalLayout>
```

同时修改background_ability_title_detail.xml文件，修改内容如下：

```
<?xml version="1.0" encoding="UTF-8" ?>
<shape xmlns:ohos="http://schemas.huawei.com/res/ohos"
    ohos:shape="rectangle">
    <corners
        ohos:radius="20"/>
    <solid
        ohos:color="#878787"/>
</shape>
```

上述代码实现效果如图9-17所示。

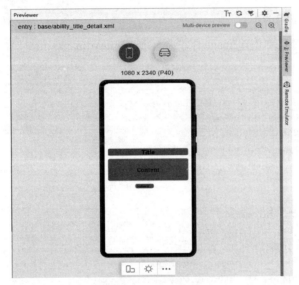

图9-17　最终效果

9.6　实战：常用显示类组件——Image

Image是用来显示图片的组件。创建一个名为Image的应用来作为演示。

9.6.1　创建Image

在Project窗口打开entry→src→main→resources→base→media，添加一个图片至media文件夹下，以waylau_616_616.jpg为例进行讲解。

创建Image主要分为两种，既可以在XML中创建Image，也可以在代码中创建Image。

1. 在 XML 中创建 Image

在XML中创建Image的方式如下：

```xml
<Image
    ohos:id="$+id:image"
    ohos:width="match_content"
    ohos:height="match_content"
    ohos:layout_alignment="center"
    ohos:image_src="$media:waylau_616_616"/>
```

2. 在代码中创建 Image

在代码中创建Image的方式如下：

```
Image image = new Image(getContext());
image.setPixelMap(ResourceTable.Media_plant);
```

3. 修改 ability_main.xml

在本例中采用在XML中创建Image的方式，修改ability_main.xml文件，内容如下：

```xml
<?xml version="1.0" encoding="utf-8"?>
<DirectionalLayout
    xmlns:ohos="http://schemas.huawei.com/res/ohos"
    ohos:height="match_parent"
    ohos:width="match_parent"
    ohos:alignment="center"
    ohos:orientation="vertical">

    <Image
        ohos:id="$+id:image"
        ohos:width="match_content"
        ohos:height="match_content"
        ohos:layout_alignment="center"
        ohos:image_src="$media:waylau_616_616"/>

</DirectionalLayout>
```

上述文件中，ohos:image_src用来配置图片的位置。最终效果如图9-18所示。

9.6.2 设置透明度

为了演示设置透明度的过程，通过DevEco Studio创建了名为AlphaAbility的Page，同时会自动创建AlphaAbility、AlphaAbilitySlice、ability_alpha.xml 和 background_ability_alpha.xml四个文件。

修改ability_alpha.xml内容如下：

```xml
<?xml version="1.0" encoding="utf-8"?>
<DirectionalLayout
```

图 9-18　界面显示效果

```xml
    xmlns:ohos="http://schemas.huawei.com/res/ohos"
    ohos:height="match_parent"
    ohos:width="match_parent"
    ohos:alignment="center"
    ohos:orientation="vertical">

    <Image
        ohos:id="$+id:image_alpha"
        ohos:width="match_content"
        ohos:height="match_content"
        ohos:layout_alignment="center"
        ohos:image_src="$media:waylau_616_616"
        ohos:alpha="0.3"/>

</DirectionalLayout>
```

上述代码中，ohos:alpha用于配置透明度。图9-19显示了预览器设置透明度之后的效果。

图9-19　设置透明度之后的效果

9.6.3　设置缩放系数

为了演示设置缩放系数的过程，通过DevEco Studio创建了名为ScaleAbility的Page，同时会自动创建4个文件。这4个文件分别是ScaleAbility、ScaleAbilitySlice、ability_scale.xml和background_ability_scale.xml。

修改ability_scale.xml内容如下：

```xml
<?xml version="1.0" encoding="utf-8"?>
<DirectionalLayout
    xmlns:ohos="http://schemas.huawei.com/res/ohos"
    ohos:height="match_parent"
    ohos:width="match_parent"
    ohos:alignment="center"
```

```
        ohos:orientation="vertical">

    <Image
        ohos:id="$+id:image_scale"
        ohos:width="match_content"
        ohos:height="match_content"
        ohos:layout_alignment="center"
        ohos:image_src="$media:waylau_616_616"
        ohos:scale_x="0.5"
        ohos:scale_y="0.5"/>

</DirectionalLayout>
```

图9-20展示了预览器显示的设置缩放系数之后的效果。

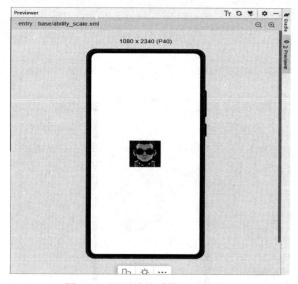

图 9-20　设置缩放系数之后的效果

9.7　实战：常用显示类组件——ProgressBar

ProgressBar用于显示内容或操作的进度。我们创建一个名为ProgressBar的应用来作为演示。

9.7.1　创建 ProgressBar

在本例中采用在XML中创建ProgressBar的方式，修改ability_main.xml文件，内容如下：

```
<?xml version="1.0" encoding="utf-8"?>
<DirectionalLayout
    xmlns:ohos="http://schemas.huawei.com/res/ohos"
    ohos:height="match_parent"
    ohos:width="match_parent"
    ohos:alignment="center"
```

```
    ohos:orientation="vertical">

    <ProgressBar
        ohos:id="$+id:progressbar"
        ohos:progress_width="20vp"
        ohos:height="60vp"
        ohos:width="260vp"
        ohos:max="100"
        ohos:min="0"
        ohos:progress="60"/>

</DirectionalLayout>
```

上述文件中，ProgressBar标签用来创建一个ProgressBar对象。其中ohos:progress用于设置当前进度，ohos:max用于设置最大值，ohos:min用于设置最小值。

最终效果如图9-21所示。

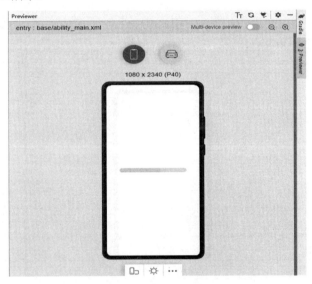

图9-21　界面显示效果

9.7.2　设置方向

默认情况下，ProgressBar的方向是水平，也可以设置为垂直。

为了演示设置方向的过程，通过DevEco Studio创建了名为OrientationAbility的Page，同时会自动创建OrientationAbility、OrientationAbilitySlice、ability_orientation.xml和background_ability_orientation.xml四个文件。

修改ability_orientation.xml内容如下：

```
<?xml version="1.0" encoding="utf-8"?>
<DirectionalLayout
    xmlns:ohos="http://schemas.huawei.com/res/ohos"
    ohos:height="match_parent"
    ohos:width="match_parent"
```

```
    ohos:alignment="center"
    ohos:orientation="vertical">

    <ProgressBar
        ohos:id="$+id:progressbar_orientation"
        ohos:progress_width="20vp"
        ohos:height="120vp"
        ohos:width="260vp"
        ohos:max="100"
        ohos:min="0"
        ohos:progress="60"
        ohos:orientation="vertical" />

</DirectionalLayout>
```

上述代码中，ohos:orientation用于配置方向，本例配置的是垂直。图9-22显示了预览器设置方向之后的效果。

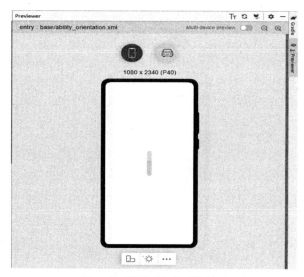

图9-22 设置方向之后的效果

9.7.3 设置颜色

为了演示设置颜色的过程，通过DevEco Studio创建了名为ElementAbility的Page，同时会自动创建4个文件。这4个文件分别是ElementAbility、ElementAbilitySlice、ability_element.xml和background_ability_element.xml。

修改ability_element.xml内容如下：

```
<?xml version="1.0" encoding="utf-8"?>
<DirectionalLayout
    xmlns:ohos="http://schemas.huawei.com/res/ohos"
    ohos:height="match_parent"
    ohos:width="match_parent"
    ohos:alignment="center"
```

```
    ohos:orientation="vertical">

    <ProgressBar
        ohos:id="$+id:progressbar_element"
        ohos:progress_width="20vp"
        ohos:height="60vp"
        ohos:width="260vp"
        ohos:max="100"
        ohos:min="0"
        ohos:progress="60"
        ohos:progress_element="#FF9900"
        ohos:background_instruct_element="#009900" />

</DirectionalLayout>
```

上述代码中,ohos:progress_element用于配置进度条颜色,而ohos:background_instruct_element用于设置ProgressBar底色。

图9-23展示了预览器显示的设置颜色之后的效果。

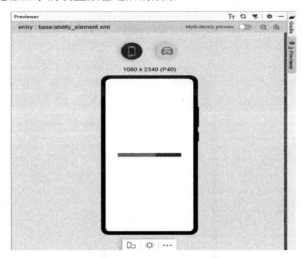

图9-23 设置颜色之后的效果

9.7.4 设置提示文字

为了演示设置提示文字的过程,通过DevEco Studio创建了名为HintAbility的Page,同时会自动创建4个文件。这4个文件分别是HintAbility、HintAbilitySlice、ability_hint.xml 和 background_ability_hint.xml。

修改ability_hint.xml内容如下:

```
<?xml version="1.0" encoding="utf-8"?>
<DirectionalLayout
    xmlns:ohos="http://schemas.huawei.com/res/ohos"
    ohos:height="match_parent"
    ohos:width="match_parent"
    ohos:alignment="center"
```

```xml
    ohos:orientation="vertical">

    <ProgressBar
        ohos:id="$+id:progressbar_hint"
        ohos:progress_width="20vp"
        ohos:height="60vp"
        ohos:width="260vp"
        ohos:max="100"
        ohos:min="0"
        ohos:progress="60"
        ohos:progress_hint_text="60%"
        ohos:progress_hint_text_color="#FFFC9F"
        />

</DirectionalLayout>
```

上述代码中，ohos:progress_hint_text用于配置提示的文字内容，而ohos:progress_hint_text_color用于设置提示文字的颜色。

图9-24展示了预览器显示的设置提示文字之后的效果。

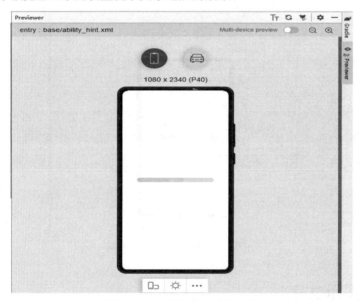

图 9-24　设置提示文字的效果

9.8　实战：常用交互类组件——Button

常用交互类组件包括Button、TextField、Checkbox、RadioButton/RadioContainer、Switch、ScrollView、Tab/TabList、ListContainer、Picker、TimePicker、DatePicker、RoundProgressBar等，这些类提供了具体场景下与用户交互响应的功能，例如Button提供了点击响应功能，Picker提供了滑动选择功能等。

Button应该是在UI界面设计中使用最为广泛的组件了。因为无论是提交表单还是执行下一页，都少不了Button组件。点击Button可以触发对应的操作。

Button通常由文本或图标组成，也可以由图标和文本共同组成。

创建一个名为Button的应用来作为演示。

9.8.1 创建 Button

以下是创建Button应用时产生的ability_main.xml文件。在该文件中增加创建Button的描述内容：

```xml
<?xml version="1.0" encoding="utf-8"?>
<DirectionalLayout
    xmlns:ohos="http://schemas.huawei.com/res/ohos"
    ohos:height="match_parent"
    ohos:width="match_parent"
    ohos:alignment="center"
    ohos:orientation="vertical">

    <Button
        ohos:id="$+id:button"
        ohos:width="match_content"
        ohos:height="match_content"
        ohos:text_size="27fp"
        ohos:text="I am Button"
        ohos:left_margin="15vp"
        ohos:bottom_margin="15vp"
        ohos:right_padding="8vp"
        ohos:left_padding="8vp"
        ohos:background_element="$graphic:background_button"
        />

    <Button
        ohos:id="$+id:button_icon"
        ohos:width="match_content"
        ohos:height="match_content"
        ohos:text_size="27fp"
        ohos:left_margin="15vp"
        ohos:bottom_margin="15vp"
        ohos:right_padding="8vp"
        ohos:left_padding="8vp"
        ohos:element_left="$graphic:ic_btn_reload"
        ohos:background_element="$graphic:background_button"
        />

    <Button
        ohos:id="$+id:button_icon_text"
        ohos:width="match_content"
        ohos:height="match_content"
        ohos:text_size="27fp"
        ohos:text="I am Button"
```

```xml
        ohos:left_margin="15vp"
        ohos:bottom_margin="15vp"
        ohos:right_padding="8vp"
        ohos:left_padding="8vp"
        ohos:element_left="$graphic:ic_btn_reload"
        ohos:background_element="$graphic:background_button"
        />

</DirectionalLayout>
```

上述文件中定义了3个Button组件。ohos:background_element可以用来配置Button的背景。上述配置引用了background_button.xml文件里面的内容，该文件放置在graphic目录下。

新增的background_button.xml文件内容如下：

```xml
<?xml version="1.0" encoding="UTF-8" ?>
<shape xmlns:ohos="http://schemas.huawei.com/res/ohos"
    ohos:shape="rectangle">
  <corners
      ohos:radius="10"/>
  <solid
      ohos:color="#007CFD"/>
</shape>
```

定义的3个Button组件中，第一个Button是纯文本的按钮，第二个Button是纯图标的按钮，第三个Button是图标加文本的按钮。图标是通过ohos:element_left配置的。图标的定义配置在ic_btn_reload.xml文件中。该文件放置在graphic目录下，文件内容如下：

```xml
<?xml version="1.0" encoding="UTF-8"?>
<vector xmlns:ohos="http://schemas.huawei.com/res/ohos" ohos:width="64vp" ohos:height="64vp" ohos:viewportWidth="1024" ohos:viewportHeight="1024">
    <path ohos:fillColor="#FF000000" ohos:pathData="M810.67,512 L952.32,512 741.12,723.2 529.92,512 724.05,512C725.33,446.29 700.59,381.01 650.24,330.67 550.4,230.83 388.27,230.83 288.43,330.67 188.59,430.51 188.59,593.07 288.43,692.91 366.93,771.41 484.69,788.05 579.41,742.83L642.13,805.55C512,882.77 341.33,865.71 227.84,753.07 94.72,619.95 95.15,404.05 228.27,270.93 362.67,137.39 577.28,136.96 710.83,270.51 777.39,337.07 810.67,424.53 810.67,512Z"></path>
</vector>
```

最终效果如图9-25所示。

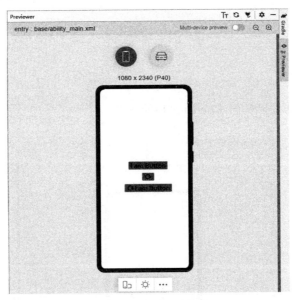

图 9-25　界面显示效果

9.8.2　设置点击事件

按钮的重要作用是当用户点击按钮时，会执行相应的操作或者界面出现相应的变化。实际上用户点击按钮时，Button对象将收到一个点击事件。开发者可以自定义响应点击事件的方法。例如，通过创建一个Component.ClickedListener对象，然后通过调用setClickedListener将其分配给按钮。这个设置与Text的点击事件类似。

为了演示点击事件的设置过程，通过DevEco Studio创建了名为ClickedListenerAbility的Page，同时会自动创建4个文件，分别是ClickedListenerAbility、ClickedListenerAbilitySlice、ability_clicked_listener.xml和background_ability_clicked_listener.xml。

修改ability_clicked_listener.xml内容如下：

```xml
<?xml version="1.0" encoding="utf-8"?>
<DirectionalLayout
    xmlns:ohos="http://schemas.huawei.com/res/ohos"
    ohos:height="match_parent"
    ohos:width="match_parent"
    ohos:alignment="center"
    ohos:orientation="vertical">

    <Button
        ohos:id="$+id:button_clicked_listener"
        ohos:width="match_content"
        ohos:height="match_content"
        ohos:text_size="27fp"
        ohos:text="I am Button"
        ohos:left_margin="15vp"
        ohos:bottom_margin="15vp"
        ohos:right_padding="8vp"
```

```
            ohos:left_padding="8vp"
            ohos:background_element="$graphic:background_button"
            />

</DirectionalLayout>
```

图9-26展示了预览器显示的创建Button之后的效果。

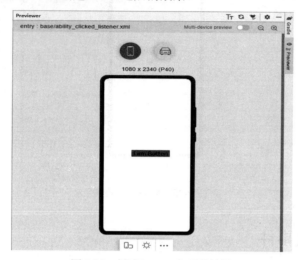

图 9-26　创建 Button 之后的效果

修改ClickedListenerAbilitySlice，代码如下：

```
package com.waylau.hmos.button.slice;

import com.waylau.hmos.button.ResourceTable;
import ohos.aafwk.ability.AbilitySlice;
import ohos.aafwk.content.Intent;
import ohos.agp.components.Button;

public class ClickedListenerAbilitySlice extends AbilitySlice {
    @Override
    public void onStart(Intent intent) {
        super.onStart(intent);
        super.setUIContent(ResourceTable.Layout_ability_clicked_listener);

        Button button =
                (Button) findComponentById(ResourceTable.Id_button_clicked_listener);

        // 为按钮设置点击事件回调
        button.setClickedListener(listener ->
                button.setText("Button was clicked!"));
    }

    @Override
    public void onActive() {
        super.onActive();
```

```
    }
    @Override
    public void onForeground(Intent intent) {
        super.onForeground(intent);
    }
}
```

上述代码中：

- 获取了Button组件。
- 在Button组件上设置了点击事件。

当点击按钮后，会将按钮上的文本修改为"Button was clicked!"。图9-27显示的就是点击Button之后的效果。

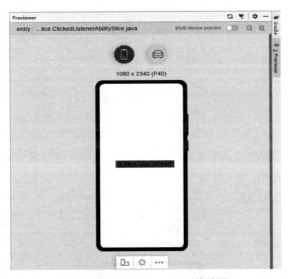

图 9-27 点击 Button 之后的效果

9.8.3 设置椭圆按钮

为了演示设置椭圆按钮的过程，通过DevEco Studio创建了名为OvalAbility的Page，同时会自动创建4个文件。这4个文件分别是OvalAbility、OvalAbilitySlice、ability_oval.xml 和 background_ability_oval.xml。

椭圆按钮是通过设置background_element来实现的，background_element的shape设置为椭圆。

修改ability_oval.xml内容如下：

```
<?xml version="1.0" encoding="utf-8"?>
<DirectionalLayout
    xmlns:ohos="http://schemas.huawei.com/res/ohos"
    ohos:height="match_parent"
    ohos:width="match_parent"
    ohos:alignment="center"
    ohos:orientation="vertical">
```

```xml
<Button
    ohos:id="$+id:button_oval"
    ohos:width="match_content"
    ohos:height="match_content"
    ohos:text_size="27fp"
    ohos:text="I am Button"
    ohos:left_margin="15vp"
    ohos:bottom_margin="15vp"
    ohos:right_padding="8vp"
    ohos:left_padding="8vp"
    ohos:background_element="$graphic:background_ability_oval"
    />

</DirectionalLayout>
```

同时修改background_ability_oval.xml文件，内容如下：

```xml
<?xml version="1.0" encoding="UTF-8" ?>
<shape xmlns:ohos="http://schemas.huawei.com/res/ohos"
    ohos:shape="oval">
    <solid
        ohos:color="#007CFD"/>
</shape>
```

上述配置将shape设置为椭圆。图9-28显示的就是设置椭圆之后的效果。

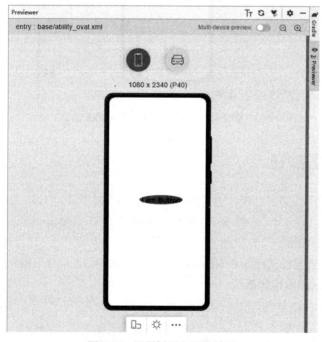

图9-28　设置椭圆之后的效果

9.8.4 设置圆形按钮

为了演示设置圆形按钮的过程，通过DevEco Studio创建了名为CircleAbility的Page，同时会自动创建4个文件。这4个文件分别是CircleAbility、CircleAbilitySlice、ability_circle.xml 和 background_ability_circle.xml。

圆形按钮和椭圆按钮的区别在于组件本身的宽度和高度需要相同。

修改ability_circle.xml内容如下：

```xml
<?xml version="1.0" encoding="utf-8"?>
<DirectionalLayout
    xmlns:ohos="http://schemas.huawei.com/res/ohos"
    ohos:height="match_parent"
    ohos:width="match_parent"
    ohos:alignment="center"
    ohos:orientation="vertical">

    <Button
        ohos:id="$+id:button_oval"
        ohos:width="200vp"
        ohos:height="200vp"
        ohos:text_size="27fp"
        ohos:text="I am Button"
        ohos:left_margin="15vp"
        ohos:bottom_margin="15vp"
        ohos:right_padding="8vp"
        ohos:left_padding="8vp"
        ohos:background_element="$graphic:background_ability_circle"
        />

</DirectionalLayout>
```

上述配置将宽、高设置成一样。

同时修改background_ability_oval.xml文件，内容如下：

```xml
<?xml version="1.0" encoding="UTF-8" ?>
<shape xmlns:ohos="http://schemas.huawei.com/res/ohos"
    ohos:shape="oval">
    <solid
        ohos:color="#007CFD"/>
</shape>
```

上述配置将shape设置为椭圆。

图9-29显示的就是设置圆形按钮之后的效果。

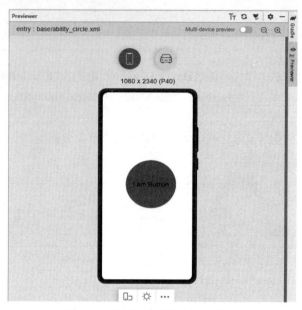

图 9-29　设置圆形按钮之后的效果

9.8.5　场景示例

接下来是一个场景示例，利用圆形按钮、胶囊按钮、文本组件绘制出电话拨号盘的UI界面。

为了演示该示例，通过DevEco Studio创建了名为DailAbility的Page，同时会自动创建4个文件。这4个文件分别是DailAbility、DailAbilitySlice、ability_dail.xml和background_ability_dail.xml。

修改ability_dail.xml内容如下：

```xml
<?xml version="1.0" encoding="utf-8"?>
<DirectionalLayout
    xmlns:ohos="http://schemas.huawei.com/res/ohos"
    ohos:height="match_parent"
    ohos:width="match_parent"
    ohos:alignment="center"
    ohos:orientation="vertical"
    ohos:background_element="$graphic:color_light_gray_element">

    <Text
        ohos:width="match_content"
        ohos:height="match_content"
        ohos:text_size="20fp"
        ohos:text="778907484"
        ohos:background_element="$graphic:green_text_element"
        ohos:text_alignment="center"
        ohos:layout_alignment="horizontal_center"
        />
    <DirectionalLayout
        ohos:width="match_parent"
        ohos:height="match_content"
        ohos:alignment="horizontal_center"
```

```xml
        ohos:orientation="horizontal"
        ohos:top_margin="5vp"
        ohos:bottom_margin="5vp">
        <Button
            ohos:width="40vp"
            ohos:height="40vp"
            ohos:text_size="15fp"
            ohos:background_element="$graphic:green_circle_button_element"
            ohos:text="1"
            ohos:text_alignment="center"
            />
        <Button
            ohos:width="40vp"
            ohos:height="40vp"
            ohos:text_size="15fp"
            ohos:background_element="$graphic:green_circle_button_element"
            ohos:text="2"
            ohos:left_margin="5vp"
            ohos:right_margin="5vp"
            ohos:text_alignment="center"
            />
        <Button
            ohos:width="40vp"
            ohos:height="40vp"
            ohos:text_size="15fp"
            ohos:background_element="$graphic:green_circle_button_element"
            ohos:text="3"
            ohos:text_alignment="center"
            />
</DirectionalLayout>
<DirectionalLayout
        ohos:width="match_parent"
        ohos:height="match_content"
        ohos:alignment="horizontal_center"
        ohos:orientation="horizontal"
        ohos:bottom_margin="5vp">
        <Button
            ohos:width="40vp"
            ohos:height="40vp"
            ohos:text_size="15fp"
            ohos:background_element="$graphic:green_circle_button_element"
            ohos:text="4"
            ohos:text_alignment="center"
            />
        <Button
            ohos:width="40vp"
            ohos:height="40vp"
            ohos:text_size="15fp"
            ohos:left_margin="5vp"
            ohos:right_margin="5vp"
            ohos:background_element="$graphic:green_circle_button_element"
```

```xml
            ohos:text="5"
            ohos:text_alignment="center"
            />
        <Button
            ohos:width="40vp"
            ohos:height="40vp"
            ohos:text_size="15fp"
            ohos:background_element="$graphic:green_circle_button_element"
            ohos:text="6"
            ohos:text_alignment="center"
            />
    </DirectionalLayout>
    <DirectionalLayout
        ohos:width="match_parent"
        ohos:height="match_content"
        ohos:alignment="horizontal_center"
        ohos:orientation="horizontal"
        ohos:bottom_margin="5vp">
        <Button
            ohos:width="40vp"
            ohos:height="40vp"
            ohos:text_size="15fp"
            ohos:background_element="$graphic:green_circle_button_element"
            ohos:text="7"
            ohos:text_alignment="center"
            />
        <Button
            ohos:width="40vp"
            ohos:height="40vp"
            ohos:text_size="15fp"
            ohos:left_margin="5vp"
            ohos:right_margin="5vp"
            ohos:background_element="$graphic:green_circle_button_element"
            ohos:text="8"
            ohos:text_alignment="center"
            />
        <Button
            ohos:width="40vp"
            ohos:height="40vp"
            ohos:text_size="15fp"
            ohos:background_element="$graphic:green_circle_button_element"
            ohos:text="9"
            ohos:text_alignment="center"
            />
    </DirectionalLayout>
    <DirectionalLayout
        ohos:width="match_parent"
        ohos:height="match_content"
        ohos:alignment="horizontal_center"
        ohos:orientation="horizontal"
        ohos:bottom_margin="5vp">
```

```xml
        <Button
            ohos:width="40vp"
            ohos:height="40vp"
            ohos:text_size="15fp"
            ohos:background_element="$graphic:green_circle_button_element"
            ohos:text="*"
            ohos:text_alignment="center"
            />
        <Button
            ohos:width="40vp"
            ohos:height="40vp"
            ohos:text_size="15fp"
            ohos:left_margin="5vp"
            ohos:right_margin="5vp"
            ohos:background_element="$graphic:green_circle_button_element"
            ohos:text="0"
            ohos:text_alignment="center"
            />
        <Button
            ohos:width="40vp"
            ohos:height="40vp"
            ohos:text_size="15fp"
            ohos:background_element="$graphic:green_circle_button_element"
            ohos:text="#"
            ohos:text_alignment="center"
            />
    </DirectionalLayout>
    <Button
        ohos:width="match_content"
        ohos:height="match_content"
        ohos:text_size="15fp"
        ohos:text="CALL"
        ohos:background_element="$graphic:green_capsule_button_element"
        ohos:bottom_margin="5vp"
        ohos:text_alignment="center"
        ohos:layout_alignment="horizontal_center"
        ohos:left_padding="10vp"
        ohos:right_padding="10vp"
        ohos:top_padding="2vp"
        ohos:bottom_padding="2vp"
        />
</DirectionalLayout>
```

上述代码采用多个DirectionalLayout进行了嵌套组合。同时新增了如下样式代码。

1. color_light_gray_element.xml

新增的color_light_gray_element.xml代码如下：

```xml
<?xml version="1.0" encoding="utf-8"?>
<shape xmlns:ohos="http://schemas.huawei.com/res/ohos"
    ohos:shape="rectangle">
    <solid
```

```xml
        ohos:color="#EDEDED"/>
</shape>
```

2. green_text_element.xml

新增的green_text_element.xml代码如下:

```xml
<?xml version="1.0" encoding="utf-8"?>
<shape xmlns:ohos="http://schemas.huawei.com/res/ohos"
    ohos:shape="rectangle">
    <corners
        ohos:radius="20"/>
    <stroke
        ohos:width="2"
        ohos:color="#006E00"/>
    <solid
        ohos:color="#EDEDED"/>
</shape>
```

3. green_circle_button_element.xml

新增的green_circle_button_element.xml代码如下:

```xml
<?xml version="1.0" encoding="utf-8"?>
<shape xmlns:ohos="http://schemas.huawei.com/res/ohos"
    ohos:shape="oval">
    <stroke
        ohos:width="5"
        ohos:color="#006E00"/>
    <solid
        ohos:color="#EDEDED"/>
</shape>
```

4. green_capsule_button_element.xml

新增的green_capsule_button_element.xml代码如下:

```xml
<?xml version="1.0" encoding="utf-8"?>
<shape xmlns:ohos="http://schemas.huawei.com/res/ohos"
    ohos:shape="rectangle">
    <corners
        ohos:radius="100"/>
    <solid
        ohos:color="#006E00"/>
</shape>
```

上述代码的实现效果如图9-30所示。

第 9 章 用 Java 开发 UI | 247

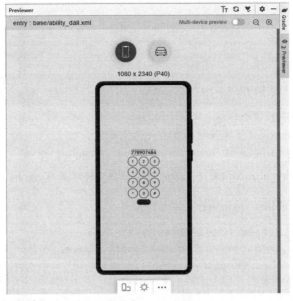

图 9-30　最终效果

9.9　实战：常用交互类组件——TextField

TextField在UI界面设计中提供了一种文本输入框。我们创建一个名为TextField的应用来作为演示。

9.9.1　创建 TextField

以下是创建TextField应用时产生的ability_main.xml文件。在该文件中增加了创建TextField组件的描述内容：

```
<?xml version="1.0" encoding="utf-8"?>
<DirectionalLayout
    xmlns:ohos="http://schemas.huawei.com/res/ohos"
    ohos:height="match_parent"
    ohos:width="match_parent"
    ohos:alignment="center"
    ohos:orientation="vertical">

    <TextField
        ohos:id="$+id:textfiled"
        ohos:height="40vp"
        ohos:width="260vp"
        ohos:left_padding="40vp"
        ohos:hint="Enter your name"
        ohos:text_alignment="vertical_center"
```

```
            ohos:background_element="$graphic:background_ability_main"
            />
</DirectionalLayout>
```

上述文件中定义了TextField组件。其中：

- ohos:hint用于设置提示文字。
- ohos:text_alignment设置了文字对齐方式是垂直居中。
- ohos:background_element可以用来配置TextField的背景。上述配置引用了background_ability_main.xml文件里面的内容，该文件放置在graphic目录下。

新增的background_ability_main.xml文件内容如下：

```
<?xml version="1.0" encoding="UTF-8" ?>
<shape xmlns:ohos="http://schemas.huawei.com/res/ohos"
       ohos:shape="rectangle">
    <corners
        ohos:radius="40"/>
    <solid
        ohos:color="#FFFF00"/>
</shape>
```

最终效果如图9-31所示。

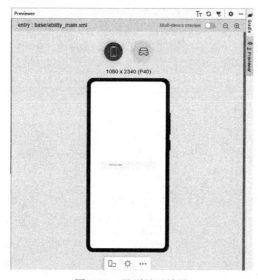

图9-31　界面显示效果

9.9.2　设置多行显示

TextField可以设置为多行显示。

为了演示多行显示的设置过程，通过DevEco Studio创建了名为MultipleLinesAbility的Page，同时会自动创建4个文件，分别是MultipleLinesAbility、MultipleLinesAbilitySlice、ability_multiple_lines.xml和background_ability_multiple_lines.xml。

修改ability_multiple_lines.xml内容如下：

```xml
<?xml version="1.0" encoding="utf-8"?>
<DirectionalLayout
    xmlns:ohos="http://schemas.huawei.com/res/ohos"
    ohos:height="match_parent"
    ohos:width="match_parent"
    ohos:alignment="center"
    ohos:orientation="vertical">

    <TextField
        ohos:id="$+id:textfiled_multiple_lines"
        ohos:height="40vp"
        ohos:width="200vp"
        ohos:left_padding="40vp"
        ohos:hint="Enter your name, your age and your country!"
        ohos:text_alignment="vertical_center"
        ohos:multiple_lines="true"
        ohos:background_element="$graphic:background_ability_multiple_lines"
        />

</DirectionalLayout>
```

上述代码中，ohos:multiple_lines即为是否启用多行显示的开关。

同时修改background_ability_multiple_lines.xml的内容如下：

```xml
<?xml version="1.0" encoding="UTF-8" ?>
<shape xmlns:ohos="http://schemas.huawei.com/res/ohos"
    ohos:shape="rectangle">
    <corners
        ohos:radius="40"/>
    <solid
        ohos:color="#FFFF00"/>
</shape>
```

图9-32显示了TextField多行显示的效果。

9.9.3 场景示例

接下来是一个场景示例，是一个常见的登录界面。

为了演示该示例，通过DevEco Studio创建了名为LoginAbility的Page，同时会自动创建4个文件。这4个文件分别是LoginAbility、LoginAbilitySlice、ability_login.xml和background_ability_login.xml。

修改ability_login.xml内容如下：

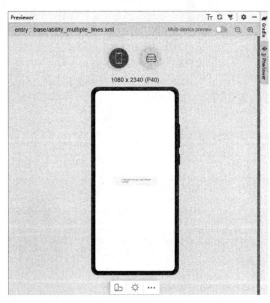

图9-32　多行显示的效果

```xml
<?xml version="1.0" encoding="utf-8"?>
<DirectionalLayout
    xmlns:ohos="http://schemas.huawei.com/res/ohos"
    ohos:height="match_parent"
    ohos:width="match_parent"
    ohos:alignment="center"
    ohos:orientation="vertical">

    <TextField
        ohos:id="$+id:textfield_name"
        ohos:height="match_content"
        ohos:width="260vp"
        ohos:background_element="$graphic:background_ability_login"
        ohos:bottom_padding="8vp"
        ohos:hint="Enter phone number"
        ohos:layout_alignment="center"
        ohos:min_height="44vp"
        ohos:multiple_lines="false"
        ohos:text_alignment="vertical_center"
        ohos:text_size="18fp"
        ohos:top_margin="10vp"
        ohos:top_padding="8vp"/>

    <TextField
        ohos:id="$+id:textfield_password"
        ohos:height="match_content"
        ohos:width="260vp"
        ohos:background_element="$graphic:background_ability_login"
        ohos:bottom_padding="8vp"
        ohos:hint="Enter password"
        ohos:layout_alignment="center"
        ohos:min_height="44vp"
        ohos:multiple_lines="false"
        ohos:text_alignment="vertical_center"
        ohos:text_size="18fp"
        ohos:top_margin="30vp"
        ohos:top_padding="8vp"/>

    <Button
        ohos:id="$+id:ensure_button"
        ohos:height="35vp"
        ohos:width="120vp"
        ohos:background_element="$graphic:background_ability_login"
        ohos:layout_alignment="horizontal_center"
        ohos:text="Log in"
        ohos:text_size="20fp"
        ohos:top_margin="40vp"/>

</DirectionalLayout>
```

同时修改background_ability_login.xml，代码如下：

```xml
<?xml version="1.0" encoding="UTF-8" ?>
<shape xmlns:ohos="http://schemas.huawei.com/res/ohos"
    ohos:shape="rectangle">
    <corners
        ohos:radius="40"/>
    <solid
        ohos:color="#FFFF00"/>
    <stroke
        ohos:color="black"
        ohos:width="6"
    />
</shape>
```

上述代码的实现效果如图9-33所示。

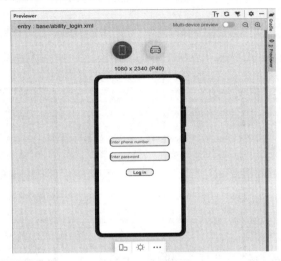

图9-33 最终效果

9.10 实战：常用交互类组件——Checkbox

Checkbox在UI界面设计中提供了一种实现选中和取消选中的功能。创建一个名为Checkbox的应用来作为演示。

9.10.1 创建 Checkbox

以下是创建Checkbox应用时产生的ability_main.xml文件。在该文件中增加了创建Checkbox组件的描述内容：

```xml
<?xml version="1.0" encoding="utf-8"?>
<DirectionalLayout
    xmlns:ohos="http://schemas.huawei.com/res/ohos"
    ohos:height="match_parent"
    ohos:width="match_parent"
```

```
    ohos:alignment="center"
    ohos:orientation="vertical"
    ohos:background_element="#FFFCCCCC">

    <Checkbox
        ohos:id="$+id:checkbox"
        ohos:height="match_content"
        ohos:width="match_content"
        ohos:text="I am Checkbox"
        ohos:text_size="40vp"/>
</DirectionalLayout>
```

上述文件中定义了Checkbox组件。初始化时，Checkbox组件如图9-34所示。

点击Checkbox组件之后，显示出了选中状态，如图9-35所示。

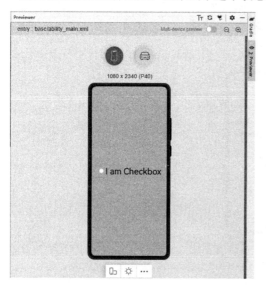
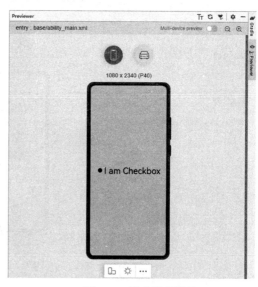

图9-34　初始化显示效果　　　　　　　　图9-35　选中显示效果

9.10.2　设置选中和取消选中时的颜色

Checkbox可以设置选中和取消选中时的颜色。

为了演示选中和取消选中时颜色的设置过程，通过DevEco Studio创建了名为OnOffAbility的Page，同时会自动创建4个文件，分别是OnOffAbility、OnOffAbilitySlice、ability_on_off.xml和background_ability_on_off.xml。

修改ability_on_off.xml内容如下：

```
<?xml version="1.0" encoding="utf-8"?>
<DirectionalLayout
    xmlns:ohos="http://schemas.huawei.com/res/ohos"
    ohos:height="match_parent"
    ohos:width="match_parent"
    ohos:alignment="center"
    ohos:orientation="vertical"
```

```
            ohos:background_element="#FFFCCCCC">

        <Checkbox
            ohos:id="$+id:checkbox_on_off"
            ohos:height="match_content"
            ohos:width="match_content"
            ohos:text="I am Checkbox"
            ohos:text_size="40vp"
            ohos:text_color_on="#00AAEE"
            ohos:text_color_off="#000000"/>
</DirectionalLayout>
```

上述代码中,ohos:text_color_on设置为选中时的颜色,而ohos:text_color_off设置为取消时的颜色。

图9-36显示了Checkbox选中时的颜色效果。

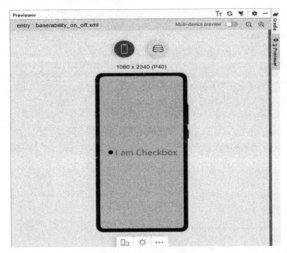

图 9-36　选中时的颜色效果

9.11　实战：常用交互类组件—— RadioButton/RadioContainer

RadioButton用于多选一的操作,需要搭配RadioContainer使用,实现单选效果。创建一个名为RadioButtonRadioContainer的应用来作为演示。

9.11.1　创建 RadioButton/RadioContainer

以下是创建RadioButtonRadioContainer应用时产生的ability_main.xml文件。在该文件中增加了创建RadioButton/RadioContainer组件的描述内容:

```
<?xml version="1.0" encoding="utf-8"?>
<DirectionalLayout
```

```xml
    xmlns:ohos="http://schemas.huawei.com/res/ohos"
    ohos:height="match_parent"
    ohos:width="match_parent"
    ohos:alignment="center"
    ohos:orientation="vertical"
    ohos:background_element="#FFFCCCCC">

    <Text
        ohos:height="match_content"
        ohos:width="match_content"
        ohos:text="最喜欢老卫哪部作品？"
        ohos:text_size="20fp"
        ohos:layout_alignment="left"
        ohos:multiple_lines="true"/>

    <RadioContainer
        ohos:id="$+id:radio_container"
        ohos:height="match_content"
        ohos:width="match_content">
        <RadioButton
            ohos:id="$+id:radio_button_1"
            ohos:height="30vp"
            ohos:width="match_content"
            ohos:text="1.《Spring Boot 企业级应用开发实战》"
            ohos:text_size="14fp"/>
        <RadioButton
            ohos:id="$+id:radio_button_2"
            ohos:height="30vp"
            ohos:width="match_content"
            ohos:text="2.《Spring Cloud 微服务架构开发实战》"
            ohos:text_size="14fp"/>
        <RadioButton
            ohos:id="$+id:radio_button_3"
            ohos:height="30vp"
            ohos:width="match_content"
            ohos:text="3.《Spring 5 开发大全》"
            ohos:text_size="14fp"/>
        <RadioButton
            ohos:id="$+id:radio_button_4"
            ohos:height="30vp"
            ohos:width="match_content"
            ohos:text="4.《Cloud Native 分布式架构原理与实践》"
            ohos:text_size="14fp"/>
        <RadioButton
            ohos:id="$+id:radio_button_5"
            ohos:height="30vp"
            ohos:width="match_content"
            ohos:text="5.《大型互联网应用轻量级架构实战》"
            ohos:text_size="14fp"/>
        <RadioButton
            ohos:id="$+id:radio_button_6"
```

```
            ohos:height="30vp"
            ohos:width="match_content"
            ohos:text="6.《Node.js企业级应用开发实战》"
            ohos:text_size="14fp"/>
        <RadioButton
            ohos:id="$+id:radio_button_7"
            ohos:height="30vp"
            ohos:width="match_content"
            ohos:text="7.《Netty原理解析与开发实战》"
            ohos:text_size="14fp"/>
    </RadioContainer>
</DirectionalLayout>
```

上述文件中定义了RadioButton/RadioContainer组件。初始化时，RadioButton/RadioContainer组件如图9-37所示。

点击RadioButton/RadioContainer组件之后，显示出了选中状态，如图9-38所示。

图9-37 初始化显示效果

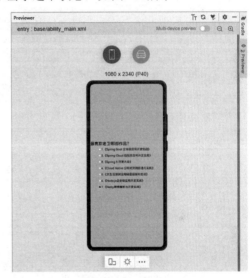

图9-38 选中显示效果

9.11.2 设置显示单选结果

可以设置响应RadioContainer状态改变的事件显示单选结果。

为了演示选中和取消选中时的颜色设置过程，通过DevEco Studio创建了名为MarkChangedAbility的Page，同时会自动创建4个文件，分别是MarkChangedAbility、MarkChangedAbilitySlice、ability_mark_changed.xml和background_ability_mark_changed.xml。

修改ability_mark_changed.xml内容如下：

```xml
<?xml version="1.0" encoding="utf-8"?>
<DirectionalLayout
    xmlns:ohos="http://schemas.huawei.com/res/ohos"
    ohos:height="match_parent"
    ohos:width="match_parent"
    ohos:alignment="center"
```

```xml
    ohos:orientation="vertical"
    ohos:background_element="#FFFCCCCC">

    <DirectionalLayout
        ohos:width="match_parent"
        ohos:height="match_content"
        ohos:orientation="horizontal">
        <Text
            ohos:id="$+id:text_question"
            ohos:height="match_content"
            ohos:width="match_content"
            ohos:text="最喜欢老卫哪部作品？答案是："
            ohos:text_size="20fp"
            ohos:layout_alignment="left"
            ohos:multiple_lines="true"/>
        <Text
            ohos:id="$+id:text_answer"
            ohos:height="match_content"
            ohos:width="match_content"
            ohos:text=""
            ohos:text_size="20fp"
            ohos:layout_alignment="left"
            ohos:multiple_lines="true"/>
    </DirectionalLayout>

    <RadioContainer
        ohos:id="$+id:radio_container"
        ohos:height="match_content"
        ohos:width="match_content">
        <RadioButton
            ohos:id="$+id:radio_button_1"
            ohos:height="30vp"
            ohos:width="match_content"
            ohos:text="1.《Spring Boot 企业级应用开发实战》"
            ohos:text_size="14fp"/>
        <RadioButton
            ohos:id="$+id:radio_button_2"
            ohos:height="30vp"
            ohos:width="match_content"
            ohos:text="2.《Spring Cloud 微服务架构开发实战》"
            ohos:text_size="14fp"/>
        <RadioButton
            ohos:id="$+id:radio_button_3"
            ohos:height="30vp"
            ohos:width="match_content"
            ohos:text="3.《Spring 5 开发大全》"
            ohos:text_size="14fp"/>
        <RadioButton
            ohos:id="$+id:radio_button_4"
            ohos:height="30vp"
            ohos:width="match_content"
```

```xml
            ohos:text="4.《Cloud Native 分布式架构原理与实践》"
            ohos:text_size="14fp"/>
        <RadioButton
            ohos:id="$+id:radio_button_5"
            ohos:height="30vp"
            ohos:width="match_content"
            ohos:text="5.《大型互联网应用轻量级架构实战》"
            ohos:text_size="14fp"/>
        <RadioButton
            ohos:id="$+id:radio_button_6"
            ohos:height="30vp"
            ohos:width="match_content"
            ohos:text="6.《Node.js企业级应用开发实战》"
            ohos:text_size="14fp"/>
        <RadioButton
            ohos:id="$+id:radio_button_7"
            ohos:height="30vp"
            ohos:width="match_content"
            ohos:text="7.《Netty原理解析与开发实战》"
            ohos:text_size="14fp"/>
    </RadioContainer>

</DirectionalLayout>
```

图9-39显示了RadioButton/RadioContainer初始化时的效果。

图 9-39　初始化时的效果

在RadioContainer上设置响应状态改变的事件。修改MarkChangedAbilitySlice代码如下：

```
package com.waylau.hmos.radiobuttonradiocontainer.slice;

import com.waylau.hmos.radiobuttonradiocontainer.ResourceTable;
import ohos.aafwk.ability.AbilitySlice;
```

```java
    import ohos.aafwk.content.Intent;
    import ohos.agp.components.RadioContainer;
    import ohos.agp.components.Text;

public class MarkChangedAbilitySlice extends AbilitySlice {
    @Override
    public void onStart(Intent intent) {
        super.onStart(intent);
        super.setUIContent(ResourceTable.Layout_ability_mark_changed);

        // 获取Text
        Text answer = (Text) findComponentById(ResourceTable.Id_text_answer);

        // 获取RadioContainer
        RadioContainer radioContainer =
            (RadioContainer)
findComponentById(ResourceTable.Id_radio_container);

        // 设置状态监听
        radioContainer.setMarkChangedListener((radioContainer1, index) -> {
            answer.setText((++index) + "");
        });
    }

    @Override
    public void onActive() {
        super.onActive();
    }

    @Override
    public void onForeground(Intent intent) {
        super.onForeground(intent);
    }
}
```

上述代码中，在RadioContainer被选中时，会获取选取到的RadioButton的索引。由于索引是从0开始的，因此需要增加1。

图9-40显示了RadioButton/RadioContainer被选中时的效果。

第 9 章 用 Java 开发 UI | 259

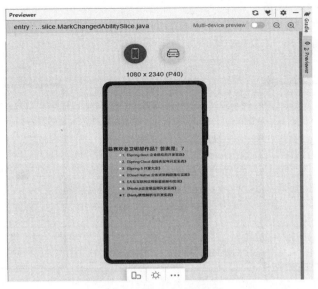

图 9-40 被选中时的效果

9.12 实战：常用交互类组件——Switch

Switch是切换单个设置开/关两种状态的组件。创建一个名为Switch的应用来作为演示。

9.12.1 创建 Switch

以下是创建Switch应用时产生的ability_main.xml文件。在该文件中增加了创建Switch组件的描述内容：

```xml
<?xml version="1.0" encoding="utf-8"?>
<DirectionalLayout
    xmlns:ohos="http://schemas.huawei.com/res/ohos"
    ohos:height="match_parent"
    ohos:width="match_parent"
    ohos:alignment="center"
    ohos:orientation="vertical">

    <Switch
        ohos:id="$+id:btn_switch"
        ohos:height="30vp"
        ohos:width="60vp"/>

</DirectionalLayout>
```

上述文件中定义了Switch组件。初始化时，Switch组件处于关闭状态，如图9-41所示。
点击Switch组件之后，Switch组件处于开启状态，如图9-42所示。

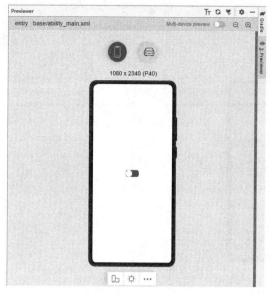
图 9-41　关闭状态

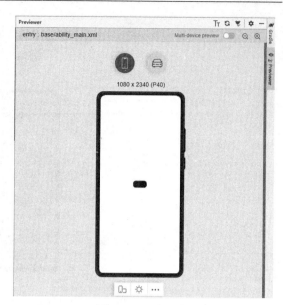
图 9-42　开启状态

9.12.2　设置文本

可以设置Switch在开启和关闭时的文本。

为了演示选中和取消选中时颜色的设置过程，通过DevEco Studio创建了名为TextStateAbility的Page，同时会自动创建4个文件，分别是TextStateAbility、TextStateAbilitySlice、ability_text_state.xml和background_ability_text_state.xml。

修改ability_text_state.xml内容如下：

```xml
<?xml version="1.0" encoding="utf-8"?>
<DirectionalLayout
    xmlns:ohos="http://schemas.huawei.com/res/ohos"
    ohos:height="match_parent"
    ohos:width="match_parent"
    ohos:alignment="center"
    ohos:orientation="vertical">

    <Switch
        ohos:id="$+id:btn_switch"
        ohos:height="30vp"
        ohos:width="60vp"
        ohos:text_state_off="OFF"
        ohos:text_state_on="ON"
        />
</DirectionalLayout>
```

图9-43显示了Switch设置文本之后的效果。

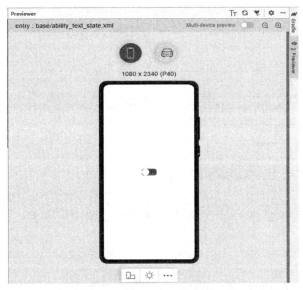

图 9-43　设置文本之后的效果

9.13　实战：常用交互类组件——ScrollView

ScrollView是一种带滚动功能的组件，它采用滑动的方式在有限的区域内显示更多的内容。这在显示面积非常受限的移动智能终端上非常实用。

我们创建一个名为ScrollView的应用来作为演示。

9.13.1　创建 ScrollView

以下是创建ScrollView应用时产生的ability_main.xml文件。在该文件中增加了创建ScrollView组件的描述内容：

```xml
<?xml version="1.0" encoding="utf-8"?>
<ScrollView
    xmlns:ohos="http://schemas.huawei.com/res/ohos"
    ohos:height="match_parent"
    ohos:width="match_parent"
    ohos:alignment="center"
    ohos:orientation="vertical">

    <Text
        ohos:id="$+id:text_lines"
        ohos:height="match_content"
        ohos:width="match_parent"
        ohos:italic="true"
        ohos:multiple_lines="true"
        ohos:text="$string:text_lines_content"
```

```
            ohos:text_color="#0000FF"
            ohos:text_font="serif"
            ohos:text_size="28fp"
            ohos:text_weight="700"/>
</ScrollView>
```

上述文件中定义了ScrollView组件。由于ScrollView本身就是继承自StackLayout，因此上述ability_main.xml文件无须再添加布局。

9.13.2 配置 Text 显示的内容

同时，在element的string.json文件中新增了text_lines_content的配置，以配置Text显示的内容。配置如下：

```
{
  "string": [
    {
      "name": "entry_MainAbility",
      "value": "entry_MainAbility"
    },
    {
      "name": "mainability_description",
      "value": "Java_Empty Ability"
    },
    {
      "name": "mainability_HelloWorld",
      "value": "Hello World"
    },
    {
      "name": "text_lines_content",
      "value": "华为HarmonyOS（鸿蒙系统）是一款"面向未来"、面向全场景（移动办公、运动健康、社交通信、媒体娱乐等）的分布式操作系统。借助HarmonyOS全场景分布式系统和设备生态，定义全新的硬件、交互和服务体验。本书采用最新的HarmonyOS 2版本作为基石，详细介绍了如何基于HarmonyOS来进行手机应用的开发，包括HarmonyOS架构、DevEco Studio、应用结构、Ability、任务调度、公共事件、通知、剪切板、Java UI、JS UI、多模输入、线程管理、视频、图像、相机、音频、媒体会话管理、媒体数据管理、安全管理、二维码、通用文字识别、蓝牙、WLAN、网络管理、电话服务、设备管理、数据管理等多个主题。本书辅以大量的实战案例，图文并茂，令读者易于理解掌握。同时，案例的选型偏重于解决实际问题，具有很强的前瞻性、应用性、趣味性。加入HarmonyOS生态，让我们一起构建万物互联的新时代！"
    }
  ]
}
```

初始化时，ScrollView组件的显示效果如图9-44所示。

上述界面无法显示完整的文本内容，因此往上拖动，以显示文本的后续内容，如图9-45所示。

图 9-44　无法显示完整的文本内容

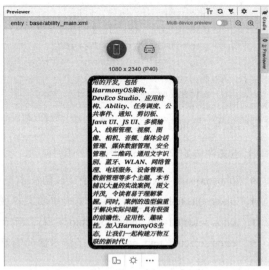

图 9-45　拖动显示

9.14　实战：常用交互类组件——Tab/TabList

TabList可以实现多个页签栏的切换，Tab为某个页签。子页签通常放在内容区上方，以展示不同的分类。页签名称应该简洁明了，清晰描述分类的内容。

创建一个名为TabList的应用来作为演示。

9.14.1　创建 TabList

以下是创建TabList应用时产生的ability_main.xml文件。在该文件中增加了创建TabList组件的描述内容：

```xml
<?xml version="1.0" encoding="utf-8"?>
<DirectionalLayout
    xmlns:ohos="http://schemas.huawei.com/res/ohos"
    ohos:height="match_parent"
    ohos:width="match_parent"
    ohos:orientation="vertical">

    <TabList
        ohos:id="$+id:tab_list"
        ohos:height="36vp"
        ohos:width="match_parent"
        ohos:layout_alignment="center"
        ohos:orientation="horizontal"
        ohos:normal_text_color="#999999"
        ohos:selected_text_color="#FF0000"
        ohos:selected_tab_indicator_color="#FF0000"
```

```xml
            ohos:selected_tab_indicator_height="2vp"
            ohos:tab_length="140vp"
            ohos:tab_margin="24vp"
            ohos:text_alignment="center"
            ohos:text_size="20fp"
            ohos:top_margin="10vp"/>

</DirectionalLayout>
```

上述文件中定义了TabList组件。其中：

- ohos:normal_text_color定义了默认的文本颜色。
- ohos:selected_text_color定义了选中的文本颜色。
- ohos:selected_tab_indicator_color定义了指示器（文本下面的横线）的颜色。
- ohos:selected_text_color定义了指示器的高度。

同时，修改MainAbilitySlice代码如下：

```java
package com.waylau.hmos.tablist.slice;

import com.waylau.hmos.tablist.ResourceTable;
import ohos.aafwk.ability.AbilitySlice;
import ohos.aafwk.content.Intent;
import ohos.agp.components.TabList;

public class MainAbilitySlice extends AbilitySlice {
    @Override
    public void onStart(Intent intent) {
        super.onStart(intent);
        super.setUIContent(ResourceTable.Layout_ability_main);

        // 获取TabList
        TabList tabList = (TabList) findComponentById(ResourceTable.Id_tab_list);

        // 在TabList中添加Tab
        TabList.Tab tab1 = tabList.new Tab(getContext());
        tab1.setText("Tab1");
        tabList.addTab(tab1);

        TabList.Tab tab2 = tabList.new Tab(getContext());
        tab2.setText("Tab2");
        tabList.addTab(tab2);

        TabList.Tab tab3 = tabList.new Tab(getContext());
        tab3.setText("Tab3");
        tabList.addTab(tab3);

        TabList.Tab tab4 = tabList.new Tab(getContext());
        tab4.setText("Tab4");
        tabList.addTab(tab4);
```

```
        // 设置FixedMode
        tabList.setFixedMode(true);

        // 初始化选中的Tab
        tabList.selectTab(tab1);
    }

    @Override
    public void onActive() {
        super.onActive();
    }

    @Override
    public void onForeground(Intent intent) {
        super.onForeground(intent);
    }
}
```

上述代码中：

- 获取了TabList。
- 在TabList中添加了4个Tab。
- 设置FixedMode为true。默认为false，该模式下，TabList的总宽度是各Tab宽度的总和，若固定了TabList的宽度，当超出可视区域时，则可通过滑动TabList来显示。如果设置为true，TabList的总宽度将与可视区域相同，各个Tab的宽度也会根据TabList的宽度而平均分配，该模式适用于Tab较少的情况。
- 初始化了选中的Tab为tab1。

初始化时，TabList组件的显示效果如图9-46所示。

图9-46　TabList组件的显示效果

9.14.2 响应焦点变化

可以在TabList中设置响应焦点变化，代码如下：

```java
package com.waylau.hmos.tablist.slice;

import com.waylau.hmos.tablist.ResourceTable;
import ohos.aafwk.ability.AbilitySlice;
import ohos.aafwk.content.Intent;
import ohos.agp.components.TabList;
import ohos.hiviewdfx.HiLog;
import ohos.hiviewdfx.HiLogLabel;

public class MainAbilitySlice extends AbilitySlice {
    private static final String TAG = MainAbilitySlice.class.getSimpleName();
    private static final HiLogLabel LABEL_LOG =
            new HiLogLabel(HiLog.LOG_APP, 0x00001, TAG);

    @Override
    public void onStart(Intent intent) {
        super.onStart(intent);
        super.setUIContent(ResourceTable.Layout_ability_main);

        // 获取TabList
        TabList tabList = (TabList) findComponentById(ResourceTable.Id_tab_list);

        // 在TabList中添加Tab
        TabList.Tab tab1 = tabList.new Tab(getContext());
        tab1.setText("Tab1");
        tabList.addTab(tab1);

        TabList.Tab tab2 = tabList.new Tab(getContext());
        tab2.setText("Tab2");
        tabList.addTab(tab2);

        TabList.Tab tab3 = tabList.new Tab(getContext());
        tab3.setText("Tab3");
        tabList.addTab(tab3);

        TabList.Tab tab4 = tabList.new Tab(getContext());
        tab4.setText("Tab4");
        tabList.addTab(tab4);

        // 设置FixedMode
        tabList.setFixedMode(true);

        // 初始化选中的Tab
        tabList.selectTab(tab1);
```

```
            // 设置响应焦点变化
            tabList.addTabSelectedListener(new TabList.TabSelectedListener() {
                @Override
                public void onSelected(TabList.Tab tab) {
                    // 当某个Tab从未选中状态变为选中状态时的回调
                    HiLog.info(LABEL_LOG, "%{public}s, onSelected",
tab.getText());
                }

                @Override
                public void onUnselected(TabList.Tab tab) {
                    // 当某个Tab从选中状态变为未选中状态时的回调
                    HiLog.info(LABEL_LOG, "%{public}s, onUnselected",
tab.getText());
                }

                @Override
                public void onReselected(TabList.Tab tab) {
                    // 当某个Tab已处于选中状态，再次被点击时的状态回调
                    HiLog.info(LABEL_LOG, "%{public}s, onReselected",
tab.getText());
                }
            });
        }

        ...
    }
```

上述代码在TabList上设置了焦点变化的事件。其中需要重写TabSelectedListener的方法：

- onSelected：当某个Tab从未选中状态变为选中状态时的回调。
- onUnselected：当某个Tab从选中状态变为未选中状态时的回调。
- onReselected：当某个Tab已处于选中状态，再次被点击时的状态回调。

运行应用，以观察回调的执行情况。初始化时，默认选中Tab1。

先点击了Tab2，控制台日志输出如下：

```
09-18 22:55:23.144 6980-6980/com.waylau.hmos.tablist I
00001/MainAbilitySlice: Tab1, onUnselected
09-18 22:55:23.144 6980-6980/com.waylau.hmos.tablist I
00001/MainAbilitySlice: Tab2, onSelected
```

再点击了Tab3，控制台日志输出如下：

```
09-18 22:55:32.321 6980-6980/com.waylau.hmos.tablist I
00001/MainAbilitySlice: Tab2, onUnselected
09-18 22:55:32.321 6980-6980/com.waylau.hmos.tablist I
00001/MainAbilitySlice: Tab3, onSelected
```

再次点击了Tab3，控制台日志输出如下：

```
09-18 22:55:33.800 6980-6980/com.waylau.hmos.tablist I
```

```
00001/MainAbilitySlice: Tab3, onReselected
```

9.15 实战：常用交互类组件——Picker

Picker提供了滑动选择器，允许用户从预定义范围中进行选择。Picker还有两个特例，分别是日期滑动选择器DatePicker和时间滑动选择器TimePicker。

创建一个名为Picker的应用来作为演示。

9.15.1 创建Picker

以下是创建Picker应用时产生的ability_main.xml文件。在该文件中增加了创建Picker组件的描述内容：

```xml
<?xml version="1.0" encoding="utf-8"?>
<DirectionalLayout
    xmlns:ohos="http://schemas.huawei.com/res/ohos"
    ohos:height="match_parent"
    ohos:width="match_parent"
    ohos:alignment="center"
    ohos:orientation="vertical">

    <Picker
        ohos:id="$+id:test_picker"
        ohos:height="match_content"
        ohos:width="300vp"
        ohos:background_element="#E1FFFF"
        ohos:layout_alignment="horizontal_center"
        ohos:normal_text_size="16fp"
        ohos:selected_text_size="16fp"/>

</DirectionalLayout>
```

上述文件中定义了Picker组件，可以选择的数字范围是0~9。Picker组件的效果显示如图9-47所示。

9.15.2 格式化Picker的显示

可以在Picker中设置格式化的显示，代码如下：

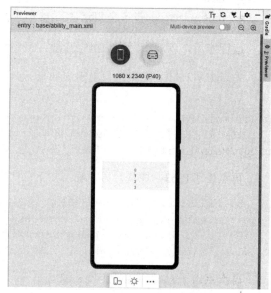

图9-47　Picker组件的显示效果

```java
package com.waylau.hmos.picke.slice;

import com.waylau.hmos.picke.ResourceTable;
import ohos.aafwk.ability.AbilitySlice;
import ohos.aafwk.content.Intent;
import ohos.agp.components.Picker;

public class MainAbilitySlice extends AbilitySlice {
    @Override
    public void onStart(Intent intent) {
        super.onStart(intent);
        super.setUIContent(ResourceTable.Layout_ability_main);

        // 获取Picker
        Picker picker = (Picker) findComponentById(ResourceTable.Id_test_picker);

        // 设置格式化
        picker.setFormatter(new Picker.Formatter() {
            @Override
            public String format(int i) {
                String value = "";
                switch (i) {
                    case 0:
                        value = "零";
                        break;
                    case 1:
                        value = "一";
                        break;
                    case 2:
                        value = "二";
                        break;
                    case 3:
                        value = "三";
                        break;
                    case 4:
                        value = "四";
                        break;
                    case 5:
                        value = "无";
                        break;
                    case 6:
                        value = "六";
                        break;
                    case 7:
                        value = "七";
                        break;
                    case 8:
                        value = "八";
                        break;
                    case 9:
```

```
                    value = "九";
                    break;
            }

            return value;
        }
    });

    ...
}
```

上述代码在Picker上设置了格式化，将数字转为了中文，Picker组件格式化后的显示效果如图9-48所示。

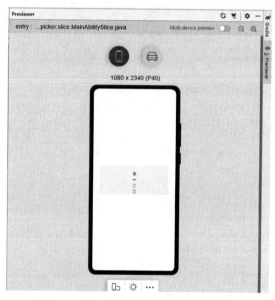

图 9-48   Picker 组件格式化后的显示效果

9.15.3　日期滑动选择器 DatePicker

为了演示日期滑动选择器DatePicker的显示效果，通过DevEco Studio创建了名为DatePickerAbility的Page，同时会自动创建4个文件，分别是DatePickerAbility、DatePickerAbilitySlice、ability_date_picker.xml和background_ability_date_picker.xml。

修改ability_date_picker.xml的内容如下：

```xml
<?xml version="1.0" encoding="utf-8"?>
<DirectionalLayout
    xmlns:ohos="http://schemas.huawei.com/res/ohos"
    ohos:height="match_parent"
    ohos:width="match_parent"
    ohos:alignment="center"
    ohos:orientation="vertical">
```

```
    <DatePicker
        ohos:id="$+id:date_pick"
        ohos:height="match_content"
        ohos:width="match_parent"
        ohos:normal_text_size="16fp"
        ohos:selected_text_size="16fp">
    </DatePicker>
</DirectionalLayout>
```

DatePicker的显示效果如图9-49所示。

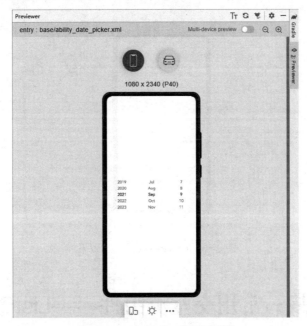

图 9-49　DatePicker 的显示效果

9.15.4　时间滑动选择器 TimePicker

为了演示日期滑动选择器 TimePicker 的显示效果，通过 DevEco Studio 创建了名为 TimePickerAbility 的 Page，同时会自动创建 4 个文件，分别是 TimePickerAbility、TimePickerAbilitySlice、ability_time_picker.xml 和 background_ability_time_picker.xml。

修改 ability_time_picker.xml 的内容如下：

```
<?xml version="1.0" encoding="utf-8"?>
<DirectionalLayout
    xmlns:ohos="http://schemas.huawei.com/res/ohos"
    ohos:height="match_parent"
    ohos:width="match_parent"
    ohos:alignment="center"
    ohos:orientation="vertical">

    <TimePicker
```

```
        ohos:id="$+id:time_picker"
        ohos:height="match_content"
        ohos:width="match_parent"
        ohos:normal_text_size="16fp"
        ohos:selected_text_size="16fp"/>

</DirectionalLayout>
```

TimePicker的显示效果如图9-50所示。

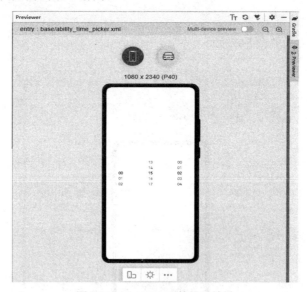

图 9-50　TimePicker 的显示效果

9.16　实战：常用交互类组件——ListContainer

ListContainer是用来呈现连续的、多行数据的组件，包含一系列相同类型的列表项。

创建一个名为ListContainer的应用来作为演示。

9.16.1　创建 ListContainer

以下是创建ListContainer应用时产生的ability_main.xml文件。在该文件中增加创建ListContainer组件的描述内容：

```
<?xml version="1.0" encoding="utf-8"?>
<DirectionalLayout
    xmlns:ohos="http://schemas.huawei.com/res/ohos"
    ohos:height="match_parent"
    ohos:width="match_parent"
    ohos:alignment="center"
    ohos:orientation="vertical">
```

```xml
<ListContainer
    ohos:id="$+id:list_container"
    ohos:height="200vp"
    ohos:width="300vp"
    ohos:layout_alignment="horizontal_center"/>
</DirectionalLayout>
```

上述文件中定义了ListContainer组件。

9.16.2　创建 ListContainer 子布局

在layout目录下新建XML文件（本例为my_item.xml）作为ListContainer的子布局，内容如下：

```xml
<?xml version="1.0" encoding="utf-8"?>
<DirectionalLayout
    xmlns:ohos="http://schemas.huawei.com/res/ohos"
    ohos:height="match_parent"
    ohos:width="match_parent"
    ohos:orientation="vertical">

    <Text
        ohos:id="$+id:item_index"
        ohos:height="match_content"
        ohos:width="match_content"
        ohos:text="Item0"
        ohos:text_size="20fp"
        ohos:layout_alignment="center"/>
</DirectionalLayout>
```

9.16.3　创建 ListContainer 数据包装类

创建MyItem.java，作为ListContainer的数据包装类，代码如下：

```java
package com.waylau.hmos.listcontainer;

public class MyItem {
    private String name;

    public MyItem(String name) {
        this.name = name;
    }

    public String getName() {
        return name;
    }

    public void setName(String name) {
        this.name = name;
    }
}
```

9.16.4 创建 ListContainer 数据提供者

ListContainer每一行可以为不同的数据，因此需要适配不同的数据结构，使其都能添加到ListContainer上。创建MyItemProvider.java，继承自RecycleItemProvider，代码如下：

```java
package com.waylau.hmos.listcontainer;

import com.waylau.hmos.listcontainer.slice.MainAbilitySlice;
import ohos.aafwk.ability.AbilitySlice;
import ohos.agp.components.*;

import java.util.List;

public class MyItemProvider extends RecycleItemProvider {
    private List<MyItem> list;
    private AbilitySlice slice;

    public MyItemProvider(List<MyItem> list, MainAbilitySlice slice) {
        this.list = list;
        this.slice = slice;
    }

    @Override
    public int getCount() {
        return list.size();
    }

    @Override
    public Object getItem(int position) {
        return list.get(position);
    }

    @Override
    public long getItemId(int position) {
        return position;
    }

    @Override
    public Component getComponent(int position,
        Component convertComponent, ComponentContainer componentContainer) {
        Component cpt = convertComponent;
        if (cpt == null) {
            cpt = LayoutScatter.getInstance(slice)
                .parse(ResourceTable.Layout_my_item, null, false);
        }
        MyItem sampleItem = list.get(position);
        Text text = (Text) cpt.findComponentById(ResourceTable.Id_item_index);
        text.setText(sampleItem.getName());
```

```
        return cpt;
    }
}
```

9.16.5 修改 MainAbilitySlice

修改MainAbilitySlice，在Java代码中添加ListContainer的数据，并适配其数据结构。代码修改如下：

```
package com.waylau.hmos.listcontainer.slice;

import com.waylau.hmos.listcontainer.MyItem;
import com.waylau.hmos.listcontainer.MyItemProvider;
import com.waylau.hmos.listcontainer.ResourceTable;
import ohos.aafwk.ability.AbilitySlice;
import ohos.aafwk.content.Intent;
import ohos.agp.components.ListContainer;

import java.util.ArrayList;
import java.util.List;

public class MainAbilitySlice extends AbilitySlice {
    @Override
    public void onStart(Intent intent) {
        super.onStart(intent);
        super.setUIContent(ResourceTable.Layout_ability_main);

        // 获取ListContainer
        ListContainer listContainer = (ListContainer) findComponentById(ResourceTable.Id_list_container);

        // 提供ListContainer的数据
        List<MyItem> list = getData();
        MyItemProvider sampleItemProvider = new MyItemProvider(list, this);
        listContainer.setItemProvider(sampleItemProvider);
    }

    private ArrayList<MyItem> getData() {
        ArrayList<MyItem> list = new ArrayList<>();
        for (int i = 0; i <= 8; i++) {
            list.add(new MyItem("Item" + i));
        }
        return list;
    }
    ...
}
```

最终显示效果如图9-51所示。

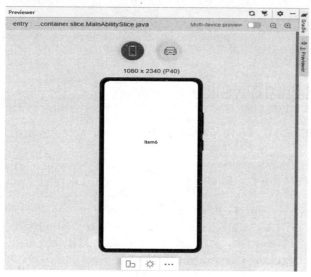

图 9-51　显示效果

9.17　实战：常用交互类组件——RoundProgressBar

RoundProgressBar继承自ProgressBar，拥有ProgressBar的属性，在设置同样的属性时用法和ProgressBar一致，用于显示环形进度。

创建一个名为RoundProgressBar的应用来作为演示。

9.17.1　创建 RoundProgressBar

以下是创建RoundProgressBar应用时产生的ability_main.xml文件。在该文件中增加创建RoundProgressBar的描述内容：

```xml
<?xml version="1.0" encoding="utf-8"?>
<DirectionalLayout
    xmlns:ohos="http://schemas.huawei.com/res/ohos"
    ohos:height="match_parent"
    ohos:width="match_parent"
    ohos:alignment="center"
    ohos:orientation="vertical">

    <RoundProgressBar
        ohos:id="$+id:round_progress_bar"
        ohos:height="200vp"
        ohos:width="200vp"
        ohos:progress_width="10vp"
        ohos:progress="20"
        ohos:progress_color="#47CC47"/>
```

```
</DirectionalLayout>
```

上述文件中定义了一个RoundProgressBar组件。ohos:progress_width可以用来配置RoundProgressBar的环的宽度，ohos:progress用来配置RoundProgressBar的进度，而ohos:progress_color用来配置环的颜色。

最终效果如图9-52所示。

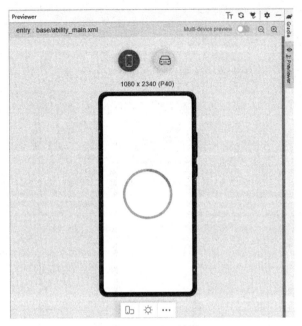

图 9-52　显示效果

9.17.2　设置开始和结束角度

可以设置环形进度条的开始和结束位置所在的角度。

为了演示设置开始和结束角度的过程，通过DevEco Studio创建了名为StartMaxAngleAbility的Page，同时会自动创建4个文件，分别是StartMaxAngleAbility、StartMaxAngleAbilitySlice、ability_start_max_angle.xml和background_ability_start_max_angle.xml。

修改ability_start_max_angle.xml内容如下：

```xml
<?xml version="1.0" encoding="utf-8"?>
<DirectionalLayout
    xmlns:ohos="http://schemas.huawei.com/res/ohos"
    ohos:height="match_parent"
    ohos:width="match_parent"
    ohos:alignment="center"
    ohos:orientation="vertical">

    <RoundProgressBar
        ohos:id="$+id:round_progress_bar_start_max_angle"
        ohos:height="200vp"
        ohos:width="200vp"
```

```
            ohos:progress_width="10vp"
            ohos:progress="20"
            ohos:progress_color="#47CC47"
            ohos:start_angle="45"
            ohos:max_angle="270"
            ohos:progress_hint_text="RoundProgressBar"
            ohos:progress_hint_text_color="#007DFF"
            />

</DirectionalLayout>
```

图9-53展示了预览器显示的设置开始和结束角度之后的效果。

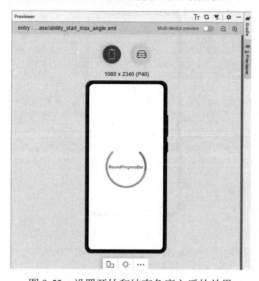

图 9-53　设置开始和结束角度之后的效果

上述代码中：

- ohos:start_angle设置了开始角度。
- ohos:max_angle设置了结束角度。
- ohos:progress_hint_text设置了提示文本内容。
- ohos:progress_hint_text_color设置了提示文本颜色。

9.18　实战：常用交互类组件——DirectionalLayout

常用的布局包括DirectionalLayout、StackLayout、DependentLayout、TableLayout等。

DirectionalLayout是Java UI中的一种非常重要的组件布局，在前面的章节中也频繁出现过。DirectionalLayout用于将一组组件按照水平或者垂直方向排布，能够方便地对齐布局内的组件。该布局和其他布局的组合可以实现更加丰富的布局方式。

创建一个名为DirectionalLayout的应用来作为演示。

9.18.1 创建 DirectionalLayout

以下是创建DirectionalLayout应用时产生的ability_main.xml文件。在该文件中增加了创建DirectionalLayout的描述内容：

```xml
<?xml version="1.0" encoding="utf-8"?>
<DirectionalLayout
    xmlns:ohos="http://schemas.huawei.com/res/ohos"
    ohos:height="match_parent"
    ohos:width="match_parent"
    ohos:alignment="center"
    ohos:orientation="vertical">

    <Button
        ohos:width="120vp"
        ohos:height="60vp"
        ohos:bottom_margin="3vp"
        ohos:left_margin="13vp"
        ohos:text_size="16vp"
        ohos:background_element="$graphic:background_ability_main"
        ohos:text="I am Button1"/>
    <Button
        ohos:width="120vp"
        ohos:height="60vp"
        ohos:bottom_margin="3vp"
        ohos:left_margin="13vp"
        ohos:text_size="16vp"
        ohos:background_element="$graphic:background_ability_main"
        ohos:text="I am Button2"/>
    <Button
        ohos:width="120vp"
        ohos:height="60vp"
        ohos:bottom_margin="3vp"
        ohos:left_margin="13vp"
        ohos:text_size="16vp"
        ohos:background_element="$graphic:background_ability_main"
        ohos:text="I am Button3"/>

</DirectionalLayout>
```

上述文件中定义了一个DirectionalLayout布局及3个Button。

DirectionalLayout布局中的ohos:orientation属性可以用来指定排列方向是水平（horizontal）还是垂直（vertical），默认为垂直排列。

Button的ohos:background_element定义在background_ability_main.xml文件中，内容如下：

```xml
<?xml version="1.0" encoding="UTF-8" ?>
<shape xmlns:ohos="http://schemas.huawei.com/res/ohos"
    ohos:shape="rectangle">
    <solid
```

```
        ohos:color="#00FFFD"/>
</shape>
```

最终效果如图9-54所示。

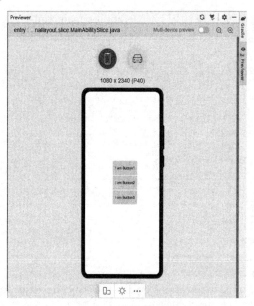

图 9-54　DirectionalLayout 的效果

9.18.2　设置水平排列

为了演示水平排列的设置过程，通过DevEco Studio创建了名为HorizontalAbility的Page，同时会自动创建4个文件，分别是HorizontalAbility、HorizontalAbilitySlice、ability_horizontal.xml和background_ability_horizontal.xml。

修改ability_horizontal.xml内容如下：

```
<?xml version="1.0" encoding="utf-8"?>
<DirectionalLayout
    xmlns:ohos="http://schemas.huawei.com/res/ohos"
    ohos:height="match_parent"
    ohos:width="match_parent"
    ohos:alignment="center"
    ohos:orientation="horizontal">

    <Button
        ohos:width="100vp"
        ohos:height="60vp"
        ohos:bottom_margin="3vp"
        ohos:left_margin="13vp"
        ohos:text_size="16vp"
        ohos:background_element="$graphic:background_ability_main"
        ohos:text="I am Button1"/>
    <Button
        ohos:width="100vp"
```

```
            ohos:height="60vp"
            ohos:bottom_margin="3vp"
            ohos:left_margin="13vp"
            ohos:text_size="16vp"
            ohos:background_element="$graphic:background_ability_main"
            ohos:text="I am Button2"/>
        <Button
            ohos:width="100vp"
            ohos:height="60vp"
            ohos:bottom_margin="3vp"
            ohos:left_margin="13vp"
            ohos:text_size="16vp"
            ohos:background_element="$graphic:background_ability_main"
            ohos:text="I am Button3"/>

</DirectionalLayout>
```

上述代码将布局的ohos:orientatio的属性改为了horizontal（水平）。

图9-55展示了预览器显示的DirectionalLayout水平排列的效果。

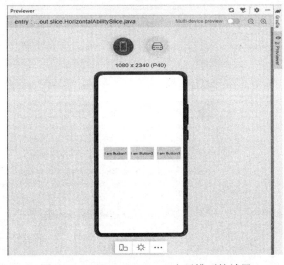

图 9-55 DirectionalLayout 水平排列的效果

9.18.3 设置权重

权重（weight）就是按比例来分配组件占用父组件的大小，在水平布局下计算公式为：

父布局可分配宽度=父布局宽度−所有子组件width之和

组件宽度=组件weight/所有组件weight之和*父布局可分配宽度

在实际使用的过程中，建议使用width=0来按比例分配父布局的宽度。

为了演示权重的设置过程，通过DevEco Studio创建了名为WeightAbility的Page，同时会自动创建4个文件，分别是WeightAbility、WeightAbilitySlice、ability_weight.xml和background_ability_weight.xml。

修改ability_weight.xml内容如下：

```xml
<?xml version="1.0" encoding="utf-8"?>
<DirectionalLayout
    xmlns:ohos="http://schemas.huawei.com/res/ohos"
    ohos:height="match_parent"
    ohos:width="match_parent"
    ohos:alignment="center"
    ohos:orientation="horizontal">

    <Button
        ohos:width="0vp"
        ohos:height="60vp"
        ohos:bottom_margin="3vp"
        ohos:left_margin="13vp"
        ohos:text_size="12vp"
        ohos:weight="1"
        ohos:background_element="$graphic:background_ability_main"
        ohos:text="I am Button1"/>
    <Button
        ohos:width="0vp"
        ohos:height="60vp"
        ohos:bottom_margin="3vp"
        ohos:left_margin="13vp"
        ohos:text_size="12vp"
        ohos:weight="1"
        ohos:background_element="$graphic:background_ability_main"
        ohos:text="I am Button2"/>
    <Button
        ohos:width="0vp"
        ohos:height="60vp"
        ohos:bottom_margin="3vp"
        ohos:left_margin="13vp"
        ohos:text_size="12vp"
        ohos:weight="2"
        ohos:background_element="$graphic:background_ability_main"
        ohos:text="I am Button3"/>

</DirectionalLayout>
```

上述代码按1∶1∶2的比例来分配父布局的宽度，效果如图9-56所示。

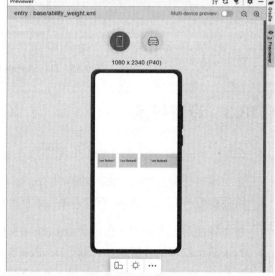

图9-56　DirectionalLayout的水平排列效果

9.19 实战：常用交互类组件——DependentLayout

DependentLayout是Java UI系统里的一种常见布局。与DirectionalLayout相比，DependentLayout拥有更多的排布方式，每个组件可以指定相对于其他同级元素的位置，或者指定相对于父组件的位置。

创建一个名为DependentLayout的应用来作为演示。

9.19.1 创建DependentLayout

以下是创建DependentLayout应用时产生的ability_main.xml文件。在该文件中增加创建DependentLayout的描述内容：

```xml
<?xml version="1.0" encoding="utf-8"?>
<DirectionalLayout
    xmlns:ohos="http://schemas.huawei.com/res/ohos"
    ohos:height="match_parent"
    ohos:width="match_parent"
    ohos:background_element="#ADEDED">
    <Text
        ohos:id="$+id:text1"
        ohos:width="match_content"
        ohos:height="match_content"
        ohos:left_margin="20vp"
        ohos:top_margin="20vp"
        ohos:bottom_margin="20vp"
        ohos:text="text1"
        ohos:text_size="20fp"
        ohos:background_element="#DDEDED"/>
    <Text
        ohos:id="$+id:text2"
        ohos:width="match_content"
        ohos:height="match_content"
        ohos:left_margin="20vp"
        ohos:top_margin="20vp"
        ohos:right_margin="20vp"
        ohos:bottom_margin="20vp"
        ohos:text="end_of_text1"
        ohos:text_size="20fp"
        ohos:background_element="#FDEDED"
        ohos:end_of="$id:text1"/>
</DirectionalLayout>
```

上述文件中定义了一个DependentLayout布局及两个Text。两个Text使用了"相对于同级组件"的布局方式，即text2处于同级组件的text1的结束侧。

最终效果如图9-57所示。

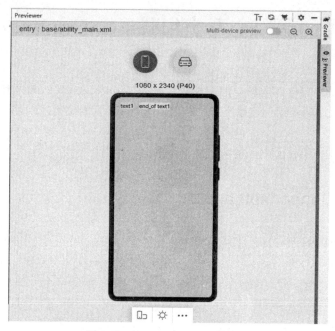

图 9-57　DependentLayout 的效果

9.19.2　相对于同级组件

上述示例使用的是"相对于同级组件"的布局方式。相对于同级组件的位置布局可配置项包括：

- above：处于同级组件的上侧。
- below：处于同级组件的下侧。
- start_of：处于同级组件的起始侧。
- end_of：处于同级组件的结束侧。
- left_of：处于同级组件的左侧。
- right_of：处于同级组件的右侧。

9.19.3　相对于父组件

相对于父组件的位置布局可配置项包括：

- align_parent_left：处于父组件的左侧。
- align_parent_right：处于父组件的右侧。
- align_parent_start：处于父组件的起始侧。
- align_parent_end：处于父组件的结束侧。
- align_parent_top：处于父组件的上侧。
- align_parent_bottom：处于父组件的下侧。
- center_in_parent：处于父组件的中间。

以上位置布局可以组合，形成处于左上角、左下角、右上角、右下角的布局。

9.19.4 场景示例

使用DependentLayout可以轻松实现内容丰富的布局。

为了演示该示例，通过DevEco Studio创建了名为FullAbility的Page，同时会自动创建4个文件，分别是FullAbility、FullAbilitySlice、ability_full.xml和background_ability_full.xml。

修改ability_full.xml内容如下：

```xml
<?xml version="1.0" encoding="utf-8"?>
<DependentLayout
    xmlns:ohos="http://schemas.huawei.com/res/ohos"
    ohos:width="match_parent"
    ohos:height="match_content"
    ohos:background_element="#ADEDED">
    <Text
        ohos:id="$+id:text1"
        ohos:width="match_parent"
        ohos:height="match_content"
        ohos:text_size="25fp"
        ohos:top_margin="15vp"
        ohos:left_margin="15vp"
        ohos:right_margin="15vp"
        ohos:background_element="#DDEDED"
        ohos:text="Title"
        ohos:text_weight="1000"
        ohos:text_alignment="horizontal_center"
        />
    <Text
        ohos:id="$+id:text2"
        ohos:width="match_content"
        ohos:height="680vp"
        ohos:text_size="10vp"
        ohos:background_element="#FDEDED"
        ohos:text="Catalog"
        ohos:top_margin="15vp"
        ohos:left_margin="15vp"
        ohos:right_margin="15vp"
        ohos:bottom_margin="15vp"
        ohos:align_parent_left="true"
        ohos:text_alignment="center"
        ohos:multiple_lines="true"
        ohos:below="$id:text1"
        ohos:text_font="serif"/>
    <Text
        ohos:id="$+id:text3"
        ohos:width="match_parent"
        ohos:height="680vp"
        ohos:text_size="25fp"
```

```xml
        ohos:background_element="#FDEDED"
        ohos:text="Content"
        ohos:top_margin="15vp"
        ohos:right_margin="15vp"
        ohos:bottom_margin="15vp"
        ohos:text_alignment="center"
        ohos:below="$id:text1"
        ohos:end_of="$id:text2"
        ohos:text_font="serif"/>
    <Button
        ohos:id="$+id:button1"
        ohos:width="70vp"
        ohos:height="match_content"
        ohos:text_size="15fp"
        ohos:background_element="#FDEDED"
        ohos:text="Previous"
        ohos:right_margin="15vp"
        ohos:bottom_margin="15vp"
        ohos:below="$id:text3"
        ohos:left_of="$id:button2"
        ohos:italic="false"
        ohos:text_weight="5"
        ohos:text_font="serif"/>
    <Button
        ohos:id="$+id:button2"
        ohos:width="70vp"
        ohos:height="match_content"
        ohos:text_size="15fp"
        ohos:background_element="#FDEDED"
        ohos:text="Next"
        ohos:right_margin="15vp"
        ohos:bottom_margin ="15vp"
        ohos:align_parent_end ="true"
        ohos:below="$id:text3"
        ohos:italic="false"
        ohos:text_weight="5"
        ohos:text_font="serif"/>
</DependentLayout>
```

上述代码的效果如图9-58所示。

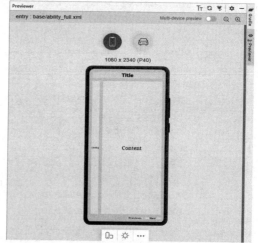

图9-58 场景示例效果

9.20 实战：常用交互类组件——StackLayout

StackLayout直接在屏幕上开辟出一块空白的区域，添加到这个布局中的视图都是以层叠的方式显示的，而它会把这些视图默认放到这块区域的左上角，第一个添加到布局中的视图显示在最底层,最后一个被放在最顶层,上一层的视图会覆盖下一层的视图。

创建一个名为StackLayout的应用来作为演示。

以下是创建StackLayout应用时产生的ability_main.xml文件。在该文件中增加创建StackLayout的描述内容:

```xml
<?xml version="1.0" encoding="utf-8"?>
<StackLayout
    xmlns:ohos="http://schemas.huawei.com/res/ohos"
    ohos:height="match_parent"
    ohos:width="match_parent">

    <Text
        ohos:text_alignment="bottom|horizontal_center"
        ohos:text_size="24fp"
        ohos:text="Layer 1"
        ohos:height="match_parent"
        ohos:width="match_parent"
        ohos:background_element="#3F56EA" />

    <Text
        ohos:text_alignment="bottom|horizontal_center"
        ohos:text_size="24fp"
        ohos:text="Layer 2"
        ohos:height="100vp"
        ohos:width="300vp"
        ohos:background_element="#00AAEE" />

    <Text
        ohos:text_alignment="center"
        ohos:text_size="24fp"
        ohos:text="Layer 3"
        ohos:height="80vp"
        ohos:width="80vp"
        ohos:background_element="#00BFC9" />

</StackLayout>
```

上述文件中定义了一个StackLayout布局及3个Text。StackLayout中组件的布局默认在区域的左上角,并且以后创建的组件都会在上层,即text1处于最下层,而text3处于最上层。

最终效果如图9-59所示。

图9-59 StackLayout 的效果

9.21 实战：常用交互类组件——TableLayout

顾名思义，TableLayout使用表格的方式划分子组件。创建一个名为TableLayout的应用来作为演示。

以下是创建TableLayout应用时产生的ability_main.xml文件。在该文件中增加创建TableLayout的描述内容：

```xml
<?xml version="1.0" encoding="utf-8"?>
<TableLayout
    xmlns:ohos="http://schemas.huawei.com/res/ohos"
    ohos:height="match_parent"
    ohos:width="match_parent"
    ohos:background_element="#87CEEB"
    ohos:layout_alignment="horizontal_center"
    ohos:padding="8vp"
    ohos:row_count="2"
    ohos:column_count="2">

    <Text
        ohos:height="60vp"
        ohos:width="60vp"
        ohos:background_element="$graphic:background_ability_main"
        ohos:margin="8vp"
        ohos:text="1"
        ohos:text_alignment="center"
        ohos:text_size="20fp"/>

    <Text
        ohos:height="60vp"
        ohos:width="60vp"
        ohos:background_element="$graphic:background_ability_main"
        ohos:margin="8vp"
        ohos:text="2"
        ohos:text_alignment="center"
        ohos:text_size="20fp"/>

    <Text
        ohos:height="60vp"
        ohos:width="60vp"
        ohos:background_element="$graphic:background_ability_main"
        ohos:margin="8vp"
        ohos:text="3"
        ohos:text_alignment="center"
        ohos:text_size="20fp"/>

    <Text
```

```
            ohos:height="60vp"
            ohos:width="60vp"
            ohos:background_element="$graphic:background_ability_main"
            ohos:margin="8vp"
            ohos:text="4"
            ohos:text_alignment="center"
            ohos:text_size="20fp"/>
</TableLayout>
```

上述文件中定义了一个TableLayout布局及4个Text。其中，ohos:row_count设置了行数，而ohos:column_count设置了列数。

Text引用的样式定义在background_ability_main.xml文件中，内容如下：

```
<?xml version="1.0" encoding="utf-8"?>
<shape xmlns:ohos="http://schemas.huawei.com/res/ohos"
    ohos:shape="rectangle">
    <corners
        ohos:radius="5vp"/>
    <stroke
        ohos:width="1vp"
        ohos:color="gray"/>
    <solid
        ohos:color="#00BFFF"/>
</shape>
```

最终效果如图9-60所示。

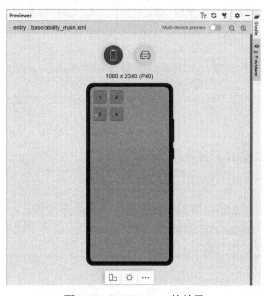

图9-60　TableLayout的效果

第 10 章

用 JS 开发 UI

在传统的前端开发中，JavaScript（简称为JS）绝对是主角。HarmonyOS提供了JS UI框架，以支持开发跨设备、高性能、声明式UI。

10.1 用 JS 开发 UI 概述

JavaScript是流行的前端开发语言，因此同样适合开发HarmonyOS应用的UI界面。

10.1.1 基础能力

HarmonyOS所提供的JS UI框架具备以下基础能力。

1. 声明式编程

JS UI框架采用类HTML和CSS声明式编程语言作为页面布局和页面样式的开发语言，页面业务逻辑则支持ECMAScript规范的JavaScript语言。JS UI框架提供的声明式编程可以让开发者避免编写UI状态切换的代码，视图配置信息更加直观。

2. 跨设备

开发框架架构上支持UI跨设备显示，运行时自动映射到不同的设备类型，开发者无感知，降低了开发者多设备的适配成本。

截至目前，JS UI框架支持包括手机（Phone）、平板（Tablet）、智慧屏（TV）、智能穿戴（Wearable）设备和轻量级智能穿戴（Lite Wearable）设备应用的开发。

3. 高性能

开发框架包含许多核心的控件，如列表、图片和各类容器组件等，针对声明式语法进行了渲染流程的优化。

10.1.2 整体架构

JS UI框架包括应用层（Application Layer）、前端框架层（Framework Layer）、引擎层（Engine Layer）和平台适配层（Porting Layer）。图10-1展示了JS UI框架的架构图。

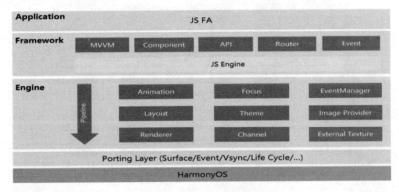

图 10-1　JS UI 框架的整体架构

1. Application

应用层表示开发者使用JS UI框架开发的FA应用，这里的FA应用特指JS FA应用。使用Java开发FA应用请参考Java UI框架。

2. Framework

前端框架层主要完成前端页面解析，以及提供MVVM（Model-View-ViewModel，模型-视图-视图模型）开发模式、页面路由机制和自定义组件等能力。

3. Engine

引擎层主要提供动画解析、文档对象模型（Document Object Model，DOM）、树构建、布局计算、渲染命令构建与绘制、事件管理等能力。

4. Porting Layer

适配层主要完成对平台层进行抽象，提供抽象接口，可以对接到系统平台，比如事件对接、渲染管线对接和系统生命周期对接等。

10.2　实战：创建 JS FA 应用

JS UI框架支持纯JavaScript、JavaScript和Java混合语言开发。JS FA指基于JavaScript或JavaScript和Java混合开发的FA。

接下来将介绍JS FA在HarmonyOS上运行时需要的基类AceAbility、加载JS FA主体的方法、JS FA开发目录。

10.2.1 创建应用

创建JS FA应用时，本例选择Empty Ability(JS)模板，如图10-2所示。

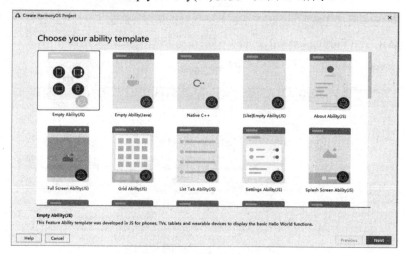

图10-2　选择"Tablet"设备类型

应用名称我们填的是"JsFa"，如图10-3所示。

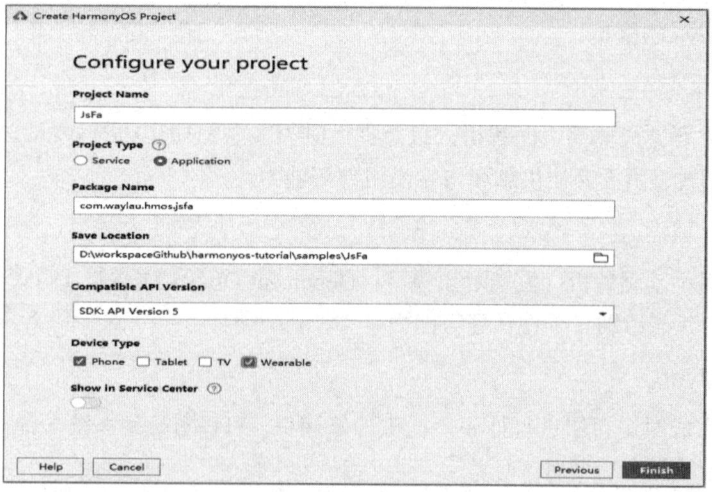

图10-3　指定应用名称

应用创建完成之后，形成的项目结构如图10-4所示。

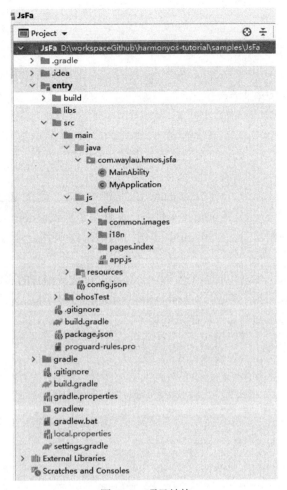

图 10-4　项目结构

10.2.2　AceAbility

AceAbility类是JS FA在HarmonyOS上运行环境的基类，继承自Ability。JsFa在初始化之后，默认生成的AceAbility代码示例如下：

```java
package com.waylau.hmos.jsfa;

import ohos.ace.ability.AceAbility;
import ohos.aafwk.content.Intent;

public class MainAbility extends AceAbility {
    @Override
    public void onStart(Intent intent) {
        super.onStart(intent);
    }

    @Override
    public void onStop() {
```

```
            super.onStop();
        }
    }
```

10.2.3 如何加载 JS FA

JS FA生命周期事件分为应用生命周期和页面生命周期，应用通过AceAbility类中的setInstanceName()接口设置该Ability的实例资源，并通过AceAbility窗口进行显示以及全局应用生命周期管理。

setInstanceName(String name) 的参数name指实例名称，实例名称与config.json文件中module.js.name的值对应。若开发者未修改实例名，而使用了默认值default，则无须调用此接口。若开发者修改了实例名，则需要在应用Ability实例的onStart()中调用此接口，并将参数name设置为修改后的实例名称。

多实例应用的module.js字段中有多个实例项，使用时可选择相应的实例名称。

setInstanceName()接口使用方法：在MainAbility的onStart()中的super.onStart()前调用此接口。以JSComponentName作为实例名称，代码示例如下：

```
public class MainAbility extends AceAbility {
    @Override
    public void onStart(Intent intent) {
        setInstanceName("JSComponentName");  // config.json配置文件中
module.js.name的标签值
        super.onStart(intent);
    }
}
```

10.2.4 JS FA 开发目录

在JS FA开发工程目录中：

- i18n下存放多语言的JSON文件。
- pages文件夹下存放多个页面，每个页面由HTML、CSS和JS文件组成。

其中，main→js→default→i18n→en-US.json文件定义了在英文模式下页面显示的变量内容，内容如下：

```
{
  "strings": {
    "hello": "Hello",
    "world": "World"
  }
}
```

同理，zh-CN.json中定义了中文模式下的页面内容。内容如下：

```
{
```

```
    "strings": {
      "hello": "您好",
      "world": "世界"
    }
  }
```

main→js→default→pages→index→index.hml文件定义了index页面的布局、index页面中用到的组件，以及这些组件的层级关系。例如，index.hml文件中包含一个text组件，内容为"Hello World"文本。

```
<div class="container">
    <text class="title">
        {{ $t('strings.hello') }} {{title}}
    </text>
</div>
```

main→js→default→pages→index→index.css文件定义了index页面的样式。例如，index.css文件定义了container和title的样式。

```css
.container {
    flex-direction: column;
    justify-content: center;
    align-items: center;
}

.title {
    font-size: 40px;
    color: #000000;
    opacity: 0.9;
}

@media screen and (device-type: tablet) and (orientation: landscape) {
    .title {
        font-size: 100px;
    }
}

@media screen and (device-type: wearable) {
    .title {
        font-size: 28px;
        color: #FFFFFF;
    }
}

@media screen and (device-type: tv) {
    .container {
        background-image: url("../../common/images/Wallpaper.png");
        background-size: cover;
        background-repeat: no-repeat;
        background-position: center;
    }
```

```css
.title {
    font-size: 100px;
    color: #FFFFFF;
}
}

@media screen and (device-type: phone) and (orientation: landscape) {
    .title {
        font-size: 60px;
    }
}
```

main→js→default→pages→index→index.js文件定义了index页面的业务逻辑，比如数据绑定、事件处理等。例如，变量title赋值为字符串World。

```javascript
export default {
    data: {
        title: ""
    },
    onInit() {
        this.title = this.$t('strings.world');
    }
}
```

10.2.5 运行

在设备模拟器中，选择Phone设备类型（P40）启动，如图10-5所示。

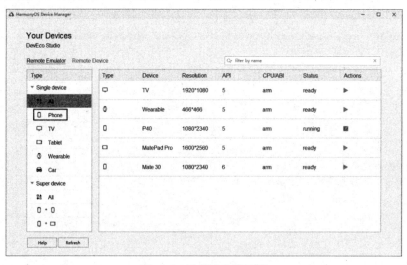

图10-5　选择设备模拟器

然后将JsFa应用在该设备模拟器中执行，可以看到如图10-6所示的效果。

当然，也可以在预览器中进行预览，可以看到如图10-7所示的效果。

图 10-6　应用运行效果

图 10-7　应用预览

10.3　组件与布局

与Java UI框架类似，JS UI框架也是有组件与布局的。

10.3.1　组件分类

根据组件的功能，可以将组件分为以下4大类：

- 基础组件：text、image、progress、rating、span、marquee、image-animator、divider、search、menu、chart。
- 容器组件：div、list、list-item、stack、swiper、tabs、tab-bar、tab-content、list-item-group、refresh、dialog。
- 媒体组件：video。
- 画布组件：canvas。

10.3.2　布局

在JS UI框架中，手机和智慧屏以720px（px指逻辑像素，非物理像素）为基准宽度，根据实际屏幕宽度进行缩放。例如，当width设为100px时，在宽度为1440物理像素的屏幕上，实际显示的宽度为200物理像素。智能穿戴设备的基准宽度为454px，换算逻辑同理。

一个页面的基本元素包含标题区域、文本区域、图片区域等，每个基本元素内还可以包含多个子元素，开发者根据需求还可以添加按钮、开关、进度条等组件。在构建页面布局时，需要对每个基本元素思考以下几个问题：

- 该元素的尺寸和排列位置是否有重叠的元素。
- 该元素的尺寸和排列位置是否需要设置对齐、内间距或者边界。
- 该元素的尺寸和排列位置是否包含子元素及其排列位置。
- 该元素的尺寸和排列位置是否需要容器组件及其类型。

将页面中的元素分解之后再对每个基本元素按顺序实现，可以减少多层嵌套造成的视觉混乱和逻辑混乱，提高代码的可读性，方便对页面做后续的调整。以图10-8为例进行分解。

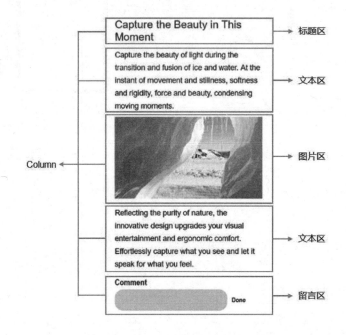

图 10-8　页面布局分解

图10-9展示的是留言区布局分解。

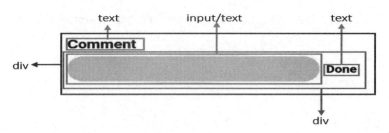

图 10-9　留言区布局分解

10.4 实战:点赞按钮

添加交互通过在组件上关联事件实现。本节将介绍如何用div、text、image组件关联click事件,实现一个点赞按钮。

10.4.1 应用概述

点赞按钮通过一个div组件关联click事件实现。div组件包含一个image组件和一个text组件:
- image组件用于显示未点赞和点赞的效果。click事件函数会交替更新点赞和未点赞图片的路径。
- text组件用于显示点赞数,点赞数会在click事件的函数中同步更新。

创建一个名为GiveLike的应用作为演示。

10.4.2 修改应用

click事件作为一个函数定义在JS文件中,可以更改isPressed的状态,从而更新显示的image组件。如果isPressed为真,则点赞数加1。该函数在HTML文件中对应的div组件上生效,点赞按钮各子组件的样式设置在CSS文件中。

对index.html文件进行修改,具体的实现示例如下:

```html
<!-- 点赞按钮 -->
<div>
  <div class="like" onclick="likeClick">
    <image class="like-img" src="{{likeImage}}" focusable="true"></image>
    <text class="like-num" focusable="true">{{total}}</text>
  </div>
</div>
```

对index.css文件进行修改,具体的实现示例如下:

```css
.like {
  width: 104px;
  height: 54px;
  border: 2px solid #bcbcbc;
  justify-content: space-between;
  align-items: center;
  margin-left: 72px;
  border-radius: 8px;
}
.like-img {
  width: 33px;
  height: 33px;
```

```
        margin-left: 14px;
    }
    .like-num {
        color: #bcbcbc;
        font-size: 20px;
        margin-right: 17px;
    }
```

对index.js文件进行修改，具体的实现示例如下：

```
export default {
    data: {
        likeImage:
'/common/images/unLike.jpg',
        isPressed: false,
        total: 20,
    },
    likeClick() {
        var temp;
        if (!this.isPressed) {
            temp = this.total + 1;
            this.likeImage =
'/common/images/like.jpg';
        } else {
            temp = this.total - 1;
            this.likeImage =
'/common/images/unLike.jpg';
        }
        this.total = temp;
        this.isPressed
= !this.isPressed;
    }
}
```

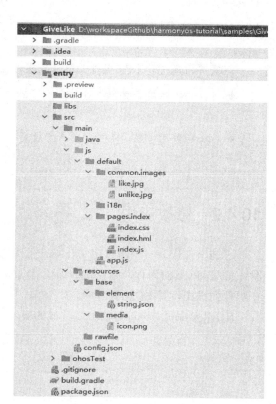

图10-10　图片资源目录

上面的代码中引用的图片资源like.jpg和unlike.jpg建议放在js→default→common→images目录下，如图10-10所示。

10.4.3　运行应用

图10-11展示的是点赞按钮未点击时的状态。图10-12展示的是点赞按钮点击后的状态。

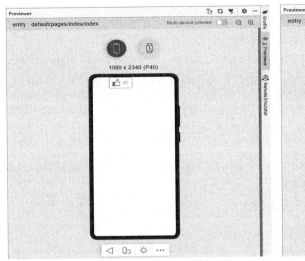
图 10-11　点赞按钮未点击

图 10-12　点赞按钮点击后

10.5　实战：JS FA 调用 PA

JS UI框架提供了JS FA调用Java PA的机制，该机制提供了一种通道来传递方法调用、数据返回以及订阅事件上报。

当前提供Ability和Internal Ability两种调用方式，开发者可以根据业务场景选择合适的调用方式进行开发。

- Ability：拥有独立的Ability生命周期，FA使用远端进程通信拉起并请求PA服务，适用于基本服务供多FA调用或者服务在后台独立运行的场景。
- Internal Ability：与FA共进程，采用内部函数调用的方式和FA进行通信，适用于对服务响应时延要求较高的场景。该方式下PA不支持其他FA访问调用。

JS端与Java端通过bundleName和abilityName进行关联。在系统收到JS调用请求后，根据开发者在JS接口中设置的参数来选择对应的处理方式。开发者在onRemoteRequest()中实现PA提供的业务逻辑。

10.5.1　FA 调用 PA 接口

FA端提供以下3个JS接口：

- FeatureAbility.callAbility(OBJECT)：调用PA能力。
- FeatureAbility.subscribeAbilityEvent(OBJECT, Function)：订阅PA能力。
- FeatureAbility.unsubscribeAbilityEvent(OBJECT)：取消订阅PA能力。

PA端提供以下两类接口：

- IRemoteObject.onRemoteRequest(int, MessageParcel, MessageParcel, MessageOption)：Ability调用方式，FA使用远端进程通信拉起并请求PA服务。
- AceInternalAbility.AceInternalAbilityHandler.onRemoteRequest(int, MessageParcel, MessageParcel, MessageOption)：Internal Ability调用方式，采用内部函数调用的方式和FA进行通信。

10.5.2 创建应用及编写FA

创建一个名为JsFaCallPa的应用作为演示。该应用会由JS FA发起运算的请求，而后将请求提交给PA做计算，最后PA将计算结果返回给JS FA。

同时，对index.js文件进行修改，具体的实现示例如下：

```js
const globalRef = Object.getPrototypeOf(global) || global
globalRef.regeneratorRuntime = require('@babel/runtime/regenerator')

const ABILITY_TYPE_EXTERNAL = 0;
const ACTION_SYNC = 0;
const ACTION_ASYNC = 1;
const ACTION_MESSAGE_CODE_PLUS = 1001;

export const playAbility = {
    sum: async function(that){
        var actionData = {};
        actionData.firstNum = 1024;
        actionData.secondNum = 2048;

        var action = {};
        action.bundleName = 'com.waylau.hmos.jsfacallpa';
        action.abilityName = 'com.waylau.hmos.jsfacallpa.PlayAbility';
        action.messageCode = ACTION_MESSAGE_CODE_PLUS;
        action.data = actionData;
        action.abilityType = ABILITY_TYPE_EXTERNAL;
        action.syncOption = ACTION_SYNC;

        var result = await FeatureAbility.callAbility(action);
        var ret = JSON.parse(result);
        if (ret.code == 0) {
            console.info('result is:' + JSON.stringify(ret.abilityResult));
            that.title = 'result is:' + JSON.stringify(ret.abilityResult);
        } else {
            console.error('plus error code:' + JSON.stringify(ret.code));
        }
    }
}

export default {
    data: {
        title: ""
    },
```

```
onInit() {
    this.title = "1024+2048=";
},
play(){
    this.title = "doing...";
    playAbility.sum(this);
}
}
```

上述代码定义了play方法,该方法最终会调用PA。

对index.html文件进行修改,具体的实现示例如下:

```
<div class="container">
    <div>
        <text class="title">
            {{title}}
        </text>
    </div>
    <div>
        <text class="title" onclick="play">
            Play
        </text>
    </div>
</div>
```

上述定义了Play的点击事件,该事件会调用play方法,最终将返回结果显示在界面上。

10.5.3 编写 PA

创建一个名为PlayAbility的Service Ability,如图10-13所示。

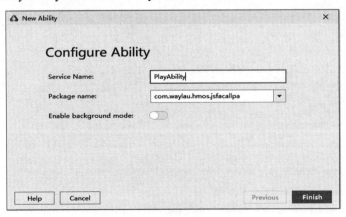

图10-13 创建 Service Ability

修改PlayAbility,代码如下:

```
package com.waylau.hmos.jsfacallpa;

import ohos.aafwk.ability.Ability;
```

```java
import ohos.aafwk.content.Intent;
import ohos.app.Context;
import ohos.hiviewdfx.HiLog;
import ohos.hiviewdfx.HiLogLabel;
import ohos.rpc.*;
import ohos.utils.zson.ZSONObject;

import java.util.HashMap;
import java.util.Map;

public class PlayAbility extends Ability {
    private static final String TAG = PlayAbility.class.getSimpleName();
    private static final HiLogLabel LABEL_LOG =
            new HiLogLabel(HiLog.LOG_APP, 0x00001, TAG);

    private static final int ERROR = -1;
    private static final int SUCCESS = 0;
    private static final int PLUS = 1001;
    private PlayRemote remote;

    @Override
    public void onStart(Intent intent) {
        HiLog.error(LABEL_LOG, "PlayAbility::onStart");
        super.onStart(intent);
    }

    @Override
    public void onBackground() {
        super.onBackground();
        HiLog.info(LABEL_LOG, "PlayAbility::onBackground");
    }

    @Override
    public void onStop() {
        super.onStop();
        HiLog.info(LABEL_LOG, "PlayAbility::onStop");
    }

    @Override
    public void onCommand(Intent intent, boolean restart, int startId) {
    }

    @Override
    public IRemoteObject onConnect(Intent intent) {
        super.onConnect(intent);
        Context c = getContext();
        remote = new PlayRemote();
        return remote.asObject();
    }
```

```java
    @Override
    public void onDisconnect(Intent intent) {
    }

    class PlayRemote extends RemoteObject implements IRemoteBroker {

        public PlayRemote() {
            super("PlayRemote 666");
        }

        @Override
        public boolean onRemoteRequest(int code, MessageParcel data,
            MessageParcel reply, MessageOption option) throws RemoteException {
            switch (code) {
                case PLUS: {

                    String zsonStr = data.readString();
                    RequestParam param = ZSONObject.stringToClass(zsonStr, RequestParam.class);

                    // 返回结果仅支持可序列化的Object类型
                    Map<String, Object> zsonResult = new HashMap<String, Object>();
                    zsonResult.put("code", SUCCESS);
                    zsonResult.put("abilityResult", param.getFirstNum() + param.getSecondNum());
                    reply.writeString(ZSONObject.toZSONString(zsonResult));
                    break;
                }
                default: {
                    reply.writeString("service not defined");
                    return false;
                }
            }
            return true;
        }

        @Override
        public IRemoteObject asObject() {
            return this;
        }
    }
}
```

上述代码实现了将调用方请求的参数做和，而后将结果返回给调用方。
其中，RequestParam的代码如下：

```java
public class RequestParam {
    private int firstNum;
```

```
    private int secondNum;

    public int getFirstNum() {
        return firstNum;
    }

    public void setFirstNum(int firstNum) {
        this.firstNum = firstNum;
    }

    public int getSecondNum() {
        return secondNum;
    }

    public void setSecondNum(int secondNum) {
        this.secondNum = secondNum;
    }
}
```

10.5.4 运行应用

图10-14展示的是点击Play前的界面状态。图10-15展示的是点击Play后的界面状态。可以看到界面显示出了PA计算的结果。

图 10-14　点击 Play 前的界面状态

图 10-15　点击 Play 后的界面状态

第 11 章

多模输入 UI 开发

HarmonyOS旨在为开发者提供NUI（Natural User Interface，自然用户界面）的交互方式，有别于传统操作系统的输入划分方式。在HarmonyOS上，将多种维度的输入整合在一起，开发者可以借助应用程序框架、系统自带的UI控件或API接口轻松地实现具有多维、自然交互特点的应用程序。

11.1 多模输入概述

目前不仅支持传统的输入交互方式，例如按键、触控、键盘、鼠标等，同时提供多模输入融合框架，可以支持语音等新型的输入交互方式。多模输入事件在不同形态产品中支持的情况如表11-1所示。

表 11-1 多模输入事件在不同形态产品中支持的情况

多模输入事件	手机	平板	智慧屏	车机	智能穿戴设备
按键输入事件	支持	支持	支持	支持	支持
触屏输入事件	支持	支持	支持	支持	支持
鼠标事件	只支持鼠标左键事件	只支持鼠标左键事件	只支持鼠标左键事件	不支持	不支持
语音事件	不支持	不支持	支持	不支持	不支持

多模输入使HarmonyOS的UI控件能够响应多种输入事件，事件来源于用户的按键、点击、触屏、语音等。需要注意的是，使用多模输入相关功能需要获取多模输入权限ohos.permission.MULTIMODAL_INTERACTIVE。目前多模输入不支持注入事件（即开发者无法模拟注入事件验证应用程序功能），仅支持KeyEvent事件的生成。

11.2 接口说明

多模输入的接口设计是基于多模事件基类（MultimodalEvent）的，派生出操作事件类（ManipulationEvent）、按键事件类（KeyEvent）、语音事件（SpeechEvent）等，另外提供创建事件类和获取输入设备信息类，如图11-1所示。

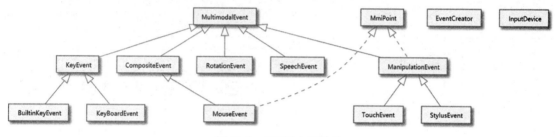

图 11-1　多模输入事件类

11.2.1　MultimodalEvent

MultimodalEvent是所有事件的基类，该类中定义了一系列高级事件类型，这些事件类型通常是对某种行为或意图的抽象。

MultimodalEvent的主要接口如下：

- getDeviceId()：获取输入设备所在的承载设备ID，如当同时有两个鼠标连接到一个机器时，该机器为这两个鼠标的承载设备。
- getInputDeviceId()：获取产生当前事件的输入设备ID，该ID是该输入设备的唯一标识，若两个鼠标同时输入，则它们会分别产生输入事件，且从事件中获取到的deviceid是不同的，开发者可以将此ID用来区分实际的输入设备源。
- getSourceDevice()：获取产生当前事件的输入设备类型。
- getOccurredTime()：获取产生当前事件的时间。
- getUuid()：获取事件的UUID。
- isSameEvent(UUID id)：判断当前事件与传入ID的事件是否为同一事件。

11.2.2　CompositeEvent

CompositeEvent处理常用设备对应的事件，目前暂时只有MouseEvent事件继承该类。

11.2.3　RotationEvent

RotationEvent处理由旋转器件产生的事件，比如智能穿戴设备上的数字表冠。

RotationEvent的主要接口有：

- getRotationValue()：获取旋转器件旋转产生的值。

11.2.4 SpeechEvent

SpeechEvent处理语音事件，开发者可以通过该类获取语音识别结果。

SpeechEvent的主要接口如下：

- createEvent(long occurTime, int action, String value)：SpeechEvent创建函数。
- getAction()：获取当前动作的类型，如打开、关闭、命中热词。
- getScene()：获取当前动作时的场景。
- getActionProperty()：获取动作所携带的属性值。
- getMatchMode()：获取识别结果的匹配模式。

11.2.5 ManipulationEvent

ManipulationEvent操作类事件主要包括手指触摸事件等事件，是对这些事件的一个抽象。该事件会持有事件发生的位置和发生的阶段等信息。通常情况下，该事件主要是作为操作回调接口的入参，开发者通过回调接口捕获及处理事件。回调接口将操作分为开始、操作过程中、结束。例如对于一次手指触控，手指接触屏幕作为操作开始，手指在屏幕上移动作为操作过程，手指抬起作为操作结束。

ManipulationEvent的主要接口如下：

- getPointerCount()：获取一次事件中触控或轨迹追踪的指针数量。
- getPointerId(int index)：获取一次事件中，指针的唯一标识ID。
- setScreenOffset(float offsetX, float offsetY)：设置相对屏幕坐标原点的偏移位置信息。
- getPointerPosition(int index)：获取一次事件中触控或轨迹追踪的某个指针相对于偏移位置的坐标信息。
- getPointerScreenPosition(int index)：获取一次事件中触控或轨迹追踪的某个指针相对屏幕坐标原点的坐标信息。
- getRadius(int index)：返回给定index手指与屏幕接触的半径值。
- getForce(int index)：获取给定index手指触控的压力值。
- getStartTime()：获取操作开始阶段时间。
- getPhase()：事件所属阶段。

11.2.6 KeyEvent

KeyEvent是对所有按键类事件的定义，该类继承MultimodalEvent类，并对按键类事件做了专属的Keycode定义以及方法封装。

KeyEvent的主要接口如下：

- getKeyCode()：获取当前按键类事件的keycode值。
- getMaxKeyCode()：获取当前定义的按键类事件的最大keycode值。
- getKeyDownDuration()：获取当前按键截止该接口被调用时被按下的时长。
- isKeyDown()：获取当前按键事件的按下状态。

11.2.7　TouchEvent

TouchEvent处理手指触控相关事件。

TouchEvent的主要接口如下：

- getAction()：获取当前触摸行为。
- getIndex()：获取发生行为的对应指针。

11.2.8　KeyBoardEvent

KeyBoardEvent处理键盘类设备的事件。

KeyBoardEvent的主要接口如下：

- enableIme()：启动输入法编辑器。
- disableIme()：关闭输入法编辑器。
- isHandledByIme()：判断输入法编辑器是否在使用。
- isNoncharacterKeyPressed(int keycode)：判定输入的单个NoncharacterKey是否处于按下状态。
- isNoncharacterKeyPressed(int keycode1, int keycode2)：判定输入的两个NoncharacterKey是否都处于按下状态。
- isNoncharacterKeyPressed(int keycode1, int keycode2, int keycode3)：判定输入的3个NoncharacterKey是否都处于按下状态。
- getUnicode()：获取按键对应的Unicode码。

11.2.9　MouseEvent

MouseEvent处理鼠标的事件。

MouseEvent的主要接口如下：

- getAction()：获取鼠标设备产生事件的行为。
- getActionButton()：获取状态发生变化的鼠标按键。
- getPressedButtons()：获取所有按下状态的鼠标按键。
- getCursor()：获取鼠标指针的位置。
- getCursorDelta(int axis)：获取鼠标指针位置相对上次的变化值。
- setCursorOffset(float offsetX, float offsetY)：设置相对屏幕的偏移位置信息。
- getScrollingDelta(int axis)：获取滚轮的滚动值。

11.2.10 MmiPoint

MmiPoint处理在指定的坐标系中的x、y和z坐标。

MmiPoint的主要接口如下：

- MmiPoint(float px, float py)：创建一个只包含x和y坐标的MmiPoint对象。
- MmiPoint(float px, float py, float pz)：创建一个包含x、y和z坐标的MmiPoint对象。
- getX()：获取x坐标值。
- getY()：获取y坐标值。
- getZ()：获取z坐标值。
- toString()：返回包含x、y、z坐标值信息的字符串。

11.2.11 EventCreator

EventCreator提供创建事件的方法，当前仅提供创建KeyEvent事件的能力。

EventCreator的主要接口如下：

- createKeyEvent(int action, int keyCode)：根据指定的action和keycode创建KeyEvent。

11.2.12 InputDevice

InputDevice提供获取承载设备上输入设备信息的方法，当前支持获取所有输入设备的ID。

InputDevice的主要接口如下：

- getAllInputDeviceID()：获取承载设备上所有输入设备的ID，例如承载设备为智慧屏，其中有键盘、鼠标和遥控器连接成功。该方法用于获取键盘、鼠标和遥控器的ID。

11.3 实战：多模输入事件

本节演示多模输入事件MultimodalEvent的使用。创建一个名为MultimodalEvent的应用作为演示。

11.3.1 修改 MainAbilitySlice

修改MainAbilitySlice，代码如下：

```
package com.waylau.hmos.multimodalevent.slice;

import com.waylau.hmos.multimodalevent.ResourceTable;
```

```java
import ohos.aafwk.ability.AbilitySlice;
import ohos.aafwk.content.Intent;
import ohos.agp.components.Component;
import ohos.agp.components.Text;
import ohos.hiviewdfx.HiLog;
import ohos.hiviewdfx.HiLogLabel;
import ohos.multimodalinput.event.KeyEvent;

public class MainAbilitySlice extends AbilitySlice {
    private static final String TAG = MainAbilitySlice.class.getSimpleName();
    private static final HiLogLabel LABEL_LOG =
            new HiLogLabel(HiLog.LOG_APP, 0x00001, TAG);

    @Override
    public void onStart(Intent intent) {
        super.onStart(intent);
        super.setUIContent(ResourceTable.Layout_ability_main);

        Text text =
                (Text) findComponentById(ResourceTable.Id_text_helloworld);

        // 为按钮设置键盘事件回调
        text.setKeyEventListener(onKeyEvent);
    }

    private Component.KeyEventListener onKeyEvent = new Component.KeyEventListener()
    {
        @Override
        public boolean onKeyEvent(Component component, KeyEvent keyEvent) {
            if (keyEvent.isKeyDown()) {
                // 检测到按键被按下,开发者根据自身需求进行实现
                HiLog.info(LABEL_LOG, "isKeyDown");
            }
            int keycode = keyEvent.getKeyCode();

            HiLog.info(LABEL_LOG, "keycode: %{public}s", keycode);

            switch (keycode) {
                case KeyEvent.KEY_DPAD_CENTER:
                    // 检测到KEY_DPAD_CENTER被按下,开发者根据自身需求进行实现
                    HiLog.info(LABEL_LOG, "KeyEvent.KEY_DPAD_CENTER");
                    break;
                case KeyEvent.KEY_DPAD_LEFT:
                    // 检测到KEY_DPAD_LEFT被按下,开发者根据自身需求进行实现
                    HiLog.info(LABEL_LOG, "KeyEvent.KEY_DPAD_LEFT");
                    break;
                case KeyEvent.KEY_DPAD_UP:
                    // 检测到KEY_DPAD_UP被按下,开发者根据自身需求进行实现
                    HiLog.info(LABEL_LOG, "KeyEvent.KEY_DPAD_UP");
```

```
                    break;
                case KeyEvent.KEY_DPAD_RIGHT:
                    // 检测到KEY_DPAD_RIGHT被按下，开发者根据自身需求进行实现
                    HiLog.info(LABEL_LOG, "KeyEvent.KEY_DPAD_RIGHT");
                    break;
                case KeyEvent.KEY_DPAD_DOWN:
                    // 检测到KEY_DPAD_DOWN被按下，开发者根据自身需求进行实现
                    HiLog.info(LABEL_LOG, "KeyEvent.KEY_DPAD_DOWN");
                    break;
                default:
                    HiLog.info(LABEL_LOG, "KeyEvent default");
                    break;
            }

            return true;
        }
    };

    @Override
    public void onActive() {
        super.onActive();
    }

    @Override
    public void onForeground(Intent intent) {
        super.onForeground(intent);
    }
}
```

上述代码定义了Component.KeyEventListener事件监听器，并通过text.setKeyEventListener(onKeyEvent)方式将事件监听器注册到text上。

11.3.2　获取多模输入权限

修改配置文件，声明多模输入权限如下：

```
// 声明权限
"reqPermissions": [
    {
        "name": "ohos.permission.MULTIMODAL_INTERACTIVE"
    }
]
```

第 12 章

线程管理

具有多线程能力的计算机因有硬件支持而能够在同一时间执行多于一个线程，实现多个任务并发执行，进而提升整体处理性能。

本章介绍HarmonyOS关于线程的管理。

12.1 线程管理概述

不同应用在各自独立的进程中运行。当应用以任何形式启动时，系统为其创建进程，该进程将持续运行。当进程完成当前任务处于等待状态且系统资源不足时，系统自动回收。

在启动应用时，系统会为该应用创建一个称为"主线程"的执行线程。该线程随着应用创建或消失，是应用的核心线程。UI界面的显示和更新等操作都是在主线程上进行的。主线程又称UI线程，默认情况下，所有的操作都是在主线程上执行的。如果需要执行比较耗时的任务（如下载文件、查询数据库），可创建其他线程来处理。

12.2 场景介绍

如果应用的业务逻辑比较复杂，可能需要创建多个线程来执行多个任务。创建的Java语言本身是支持创建多线程的。

12.2.1 传统 Java 多线程管理

比如下面这个多线程的示例（该示例选自《Java核心编程》）。SimpleThreads示例有两个线程。第一个线程是每个Java应用程序都有的主线程。主线程创建Runnable对象的MessageLoop，并等待它完成。如果MessageLoop需要很长时间才能完成，主线程就中断它。

MessageLoop线程打印出一系列消息。如果中断之前就已经打印了所有消息，则MessageLoop线程打印一条消息并退出。

```java
class SimpleThreads {

    // 显示当前执行线程的名称、信息
    static void threadMessage(String message) {
        String threadName =
            Thread.currentThread().getName();
        System.out.format("%s: %s%n",
                    threadName,
                    message);
    }

    private static class MessageLoop
        implements Runnable {
        public void run() {
            String importantInfo[] = {
                "Mares eat oats",
                "Does eat oats",
                "Little lambs eat ivy",
                "A kid will eat ivy too"
            };
            try {
                for (int i = 0; i < importantInfo.length; i++) {

                    // 暂停4秒
                    Thread.sleep(4000);

                    // 打印消息
                    threadMessage(importantInfo[i]);
                }
            } catch (InterruptedException e) {
                threadMessage("I wasn't done!");
            }
        }
    }

    public static void main(String args[])
        throws InterruptedException {

        // 在中断MessageLoop线程（默认为1小时）前先延迟一段时间（单位是毫秒）
```

```
        long patience = 1000 * 60 * 60;

// 如果命令行参数出现
// 设置patience的时间值
// 单位是秒
        if (args.length > 0) {
            try {
                patience = Long.parseLong(args[0]) * 1000;
            } catch (NumberFormatException e) {
                System.err.println("Argument must be an integer.");
                System.exit(1);
            }
        }

        threadMessage("Starting MessageLoop thread");
        long startTime = System.currentTimeMillis();
        Thread t = new Thread(new MessageLoop());
        t.start();

        threadMessage("Waiting for MessageLoop thread to finish");

// 循环直到MessageLoop线程退出
        while (t.isAlive()) {
            threadMessage("Still waiting...");

            // 最长等待1秒
            // 由MessageLoop线程来完成
            t.join(1000);
            if (((System.currentTimeMillis() - startTime) > patience)
                && t.isAlive()) {
                threadMessage("Tired of waiting!");
                t.interrupt();

                // 等待
                t.join();
            }
        }
        threadMessage("Finally!");
    }
}
```

如果线程的数量再增多的话，则需要引入 Executor 框架。Executor 框架核心的类是 ThreadPoolExecutor，它是线程池的实现类。通过线程池可以使线程得到复用，避免创建过多的线程对象。

有关Java线程池的示例不再赘述，有兴趣的读者可以自行参阅相关资料。

12.2.2 HarmonyOS 多线程管理

通过Java语言自行实现线程管理是复杂的，代码复杂难以维护，任务与线程的交互也会更加繁杂。为了解决此问题，HarmonyOS提供了TaskDispatcher（任务分发器），开发者通过TaskDispatcher可以分发不同的任务。

接下来将详细介绍TaskDispatcher接口。

12.3 接口说明

TaskDispatcher是Ability分发任务的基本接口，隐藏任务所在线程的实现细节。

为了保证应用有更好的响应性，需要设计任务的优先级。在UI线程上运行的任务默认以高优先级运行，如果某个任务无须等待结果，则可以用低优先级运行。

HarmonyOS线程的优先级分类如下：

- HIGH：最高任务优先级，比默认优先级、低优先级的任务有更高的概率得到执行。
- DEFAULT：默认任务优先级，比低优先级的任务有更高的概率得到执行。
- LOW：低任务优先级，比高优先级、默认优先级的任务有更低的概率得到执行。

TaskDispatcher具有多种实现，每种实现对应不同的任务分发器。在分发任务时可以指定任务的优先级，由同一个任务分发器分发出的任务具有相同的优先级。系统提供的任务分发器有GlobalTaskDispatcher、ParallelTaskDispatcher、SerialTaskDispatcher、SpecTaskDispatcher。

12.3.1 GlobalTaskDispatcher

全局并发任务分发器由Ability执行getGlobalTaskDispatcher()获取，适用于任务之间没有联系的情况。一个应用只有一个GlobalTaskDispatcher，它在程序结束时才被销毁。

代码示例如下：

```
TaskDispatcher globalTaskDispatcher =
getGlobalTaskDispatcher(TaskPriority.DEFAULT);
```

12.3.2 ParallelTaskDispatcher

并发任务分发器由Ability执行createParallelTaskDispatcher()创建并返回。与GlobalTaskDispatcher不同的是，ParallelTaskDispatcher不具有全局唯一性，可以创建多个。开发者在创建或销毁dispatcher时，需要持有对应的对象引用。

代码示例如下：

```
String dispatcherName = "parallelTaskDispatcher";
```

```
TaskDispatcher parallelTaskDispatcher =
    createParallelTaskDispatcher(dispatcherName, TaskPriority.DEFAULT);
```

12.3.3　SerialTaskDispatcher

串行任务分发器由Ability执行createSerialTaskDispatcher()创建并返回。由该分发器分发的所有任务都是按顺序执行的，但是执行这些任务的线程并不是固定的。如果要执行并行任务，应使用ParallelTaskDispatcher或者GlobalTaskDispatcher，而不是创建多个SerialTaskDispatcher。如果任务之间没有依赖，应使用GlobalTaskDispatcher来实现。它的创建和销毁由开发者自己管理，开发者在使用期间需要持有该对象引用。

代码示例如下：

```
String dispatcherName = "serialTaskDispatcher";
TaskDispatcher serialTaskDispatcher =
    createSerialTaskDispatcher(dispatcherName, TaskPriority.DEFAULT);
```

12.3.4　SpecTaskDispatcher

专有任务分发器是绑定到专有线程上的任务分发器。目前已有的专有线程是主线程。UITaskDispatcher和MainTaskDispatcher都属于SpecTaskDispatcher。建议使用UITaskDispatcher。

1. UITaskDispatcher

UITaskDispatcher是绑定到应用主线程的专有任务分发器，由Ability执行getUITaskDispatcher()创建并返回。由该分发器分发的所有任务都是在主线程上按顺序执行的，它在应用程序结束时被销毁。

代码示例如下：

```
TaskDispatcher uiTaskDispatcher = getUITaskDispatcher();
```

2. MainTaskDispatcher

MainTaskDispatcher是由Ability执行getMainTaskDispatcher()创建并返回的。

代码示例如下：

```
TaskDispatcher mainTaskDispatcher= getMainTaskDispatcher()
```

12.4　实战：线程管理示例

创建一个名为ParallelTaskDispatcher的应用，用于演示ParallelTaskDispatcher任务分发器派发任务的使用。

12.4.1 修改 ability_main.xml

修改ability_main.xml内容如下：

```xml
<?xml version="1.0" encoding="utf-8"?>
<DirectionalLayout
    xmlns:ohos="http://schemas.huawei.com/res/ohos"
    ohos:height="match_parent"
    ohos:width="match_parent"
    ohos:alignment="center"
    ohos:orientation="vertical">

    <Text
        ohos:id="$+id:text_start_parallel_task_dispatcher"
        ohos:height="match_content"
        ohos:width="match_content"
        ohos:background_element="$graphic:background_ability_main"
        ohos:layout_alignment="horizontal_center"
        ohos:text="Start ParallelTaskDispatcher"
        ohos:text_size="20vp"
        />

</DirectionalLayout>
```

显示界面效果如图12-1所示。

图 12-1　界面效果

12.4.2 自定义任务

MyTask是一个自定义的任务。该任务逻辑比较简单，只是模拟了一个耗时的操作。

```java
package com.waylau.hmos.paralleltaskdispatcher;

import ohos.hiviewdfx.HiLog;
import ohos.hiviewdfx.HiLogLabel;

import java.util.concurrent.TimeUnit;

public class MyTask implements Runnable {
    private static final String TAG = MyTask.class.getSimpleName();
    private static final HiLogLabel LABEL_LOG =
            new HiLogLabel(HiLog.LOG_APP, 0x00001, TAG);

    private String taskName;

    public MyTask(String taskName) {
        this.taskName = taskName;
    }

    @Override
    public void run() {
        HiLog.info(LABEL_LOG, "before %{public}s run", taskName);
        int task1Result = getRandomInt();
        try {
            // 模拟一个耗时的操作
            TimeUnit.MILLISECONDS.sleep(task1Result);
        } catch (InterruptedException e) {
            e.printStackTrace();
        }

        HiLog.info(LABEL_LOG, "after %{public}s run, result is: %{public}s", taskName, task1Result);
    }

    // 返回随机整数
    private int getRandomInt() {
        // 获取[0, 1000)的int整数，方法如下
        double a = Math.random();
        int result = (int) (a * 1000);
        return result;
    }
}
```

这个耗时操作是通过获取一个随机数，而后根据随机数执行线程sleep实现的。

12.4.3　执行任务派发

修改MainAbilitySlice，增加任务派发器相关的逻辑。

```java
package com.waylau.hmos.paralleltaskdispatcher.slice;

import com.waylau.hmos.paralleltaskdispatcher.MyTask;
import com.waylau.hmos.paralleltaskdispatcher.ResourceTable;
import ohos.aafwk.ability.AbilitySlice;
import ohos.aafwk.content.Intent;
import ohos.agp.components.Text;
import ohos.app.dispatcher.Group;
import ohos.app.dispatcher.TaskDispatcher;
import ohos.app.dispatcher.task.TaskPriority;
import ohos.hiviewdfx.HiLog;
import ohos.hiviewdfx.HiLogLabel;

public class MainAbilitySlice extends AbilitySlice {
    private static final String TAG = MainAbilitySlice.class.getSimpleName();
    private static final HiLogLabel LABEL_LOG =
            new HiLogLabel(HiLog.LOG_APP, 0x00001, TAG);

    @Override
    public void onStart(Intent intent) {
        super.onStart(intent);
        super.setUIContent(ResourceTable.Layout_ability_main);

        // 添加点击事件来触发
        Text textStartDispatcher =
                (Text) findComponentById(ResourceTable.Id_text_start_parallel_task_dispatcher);
        textStartDispatcher.setClickedListener(listener -> startDispatcher());
    }

    // 指定任务派发
    private void startDispatcher() {
        String dispatcherName = "MyDispatcher";

        TaskDispatcher dispatcher =
                this.getContext().createParallelTaskDispatcher(dispatcherName, TaskPriority.DEFAULT);

        // 创建任务组
        Group group = dispatcher.createDispatchGroup();

        // 将任务1加入任务组
```

```
            dispatcher.asyncGroupDispatch(group, new MyTask("task1"));

            // 将与任务1相关联的任务2加入任务组
            dispatcher.asyncGroupDispatch(group, new MyTask("task2"));

            // task3必须要等任务组中的所有任务执行完成后才会执行
            dispatcher.groupDispatchNotify(group, new MyTask("task3"));
        }

    @Override
    public void onActive() {
        super.onActive();
    }

    @Override
    public void onForeground(Intent intent) {
        super.onForeground(intent);
    }
}
```

上述代码中：

- Text增加了点击事件，以触发startDispatcher任务。
- startDispatcher方法中创建了ParallelTaskDispatcher任务派发器。
- 创建了3个MyTask任务实例，这些任务都是一个任务组。
- task1和task2通过asyncGroupDispatch方式异步派发。
- task3通过groupDispatchNotify方式派发。groupDispatchNotify方式需要等任务组中的所有任务执行完成后才会执行指定任务。

12.4.4　运行

运行应用后，点击两次界面上的文本Start ParallelTaskDispatcher以触发任务派发。此时，控制台输出如下：

```
    09-11 11:29:18.180 15991-16685/com.waylau.hmos.paralleltaskdispatcher I
00001/MyTask: before task1 run
    09-11 11:29:18.181 15991-16686/com.waylau.hmos.paralleltaskdispatcher I
00001/MyTask: before task2 run
    09-11 11:29:18.338 15991-16686/com.waylau.hmos.paralleltaskdispatcher I
00001/MyTask: after task2 run, result is: 157
    09-11 11:29:18.974 15991-16685/com.waylau.hmos.paralleltaskdispatcher I
00001/MyTask: after task1 run, result is: 793
    09-11 11:29:18.976 15991-16744/com.waylau.hmos.paralleltaskdispatcher I
00001/MyTask: before task3 run
    09-11 11:29:19.248 15991-16744/com.waylau.hmos.paralleltaskdispatcher I
00001/MyTask: after task3 run, result is: 269
```

```
    09-11 11:29:22.499 15991-16946/com.waylau.hmos.paralleltaskdispatcher I
00001/MyTask: before task1 run
    09-11 11:29:22.500 15991-16947/com.waylau.hmos.paralleltaskdispatcher I
00001/MyTask: before task2 run
    09-11 11:29:22.666 15991-16947/com.waylau.hmos.paralleltaskdispatcher I
00001/MyTask: after task2 run, result is: 166
    09-11 11:29:23.203 15991-16946/com.waylau.hmos.paralleltaskdispatcher I
00001/MyTask: after task1 run, result is: 704
    09-11 11:29:23.204 15991-16995/com.waylau.hmos.paralleltaskdispatcher I
00001/MyTask: before task3 run
    09-11 11:29:23.750 15991-16995/com.waylau.hmos.paralleltaskdispatcher I
00001/MyTask: after task3 run, result is: 545
```

分别执行了两次，可以看到task1和taks2先后顺序是随机的，但task3一定是在task1和taks2完成之后才会执行。

12.5 线程间通信概述

读者如果开发过Netty或者Node.js应用，那么对于事件循环器应该不太陌生，事件循环器是高并发非阻塞的"秘笈"。在HarmonyOS中，事件循环器的实现方式就是EventHandler机制。当前线程中处理较为耗时的操作时，如果不希望当前的线程受到阻塞，就可以使用EventHandler机制。EventHandler是HarmonyOS用于处理线程间通信的一种机制，可以通过EventRunner创建新线程，将耗时的操作放到新线程上执行。这样既不阻塞原来的线程，任务又可以得到合理的处理。比如，主线程使用EventHandler创建子线程，子线程做耗时的下载图片操作，下载完成后，子线程通过EventHandler通知主线程，主线程再更新UI。

Netty或者Node.js方面的内容可以参见本书最后的"参考文献"部分。

12.5.1 基本概念

EventRunner是一种事件循环器，循环处理从EventRunner创建的新线程的事件队列中获取InnerEvent事件或者Runnable任务。InnerEvent是EventHandler投递的事件。

EventHandler是一种用户在当前线程上投递InnerEvent事件或者Runnable任务到异步线程上处理的机制。每一个EventHandler和指定的EventRunner所创建的新线程绑定，并且该新线程内部有一个事件队列。EventHandler可以投递指定的InnerEvent事件或Runnable任务到这个事件队列。EventRunner从事件队列中循环地取出事件，如果取出的事件是InnerEvent事件，将在EventRunner所在线程执行processEvent回调；如果取出的事件是Runnable任务，将在EventRunner所在线程执行Runnable的run回调。一般，EventHandler有两个主要作用：

- 在不同线程间分发和处理InnerEvent事件或Runnable任务。
- 延迟处理InnerEvent事件或Runnable任务。

12.5.2 运作机制

EventHandler的运作机制如图12-2所示。

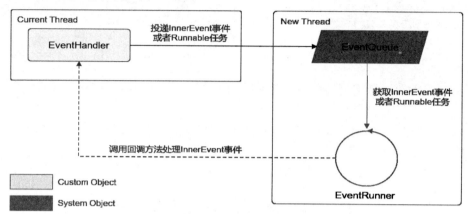

图 12-2　EventHandler 的运作机制

使用EventHandler实现线程间通信的主要流程如下：

- EventHandler投递具体的InnerEvent事件或者Runnable任务到EventRunner所创建的线程的事件队列。
- EventRunner循环从事件队列中获取InnerEvent事件或者Runnable任务。

处理事件或任务：

- 如果EventRunner取出的事件为InnerEvent事件，则触发EventHandler的回调方法并触发EventHandler的处理方法，在新线程上处理该事件。
- 如果EventRunner取出的事件为Runnable任务，则EventRunner直接在新线程上处理Runnable任务。

12.5.3 约束限制

在进行线程间通信的时候，EventHandler只能和EventRunner所创建的线程进行绑定，EventRunner创建时需要判断是否创建成功，只有确保获取的EventRunner实例非空时，才可以使用EventHandler绑定EventRunner。

一个EventHandler只能同时与一个EventRunner绑定，一个EventRunner上可以创建多个EventHandler。

12.6　实战：线程间通信示例

创建一个名为EventHandler的应用，用于演示EventHandler处理线程间通信的使用。

12.6.1 修改 ability_main.xml

修改ability_main.xml内容如下:

```xml
<?xml version="1.0" encoding="utf-8"?>
<DirectionalLayout
    xmlns:ohos="http://schemas.huawei.com/res/ohos"
    ohos:height="match_parent"
    ohos:width="match_parent"
    ohos:alignment="center"
    ohos:orientation="vertical">

    <Text
        ohos:id="$+id:text_send_event"
        ohos:height="match_content"
        ohos:width="match_content"
        ohos:background_element="$graphic:background_ability_main"
        ohos:layout_alignment="horizontal_center"
        ohos:text="Send Event"
        ohos:text_size="40vp"
        />

</DirectionalLayout>
```

显示界面效果如图12-3所示。

图 12-3　界面效果

12.6.2 自定义事件处理器

MyEventHandler是自定义的一个事件处理器。该事件处理器的逻辑比较简单,只是模拟了一个耗时的操作。

```java
package com.waylau.hmos.eventhandler;

import ohos.eventhandler.EventHandler;
import ohos.eventhandler.EventRunner;
import ohos.eventhandler.InnerEvent;
import ohos.hiviewdfx.HiLog;
import ohos.hiviewdfx.HiLogLabel;

import java.util.concurrent.TimeUnit;

public class MyEventHandler extends EventHandler {
    private static final String TAG = MyEventHandler.class.getSimpleName();
    private static final HiLogLabel LABEL_LOG =
            new HiLogLabel(HiLog.LOG_APP, 0x00001, TAG);

    public MyEventHandler(EventRunner runner) throws IllegalArgumentException {
        super(runner);
    }

    @Override
    public void processEvent(InnerEvent event) {
        super.processEvent(event);
        if (event == null) {
            HiLog.info(LABEL_LOG, "before processEvent event is null");
            return;
        }
        int eventId = event.eventId;
        HiLog.info(LABEL_LOG, "before processEvent eventId: %{public}s", eventId);

        int task1Result = getRandomInt();
        try {
            // 模拟一个耗时的操作
            TimeUnit.MILLISECONDS.sleep(task1Result);
        } catch (InterruptedException e) {
            e.printStackTrace();
        }

        HiLog.info(LABEL_LOG, "after processEvent eventId %{public}s", eventId);
    }

    // 返回随机整数
    private int getRandomInt() {
        // 获取[0, 1000)的int整数,方法如下
        double a = Math.random();
        int result = (int) (a * 1000);
        return result;
    }
```

}
```

这个耗时操作是通过获取一个随机数，而后根据随机数执行线程sleep实现的。

## 12.6.3 执行事件发送

修改MainAbilitySlice，增加事件发送相关的逻辑。

```
package com.waylau.hmos.eventhandler.slice;

import com.waylau.hmos.eventhandler.MyEventHandler;
import com.waylau.hmos.eventhandler.ResourceTable;
import ohos.aafwk.ability.AbilitySlice;
import ohos.aafwk.content.Intent;
import ohos.agp.components.Text;
import ohos.eventhandler.EventRunner;
import ohos.hiviewdfx.HiLog;
import ohos.hiviewdfx.HiLogLabel;

public class MainAbilitySlice extends AbilitySlice {
    private static final String TAG = MainAbilitySlice.class.getSimpleName();
    private static final HiLogLabel LABEL_LOG =
            new HiLogLabel(HiLog.LOG_APP, 0x00001, TAG);

    // 创建EventRunner
    private EventRunner eventRunner = EventRunner.create("MyEventRunner"); // 内部会新建一个线程
    private int eventId = 0; // 事件ID，递增的序列

    @Override
    public void onStart(Intent intent) {
        super.onStart(intent);
        super.setUIContent(ResourceTable.Layout_ability_main);

        // 添加点击事件来触发
        Text textSendEvent =
                (Text) findComponentById(ResourceTable.Id_text_send_event);
        textSendEvent.setClickedListener(listener -> sendEvent());
    }

    private void sendEvent() {
        HiLog.info(LABEL_LOG, "before sendEvent");

        // 创建MyEventHandler实例
        MyEventHandler handler = new MyEventHandler(eventRunner);

        eventId++;

        // 向EventRunner发送事件
        handler.sendEvent(eventId);
```
```

```
        HiLog.info(LABEL_LOG, "end sendEvent eventId: %{public}s", eventId);
    }

    @Override
    public void onActive() {
        super.onActive();
    }

    @Override
    public void onForeground(Intent intent) {
        super.onForeground(intent);
    }
}
```

上述代码中：

- Text增加了点击事件，以触发sendEvent任务。
- sendEvent方法中创建了EventHandler事件处理器。
- 通过EventHandler的sendEvent方法发送了一个事件ID。事件ID是自增的序列。

12.6.4 运行

运行应用后，点击3次界面上的文本Start Event以触发事件派发。此时，控制台输出如下：

```
    09-11 14:47:05.056 3943-3943/com.waylau.hmos.eventhandler I
00001/MainAbilitySlice: before sendEvent
    09-11 14:47:05.057 3943-3943/com.waylau.hmos.eventhandler I
00001/MainAbilitySlice: end sendEvent eventId: 1
    09-11 14:47:05.058 3943-4024/com.waylau.hmos.eventhandler I
00001/MyEventHandler: before processEvent eventId: 1
    09-11 14:47:05.196 3943-4024/com.waylau.hmos.eventhandler I
00001/MyEventHandler: after processEvent eventId 1

    09-11 14:47:07.731 3943-3943/com.waylau.hmos.eventhandler I
00001/MainAbilitySlice: before sendEvent
    09-11 14:47:07.732 3943-4024/com.waylau.hmos.eventhandler I
00001/MyEventHandler: before processEvent eventId: 2
    09-11 14:47:07.732 3943-3943/com.waylau.hmos.eventhandler I
00001/MainAbilitySlice: end sendEvent eventId: 2
    09-11 14:47:08.104 3943-4024/com.waylau.hmos.eventhandler I
00001/MyEventHandler: after processEvent eventId 2

    09-11 14:47:09.149 3943-3943/com.waylau.hmos.eventhandler I
00001/MainAbilitySlice: before sendEvent
    09-11 14:47:09.151 3943-3943/com.waylau.hmos.eventhandler I
00001/MainAbilitySlice: end sendEvent eventId: 3
    09-11 14:47:09.159 3943-4024/com.waylau.hmos.eventhandler I
00001/MyEventHandler: before processEvent eventId: 3
```

```
    09-11 14:47:09.300 3943-4024/com.waylau.hmos.eventhandler I
00001/MyEventHandler: after processEvent eventId 3
```

从上述日志可以看出，MainAbilitySlice先发送了3次事件，而后MyEventHandler处理了这些事件。

第 13 章

视 频

HarmonyOS视频模块支持视频业务的开发和生态开放,开发者可以通过已开放的接口很容易地实现视频媒体的播放、操作和新功能的开发。

13.1 视频概述

视频媒体的常见操作有视频编解码、视频合成、视频提取、视频播放以及视频录制等。视频媒体常见的概念包括:

- 编码:编码是信息从一种形式或格式转换为另一种形式的过程。用预先规定的方法将文字、数字或其他对象编成数码,或将信息、数据转换成规定的电脉冲信号。在本模块中,编码是指编码器将原始的视频信息压缩为另一种格式的过程。
- 解码:解码是一种用特定方法把数码还原成它所代表的内容或将电脉冲信号、光信号、无线电波等转换成它所代表的信息、数据等的过程。在本模块中,解码是指解码器将接收到的数据还原为视频信息的过程,与编码过程相对应。
- 帧率:帧率是以帧称为单位的位图图像连续出现在显示器上的频率(速率),以赫兹(Hz)为单位。该术语同样适用于胶片、摄像机、计算机图形和动作捕捉系统。

13.2 实战:媒体编解码能力查询

媒体编解码能力查询主要指查询设备所支持的编解码器的MIME(Multipurpose Internet Mail

Extensions，媒体类型）列表，并判断设备是否支持指定MIME对应的编码器/解码器。

13.2.1 接口说明

媒体编解码能力查询类CodecDescriptionList的主要接口有：

- getSupportedMimes()：获取某设备所支持的编解码器的MIME列表。
- isDecodeSupportedByMime(String mime)：判断某设备是否支持指定MIME对应的解码器。
- isEncodeSupportedByMime(String mime)：判断某设备是否支持指定MIME对应的编码器。
- isDecoderSupportedByFormat(Format format)：判断某设备是否支持指定媒体格式对应的解码器。
- isEncoderSupportedByFormat(Format format)：判断某设备是否支持指定媒体格式对应的编码器。

13.2.2 创建应用

为了演示媒体编解码能力查询的功能，创建一个名为CodecDescriptionList的应用。

在应用的界面上，通过点击按钮来触发获取查询媒体编解码能力，并将能力文本信息打印到界面上。

13.2.3 修改 ability_main.xml

修改ability_main.xml内容如下：

```
<?xml version="1.0" encoding="utf-8"?>
<DirectionalLayout
    xmlns:ohos="http://schemas.huawei.com/res/ohos"
    ohos:height="match_parent"
    ohos:width="match_parent"
    ohos:alignment="center"
    ohos:orientation="vertical">

    <Button
        ohos:id="$+id:button"
        ohos:height="match_content"
        ohos:width="match_content"
        ohos:background_element="#F76543"
        ohos:text="Get"
        ohos:text_size="27fp"
        />

    <Text
        ohos:id="$+id:text"
        ohos:height="match_content"
```

```
            ohos:width="match_content"
            ohos:background_element="$graphic:background_ability_main"
            ohos:layout_alignment="horizontal_center"
            ohos:multiple_lines="true"
            ohos:text="Hello World"
            ohos:text_size="20vp"
            />

</DirectionalLayout>
```

界面预览效果如图13-1所示。

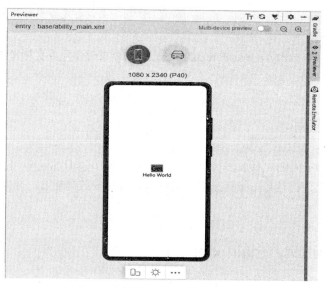

图 13-1　界面效果

13.2.4　修改 MainAbilitySlice

修改MainAbilitySlice内容如下：

```
package com.waylau.hmos.codecdescriptionlist.slice;

import com.waylau.hmos.codecdescriptionlist.ResourceTable;
import ohos.aafwk.ability.AbilitySlice;
import ohos.aafwk.content.Intent;
import ohos.agp.components.Button;
import ohos.agp.components.Text;
import ohos.media.codec.CodecDescriptionList;
import ohos.media.common.Format;

import java.util.List;

public class MainAbilitySlice extends AbilitySlice {
    @Override
    public void onStart(Intent intent) {
```

```java
        super.onStart(intent);
        super.setUIContent(ResourceTable.Layout_ability_main);

        Button button =
                (Button) findComponentById(ResourceTable.Id_button);

        Text text =
                (Text) findComponentById(ResourceTable.Id_text);

        // 为按钮设置点击事件回调
        button.setClickedListener(listener -> {
                showInfo(text);
            }
        );
    }

    private void showInfo(Text text) {
        // 获取某设备所支持的编解码器的MIME列表
        List<String> mimes = CodecDescriptionList.getSupportedMimes();
        text.setText("mimes:" + mimes);

        // 判断某设备是否支持指定MIME对应的解码器,支持返回true,否则返回false
        boolean isDecodeSupportedByMime =
            CodecDescriptionList.isDecodeSupportedByMime(Format.VIDEO_VP9);
        text.insert("isDecodeSupportedByMime:" + isDecodeSupportedByMime);

        // 判断某设备是否支持指定MIME对应的编码器,支持返回true,否则返回false
        boolean isEncodeSupportedByMime =
            CodecDescriptionList.isEncodeSupportedByMime(Format.AUDIO_FLAC);
        text.insert("isEncodeSupportedByMime:" + isEncodeSupportedByMime);

        // 判断某设备是否支持指定Format的编解码器,支持返回true,否则返回false
        Format format = new Format();
        format.putStringValue(Format.MIME, Format.VIDEO_AVC);
        format.putIntValue(Format.WIDTH, 2560);
        format.putIntValue(Format.HEIGHT, 1440);
        format.putIntValue(Format.FRAME_RATE, 30);
        format.putIntValue(Format.FRAME_INTERVAL, 1);
        boolean isDecoderSupportedByFormat =
            CodecDescriptionList.isDecoderSupportedByFormat(format);
        text.insert("isDecoderSupportedByFormat:" +
isDecoderSupportedByFormat);
        boolean isEncoderSupportedByFormat =
            CodecDescriptionList.isEncoderSupportedByFormat(format);
        text.insert("isEncoderSupportedByFormat:" +
isEncoderSupportedByFormat);
    }

    @Override
    public void onActive() {
```

```
    super.onActive();
}

@Override
public void onForeground(Intent intent) {
    super.onForeground(intent);
}
}
```

上述代码中：

- 在按钮上设置了点击事件，以触发获取查询媒体编解码能力。
- 将获取到的查询媒体编解码能力以文本信息打印到界面上。

13.2.5 运行

运行应用后，界面效果如图13-2所示。

图 13-2　界面效果

可以看到媒体编解码能力以文本信息打印到界面上。

13.3　实战：视频编解码

视频编解码的主要工作是将视频进行编码和解码。

13.3.1 接口说明

视频编解码类Codec的主要接口有：

- createDecoder()：创建解码器Codec实例。
- createEncoder()：创建编码器Codec实例。
- registerCodecListener(ICodecListener listener)：注册侦听器用来异步接收编码或解码后的数据。
- setSource(Source source, TrackInfo trackInfo)：根据解码器的源轨道信息设置数据源，对于编码器trackInfo无效。
- setSourceFormat(Format format)：编码器的管道模式下，设置编码器的编码格式。
- setCodecFormat(Format format)：普通模式设置编/解码器参数。
- setVideoSurface(Surface surface)：设置解码器的Surface。
- getAvailableBuffer(long timeout)：普通模式获取可用ByteBuffer。
- writeBuffer(ByteBuffer buffer, BufferInfo info)：推送源数据给Codec。
- getBufferFormat(ByteBuffer buffer)：获取输出Buffer数据格式。
- start()：启动编/解码。
- stop()：停止编/解码。
- release()：释放所有资源。

13.3.2 创建应用

为了演示视频编解码的功能，创建一个名为Codec的应用。
在应用的界面上分别设置编码和解码按钮，通过触发按钮来执行相应的编码和解码操作。
同时，在应用的resources/rawfile目录下，增加big_buck_bunny.mp4视频文件。

13.3.3 修改ability_main.xml

修改ability_main.xml内容如下：

```xml
<?xml version="1.0" encoding="utf-8"?>
<DirectionalLayout
    xmlns:ohos="http://schemas.huawei.com/res/ohos"
    ohos:height="match_parent"
    ohos:width="match_parent"
    ohos:alignment="center"
    ohos:orientation="vertical">

    <Button
        ohos:id="$+id:button_encode"
        ohos:height="match_content"
        ohos:width="match_content"
        ohos:background_element="#F76543"
```

```xml
            ohos:text="Encode"
            ohos:text_size="27fp"
            />

    <Button
        ohos:id="$+id:button_decode"
        ohos:height="match_content"
        ohos:width="match_content"
        ohos:background_element="#F76543"
        ohos:text="Decode"
        ohos:text_size="27fp"
        />

</DirectionalLayout>
```

界面预览效果如图13-3所示。

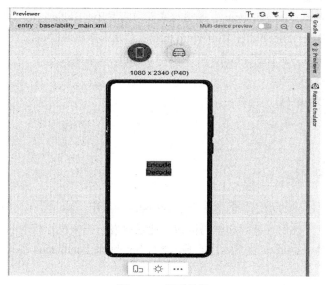

图 13-3　界面效果

13.3.4　修改 MainAbilitySlice

修改MainAbilitySlice内容如下：

```java
package com.waylau.hmos.codec.slice;

import com.waylau.hmos.codec.ResourceTable;
import ohos.aafwk.ability.AbilitySlice;
import ohos.aafwk.content.Intent;
import ohos.agp.components.Button;
import ohos.global.resource.RawFileEntry;
import ohos.hiviewdfx.HiLog;
import ohos.hiviewdfx.HiLogLabel;
import ohos.media.codec.Codec;
```

```java
import ohos.media.codec.TrackInfo;
import ohos.media.common.BufferInfo;
import ohos.media.common.Format;
import ohos.media.common.Source;

import java.io.FileDescriptor;
import java.io.IOException;
import java.nio.ByteBuffer;

public class MainAbilitySlice extends AbilitySlice {
    private static final String TAG = MainAbilitySlice.class.getSimpleName();
    private static final HiLogLabel LABEL_LOG =
            new HiLogLabel(HiLog.LOG_APP, 0x00001, TAG);

    @Override
    public void onStart(Intent intent) {
        super.onStart(intent);
        super.setUIContent(ResourceTable.Layout_ability_main);

        Button buttonEncode =
                (Button) findComponentById(ResourceTable.Id_button_encode);

        Button buttonDecode =
                (Button) findComponentById(ResourceTable.Id_button_decode);

        // 为按钮设置点击事件回调
        buttonEncode.setClickedListener(listener -> {
            try {
                encode();
            } catch (IOException e) {
                HiLog.error(LABEL_LOG, "exception : %{public}s", e);
            }
        });

        buttonDecode.setClickedListener(listener -> {
            try {
                decode();
            } catch (IOException e) {
                HiLog.error(LABEL_LOG, "exception : %{public}s", e);
            }
        });
    }

    private void decode() throws IOException {
        HiLog.info(LABEL_LOG, "before decode()");

        //1.创建解码Codec实例，可调用createDecoder()创建
        final Codec decoder = Codec.createDecoder();

        //2.调用setSource()设置数据源，支持设定文件路径或者文件File Descriptor
```

```java
        RawFileEntry rawFileEntry = this.getResourceManager().
getRawFileEntry("resources/rawfile/big_buck_bunny.mp4");
        FileDescriptor fd =
rawFileEntry.openRawFileDescriptor().getFileDescriptor();
        decoder.setSource(new Source(fd), new TrackInfo());

        //3.构造数据源格式或者从Extractor中读取数据源格式,并设置给Codec实例,调用
setSourceFormat()
        Format fmt = new Format();
        fmt.putStringValue(Format.MIME, Format.VIDEO_AVC);
        fmt.putIntValue(Format.WIDTH, 1920);
        fmt.putIntValue(Format.HEIGHT, 1080);
        fmt.putIntValue(Format.BIT_RATE, 392000);
        fmt.putIntValue(Format.FRAME_RATE, 30);
        fmt.putIntValue(Format.FRAME_INTERVAL, -1);
        decoder.setSourceFormat(fmt);

        //4.设置监听器
        decoder.registerCodecListener(listener);

        //5.开始编码
        decoder.start();

        //6.停止编码
        decoder.stop();

        //7.释放资源
        decoder.release();

        HiLog.info(LABEL_LOG, "end decode()");
    }

    private void encode() throws IOException {
        HiLog.info(LABEL_LOG, "before encode()");

        //1.创建编码Codec实例,可调用createEncoder()创建
        final Codec encoder = Codec.createEncoder();

        //2.调用setSource()设置数据源,支持设定文件路径或者文件File Descriptor
        // 获取源文件
        RawFileEntry rawFileEntry = this.getResourceManager().
getRawFileEntry("resources/rawfile/big_buck_bunny.mp4");
        FileDescriptor fd =
rawFileEntry.openRawFileDescriptor().getFileDescriptor();
        encoder.setSource(new Source(fd), new TrackInfo());

        //3.构造数据源格式或者从Extractor中读取数据源格式,并设置给Codec实例,调用
setSourceFormat()
        Format fmt = new Format();
        fmt.putStringValue(Format.MIME, Format.VIDEO_AVC);
```

```
            fmt.putIntValue(Format.WIDTH, 1920);
            fmt.putIntValue(Format.HEIGHT, 1080);
            fmt.putIntValue(Format.BIT_RATE, 392000);
            fmt.putIntValue(Format.FRAME_RATE, 30);
            fmt.putIntValue(Format.FRAME_INTERVAL, -1);
            encoder.setSourceFormat(fmt);

            //4.设置监听器
            encoder.registerCodecListener(listener);

            //5.开始编码
            encoder.start();

            //6.停止编码
            encoder.stop();

            //7.释放资源
            encoder.release();

            HiLog.info(LABEL_LOG, "end encode()");
        }

        private Codec.ICodecListener listener = new Codec.ICodecListener() {
            @Override
            public void onReadBuffer(ByteBuffer byteBuffer, BufferInfo bufferInfo, int trackId) {
                HiLog.info(LABEL_LOG, "onReadBuffer trackId: %{public}s", trackId);
            }

            @Override
            public void onError(int errorCode, int act, int trackId) {
                throw new RuntimeException();
            }
        };

        @Override
        public void onActive() {
            super.onActive();
        }

        @Override
        public void onForeground(Intent intent) {
            super.onForeground(intent);
        }
    }
```

上述代码中：

- 在按钮上设置了点击事件，以触发编码和解码。
- 创建解码Codec实例，可调用createDecoder()创建。

- 创建编码Codec实例，可调用createEncoder()创建。
- 通过registerCodecListener方法分别给解码Codec实例、编码Codec实例注册监听器。该监听器是Codec.ICodecListener的实例。

13.3.5　运行

运行应用，分别点击Encode、Decode按钮，可以看到控制台输出内容如下：

```
09-12 09:52:18.949 27137-27137/com.waylau.hmos.codec I
00001/MainAbilitySlice: before encode()
09-12 09:52:18.994 27137-27137/com.waylau.hmos.codec I
00001/MainAbilitySlice: end encode()
09-12 09:52:20.804 27137-27137/com.waylau.hmos.codec I
00001/MainAbilitySlice: before decode()
09-12 09:52:20.831 27137-27137/com.waylau.hmos.codec I
00001/MainAbilitySlice: end decode()
```

13.4　实战：视频播放

视频播放包括播放控制、播放设置和播放查询，如播放的开始/停止、播放速度设置和是否循环播放等。

13.4.1　接口说明

视频播放类Player的主要接口有：

- Player(Context context)：创建Player实例。
- setSource(Source source)：设置媒体源。
- prepare()：准备播放。
- play()：开始播放。
- pause()：暂停播放。
- stop()：停止播放。
- rewindTo(long microseconds)：拖曳播放。
- setVolume(float volume)：调节播放音量。
- setVideoSurface(Surface surface)：设置视频播放的窗口。
- enableSingleLooping(boolean looping)：设置为单曲循环。
- isSingleLooping()：检查是否单曲循环播放。
- isNowPlaying()：检查是否播放。
- getCurrentTime()：获取当前播放位置。
- getDuration()：获取媒体文件总时长。

- getVideoWidth()：获取视频宽度。
- getVideoHeight()：获取视频高度。
- setPlaybackSpeed(float speed)：设置播放速度。
- getPlaybackSpeed()：获取播放速度。
- setAudioStreamType(int type)：设置音频类型。
- getAudioStreamType()：获取音频类型。
- setNextPlayer(Player next)：设置当前播放结束后的下一个播放器。
- reset()：重置播放器。
- release()：释放播放资源。
- setPlayerCallback(IPlayerCallback callback)：注册回调，接收播放器的事件通知或异常通知。

13.4.2 创建应用

为了演示视频播放的功能，创建一个名为Player的应用。

在应用的界面上，通过点击按钮来触发播放视频。

在rawfile目录下放置一个big_buck_bunny.mp4视频文件，以备测试。

13.4.3 修改 ability_main.xml

修改ability_main.xml内容如下：

```xml
<?xml version="1.0" encoding="utf-8"?>
<DirectionalLayout
    xmlns:ohos="http://schemas.huawei.com/res/ohos"
    ohos:height="match_parent"
    ohos:width="match_parent"
    ohos:alignment="center"
    ohos:orientation="vertical">

    <DirectionalLayout
        ohos:height="60vp"
        ohos:width="match_content"
        ohos:orientation="horizontal">

        <Button
            ohos:id="$+id:button_play"
            ohos:height="match_content"
            ohos:width="match_content"
            ohos:background_element="#F76543"
            ohos:margin="10vp"
            ohos:padding="10vp"
            ohos:text="Play"
            ohos:text_size="20vp"
            />
```

```xml
        <Button
            ohos:id="$+id:button_pause"
            ohos:height="match_content"
            ohos:width="match_content"
            ohos:background_element="#F76543"
            ohos:margin="10vp"
            ohos:padding="10vp"
            ohos:text="Pause"
            ohos:text_size="20vp"
            />

        <Button
            ohos:id="$+id:button_stop"
            ohos:height="match_content"
            ohos:width="match_content"
            ohos:background_element="#F76543"
            ohos:margin="10vp"
            ohos:padding="10vp"
            ohos:text="Stop"
            ohos:text_size="20vp"
            />
    </DirectionalLayout>

    <DependentLayout
        ohos:id="$+id:layout_surface_provider"
        ohos:height="match_parent"
        ohos:width="match_content"
        ohos:orientation="horizontal"/>

</DirectionalLayout>
```

界面预览效果如图13-4所示。

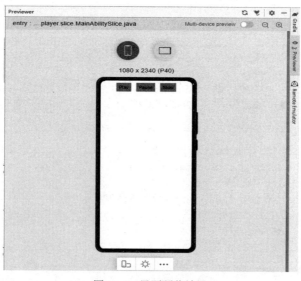

图13-4　界面预览效果

13.4.4 修改 MainAbilitySlice

修改MainAbilitySlice内容如下：

```java
package com.waylau.hmos.player.slice;

import com.waylau.hmos.player.ResourceTable;
import ohos.aafwk.ability.AbilitySlice;
import ohos.aafwk.content.Intent;
import ohos.agp.components.Button;
import ohos.agp.components.ComponentContainer;
import ohos.agp.components.DependentLayout;
import ohos.agp.components.surfaceprovider.SurfaceProvider;
import ohos.agp.graphics.Surface;
import ohos.agp.graphics.SurfaceOps;
import ohos.global.resource.RawFileDescriptor;
import ohos.hiviewdfx.HiLog;
import ohos.hiviewdfx.HiLogLabel;
import ohos.media.common.Source;
import ohos.media.player.Player;

public class MainAbilitySlice extends AbilitySlice {
    private static final String TAG = MainAbilitySlice.class.getSimpleName();
    private static final HiLogLabel LABEL_LOG =
            new HiLogLabel(HiLog.LOG_APP, 0x00001, TAG);

    private static Player player;

    private SurfaceProvider surfaceProvider;

    @Override
    public void onStart(Intent intent) {
        super.onStart(intent);
        super.setUIContent(ResourceTable.Layout_ability_main);

        Button buttonPlay =
                (Button) findComponentById(ResourceTable.Id_button_play);

        // 为按钮设置点击事件回调
        buttonPlay.setClickedListener(listener -> player.play());

        Button buttonPause =
                (Button) findComponentById(ResourceTable.Id_button_pause);

        // 为按钮设置点击事件回调
        buttonPause.setClickedListener(listener -> player.pause());

        Button buttonStop =
                (Button) findComponentById(ResourceTable.Id_button_stop);
```

```java
        // 为按钮设置点击事件回调
        buttonStop.setClickedListener(listener -> player.stop());
    }

    @Override
    public void onForeground(Intent intent) {
        super.onForeground(intent);
    }


    @Override
    protected void onActive() {
        super.onActive();
        initSurfaceProvider();
    }

    private void initSurfaceProvider() {
        HiLog.info(LABEL_LOG, "before initSurfaceProvider");

        player = new Player(this);

        surfaceProvider = new SurfaceProvider(this);
        surfaceProvider.getSurfaceOps().get().addCallback(new VideoSurfaceCallback());
        surfaceProvider.pinToZTop(true);
        surfaceProvider.setWidth(ComponentContainer.LayoutConfig.MATCH_CONTENT);
        surfaceProvider.setHeight(ComponentContainer.LayoutConfig.MATCH_PARENT);

        DependentLayout layout
                = (DependentLayout) findComponentById(ResourceTable.Id_layout_surface_provider);
        layout.addComponent(surfaceProvider);

        HiLog.info(LABEL_LOG, "end initSurfaceProvider");
    }

    class VideoSurfaceCallback implements SurfaceOps.Callback {
        @Override
        public void surfaceCreated(SurfaceOps surfaceOps) {
            if (surfaceProvider.getSurfaceOps().isPresent()) {
                Surface surface = surfaceProvider.getSurfaceOps().get().getSurface();
                playLocalFile(surface);
            }

            HiLog.info(LABEL_LOG, "surfaceCreated");
        }
```

```java
        @Override
        public void surfaceChanged(SurfaceOps surfaceOps, int i, int i1, int i2) {
            HiLog.info(LABEL_LOG,
"surfaceChanged, %{public}s, %{public}s, %{public}s",
                    i, i1, i2);
        }

        @Override
        public void surfaceDestroyed(SurfaceOps surfaceOps) {
            HiLog.info(LABEL_LOG, "surfaceDestroyed");
        }
    }

    private void playLocalFile(Surface surface) {
        HiLog.info(LABEL_LOG, "before playLocalFile");

        try {
            RawFileDescriptor filDescriptor =
                    getResourceManager()
                    .getRawFileEntry("resources/rawfile/big_buck_bunny.mp4")
                    .openRawFileDescriptor();

            Source source = new Source(filDescriptor.getFileDescriptor(),
                    filDescriptor.getStartPosition(),
filDescriptor.getFileSize());
            player.setSource(source);
            player.setVideoSurface(surface);
            player.setPlayerCallback(new VideoPlayerCallback());
            player.prepare();

            surfaceProvider.setTop(0);
        } catch (Exception e) {
            e.printStackTrace();
        }

        HiLog.info(LABEL_LOG, "before playLocalFile");
    }

    @Override
    protected void onStop() {
        super.onStop();
        if (player != null) {
            player.stop();
        }
        surfaceProvider.removeFromWindow();
    }

    private class VideoPlayerCallback implements Player.IPlayerCallback {
```

```java
            @Override
            public void onPrepared() {
                HiLog.info(LABEL_LOG, "onPrepared");
            }

            @Override
            public void onMessage(int i, int i1) {
                HiLog.info(LABEL_LOG, "onMessage, %{public}s, %{public}s", i, i1);
            }

            @Override
            public void onError(int i, int i1) {
                HiLog.info(LABEL_LOG, "onError, %{public}s, %{public}s", i, i1);
            }

            @Override
            public void onResolutionChanged(int i, int i1) {
                HiLog.info(LABEL_LOG,
"onResolutionChanged, %{public}s, %{public}s", i, i1);
            }

            @Override
            public void onPlayBackComplete() {
                HiLog.info(LABEL_LOG, "onPlayBackComplete");
            }

            @Override
            public void onRewindToComplete() {
                HiLog.info(LABEL_LOG, "onRewindToComplete");
            }

            @Override
            public void onBufferingChange(int i) {
                HiLog.info(LABEL_LOG, "onBufferingChange, %{public}s", i);
            }

            @Override
            public void onNewTimedMetaData(Player.MediaTimedMetaData
mediaTimedMetaData) {
                HiLog.info(LABEL_LOG, "onNewTimedMetaData");
            }

            @Override
            public void onMediaTimeIncontinuity(Player.MediaTimeInfo
mediaTimeInfo) {
                HiLog.info(LABEL_LOG, "onMediaTimeIncontinuity");
            }
        }
    }
```

上述代码中：

- 在按钮上设置了点击事件，以触发Player的播放、暂停和停止操作。
- 重写了onActive方法，以初始化SurfaceProvider。
- SurfaceProvider会设置VideoSurfaceCallback的回调。
- 当VideoSurfaceCallback的surfaceCreated回调执行时，Player会准备好视频源以备后续操作。
- 重写了onStop方法，以关闭Player。

13.4.5 运行

运行应用后，界面效果如图13-5所示。

图13-5　界面效果

13.5　实战：视频录制

视频录制的主要工作是选择视频/音频来源后，录制并生成视频/音频文件。

13.5.1　接口说明

视频录制类Recorder的主要接口有：

- Recorder()：创建Recorder实例。
- setSource(Source source)：设置音视频源。
- setAudioProperty(AudioProperty property)：设置音频属性。
- setVideoProperty(VideoProperty property)：设置视频属性。
- setStorageProperty(StorageProperty property)：设置音视频存储属性。

- prepare()：准备录制资源。
- start()：开始录制。
- stop()：停止录制。
- pause()：暂停录制。
- resume()：恢复录制。
- reset()：重置录制。
- setRecorderLocation(float latitude, float longitude)：设置视频的经纬度。
- setOutputFormat(int outputFormat)：设置输出文件格式。
- getVideoSurface()：获取视频窗口。
- setRecorderProfile(RecorderProfile profile)：设置媒体录制配置信息。
- registerRecorderListener(IRecorderListener listener)：注册媒体录制回调。
- release()：释放媒体录制资源。

13.5.2 创建应用

为了演示视频播放的功能，创建一个名为Recorder的应用。

在应用的界面上，通过点击按钮来触发录制视频。

在rawfile目录下放置一个big_buck_bunny.mp4视频文件，以备测试。

13.5.3 修改 ability_main.xml

修改ability_main.xml内容如下：

```xml
<?xml version="1.0" encoding="utf-8"?>
<DirectionalLayout
    xmlns:ohos="http://schemas.huawei.com/res/ohos"
    ohos:height="match_parent"
    ohos:width="match_parent"
    ohos:alignment="center"
    ohos:orientation="vertical">

    <DirectionalLayout
        ohos:height="60vp"
        ohos:width="match_content"
        ohos:orientation="horizontal">

        <Button
            ohos:id="$+id:button_start"
            ohos:height="match_content"
            ohos:width="match_content"
            ohos:background_element="#F76543"
            ohos:margin="10vp"
            ohos:padding="10vp"
            ohos:text="Start"
```

```xml
            ohos:text_size="27fp"
            />

        <Button
            ohos:id="$+id:button_pause"
            ohos:height="match_content"
            ohos:width="match_content"
            ohos:background_element="#F76543"
            ohos:margin="10vp"
            ohos:padding="10vp"
            ohos:text="Pause"
            ohos:text_size="27fp"
            />

        <Button
            ohos:id="$+id:button_stop"
            ohos:height="match_content"
            ohos:width="match_content"
            ohos:background_element="#F76543"
            ohos:margin="10vp"
            ohos:padding="10vp"
            ohos:text="Stop"
            ohos:text_size="27fp"
            />
    </DirectionalLayout>

    <Text
        ohos:id="$+id:text_helloworld"
        ohos:height="match_parent"
        ohos:width="match_content"
        ohos:background_element="$graphic:background_ability_main"
        ohos:layout_alignment="horizontal_center"
        ohos:text="Hello World"
        ohos:text_size="50"
        />

</DirectionalLayout>
```

界面预览效果如图13-6所示。

图13-6　界面预览效果

13.5.4 修改 MainAbilitySlice

修改MainAbilitySlice内容如下：

```java
package com.waylau.hmos.recorder.slice;

import com.waylau.hmos.recorder.ResourceTable;
import ohos.aafwk.ability.AbilitySlice;
import ohos.aafwk.content.Intent;
import ohos.agp.components.Button;
import ohos.agp.components.Text;
import ohos.global.resource.RawFileDescriptor;
import ohos.media.common.Source;
import ohos.media.common.StorageProperty;
import ohos.media.recorder.Recorder;

import java.io.IOException;

public class MainAbilitySlice extends AbilitySlice {
    private final Recorder recorder = new Recorder();

    @Override
    public void onStart(Intent intent) {
        super.onStart(intent);
        super.setUIContent(ResourceTable.Layout_ability_main);

        Text text =
                (Text) findComponentById(ResourceTable.Id_text_helloworld);

        Button buttonStart =
                (Button) findComponentById(ResourceTable.Id_button_start);

        // 为按钮设置点击事件回调
        buttonStart.setClickedListener(listener -> {
            recorder.start();
            text.setText("Start");
        });

        Button buttonPause =
                (Button) findComponentById(ResourceTable.Id_button_pause);

        // 为按钮设置点击事件回调
        buttonPause.setClickedListener(listener -> {
            recorder.pause();
            text.setText("Pause");
        });

        Button buttonStop =
```

```java
        (Button) findComponentById(ResourceTable.Id_button_stop);

    // 为按钮设置点击事件回调
    buttonStop.setClickedListener(listener -> {
        recorder.stop();
        text.setText("Stop");
    });
}

@Override
public void onActive() {
    super.onActive();

    RawFileDescriptor filDescriptor = null;
    try {
        filDescriptor = getResourceManager()
                .getRawFileEntry("resources/rawfile/big_buck_bunny.mp4")
                .openRawFileDescriptor();
    } catch (IOException e) {
        e.printStackTrace();
    }

    // 设置媒体源
    Source source = new Source(filDescriptor.getFileDescriptor(),
            filDescriptor.getStartPosition(),
filDescriptor.getFileSize());
    source.setRecorderAudioSource(Recorder.AudioSource.DEFAULT);
    recorder.setSource(source);

    // 设置存储属性
    String path = "record.mp4";
    StorageProperty storageProperty = new StorageProperty.Builder()
            .setRecorderPath(path)
            .setRecorderMaxDurationMs(-1)
            .setRecorderMaxFileSizeBytes(-1)
            .build();
    recorder.setStorageProperty(storageProperty);

    // 准备
    recorder.prepare();
}

@Override
protected void onStop() {
    super.onStop();
    if (recorder != null) {
        recorder.stop();
    }
}
```

```
    @Override
    public void onForeground(Intent intent) {
        super.onForeground(intent);
    }
}
```

上述代码中：

- 调用Recorder()方法创建Recorder实例。
- 在按钮上设置了点击事件，以触发Recorder的开始、暂停和停止操作，并将状态信息回写到界面的Text。
- 重写了onActive方法，以设置Recorder实例的属性。这些属性包括媒体源、存储属性等。
- 重写了onStop方法，以关闭Player。

13.5.5 运行

运行应用后，界面效果如图13-7所示。

图 13-7 界面效果

第 14 章

图 像

HarmonyOS图像模块支持图像业务的开发，常见功能如图像解码、图像编码、基本的位图操作、图像编辑等。

14.1 图像概述

HarmonyOS图像模块支持图像业务的开发，常见功能如图像解码、图像编码、基本的位图操作、图像编辑等。当然，也支持通过接口组合来实现更复杂的图像处理逻辑。

14.1.1 基本概念

图像业务包含以下基本概念：

- 图像解码：图像解码就是不同的存档格式图片（如JPEG、PNG等）解码为无压缩的位图格式，以方便在应用或者系统中进行相应的处理。
- PixelMap：PixelMap是图像解码后无压缩的位图格式，用于图像显示或者进一步的处理。
- 渐进式解码：渐进式解码是在无法一次性提供完整图像文件数据的场景下，随着图像文件数据的逐步增加，通过多次增量解码逐步完成图像解码的模式。
- 预乘：预乘时，RGB各通道的值被替换为原始值乘以Alpha通道不透明的比例（0~1）后的值，方便后期直接合成叠加。不预乘指RGB各通道的数值是图像的原始值，与Alpha通道的值无关。
- 图像编码：图像编码就是将无压缩的位图格式编码成不同格式的存档格式图片（JPEG、PNG等），以方便在应用或者系统中进行相应的处理。

14.1.2 约束与限制

为了及时释放本地资源,建议在图像解码的ImageSource对象、位图图像PixelMap对象或图像编码的ImagePacker对象使用完成后,主动调用ImageSource、PixelMap和ImagePacker的release()方法。

14.2 实战:图像解码和编码

图像解码就是将所支持格式的存档图片解码成统一的PixelMap图像,用于后续图像显示或其他处理,比如旋转、缩放、裁剪等。当前支持的格式包括JPEG、PNG、GIF、HEIF、WebP、BMP。

14.2.1 接口说明

ImageSource主要用于图像解码。ImageSource的主要接口有:

- create(String pathName, SourceOptions opts):从图像文件路径创建图像数据源。
- create(InputStream is, SourceOptions opts):从输入流创建图像数据源。
- create(byte[] data, SourceOptions opts):从字节数组创建图像源。
- create(byte[] data, int offset, int length, SourceOptions opts):从字节数组指定范围创建图像源。
- create(File file, SourceOptions opts):从文件对象创建图像数据源。
- create(FileDescriptor fd, SourceOptions opts):从文件描述符创建图像数据源。
- createIncrementalSource(SourceOptions opts):创建渐进式图像数据源。
- createIncrementalSource(IncrementalSourceOptions opts):创建渐进式图像数据源,支持设置渐进式数据更新模式。
- createPixelmap(DecodingOptions opts):从图像数据源解码并创建PixelMap图像。
- createPixelmap(int index, DecodingOptions opts):从图像数据源解码并创建PixelMap图像,如果图像数据源支持多张图片,则支持指定图像索引。
- updateData(byte[] data, boolean isFinal):更新渐进式图像源数据。
- updateData(byte[] data, int offset, int length, boolean isFinal):更新渐进式图像源数据,支持设置输入数据的有效数据范围。
- getImageInfo():获取图像基本信息。
- getImageInfo(int index):根据特定的索引获取图像基本信息。
- getSourceInfo():获取图像源信息。
- release():释放对象关联的本地资源。

14.2.2 创建应用

为了演示图像编解码的功能,创建一个名为ImageCodec的应用。

在应用的界面上,通过点击按钮来触发解码或者编码图片的操作。

在media目录下放置一张用于测试的图片waylau_616_616.jpeg。

14.2.3 修改 ability_main.xml

修改ability_main.xml内容如下:

```xml
<?xml version="1.0" encoding="utf-8"?>
<DirectionalLayout
    xmlns:ohos="http://schemas.huawei.com/res/ohos"
    ohos:height="match_parent"
    ohos:width="match_parent"
    ohos:orientation="horizontal">

    <DirectionalLayout
        ohos:height="match_content"
        ohos:width="0vp"
        ohos:orientation="vertical"
        ohos:weight="1">

        <Button
            ohos:id="$+id:button_decode"
            ohos:height="match_content"
            ohos:width="match_content"
            ohos:background_element="#F76543"
            ohos:left_padding="10vp"
            ohos:text="Decode"
            ohos:text_size="27fp"
            />

        <Image
            ohos:id="$+id:image_decode"
            ohos:height="100vp"
            ohos:width="100vp"/>

    </DirectionalLayout>

    <DirectionalLayout
        ohos:height="match_content"
        ohos:width="0vp"
        ohos:orientation="vertical"
        ohos:weight="1">

        <Button
```

```
            ohos:id="$+id:button_encode"
            ohos:height="match_content"
            ohos:width="match_content"
            ohos:background_element="#F76543"
            ohos:left_padding="10vp"
            ohos:text="Encode"
            ohos:text_size="27fp"
            />

    <Image
        ohos:id="$+id:image_encode"
        ohos:height="100vp"
        ohos:width="100vp"/>
    </DirectionalLayout>
</DirectionalLayout>
```

界面预览效果如图14-1所示。

图 14-1　界面预览效果

14.2.4　修改 MainAbilitySlice

修改MainAbilitySlice内容如下：

```
package com.waylau.hmos.imagedecoder.slice;

import com.waylau.hmos.imagedecoder.ResourceTable;
import ohos.aafwk.ability.AbilitySlice;
import ohos.aafwk.content.Intent;
import ohos.agp.components.Button;
import ohos.agp.components.Image;
import ohos.global.resource.NotExistException;
import ohos.hiviewdfx.HiLog;
```

```java
import ohos.hiviewdfx.HiLogLabel;
import ohos.media.image.ImagePacker;
import ohos.media.image.ImageSource;
import ohos.media.image.PixelMap;
import ohos.media.image.common.Rect;
import ohos.media.image.common.Size;

import java.io.*;
import java.nio.file.Paths;

public class MainAbilitySlice extends AbilitySlice {
    private static final String TAG = MainAbilitySlice.class.getSimpleName();
    private static final HiLogLabel LABEL_LOG =
            new HiLogLabel(HiLog.LOG_APP, 0x00001, TAG);

    private ImageSource imageSource;
    private PixelMap pixelMap;

    @Override
    public void onStart(Intent intent) {
        super.onStart(intent);
        super.setUIContent(ResourceTable.Layout_ability_main);

        Button buttonDecode =
                (Button) findComponentById(ResourceTable.Id_button_decode);

        // 为按钮设置点击事件回调
        buttonDecode.setClickedListener(listener -> {
            try {
                decode();
            } catch (IOException e) {
                e.printStackTrace();
            } catch (NotExistException e) {
                e.printStackTrace();
            }
        });

        Button buttonEncode =
                (Button) findComponentById(ResourceTable.Id_button_encode);

        // 为按钮设置点击事件回调
        buttonEncode.setClickedListener(listener -> {
            try {
                encode();
            } catch (IOException e) {
                e.printStackTrace();
            }
        });
    }
```

```java
        private void decode() throws IOException, NotExistException {
            // 获取图片流
            InputStream drawableInputStream =
getResourceManager().getResource(ResourceTable.Media_waylau_616_616);

            // 创建图像数据源ImageSource对象
            imageSource = ImageSource.create(drawableInputStream,
this.getSourceOptions());

            // 普通解码叠加旋转、缩放、裁剪
            pixelMap = imageSource.createPixelmap(this.getDecodingOptions());

            Image imageDecode =
                    (Image) findComponentById(ResourceTable.Id_image_decode);

            imageDecode.setPixelMap(pixelMap);
        }

        private void encode() throws IOException {
            HiLog.info(LABEL_LOG, "before encode()");

            // 创建图像编码ImagePacker对象
            ImagePacker imagePacker = ImagePacker.create();

            // 获取数据目录
            File dataDir = new File(this.getDataDir().toString());
            if(!dataDir.exists()){
                dataDir.mkdirs();
            }

            // 文件路径
            String filePath =
Paths.get(dataDir.toString(),"test.jpeg").toString();

            // 构建目标文件
            File targetFile = new File(filePath);

            // 设置编码输出流和编码参数
            FileOutputStream outputStream = new FileOutputStream(targetFile);

            // 初始化打包
            imagePacker.initializePacking(outputStream,
this.getPackingOptions());

            // 添加需要编码的PixelMap对象，进行编码操作
            imagePacker.addImage(pixelMap);

            // 完成图像打包任务
            imagePacker.finalizePacking();
```

```java
        Image imageEncode =
                (Image) findComponentById(ResourceTable.Id_image_encode);

        // 文件转成图像
        imageSource = ImageSource.create(targetFile,this.getSourceOptions());

        pixelMap = imageSource.createPixelmap(this.getDecodingOptions());

        imageEncode.setPixelMap(pixelMap);

        HiLog.info(LABEL_LOG, "end encode()");
    }


    // 设置打包格式
    private ImagePacker.PackingOptions getPackingOptions() {
        ImagePacker.PackingOptions packingOptions = new ImagePacker.PackingOptions();
        packingOptions.format = "image/jpeg"; // 设置format为编码的图像格式，当前支持JPEG格式
        packingOptions.quality = 90; // 设置quality为图像质量，范围为0~100，100为最佳质量
        return packingOptions;
    }

    // 设置解码格式
    private ImageSource.DecodingOptions getDecodingOptions() {
        ImageSource.DecodingOptions decodingOpts = new ImageSource.DecodingOptions();
        decodingOpts.desiredSize = new Size(600, 600);
        decodingOpts.desiredRegion = new Rect(0, 0, 600, 600);
        decodingOpts.rotateDegrees = 90;

        return decodingOpts;
    }

    // 设置数据源的格式信息
    private ImageSource.SourceOptions getSourceOptions() {
        ImageSource.SourceOptions sourceOptions = new ImageSource.SourceOptions();
        sourceOptions.formatHint = "image/jpeg";

        return sourceOptions;
    }

    @Override
    protected void onStop() {
        super.onStop();
        if (imageSource != null) {
            // 释放资源
```

```
            imageSource.release();
        }
        if (pixelMap != null) {
            // 释放资源
            pixelMap.release();
        }
    }

    @Override
    public void onForeground(Intent intent) {
        super.onForeground(intent);
    }
}
```

上述代码在按钮上设置了点击事件,以触发解码和编码操作。

14.2.5 解码操作说明

解码的过程是这样的:

- 在decode()方法内部,通过getResourceManager().getResource方法获取测试图片,并转为输入流。
- 将上述输入流作为创建图像数据源ImageSource对象的参数之一。
- 通过imageSource.createPixelmap方法将ImageSource转为PixelMap。
- 通过imageDecode.setPixelMap方法将图像信息设置到Image组件上,从而在界面上显示图片。

运行应用后,点击Decode按钮,界面效果如图14-2所示。

图14-2 界面效果

14.2.6 编码操作说明

编码的过程是这样的：

- 在encode()方法内部，通过ImagePacker.create()方法来创建图像编码ImagePacker对象。
- 构建目标文件targetFile。
- 执行initializePacking、addImage、finalizePacking等系列操作后完成打包。
- 最后将图像信息设置到Image组件上，从而在界面上显示图片。

运行应用后，点击Encode按钮，界面效果如图14-3所示。

图 14-3　界面效果

14.3　实战：位图操作

位图操作是指对PixelMap图像进行相关的操作，比如创建、查询信息、读写像素数据等。

14.3.1　接口说明

位图操作类PixelMap的主要接口有：

- create(InitializationOptions opts)：根据图像大小、像素格式、alpha类型等初始化选项创建PixelMap。
- create(int[] colors, InitializationOptions opts)：根据图像大小、像素格式、alpha类型等初始化选项，以像素颜色数组为数据源创建PixelMap。

- create(int[] colors, int offset, int stride, InitializationOptions opts)：根据图像大小、像素格式、alpha类型等初始化选项，以像素颜色数组、起始偏移量、行像素大小描述的数据源创建PixelMap。
- create(PixelMap source, InitializationOptions opts)：根据图像大小、像素格式、alpha类型等初始化选项，以源PixelMap为数据源创建PixelMap。
- create(PixelMap source, Rect srcRegion, InitializationOptions opts)：根据图像大小、像素格式、alpha类型等初始化选项，以源PixelMap、源裁剪区域描述的数据源创建PixelMap。
- getBytesNumberPerRow()：获取每行像素数据占用的字节数。
- getPixelBytesCapacity()：获取存储Pixelmap像素数据的内存容量。
- isEditable()：判断PixelMap是否允许修改。
- isSameImage(PixelMap other)：判断两个图像是否相同，包括ImageInfo属性信息和像素数据。
- readPixel(Position pos)：读取指定位置像素的颜色值，返回的颜色格式为PixelFormat.ARGB_8888。
- readPixels(int[] pixels, int offset, int stride, Rect region)：读取指定区域像素的颜色值，输出到以起始偏移量、行像素大小描述的像素数组，返回的颜色格式为PixelFormat.ARGB_8888。
- readPixels(Buffer dst)：读取像素的颜色值到缓冲区，返回的数据是PixelMap中像素数据的原样复制，即返回的颜色数据格式与PixelMap中的像素格式一致。
- resetConfig(Size size, PixelFormat pixelFormat)：重置PixelMap的大小和像素格式配置，但不会改变原有的像素数据，也不会重新分配像素数据的内存，重置后图像数据的字节数不能超过PixelMap的内存容量。
- setAlphaType(AlphaType alphaType)：设置PixelMap的Alpha类型。
- writePixel(Position pos, int color)：向指定位置像素写入颜色值，写入颜色格式为PixelFormat.ARGB_8888。
- writePixels(int[] pixels, int offset, int stride, Rect region)：将像素颜色数组、起始偏移量、行像素的个数描述的源像素数据写入PixelMap的指定区域，写入颜色格式为PixelFormat.ARGB_8888。
- writePixels(Buffer src)：将缓冲区描述的源像素数据写入PixelMap，写入的数据将原样覆盖PixelMap中的像素数据，即写入数据的颜色格式应与PixelMap的配置兼容。
- writePixels(int color)：将所有像素都填充为指定的颜色值，写入颜色格式为PixelFormat.ARGB_8888。
- getPixelBytesNumber()：获取全部像素数据包含的字节数。
- setBaseDensity(int baseDensity)：设置PixelMap的基础像素密度值。
- getBaseDensity()：获取PixelMap的基础像素密度值。
- setUseMipmap(boolean useMipmap)：设置PixelMap渲染是否使用mipmap。
- useMipmap()：获取PixelMap渲染是否使用mipmap。
- getNinePatchChunk()：获取图像的NinePatchChunk数据。
- getFitDensitySize(int targetDensity)：获取适应目标像素密度的图像缩放的尺寸。
- getImageInfo()：获取图像基本信息。
- release()：释放对象关联的本地资源。

14.3.2 创建应用

为了演示位图操作的功能，创建一个名为PixelMap的应用。

在应用的界面上，通过点击按钮来触发位图操作。

14.3.3 修改 ability_main.xml

修改ability_main.xml内容如下：

```xml
<?xml version="1.0" encoding="utf-8"?>
<DirectionalLayout
    xmlns:ohos="http://schemas.huawei.com/res/ohos"
    ohos:height="match_parent"
    ohos:width="match_parent"
    ohos:alignment="center"
    ohos:orientation="vertical">

    <Button
        ohos:id="$+id:button_create"
        ohos:height="match_content"
        ohos:width="match_content"
        ohos:background_element="#F76543"
        ohos:text="Create"
        ohos:text_size="27fp"
        />

    <Button
        ohos:id="$+id:button_get"
        ohos:height="match_content"
        ohos:width="match_content"
        ohos:background_element="#F76543"
        ohos:text="Get Info"
        ohos:text_size="27fp"
        />

    <Button
        ohos:id="$+id:button_read_write"
        ohos:height="match_content"
        ohos:width="match_content"
        ohos:background_element="#F76543"
        ohos:text="Read and write"
        ohos:text_size="27fp"
        />

</DirectionalLayout>
```

界面预览效果如图14-4所示。

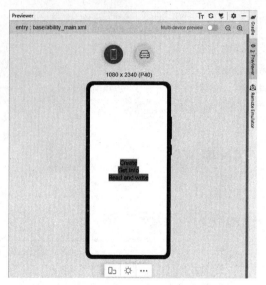

图 14-4　界面预览效果

14.3.4　修改 MainAbilitySlice

修改MainAbilitySlice内容如下：

```java
package com.waylau.hmos.pixelmap.slice;

import com.waylau.hmos.pixelmap.ResourceTable;
import ohos.aafwk.ability.AbilitySlice;
import ohos.aafwk.content.Intent;
import ohos.agp.components.Button;
import ohos.hiviewdfx.HiLog;
import ohos.hiviewdfx.HiLogLabel;
import ohos.media.image.PixelMap;
import ohos.media.image.common.PixelFormat;
import ohos.media.image.common.Position;
import ohos.media.image.common.Rect;
import ohos.media.image.common.Size;

import java.nio.IntBuffer;

public class MainAbilitySlice extends AbilitySlice {
    private static final String TAG = MainAbilitySlice.class.getSimpleName();
    private static final HiLogLabel LABEL_LOG =
            new HiLogLabel(HiLog.LOG_APP, 0x00001, TAG);

    private PixelMap pixelMap1;
    private PixelMap pixelMap2;

    @Override
    public void onStart(Intent intent) {
```

```
        super.onStart(intent);
        super.setUIContent(ResourceTable.Layout_ability_main);

        Button buttonCreate =
                (Button) findComponentById(ResourceTable.Id_button_create);
        buttonCreate.setClickedListener(listener -> createPixelMap());

        Button buttonGet =
                (Button) findComponentById(ResourceTable.Id_button_get);
        buttonGet.setClickedListener(listener -> getPixelMapInfo());

        Button buttonReadWrite =
                (Button)
findComponentById(ResourceTable.Id_button_read_write);
        buttonReadWrite.setClickedListener(listener -> readWritePixels());
    }
    ...
}
```

上述代码在按钮上设置了点击事件，以触发位图操作。

14.3.5 创建 PixelMap 操作说明

createPixelMap方法用于创建PixelMap操作，代码如下：

```
private void createPixelMap() {
    HiLog.info(LABEL_LOG, "before createPixelMap");

    // 像素颜色数组
    int[] defaultColors = new int[]{5, 5, 5, 5, 6, 6, 3, 3, 3, 0};

    // 初始化选项
    PixelMap.InitializationOptions initializationOptions = new
PixelMap.InitializationOptions();
    initializationOptions.size = new Size(3, 2);
    initializationOptions.pixelFormat = PixelFormat.ARGB_8888;

    // 根据像素颜色数组、初始化选项创建位图对象PixelMap
    pixelMap1 = PixelMap.create(defaultColors, initializationOptions);

    // 根据PixelMap作为数据源创建
    pixelMap2 = PixelMap.create(pixelMap1, initializationOptions);

    HiLog.info(LABEL_LOG, "end createPixelMap");
}
```

上述方法中：

- pixelMap1根据像素颜色数组、初始化选项创建位图对象PixelMap。

- pixelMap2以PixelMap作为数据源创建。

运行应用后，点击Create按钮，控制台输出内容如下：

```
02-08 14:40:18.664 2717-2717/com.waylau.hmos.pixelmap I
00001/MainAbilitySlice: before createPixelMap
02-08 14:40:18.669 2717-2717/com.waylau.hmos.pixelmap I
00001/MainAbilitySlice: end createPixelMap
```

14.3.6　从位图对象中获取信息操作说明

getPixelMapInfo方法用于从位图对象中获取信息，代码如下：

```java
private void getPixelMapInfo() {
    //从位图对象中获取信息
    long capacity = pixelMap1.getPixelBytesCapacity();
    long bytesNumber = pixelMap1.getPixelBytesNumber();
    int rowBytes = pixelMap1.getBytesNumberPerRow();
    byte[] ninePatchData = pixelMap1.getNinePatchChunk();

    HiLog.info(LABEL_LOG, "capacity: %{public}s", capacity);
    HiLog.info(LABEL_LOG, "bytesNumber: %{public}s", bytesNumber);
    HiLog.info(LABEL_LOG, "rowBytes: %{public}s", rowBytes);
    HiLog.info(LABEL_LOG, "ninePatchData: %{public}s", ninePatchData);
}
```

上述方法从pixelMap1中获取位图的信息。

运行应用后，点击Get Info按钮，控制台输出内容如下：

```
02-08 14:42:52.853 2717-2717/com.waylau.hmos.pixelmap I
00001/MainAbilitySlice: capacity: 24
02-08 14:42:52.853 2717-2717/com.waylau.hmos.pixelmap I
00001/MainAbilitySlice: bytesNumber: 24
02-08 14:42:52.853 2717-2717/com.waylau.hmos.pixelmap I
00001/MainAbilitySlice: rowBytes: 12
02-08 14:42:52.853 2717-2717/com.waylau.hmos.pixelmap I
00001/MainAbilitySlice: ninePatchData: null
```

14.3.7　读取和写入像素操作说明

readWritePixels方法用于读取和写入像素，代码如下：

```java
private void readWritePixels() {
    // 读取指定位置像素
    int color = pixelMap1.readPixel(new Position(1, 1));
    HiLog.info(LABEL_LOG, "readPixel color: %{public}s", color);

    // 读取指定区域像素
    int[] pixelArray = new int[50];
```

```
    Rect region = new Rect(0, 0, 3, 2);
    pixelMap1.readPixels(pixelArray, 0, 10, region);
    HiLog.info(LABEL_LOG, "readPixel pixelArray: %{public}s", pixelArray);

    // 读取像素到Buffer
    IntBuffer pixelBuf = IntBuffer.allocate(50);
    pixelMap1.readPixels(pixelBuf);
    HiLog.info(LABEL_LOG, "readPixel pixelBuf: %{public}s", pixelBuf);

    // 在指定位置写入像素
    pixelMap1.writePixel(new Position(1, 1), 0xFF112233);

    // 在指定区域写入像素
    pixelMap1.writePixels(pixelArray, 0, 10, region);

    // 写入Buffer中的像素
    pixelMap1.writePixels(pixelBuf);
}
```

上述方法分别执行readPixels和writePixels来进行像素的读取和写入。

运行应用后，点击Read and write按钮，控制台输出内容如下：

```
02-08 14:47:05.295 2717-2717/com.waylau.hmos.pixelmap I
00001/MainAbilitySlice: readPixel color: 0
02-08 14:47:05.296 2717-2717/com.waylau.hmos.pixelmap I
00001/MainAbilitySlice: readPixel pixelArray: [I@21e5496
02-08 14:47:05.296 2717-2717/com.waylau.hmos.pixelmap I
00001/MainAbilitySlice: readPixel pixelBuf: java.nio.HeapIntBuffer[pos=6 lim=50
cap=50]
```

14.3.8 释放资源

在onStop方法中释放资源，代码如下：

```
@Override
protected void onStop() {
    super.onStop();
    if (pixelMap1 != null) {
        // 释放资源
        pixelMap1.release();
    }
    if (pixelMap2 != null) {
        // 释放资源
        pixelMap2.release();
    }
}
```

14.4 实战：图像属性解码

图像属性解码就是获取图像中包含的属性信息，比如EXIF属性。

14.4.1 接口说明

图像属性解码的功能主要由ImageSource和ExifUtils提供。
ImageSource的主要接口有：

- getThumbnailInfo()：获取嵌入图像文件的缩略图的基本信息。
- getImageThumbnailBytes()：获取嵌入图像文件的缩略图的原始数据。
- getThumbnailFormat()：获取嵌入图像文件的缩略图的格式。

ExifUtils的主要接口有：

- getLatLong(ImageSource imageSource)：获取嵌入图像文件的经纬度信息。
- getAltitude(ImageSource imageSource, double defaultValue)：获取嵌入图像文件的海拔信息。

14.4.2 创建应用

为了演示图像属性解码的功能，创建一个名为ImageSourceExifUtils的应用。
在应用的界面上，通过点击按钮来触发图像属性解码的操作。
在media目录下放置一张用于测试的照片IMG_20210219_175445.jpg。需要注意，测试照片需要包含EXIF属性信息。

14.4.3 修改 ability_main.xml

修改ability_main.xml内容如下：

```xml
<?xml version="1.0" encoding="utf-8"?>
<DirectionalLayout
    xmlns:ohos="http://schemas.huawei.com/res/ohos"
    ohos:height="match_parent"
    ohos:width="match_parent"
    ohos:alignment="center"
    ohos:orientation="vertical">

    <Button
        ohos:id="$+id:button_get_info"
        ohos:height="match_content"
        ohos:width="match_content"
```

```
            ohos:background_element="#F76543"
            ohos:left_padding="10vp"
            ohos:text="Get Info"
            ohos:text_size="27fp"
            />

</DirectionalLayout>
```

界面预览效果如图14-5所示。

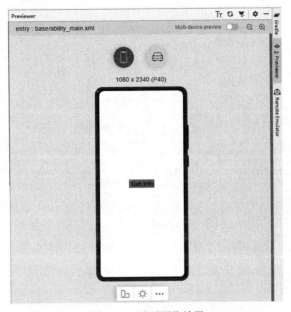

图 14-5　界面预览效果

14.4.4　修改 MainAbilitySlice

修改MainAbilitySlice内容如下：

```
package com.waylau.hmos.imagesourceexifutils.slice;

import com.waylau.hmos.imagesourceexifutils.ResourceTable;
import ohos.aafwk.ability.AbilitySlice;
import ohos.aafwk.content.Intent;
import ohos.agp.components.Button;
import ohos.global.resource.RawFileEntry;
import ohos.hiviewdfx.HiLog;
import ohos.hiviewdfx.HiLogLabel;
import ohos.media.image.ExifUtils;
import ohos.media.image.ImageSource;
import ohos.media.image.PixelMap;
import ohos.media.image.common.ImageInfo;
import ohos.utils.Pair;
```

```java
import java.io.*;

public class MainAbilitySlice extends AbilitySlice {
    private static final String TAG = MainAbilitySlice.class.getSimpleName();
    private static final HiLogLabel LABEL_LOG =
            new HiLogLabel(HiLog.LOG_APP, 0x00001, TAG);

    private ImageSource imageSource;
    private PixelMap pixelMap;

    @Override
    public void onStart(Intent intent) {
        super.onStart(intent);
        super.setUIContent(ResourceTable.Layout_ability_main);

        Button buttonGetInfo =
                (Button) findComponentById(ResourceTable.Id_button_get_info);

        // 为按钮设置点击事件回调
        buttonGetInfo.setClickedListener(listener -> {
            try {
                getInfo();
            } catch (IOException e) {
                e.printStackTrace();
            }
        });
    }

    private void getInfo() throws IOException {
        // 获取图片
        RawFileEntry fileEntry = getResourceManager().
getRawFileEntry("resources/base/media/IMG_20210219_175445.jpg");

        //获取文件大小
        int fileSize = (int) fileEntry.openRawFileDescriptor().getFileSize();

        //定义读取文件的字节
        byte[] fileData = new byte[fileSize];

        //读取文件字节
        fileEntry.openRawFile().read(fileData);

        imageSource = ImageSource.create(fileData, this.getSourceOptions());
        // 获取嵌入图像文件的缩略图的基本信息
        ImageInfo imageInfo = imageSource.getThumbnailInfo();
        HiLog.info(LABEL_LOG, "imageInfo: %{public}s", imageInfo);

        // 获取嵌入图像文件的缩略图的原始数据
```

```
            byte[] imageThumbnailBytes = imageSource.getImageThumbnailBytes();
            HiLog.info(LABEL_LOG, "imageThumbnailBytes: %{public}s",
imageThumbnailBytes);

            // 获取嵌入图像文件的缩略图的格式
            int thumbnailFormat = imageSource.getThumbnailFormat();
            HiLog.info(LABEL_LOG, "thumbnailFormat: %{public}s",
thumbnailFormat);

            // 获取嵌入图像文件的经纬度信息
            Pair<Float, Float> lat = ExifUtils.getLatLong(imageSource);
            HiLog.info(LABEL_LOG, "lat first: %{public}s", lat.f);
            HiLog.info(LABEL_LOG, "lat second: %{public}s", lat.s);

            // 获取嵌入图像文件的海拔信息
            double defaultValue = 100;
            double altitude = ExifUtils.getAltitude(imageSource, defaultValue);
            HiLog.info(LABEL_LOG, "altitude: %{public}s", altitude);
        }

        // 设置数据源的格式信息
        private ImageSource.SourceOptions getSourceOptions() {
            ImageSource.SourceOptions sourceOptions = new
ImageSource.SourceOptions();
            sourceOptions.formatHint = "image/jpeg";

            return sourceOptions;
        }

        @Override
        protected void onStop() {
            super.onStop();
            if (imageSource != null) {
                // 释放资源
                imageSource.release();
            }
        }

        @Override
        public void onForeground(Intent intent) {
            super.onForeground(intent);
        }
    }
```

上述代码中：

- 在按钮上设置了点击事件，以触发获取图像属性信息的操作。
- 通过ImageSource的主要接口来获取嵌入图像文件的缩略图的基本信息、原始数据和格式。
- 通过ExifUtils的主要接口获取嵌入图像文件的经纬度信息、海拔信息。

14.4.5 运行

运行应用，点击Get Info按钮，可以看到控制台输出内容如下：

```
02-20 16:01:44.045 2947-2947/com.waylau.hmos.imagesourceexifutils I
00001/MainAbilitySlice: imageInfo: ohos.media.image.common.ImageInfo@b982677
02-20 16:01:44.045 2947-2947/com.waylau.hmos.imagesourceexifutils I
00001/MainAbilitySlice: imageThumbnailBytes: [B@28ad2e4
02-20 16:01:44.046 2947-2947/com.waylau.hmos.imagesourceexifutils I
00001/MainAbilitySlice: thumbnailFormat: 3
02-20 16:01:44.047 2947-2947/com.waylau.hmos.imagesourceexifutils I
00001/MainAbilitySlice: lat first: 23.081373
02-20 16:01:44.047 2947-2947/com.waylau.hmos.imagesourceexifutils I
00001/MainAbilitySlice: lat second: 114.40377
02-20 16:01:44.047 2947-2947/com.waylau.hmos.imagesourceexifutils I
00001/MainAbilitySlice: altitude: 100.0
```

第15章

相 机

HarmonyOS相机模块支持相机业务的开发，开发者可以通过已开放的接口实现相机硬件的访问、操作和新功能开发。

15.1 相机概述

HarmonyOS相机模块支持相机业务的开发，开发者可以通过已开放的接口实现相机硬件的访问、操作和新功能开发，常见的操作有预览、拍照、连拍和录像等。

15.1.1 基本概念

在相机业务的开发中，基本的概念包括：

- 相机静态能力：用于描述相机的固有能力的一系列参数，比如朝向、支持的分辨率等信息。
- 物理相机：物理相机就是独立的实体摄像头设备。物理相机ID是用于标志每个物理摄像头的唯一字串。
- 逻辑相机：逻辑相机是多个物理相机组合出来的抽象设备，逻辑相机通过同时控制多个物理相机设备来完成相机的某些功能，如大光圈、变焦等功能。逻辑相机ID是一个唯一的字符串，标识多个物理相机的抽象能力。
- 帧捕获：相机启动后对帧的捕获动作统称为帧捕获，主要包含单帧捕获、多帧捕获、循环帧捕获。
- 单帧捕获：指的是相机启动后，在帧数据流中捕获一帧数据，常用于普通拍照。
- 多帧捕获：指的是相机启动后，在帧数据流中连续捕获多帧数据，常用于连拍。

- 循环帧捕获：指的是相机启动后，在帧数据流中一直捕获帧数据，常用于预览和录像。

15.1.2 约束与限制

开发相机业务主要的约束与限制包括：

- 在同一时刻只能有一个相机应用在运行中。
- 相机模块内部有状态控制，开发者必须按照指导文档中的流程对接口进行顺序调用，否则可能会出现调用失败等问题。
- 为了开发的相机应用拥有更好的兼容性，在创建相机对象或者设置相关参数前务必进行能力查询。

15.1.3 相机开发流程

相机模块的主要工作是给相机应用开发者提供基本的相机API接口，用于使用相机系统的功能，进行相机硬件的访问、操作和新功能开发。相机的开发流程如图15-1所示。

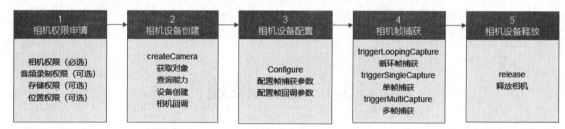

图 15-1 相机的开发流程

15.1.4 核心接口

相机模块为相机应用开发者提供了3个包的内容，包括方法、枚举以及常量/变量，方便开发者更容易地实现相机功能。接口如下：

- ohos.media.camera.CameraKit：相机功能入口类。获取当前支持的相机列表及其静态能力信息，创建相机对象。
- ohos.media.camera.device：相机设备操作类。提供相机能力查询、相机配置、相机帧捕获、相机状态回调等功能。
- ohos.media.camera.params：相机参数类。提供相机属性、参数和操作结果的定义。

15.1.5 相机权限

在使用相机之前，需要申请相机的相关权限，保证应用拥有相机硬件及其他功能权限。

相机涉及的权限如下：

- 相机权限：ohos.permission.CAMERA，必选。
- 录音权限：ohos.permission.MICROPHONE，可选（需要录像时申请）。
- 存储权限：ohos.permission.WRITE_MEDIA，可选（需要保存图像及视频到设备的外部存储时申请）。
- 位置权限：ohos.permission.MEDIA_LOCATION，可选（需要保存图像及视频位置信息时申请）。

15.2 实战：相机设备创建

本节将演示如何创建相机设备。

15.2.1 接口说明

CameraKit类是相机的入口API类，用于获取相机设备特性、打开相机，其接口如下：

- createCamera(String cameraId, CameraStateCallback callback, EventHandler handler)：创建相机对象。
- getCameraAbility(String cameraId)：获取指定逻辑相机或物理相机的静态能力。
- getCameraIds()：获取当前逻辑相机列表。
- getCameraInfo(String cameraId)：获取指定逻辑相机的信息。
- getInstance(Context context)：获取CameraKit实例。
- registerCameraDeviceCallback(CameraDeviceCallback callback, EventHandler handler)：注册相机使用状态回调。
- unregisterCameraDeviceCallback(CameraDeviceCallback callback)：注销相机使用状态回调。

15.2.2 创建应用

为了演示相机设备的功能，创建一个名为CameraKit的应用。
在应用的界面上，通过点击按钮来触发相机设备的操作。

15.2.3 声明相机权限

修改配置文件，声明相机权限如下：

```
// 声明权限
"reqPermissions": [
  {
  "name": "ohos.permission.CAMERA"
  }
```

    ]

同时，在应用启动时，显式声明相机权限。其代码如下：

```java
package com.waylau.hmos.camerakit;

import com.waylau.hmos.camerakit.slice.MainAbilitySlice;
import ohos.aafwk.ability.Ability;
import ohos.aafwk.content.Intent;

import java.util.ArrayList;
import java.util.List;

public class MainAbility extends Ability {
    @Override
    public void onStart(Intent intent) {
        super.onStart(intent);
        super.setMainRoute(MainAbilitySlice.class.getName());

        // 显式声明需要使用的权限
        requestPermission();
    }

    // 显式声明需要使用的权限
    private void requestPermission() {
        String[] permission = {
                "ohos.permission.CAMERA"};
        List<String> applyPermissions = new ArrayList<>();
        for (String element : permission) {
            if (verifySelfPermission(element) != 0) {
                if (canRequestPermission(element)) {
                    applyPermissions.add(element);
                }
            }
        }
        requestPermissionsFromUser(applyPermissions.toArray(new String[0]), 0);
    }
}
```

15.2.4　修改 ability_main.xml

修改ability_main.xml内容如下：

```xml
<?xml version="1.0" encoding="utf-8"?>
<DirectionalLayout
    xmlns:ohos="http://schemas.huawei.com/res/ohos"
    ohos:id="$+id:layout"
    ohos:height="match_parent"
    ohos:width="match_parent"
```

```
    ohos:orientation="vertical">

    <Button
        ohos:id="$+id:button_create_camera"
        ohos:height="match_content"
        ohos:width="match_content"
        ohos:background_element="#F76543"
        ohos:layout_alignment="horizontal_center"
        ohos:text="Get Camera"
        ohos:text_size="40vp"
        />

</DirectionalLayout>
```

界面预览效果如图15-2所示。

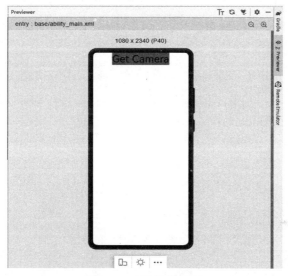

图 15-2　界面预览效果

上述代码设置了Get Camera按钮，以备设置点击事件，以触发创建相机的操作。

15.2.5　修改 MainAbilitySlice

修改MainAbilitySlice内容如下：

```
package com.waylau.hmos.camerakit.slice;

import com.waylau.hmos.camerakit.ResourceTable;
import ohos.aafwk.ability.AbilitySlice;
import ohos.aafwk.content.Intent;
import ohos.agp.components.Button;
import ohos.eventhandler.EventHandler;
import ohos.eventhandler.EventRunner;
import ohos.hiviewdfx.HiLog;
```

```java
    import ohos.hiviewdfx.HiLogLabel;
    import ohos.media.camera.CameraKit;
    import ohos.media.camera.device.*;
    import ohos.utils.zson.ZSONObject;

    public class MainAbilitySlice extends AbilitySlice {
        private static final String TAG = MainAbilitySlice.class.getSimpleName();
        private static final HiLogLabel LABEL_LOG =
                new HiLogLabel(HiLog.LOG_APP, 0x00001, TAG);

        // 相机
        private Camera cameraDevice;
        // 相机创建和相机运行时的回调
        private CameraStateCallbackImpl cameraStateCallback = new
CameraStateCallbackImpl();
        // 执行回调的EventHandler
        private EventHandler eventHandler = new
EventHandler(EventRunner.create("CameraCb"));

        @Override
        public void onStart(Intent intent) {
            super.onStart(intent);
            super.setUIContent(ResourceTable.Layout_ability_main);

            Button buttonCreateCamera =
                    (Button)
findComponentById(ResourceTable.Id_button_create_camera);

            // 为按钮设置点击事件回调
            buttonCreateCamera.setClickedListener(listener -> createCamera());
        }

        private void createCamera() {
            // 获取CameraKit对象
            openCamera();
        }

        private void openCamera() {
            // 获取CameraKit对象
            CameraKit cameraKit = CameraKit.getInstance(this);
            if (cameraKit != null) {
                try {
                    // 获取当前设备的逻辑相机列表
                    String[] cameraIds = cameraKit.getCameraIds();
                    if (cameraIds.length <= 0) {
                        HiLog.error(LABEL_LOG, "cameraIds size is 0");
                    } else {
                        for (String cameraId : cameraIds) {
                            CameraInfo cameraInfo =
cameraKit.getCameraInfo(cameraId);
```

```java
                        HiLog.info(LABEL_LOG,
                                "cameraId: %{public}s, CameraInfo: %{public}s",
                                cameraId, ZSONObject.toZSONString(cameraInfo));

                        CameraAbility cameraAbility =
cameraKit.getCameraAbility(cameraId);
                        HiLog.info(LABEL_LOG,
                                "cameraId: %{public}s, CameraAbility: %{public}s",
                                cameraId,
ZSONObject.toZSONString(cameraAbility));
                    }
                }

                // 相机创建和相机运行时的回调
                if (cameraStateCallback == null) {
                    HiLog.error(LABEL_LOG, "cameraStateCallback is null");
                }

                // 执行回调的EventHandler
                if (eventHandler == null) {
                    HiLog.error(LABEL_LOG, "eventHandler is null");
                }

                // 创建相机设备
                cameraKit.createCamera(cameraIds[1], cameraStateCallback,
eventHandler);
            } catch (IllegalStateException e) {
                // 处理异常
                HiLog.error(LABEL_LOG, "exception %{public}s", e.getMessage());

            }
        }
    }

    @Override
    protected void onStop() {
        super.onStop();

        // 释放资源
        releaseCamera();
    }

    @Override
    public void onActive() {
        super.onActive();
    }

    @Override
    public void onForeground(Intent intent) {
```

```java
        super.onForeground(intent);
    }

    private final class CameraStateCallbackImpl extends CameraStateCallback {
        @Override
        public void onCreated(Camera camera) {
            // 创建相机设备
            HiLog.info(LABEL_LOG, "Camera onCreated");
        }

        @Override
        public void onConfigured(Camera camera) {
            // 配置相机设备
            HiLog.info(LABEL_LOG, "Camera onConfigured");
        }

        @Override
        public void onPartialConfigured(Camera camera) {
            // 当使用addDeferredSurfaceSize配置了相机，会接到此回调
            HiLog.info(LABEL_LOG, "Camera onPartialConfigured");
        }

        @Override
        public void onReleased(Camera camera) {
            // 释放相机设备
            HiLog.info(LABEL_LOG, "Camera onReleased");
        }

        @Override
        public void onFatalError(Camera camera, int errorCode) {
            HiLog.info(LABEL_LOG, "Camera onFatalError, errorCode: %{public}s",
errorCode);
        }

        @Override
        public void onCreateFailed(String cameraId, int errorCode) {
            HiLog.info(LABEL_LOG, "Camera onCreateFailed, 
errorCode: %{public}s", errorCode);
        }

    }

    private void releaseCamera() {
        if (cameraDevice != null) {
            // 关闭相机和释放资源
            cameraDevice.release();
            cameraDevice = null;
        }
    }
}
```

上述代码中：

- 在Button上设置了点击事件，以触发createCamera方法的执行。
- 在openCamera方法中，通过CameraKit.getInstance(Context context)方法获取唯一的CameraKit对象。如果此步骤操作失败，相机可能被占用或无法使用。如果被占用，则必须等到相机释放后才能重新获取CameraKit对象。
- 通过getCameraIds()方法获取当前使用的设备支持的逻辑相机列表。逻辑相机列表中存储了当前设备拥有的所有逻辑相机ID，如果列表不为空，则列表中的每个ID都支持独立创建相机对象；否则，说明正在使用的设备无可用的相机，不能继续后续的操作。
- 遍历cameraIds数组，并在日志中记录相机的信息。这些信息对象通过ZSONObject转化为JSON格式的字符串，方便在日志中查看。
- 通过createCamera(String cameraId, CameraStateCallback callback, EventHandler handler)方法创建相机对象，此步骤执行成功意味着相机系统的硬件已经完成了上电。其中，第一个参数cameraId可以是上一步获取的逻辑相机列表中的任何一个相机ID。第二个和第三个参数负责相机创建和相机运行时的数据和状态检测，务必保证在整个相机运行周期内有效。

至此，相机设备的创建已经完成。相机设备创建成功会在CameraStateCallback中触发onCreated(Camera camera)回调。在进入相机设备配置前，请确保相机设备已经创建成功，否则会触发相机设备创建失败的回调，并返回错误码，需要进行错误处理后，重新执行相机设备的创建。

15.2.6 运行

运行应用后，点击Get Camera按钮，控制台输出内容如下：

```
    09-12 16:41:28.850 15364-15364/com.waylau.hmos.camerakit I
00001/MainAbilitySlice: cameraId: 0, CameraInfo: {"facingType":1,
"logicalCameraAvailable":true,"logicalId":"0","physicalIdList":["0"]}
    09-12 16:41:28.864 15364-15364/com.waylau.hmos.camerakit I
00001/MainAbilitySlice: cameraId: 0, CameraAbility:
{"cameraId":"0","logicalCamera":false,"physicalCameraIds":[],"supportedAeMode"
:[0,1],"supportedAfMode":[2,1],"supportedAwbMode":[0,1],"supportedFaceDetectio
n":[],"supportedFlashMode":[],"supportedFormats":[2,3],"supportedParameters":[
{"name":"ohos.camera.imageCompressionQuality"},{"name":"ohos.camera.imageMirro
r"},{"name":"ohos.camera.videoStabilization"},{"name":"ohos.camera.exposureFps
Range"},{"name":"ohos.camera.vendorCustom"},{"name":"ohos.camera.zoom"},{"name
":"ohos.camera.aeMode"},{"name":"ohos.camera.aeRegion"},{"name":"ohos.camera.a
eTrigger"},{"name":"ohos.camera.afMode"},{"name":"ohos.camera.afRegion"},{"nam
e":"ohos.camera.afTrigger"},{"name":"ohos.camera.awbMode"},{"name":"ohos.camer
a.awbRegion"},{"name":"ohos.camera.flashMode"},{"name":"ohos.camera.faceDetect
ionType"},{"name":"ohos.camera.imageRotation"},{"name":"ohos.camera.location"}
,{"name":"ohos.camera.captureMirrorMode"},{"name":"ohos.camera.faceBeautyMode"
},{"name":"ohos.camera.faceBeautyLevel"},{"name":"ohos.cam
    09-12 16:41:28.865 15364-15364/com.waylau.hmos.camerakit I
00001/MainAbilitySlice: cameraId: 1, CameraInfo: {"facingType":0,
"logicalCameraAvailable":true,"logicalId":"1","physicalIdList":["1"]}
```

```
    09-12 16:41:28.869 15364-15364/com.waylau.hmos.camerakit I
00001/MainAbilitySlice: cameraId: 1, CameraAbility:
{"cameraId":"1","logicalCamera":false,"physicalCameraIds":[],"supportedAeMode"
:[0,1],"supportedAfMode":[],"supportedAwbMode":[0,1],"supportedFaceDetection":
[],"supportedFlashMode":[],"supportedFormats":[2,3],"supportedParameters":[{"n
ame":"ohos.camera.imageCompressionQuality"},{"name":"ohos.camera.imageMirror"}
,{"name":"ohos.camera.videoStabilization"},{"name":"ohos.camera.exposureFpsRan
ge"},{"name":"ohos.camera.vendorCustom"},{"name":"ohos.camera.zoom"},{"name":"
ohos.camera.aeMode"},{"name":"ohos.camera.aeRegion"},{"name":"ohos.camera.aeTr
igger"},{"name":"ohos.camera.afMode"},{"name":"ohos.camera.afRegion"},{"name":
"ohos.camera.afTrigger"},{"name":"ohos.camera.awbMode"},{"name":"ohos.camera.a
wbRegion"},{"name":"ohos.camera.flashMode"},{"name":"ohos.camera.faceDetection
Type"},{"name":"ohos.camera.imageRotation"},{"name":"ohos.camera.location"},{"
name":"ohos.camera.captureMirrorMode"},{"name":"ohos.camera.faceBeautyMode"},{
"name":"ohos.camera.faceBeautyLevel"},{"name":"ohos.camera
    09-12 16:41:28.948 15364-16070/com.waylau.hmos.camerakit I
00001/MainAbilitySlice: Camera onCreated
```

15.3 实战：相机设备配置

创建相机设备成功后，在CameraStateCallback中会触发onCreated(Camera camera)回调，并且带回Camera对象，用于执行相机设备的操作。

当一个新的相机设备成功创建后，首先需要对相机进行配置，调用configure(CameraConfig)方法实现配置。相机配置主要是设置预览、拍照、录像用到的Surface（详见ohos.agp.graphics.Surface），没有配置过Surface，相应的功能就不能使用。

为了检测相机帧捕获结果的数据和状态，还需要在相机配置时调用setFrameStateCallback(FrameStateCallback, EventHandler)方法设置帧回调。

15.3.1 添加类变量

修改MainAbilitySlice，添加如下类变量：

```
// 图像帧数据接收处理对象
private ImageReceiver imageReceiver;

// 配置预览的Surface
private Surface previewSurface;
private FrameConfig.Builder frameConfigBuilder;
private CameraConfig.Builder cameraConfigBuilder;
private FrameStateCallbackImpl frameStateCallbackImpl = new
FrameStateCallbackImpl();
private SurfaceProvider surfaceProvider;
```

其中，FrameStateCallbackImpl为内部类。

15.3.2 新增FrameStateCallbackImpl

为了检测相机帧捕获结果的数据和状态，新增FrameStateCallbackImpl内部类。该类继承自FrameStateCallback，代码如下：

```
private final class FrameStateCallbackImpl extends FrameStateCallback {
    @Override
    public void onFrameStarted(Camera camera, FrameConfig frameConfig, long frameNumber, long timestamp) {
        // 开始帧捕获时，触发回调
        HiLog.info(LABEL_LOG, "onFrameStarted");
    }

    @Override
    public void onFrameFinished(Camera camera, FrameConfig frameConfig, FrameResult frameResult) {
        // 当帧捕获完成并且所有结果都可用时调用
        HiLog.info(LABEL_LOG, "onFrameFinished");
    }

    @Override
    public void onFrameProgressed(Camera camera, FrameConfig frameConfig, FrameResult frameResult) {
        // 在帧捕获期间有部分结果时调用
        HiLog.info(LABEL_LOG, "onFrameProgressed");
    }

    @Override
    public void onFrameError(Camera camera, FrameConfig frameConfig, int errorCode, FrameResult frameResult) {
        // 在帧捕获过程中发生错误时调用
        HiLog.info(LABEL_LOG, "onFrameError");
    }

    @Override
    public void onCaptureTriggerStarted(Camera camera, int captureTriggerId, long firstFrameNumber) {
        // 在启动触发器的帧捕获时调用
        HiLog.info(LABEL_LOG, "onCaptureTriggerStarted");
    }

    @Override
    public void onCaptureTriggerFinished(Camera camera, int captureTriggerId, long lastFrameNumber) {
        // 当触发的帧捕获动作完成时调用
        HiLog.info(LABEL_LOG, "onCaptureTriggerFinished");
    }
```

```
        @Override
        public void onCaptureTriggerInterrupted(Camera camera, int
captureTriggerId) {
            // 当已触发的帧捕获动作提前停止时调用
            HiLog.info(LABEL_LOG, "onCaptureTriggerInterrupted");
        }
    }
```

15.3.3 修改 onStart 方法

修改onStart方法，新增了初始化预览的Surface的方法initSurface()，代码如下：

```
@Override
public void onStart(Intent intent) {
    super.onStart(intent);
    super.setUIContent(ResourceTable.Layout_ability_main);

    // 初始化预览的Surface
    initSurface();

    Button buttonCreateCamera =
            (Button) findComponentById(ResourceTable.Id_button_create_camera);

    // 为按钮设置点击事件回调
    buttonCreateCamera.setClickedListener(listener -> createCamera());
}
```

initSurface()方法实现如下：

```
private void initSurface() {
    // 获取组件对象
    DirectionalLayout stackLayout = (DirectionalLayout) findComponentById(ResourceTable.Id_layout);
    surfaceProvider = new SurfaceProvider(getContext());
    // 放在AGP容器组件的顶层
    surfaceProvider.pinToZTop(true);
    // 设置样式
    DirectionalLayout.LayoutConfig layoutConfig = new DirectionalLayout.LayoutConfig();
    layoutConfig.alignment = LayoutAlignment.CENTER;
    layoutConfig.setMarginBottom(55);
    // 根据比例设置相机长宽值
    layoutConfig.width = 1000;
    layoutConfig.height = 1000;
    surfaceProvider.setLayoutConfig(layoutConfig);
    // 将组件添加到容器中
    stackLayout.addComponent(surfaceProvider);
    // 获取SurfaceOps对象
    SurfaceOps surfaceOps = surfaceProvider.getSurfaceOps().get();
```

```
    // 设置像素格式
    surfaceOps.setFormat(ImageFormat.JPEG);
    // 设置屏幕一直打开
    surfaceOps.setKeepScreenOn(true);
    // 添加回调
    surfaceOps.addCallback(new SurfaceOpsCallBack());

    // 创建ImageReceiver对象，注意create函数中的宽度要大于高度。5为最大支持的图像数，
请根据实际设置
    imageReceiver = ImageReceiver.create(100,100, ImageFormat.JPEG, 5);

}
```

上述代码添加了回调SurfaceOpsCallBack。SurfaceOpsCallBack方法如下：

```
private class SurfaceOpsCallBack implements SurfaceOps.Callback {

    @Override
    public void surfaceCreated(SurfaceOps surfaceOps) {
        HiLog.info(LABEL_LOG, "surfaceCreated");

        // 获取surface对象
        previewSurface = surfaceOps.getSurface();
    }

    @Override
    public void surfaceChanged(SurfaceOps surfaceOps, int i, int i1, int i2) {
        HiLog.info(LABEL_LOG, "surfaceChanged");
    }

    @Override
    public void surfaceDestroyed(SurfaceOps surfaceOps) {
        HiLog.info(LABEL_LOG, "surfaceDestroyed");
    }
}
```

上述回调用于获取配置预览的Surface对象。

15.3.4 修改 onCreated 方法

修改CameraStateCallbackImpl的onCreated方法，代码如下：

```
@Override
public void onCreated(Camera camera) {
    // 创建相机设备
    HiLog.info(LABEL_LOG, "Camera onCreated");

    cameraConfigBuilder = camera.getCameraConfigBuilder();
    if (cameraConfigBuilder == null) {
```

```
        HiLog.error(LABEL_LOG, "onCreated cameraConfigBuilder is null");
        return;
    }

    // 配置预览的Surface
    cameraConfigBuilder.addSurface(previewSurface);
    // 配置拍照的Surface
    cameraConfigBuilder.addSurface(imageReceiver.getRecevingSurface());
    // 配置帧结果的回调
    cameraConfigBuilder.setFrameStateCallback(frameStateCallbackImpl,
eventHandler);

    try {
        // 相机设备配置
        camera.configure(cameraConfigBuilder.build());
    } catch (IllegalArgumentException e) {
        HiLog.error(LABEL_LOG, "Argument Exception");
    } catch (IllegalStateException e) {
        HiLog.error(LABEL_LOG, "State Exception");
    }
}
```

上述代码在相机创建之后需要配置预览的Surface、拍照的Surface以及帧结果的回调。

15.3.5 运行

运行应用后,可以看到如图15-3所示的界面运行效果。

图15-3　界面运行效果

在上述界面中可以看到预览的Surface。

15.4 实战：相机帧捕获

Camera操作类包括相机预览、录像、拍照等功能接口。

15.4.1 接口说明

Camera的主要接口有：

- triggerSingleCapture(FrameConfig frameConfig)：启动相机帧的单帧捕获。
- triggerMultiCapture(List frameConfigs)：启动相机帧的多帧捕获。
- configure(CameraConfig config)：配置相机。
- flushCaptures()：停止并清除相机帧的捕获，包括循环帧、单帧、多帧捕获。
- getCameraConfigBuilder()：获取相机配置构造器对象。
- getCameraId()：获取当前相机的ID。
- getFrameConfigBuilder(int type)：获取指定类型的相机帧配置构造器对象。
- release()：释放相机对象及资源。
- triggerLoopingCapture(FrameConfig frameConfig)：启动或者更新相机帧的循环捕获。
- stopLoopingCapture()：停止当前相机帧的循环捕获。

15.4.2 启动预览（循环帧捕获）

用户一般都是先看见预览画面才执行拍照或者其他功能，所以对于一个普通的相机应用，预览是必不可少的。启动预览的步骤如下：

- 通过getFrameConfigBuilder(FRAME_CONFIG_PREVIEW)方法获取预览配置模板。
- 通过triggerLoopingCapture(FrameConfig)方法实现循环帧捕获（如预览、录像）。

新增类变量，代码如下：

```
private FrameConfig previewFrameConfig;
```

修改onConfigured方法，代码如下：

```
import static ohos.media.camera.device.Camera.FrameConfigType.FRAME_CONFIG_PREVIEW;

@Override
public void onConfigured(Camera camera) {
    // 配置相机设备
    HiLog.info(LABEL_LOG, "Camera onConfigured");

    // 获取预览配置模板
```

```
        frameConfigBuilder = camera.getFrameConfigBuilder(FRAME_CONFIG_PREVIEW);
        // 配置预览Surface
        frameConfigBuilder.addSurface(previewSurface);
        previewFrameConfig = frameConfigBuilder.build();
        try {
            // 启动循环帧捕获
            camera.triggerLoopingCapture(previewFrameConfig);

            // 设置类变量
            cameraDevice = camera;
        } catch (IllegalArgumentException e) {
            HiLog.error(LABEL_LOG, "Argument Exception");
        } catch (IllegalStateException e) {
            HiLog.error(LABEL_LOG, "State Exception");
        }
    }
```

15.4.3 释放资源

最后，为了能释放资源，修改releaseCamera方法如下：

```
private void releaseCamera() {
    if (camera != null) {
        // 关闭相机和释放资源
        camera.release();
        camera = null;
    }

    // 拍照配置模板置空
    frameConfigBuilder = null;
    // 预览配置模板置空
    previewFrameConfig = null;
}
```

15.4.4 运行

经过以上操作，相机应用已经可以正常进行实时预览了。运行应用后，点击Get Camera按钮，可以看到如图15-4所示的界面运行效果。

在上述界面中可以看到相机预览的画面。

15.4.5 实现拍照（单帧捕获）

拍照的步骤如下：

- 通过getFrameConfigBuilder(FRAME_CONFIG_PICTURE)方法获取拍照配置模板，并且设置

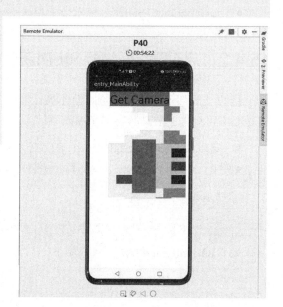

图15-4　界面运行效果

拍照帧配置。
- 拍照前准备图像帧数据的接收实现。
- 通过triggerSingleCapture(FrameConfig)方法实现单帧捕获（如拍照）。

修改ability_main.xml，增加如下内容：

```xml
<Button
    ohos:id="$+id:button_single_capture"
    ohos:height="match_content"
    ohos:width="match_content"
    ohos:background_element="#F76543"
    ohos:layout_alignment="horizontal_center"
    ohos:text="Single Capture"
    ohos:text_size="40vp"
    />
```

Single Capture按钮用于触发实现单帧捕获的方法。

修改onStart方法，增加Single Capture按钮的事件处理，代码如下：

```java
@Override
public void onStart(Intent intent) {
    super.onStart(intent);
    super.setUIContent(ResourceTable.Layout_ability_main);

    // 初始化预览的Surface
    initSurface();

    Button buttonCreateCamera =
            (Button) findComponentById(ResourceTable.Id_button_create_camera);

    // 为按钮设置点击事件回调
    buttonCreateCamera.setClickedListener(listener -> createCamera());

    Button buttonSingleCapture =
            (Button) findComponentById(ResourceTable.Id_button_single_capture);

    // 为按钮设置点击事件回调
    buttonSingleCapture.setClickedListener(listener -> capture());
}
```

capture方法实现如下：

```java
import static ohos.media.camera.device.Camera.FrameConfigType.FRAME_CONFIG_PICTURE;

    private FrameConfig.Builder framePictureConfigBuilder;

    private void capture() {
        // 获取照片前的准备
```

```
        captureInit();

        // 获取拍照配置模板
        framePictureConfigBuilder =
cameraDevice.getFrameConfigBuilder(FRAME_CONFIG_PICTURE);
        // 配置拍照Surface
framePictureConfigBuilder.addSurface(imageReceiver.getRecevingSurface());
        // 配置拍照其他参数
        framePictureConfigBuilder.setImageRotation(90);
        try {
            // 启动单帧捕获(拍照)
cameraDevice.triggerSingleCapture(framePictureConfigBuilder.build());
        } catch (IllegalArgumentException e) {
            HiLog.error(LABEL_LOG, "Argument Exception");
        } catch (IllegalStateException e) {
            HiLog.error(LABEL_LOG, "State Exception");
        }
    }
```

其中，captureInit方法如下：

```
    private void captureInit() {
        imageReceiver.setImageArrivalListener(imageArrivalListener);
    }

    // 单帧捕获生成图像回调Listener
    private final ImageReceiver.IImageArrivalListener imageArrivalListener = new
ImageReceiver.IImageArrivalListener() {
        @Override
        public void onImageArrival(ImageReceiver imageReceiver) {
            // 创建相机设备
            HiLog.info(LABEL_LOG, "onImageArrival");

            StringBuffer fileName = new StringBuffer("picture_");
            fileName.append(UUID.randomUUID()).append(".jpg"); // 定义生成的图片文
件名

            // 获取数据目录
            File dataDir = new File(getExternalCacheDir().toString());
            if (!dataDir.exists()) {
                dataDir.mkdirs();
            }

            // 构建目标文件
            String dirFile = dataDir.toString();

            File myFile = new File(dirFile, fileName.toString()); // 创建图片文件
            ImageSaver imageSaver = new ImageSaver(imageReceiver.readNextImage(),
myFile); // 创建一个读写线程任务用于保存图片
```

```
            eventHandler.postTask(imageSaver); // 执行读写线程任务生成图片
        }
    };

    // 保存图片、图片数据读写及图像生成见run方法
    class ImageSaver implements Runnable {
        private final Image myImage;
        private final File myFile;

        ImageSaver(Image image, File file) {
            myImage = image;
            myFile = file;
        }

        @Override
        public void run() {
            Image.Component component =
myImage.getComponent(ImageFormat.ComponentType.JPEG);
            byte[] bytes = new byte[component.remaining()];
            component.read(bytes);
            FileOutputStream output = null;
            try {
                output = new FileOutputStream(myFile);
                output.write(bytes); // 写图像数据
                output.flush();

                File dataDir = new File(getExternalCacheDir().toString());
                HiLog.info(LABEL_LOG, "dateDir list:", dataDir.list());
            } catch (IOException e) {
                HiLog.error(LABEL_LOG, "save picture occur exception!");
            } finally {
                myImage.release();
                if (output != null) {
                    try {
                        output.close(); // 关闭流
                    } catch (IOException e) {
                        HiLog.error(LABEL_LOG, "image release occur exception!");
                    }
                }
            }
        }
    }
}
```

上述方法主要设置了单帧捕获生成图像回调监听器，用于创建一个读写线程任务用于保存图片。

15.4.6 声明存储权限

修改配置文件，声明存储图片的权限，代码如下：

```
// 声明权限
"reqPermissions": [
  {
    "name": "ohos.permission.CAMERA"
  },
  {
    "name": "ohos.permission.WRITE_USER_STORAGE"
  }
]
```

同时，在应用启动时，显式声明存储图片的权限，代码如下：

```java
package com.waylau.hmos.camerakit;

import com.waylau.hmos.camerakit.slice.MainAbilitySlice;
import ohos.aafwk.ability.Ability;
import ohos.aafwk.content.Intent;

import java.util.ArrayList;
import java.util.List;

public class MainAbility extends Ability {
    @Override
    public void onStart(Intent intent) {
        super.onStart(intent);
        super.setMainRoute(MainAbilitySlice.class.getName());

        // 显式声明需要使用的权限
        requestPermission();
    }

    // 显式声明需要使用的权限
    private void requestPermission() {
        String[] permission = {
                "ohos.permission.CAMERA",
                "ohos.permission.WRITE_USER_STORAGE"};
        List<String> applyPermissions = new ArrayList<>();
        for (String element : permission) {
            if (verifySelfPermission(element) != 0) {
                if (canRequestPermission(element)) {
                    applyPermissions.add(element);
                }
            }
        }
        requestPermissionsFromUser(applyPermissions.toArray(new String[0]), 0);
    }
}
```

15.4.7 运行

运行应用后,先点击Get Camera按钮,再点击Single Capture按钮,可以看到如图15-5所示的界面运行效果。

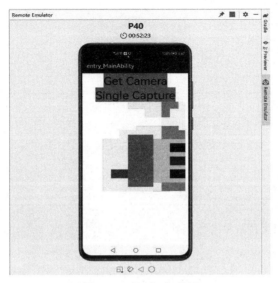

图 15-5　界面运行效果

在上述界面中可以看到相机预览和拍照的画面。

第 16 章

音 频

HarmonyOS音频模块支持音频业务的开发,提供音频相关的功能。

16.1 音频概述

HarmonyOS音频模块支持音频业务的开发,提供音频相关的功能,主要包括音频播放、音频采集、音量管理和短音播放等。

16.1.1 基本概念

音频业务的开发主要涉及以下核心概念:

- 采样:采样是指将连续时域上的模拟信号按照一定的时间间隔采样,获取到离散时域上离散信号的过程。
- 采样率:采样率为每秒从连续信号中提取并组成离散信号的采样次数,单位用赫兹(Hz)来表示。通常人耳能听到频率范围在20Hz~20kHz的声音。常用的音频采样频率有8kHz、11.025kHz、22.05kHz、16kHz、37.8kHz、44.1kHz、48kHz、96kHz、192kHz等。
- 声道:声道是指声音在录制或播放时在不同空间位置采集或回放的相互独立的音频信号,所以声道数也就是声音录制时的音源数量或回放时相应的扬声器数量。
- 音频帧:音频数据是流式的,本身没有明确的一帧帧的概念,在实际应用中,为了音频算法处理/传输的方便,一般约定俗成取2.5ms~60ms为单位的数据量为一帧音频。这个时间被称为"采样时间",其长度没有特别的标准,它是根据编解码器和具体应用的需求来决定的。
- PCM:Pulse Code Modulation(脉冲编码调制),是一种将模拟信号数字化的方法,是将时间连续、取值连续的模拟信号转换成时间离散、抽样值离散的数字信号的过程。
- 短音：使用源于应用程序包内的资源或者文件系统中的文件为样本，将其解码成一个16bit的单声道或者立体声的PCM流并加载到内存中，这使得应用程序可以直接用压缩数据流同时摆脱CPU加载数据的压力和播放时重解压的延迟。
- tone音：根据特定频率生成的波形，比如拨号盘的声音。
- 系统音：系统预置的短音，比如按键音、删除音等。

16.1.2 约束与限制

开发音频业务主要的约束与限制包括：
- 在使用完AudioRenderer音频播放类和AudioCapturer音频采集类后，需要调用release()方法进行资源释放。
- 音频采集所使用的最终采样率与采样格式取决于输入设备，不同设备支持的格式及采样率范围不同，可以通过AudioManager类的getDevices接口查询。
- 在进行音频采集之前，需要申请麦克风权限ohos.permission.MICROPHONE。如果是要写入、读取外部存储中的媒体文件，则还需要申请ohos.permission.WRITE_MEDIA、ohos.permission.READ_MEDIA权限。

16.2 实战：音频播放

本节将演示如何进行音频播放。音频播放的主要工作是将音频数据转码为可听见的音频模拟信号并通过输出设备进行播放，同时对播放任务进行管理。

16.2.1 接口说明

音频播放类AudioRenderer的主要接口有：
- AudioRenderer(AudioRendererInfo audioRendererInfo, PlayMode pm)：构造函数，设置播放相关音频的参数和播放模式，使用默认播放设备。
- AudioRenderer(AudioRendererInfo audioRendererInfo, PlayMode pm, AudioDeviceDescriptor outputDevice)：构造函数，设置播放相关音频的参数、播放模式和播放设备。
- start()：播放音频流。
- write(byte[] data, int offset, int size)：将音频数据以byte流写入音频接收器以进行播放。
- write(short[] data, int offset, int size)：将音频数据以short流写入音频接收器以进行播放。
- write(float[] data, int offset, int size)：将音频数据以float流写入音频接收器以进行播放。
- write(java.nio.ByteBuffer data, int size)：将音频数据以ByteBuffer流写入音频接收器以进行播放。

- pause()：暂停播放音频流。
- stop()：停止播放音频流。
- release()：释放播放资源。
- getCurrentDevice()：获取当前工作的音频播放设备。
- setPlaybackSpeed(float speed)：设置播放速度。
- setPlaybackSpeed(AudioRenderer.SpeedPara speedPara)：设置播放速度与音调。
- setVolume(ChannelVolume channelVolume)：设置指定声道上的输出音量。
- setVolume(float vol)：设置所有声道上的输出音量。
- getMinBufferSize(int sampleRate, AudioStreamInfo.EncodingFormat format, AudioStreamInfo.ChannelMask channelMask)：获取Stream播放模式所需的buffer大小。
- getState()：获取音频播放的状态。
- getRendererSessionId()：获取音频播放的session ID。
- getSampleRate()：获取采样率。
- getPosition()：获取音频播放的帧数位置。
- setPosition(int position)：设置起始播放帧位置。
- getRendererInfo()：获取音频渲染信息。
- duckVolume()：降低音量并将音频与另一个拥有音频焦点的应用程序混合。
- unduckVolume()：恢复音量。
- getPlaybackSpeed()：获取播放速度、音调参数。
- setSpeed(SpeedPara speedPara)：设置播放速度、音调参数。
- getAudioTime()：获取播放时间戳信息。
- flush()：刷新当前的播放流数据队列。
- getMaxVolume()：获取播放流可设置的最大音量。
- getMinVolume()：获取播放流可设置的最小音量。
- getStreamType()：获取播放流的音频流类型。

16.2.2 创建应用

为了演示音频播放的功能，创建一个名为AudioRenderer的应用。

在应用的界面上，通过点击按钮来触发音频的操作。

在rawfile目录下放置一个test.wav音频文件，以备测试。

16.2.3 修改 ability_main.xml

修改ability_main.xml内容如下：

```
<?xml version="1.0" encoding="utf-8"?>
<DirectionalLayout
    xmlns:ohos="http://schemas.huawei.com/res/ohos"
    ohos:height="match_parent"
```

```xml
        ohos:width="match_parent"
        ohos:alignment="center"
        ohos:orientation="vertical">

    <Button
        ohos:id="$+id:button_start"
        ohos:height="match_content"
        ohos:width="match_content"
        ohos:background_element="#F76543"
        ohos:margin="10vp"
        ohos:padding="10vp"
        ohos:text="Start"
        ohos:text_size="40vp"
        />

    <Button
        ohos:id="$+id:button_pause"
        ohos:height="match_content"
        ohos:width="match_content"
        ohos:background_element="#F76543"
        ohos:margin="10vp"
        ohos:padding="10vp"
        ohos:text="Pause"
        ohos:text_size="40vp"
        />

    <Button
        ohos:id="$+id:button_stop"
        ohos:height="match_content"
        ohos:width="match_content"
        ohos:background_element=
"#F76543"
        ohos:margin="10vp"
        ohos:padding="10vp"
        ohos:text="Stop"
        ohos:text_size="40vp"
        />
</DirectionalLayout>
```

界面预览效果如图16-1所示。

上述代码设置了Start、Pause、Stop按钮，以备设置点击事件，以触发音频播放相关的操作。

16.2.4 修改 MainAbilitySlice

修改MainAbilitySlice内容如下：

```
package com.waylau.hmos.audiorenderer.slice;
```

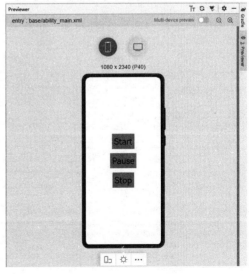

图 16-1 界面预览效果

```java
import com.waylau.hmos.audiorenderer.ResourceTable;
import ohos.aafwk.ability.AbilitySlice;
import ohos.aafwk.content.Intent;
import ohos.agp.components.Button;
import ohos.global.resource.NotExistException;
import ohos.global.resource.RawFileEntry;
import ohos.global.resource.Resource;
import ohos.hiviewdfx.HiLog;
import ohos.hiviewdfx.HiLogLabel;
import ohos.media.audio.AudioRenderer;
import ohos.media.audio.AudioRendererInfo;
import ohos.media.audio.AudioStreamInfo;

import java.io.*;

public class MainAbilitySlice extends AbilitySlice {
    private static final String TAG = MainAbilitySlice.class.getSimpleName();
    private static final HiLogLabel LABEL_LOG =
            new HiLogLabel(HiLog.LOG_APP, 0x00001, TAG);

    // 音频播放类
    private AudioRenderer audioRenderer;

    @Override
    public void onStart(Intent intent) {
        super.onStart(intent);
        super.setUIContent(ResourceTable.Layout_ability_main);

        // 初始化AudioRenderer
        initAudioRenderer();

        // 为按钮设置点击事件回调
        Button buttonStart =
                (Button) findComponentById(ResourceTable.Id_button_start);
        buttonStart.setClickedListener(listener ->
        {
            try {
                start();
            } catch (IOException | NotExistException e) {
                e.printStackTrace();
            }
        });

        Button buttonPause =
                (Button) findComponentById(ResourceTable.Id_button_pause);
        buttonPause.setClickedListener(listener -> audioRenderer.pause());

        Button buttonStop =
                (Button) findComponentById(ResourceTable.Id_button_stop);
```

```java
            buttonStop.setClickedListener(listener -> audioRenderer.stop());
        }

        private void start() throws IOException, NotExistException {
            HiLog.info(LABEL_LOG, "before getFile");

            RawFileEntry rawFileEntry =
this.getResourceManager().getRawFileEntry("resources/rawfile/test.wav");
            Resource soundInputStream = rawFileEntry.openRawFile();

            int bufSize = audioRenderer.getBufferFrameSize();

            HiLog.info(LABEL_LOG, "bufSize: %{public}s", bufSize);

            if (bufSize <=0) {
                HiLog.info(LABEL_LOG, "bufSize is empty");
                return;
            }

            byte[] buffer = new byte[4096];
            try {
                audioRenderer.start();

                int count = 0;

                // 源文件内容写入目标文件
                while((count = soundInputStream.read(buffer)) >= 0){
                    audioRenderer.write(buffer,0,count);

                    HiLog.info(LABEL_LOG, "getFile buffer.length: %{public}s,
state:%{public}s",
                        buffer.length, getState().getValue());
                }

                soundInputStream.close();
            } catch (Exception e) {
                e.printStackTrace();
            }

            HiLog.info(LABEL_LOG, "end getFile");
        }


        @Override
        public void onStop() {
            super.onStop();

            // 关闭、释放资源
            release();
        }
```

```java
        private void initAudioRenderer() {
            HiLog.info(LABEL_LOG, "before initAudioRenderer");

            AudioStreamInfo audioStreamInfo = new 
AudioStreamInfo.Builder().sampleRate(44100) // 44.1kHz
                    .audioStreamFlag(AudioStreamInfo.AudioStreamFlag.AUDIO_STREAM
_FLAG_MAY_DUCK) // 混音
                    .encodingFormat(AudioStreamInfo.EncodingFormat.ENCODING_PCM_1
6BIT) // 16-bit PCM
                    .channelMask(AudioStreamInfo.ChannelMask.CHANNEL_OUT_STEREO)
// 双声道输出
                    .streamUsage(AudioStreamInfo.StreamUsage.STREAM_USAGE_MEDIA)
// 媒体类音频
                    .build();

            AudioRendererInfo audioRendererInfo = new 
AudioRendererInfo.Builder().audioStreamInfo(audioStreamInfo)
                    .audioStreamOutputFlag(AudioRendererInfo.
AudioStreamOutputFlag.AUDIO_STREAM_OUTPUT_FLAG_DIRECT_PCM) // pcm格式的输出流
                    .bufferSizeInBytes(1024)
                    .isOffload(false) // false表示分段传输buffer并播放，true表示整个音
频流一次性传输到HAL层播放
                    .build();

            audioRenderer = new AudioRenderer(audioRendererInfo,
AudioRenderer.PlayMode.MODE_STREAM);
            audioRenderer.setVolume(90f); // 音量

            HiLog.info(LABEL_LOG, "end initAudioRenderer");
        }

        private void release() {
            HiLog.info(LABEL_LOG, "release");

            if (audioRenderer != null) {
                // 关闭、释放资源
                audioRenderer.release();
                audioRenderer = null;
            }
        }

        @Override
        public void onActive() {
            super.onActive();
        }

        @Override
        public void onForeground(Intent intent) {
```

```
        super.onForeground(intent);
    }
}
```

上述代码中：

- 在Button上设置了点击事件。
- initAudioRenderer方法用于初始化AudioRenderer对象。
- start方法用于读取音频文件，并通过分段的方式将字节流写入audioRenderer。
- pause和stop方法比较简单，只是单纯执行audioRenderer的pause和stop方法。
- release方法用于释放资源。

16.2.5 运行

运行应用后，分别点击Start、Pause、Stop按钮，控制台输出内容如下：

```
   09-12 18:24:39.489 21946-21946/com.waylau.hmos.audiorenderer I
00001/MainAbilitySlice: before initAudioRenderer
   09-12 18:24:39.506 21946-21946/com.waylau.hmos.audiorenderer I
00001/MainAbilitySlice: end initAudioRenderer
   09-12 18:24:42.620 21946-21946/com.waylau.hmos.audiorenderer I
00001/MainAbilitySlice: before getFile
   09-12 18:24:42.624 21946-21946/com.waylau.hmos.audiorenderer I
00001/MainAbilitySlice: bufSize: 2124
   09-12 18:24:42.636 21946-21946/com.waylau.hmos.audiorenderer I
00001/MainAbilitySlice: getFile buffer.length: 4096, state:2
   09-12 18:24:42.990 21946-21946/com.waylau.hmos.audiorenderer I
00001/MainAbilitySlice: getFile buffer.length: 4096, state:2
   09-12 18:24:42.991 21946-21946/com.waylau.hmos.audiorenderer I
00001/MainAbilitySlice: getFile buffer.length: 4096, state:2
   09-12 18:24:43.567 21946-21946/com.waylau.hmos.audiorenderer I
00001/MainAbilitySlice: getFile buffer.length: 4096, state:2
   09-12 18:24:43.614 21946-21946/com.waylau.hmos.audiorenderer I
00001/MainAbilitySlice: getFile buffer.length: 4096, state:2
   09-12 18:24:44.046 21946-21946/com.waylau.hmos.audiorenderer I
00001/MainAbilitySlice: getFile buffer.length: 4096, state:2
   09-12 18:24:44.574 21946-21946/com.waylau.hmos.audiorenderer I
00001/MainAbilitySlice: getFile buffer.length: 4096, state:2
   09-12 18:24:44.622 21946-21946/com.waylau.hmos.audiorenderer I
00001/MainAbilitySlice: getFile buffer.length: 4096, state:2
   09-12 18:24:45.102 21946-21946/com.waylau.hmos.audiorenderer I
00001/MainAbilitySlice: getFile buffer.length: 4096, state:2
   09-12 18:24:45.390 21946-21946/com.waylau.hmos.audiorenderer I
00001/MainAbilitySlice: getFile buffer.length: 4096, state:2
   09-12 18:24:45.438 21946-21946/com.waylau.hmos.audiorenderer I
00001/MainAbilitySlice: getFile buffer.length: 4096, state:2
   09-12 18:24:45.591 21946-21946/com.waylau.hmos.audiorenderer I
00001/MainAbilitySlice: getFile buffer.length: 4096, state:2
   09-12 18:24:45.642 21946-21946/com.waylau.hmos.audiorenderer I
```

```
00001/MainAbilitySlice: getFile buffer.length: 4096, state:2
    09-12 18:24:46.122 21946-21946/com.waylau.hmos.audiorenderer I
00001/MainAbilitySlice: getFile buffer.length: 4096, state:2
    09-12 18:24:46.794 21946-21946/com.waylau.hmos.audiorenderer I
00001/MainAbilitySlice: getFile buffer.length: 4096, state:2
    09-12 18:24:46.794 21946-21946/com.waylau.hmos.audiorenderer I
00001/MainAbilitySlice: end getFile
```

16.3 实战：音频采集

本节将演示如何进行音频采集。音频采集的主要工作是通过输入设备将声音采集并转码为音频数据，同时对采集任务进行管理。

16.3.1 接口说明

音频采集类AudioCapturer的主要接口有：

- AudioCapturer(AudioCapturerInfo audioCapturerInfo) throws IllegalArgumentException：构造函数，设置录音相关的音频参数，使用默认录音设备。
- AudioCapturer(AudioCapturerInfo audioCapturerInfo, AudioDeviceDescriptor devInfo) throws IllegalArgumentException：构造函数，设置录音相关音频的参数并指定录音设备。
- getMinBufferSize(int sampleRate, int channelCount, int audioFormat)：获取指定参数条件下所需的最小缓冲区大小。
- addSoundEffect(UUID type, String packageName)：增加录音的音频音效。
- start()：开始录音。
- read(byte[] data, int offset, int size)：读取音频数据。
- read(byte[] data, int offset, int size, boolean isBlocking)：读取音频数据并写入传入的byte数组中。
- read(float[] data, int offsetInFloats, int sizeInFloats)：阻塞式读取音频数据并写入传入的float数组中。
- read(float[] data, int offsetInFloats, int sizeInFloats, boolean isBlocking)：读取音频数据并写入传入的float数组中。
- read(short[] data, int offsetInShorts, int sizeInShorts)：阻塞式读取音频数据并写入传入的short数组中。
- read(short[] data, int offsetInShorts, int sizeInShorts, boolean isBlocking)：读取音频数据并写入传入的short数组中。
- read(java.nio.ByteBuffer buffer, int sizeInBytes)：阻塞式读取音频数据并写入传入的ByteBuffer对象中。
- read(java.nio.ByteBuffer buffer, int sizeInBytes, boolean isBlocking)：读取音频数据并写入传入的ByteBuffer对象中。

- stop()：停止录音。
- release()：释放录音资源。
- getSelectedDevice()：获取输入设备信息。
- getCurrentDevice()：获取当前正在录制音频的设备信息。
- getCapturerSessionId()：获取录音的session ID。
- getSoundEffects()：获取已经激活的音频音效列表。
- getState()：获取音频采集状态。
- getSampleRate()：获取采样率。
- getAudioInputSource()：获取录音的输入设备信息。
- getBufferFrameCount()：获取以帧为单位的缓冲区大小。
- getChannelCount()：获取音频采集通道数。
- getEncodingFormat()：获取音频采集的音频编码格式。
- getAudioTime(Timestamp timestamp, Timestamp.Timebase timebase)：获取一个即时的捕获时间戳。

16.3.2 创建应用

为了演示音频采集的功能，创建一个名为AudioCapturer的应用。

在应用的界面上，通过点击按钮，来触发音频采集的操作。

16.3.3 声明麦克风权限

修改配置文件，声明麦克风权限如下：

```
// 声明权限
"reqPermissions": [
  {
    "name": "ohos.permission.MICROPHONE"
  }
]
```

同时，在应用启动时，显式声明麦克风权限，代码如下：

```java
package com.waylau.hmos.audiocapturer;

import com.waylau.hmos.audiocapturer.slice.MainAbilitySlice;
import ohos.aafwk.ability.Ability;
import ohos.aafwk.content.Intent;

import java.util.ArrayList;
import java.util.List;

public class MainAbility extends Ability {
    @Override
```

```
public void onStart(Intent intent) {
    super.onStart(intent);
    super.setMainRoute(MainAbilitySlice.class.getName());

    // 显式声明需要使用的权限
    requestPermission();
}

// 显式声明需要使用的权限
private void requestPermission() {
    String[] permission = {
            "ohos.permission.MICROPHONE"};
    List<String> applyPermissions = new ArrayList<>();
    for (String element : permission) {
        if (verifySelfPermission(element) != 0) {
            if (canRequestPermission(element)) {
                applyPermissions.add(element);
            }
        }
    }
    requestPermissionsFromUser(applyPermissions.toArray(new String[0]), 0);
}
```

16.3.4 修改 ability_main.xml

修改ability_main.xml内容如下：

```xml
<?xml version="1.0" encoding="utf-8"?>
<DirectionalLayout
    xmlns:ohos="http://schemas.huawei.com/res/ohos"
    ohos:height="match_parent"
    ohos:width="match_parent"
    ohos:alignment="center"
    ohos:orientation="vertical">

    <Button
        ohos:id="$+id:button_start"
        ohos:height="match_content"
        ohos:width="match_content"
        ohos:background_element="#F76543"
        ohos:margin="10vp"
        ohos:padding="10vp"
        ohos:text="Pause"
        ohos:text_size="40vp"
        />

    <Button
```

```xml
        ohos:id="$+id:button_stop"
        ohos:height="match_content"
        ohos:width="match_content"
        ohos:background_element="#F76543"
        ohos:margin="10vp"
        ohos:padding="10vp"
        ohos:text="Stop"
        ohos:text_size="40vp"
        />

</DirectionalLayout>
```

界面预览效果如图16-2所示。

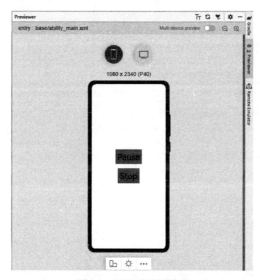

图 16-2　界面预览效果

上述代码设置了Start、Stop按钮，以备设置点击事件，以触发音频采集相关的操作。

16.3.5　修改 MainAbilitySlice

修改MainAbilitySlice内容如下：

```java
package com.waylau.hmos.audiocapturer.slice;

import com.waylau.hmos.audiocapturer.ResourceTable;
import ohos.aafwk.ability.AbilitySlice;
import ohos.aafwk.content.Intent;
import ohos.agp.components.Button;
import ohos.global.resource.NotExistException;
import ohos.hiviewdfx.HiLog;
import ohos.hiviewdfx.HiLogLabel;
import ohos.media.audio.*;

import java.io.*;
```

```java
public class MainAbilitySlice extends AbilitySlice {
    private static final String TAG = MainAbilitySlice.class.getSimpleName();
    private static final HiLogLabel LABEL_LOG =
            new HiLogLabel(HiLog.LOG_APP, 0x00001, TAG);

    // 音频采集类
    private AudioCapturer audioCapturer;

    @Override
    public void onStart(Intent intent) {
        super.onStart(intent);
        super.setUIContent(ResourceTable.Layout_ability_main);

        // 初始化AudioCapturer
        initAudioCapturer();

        // 为按钮设置点击事件回调
        Button buttonStart =
                (Button) findComponentById(ResourceTable.Id_button_start);
        buttonStart.setClickedListener(listener ->
        {
            try {
                start();
            } catch (IOException | NotExistException e) {
                e.printStackTrace();
            }
        });

        Button buttonStop =
                (Button) findComponentById(ResourceTable.Id_button_stop);
        buttonStop.setClickedListener(listener -> audioCapturer.stop());
    }

    private void start() throws IOException, NotExistException {
        HiLog.info(LABEL_LOG, "before start");

        audioCapturer.start();

        HiLog.info(LABEL_LOG, "end start");
    }


    @Override
    public void onStop() {
        super.onStop();

        // 关闭、释放资源
        release();
    }
```

```java
    private void initAudioCapturer() {
        HiLog.info(LABEL_LOG, "before initAudioCapturer");

        AudioStreamInfo audioStreamInfo = new
AudioStreamInfo.Builder().encodingFormat(
                AudioStreamInfo.EncodingFormat.ENCODING_PCM_16BIT) // 16-bit PCM
                .channelMask(AudioStreamInfo.ChannelMask.CHANNEL_IN_STEREO) // 双声道输入
                .sampleRate(44100) // 44.1kHz
                .build();

        AudioCapturerInfo audioCapturerInfo = new AudioCapturerInfo.Builder()
                .audioStreamInfo(audioStreamInfo)
                .build();

        audioCapturer = new AudioCapturer(audioCapturerInfo);

        HiLog.info(LABEL_LOG, "end initAudioCapturer");
    }

    private void release() {
        HiLog.info(LABEL_LOG, "release");

        if (audioCapturer != null) {
            // 关闭、释放资源
            audioCapturer.release();
            audioCapturer = null;
        }
    }

    @Override
    public void onActive() {
        super.onActive();
    }

    @Override
    public void onForeground(Intent intent) {
        super.onForeground(intent);
    }
}
```

上述代码中:

- 在Button上设置了点击事件。
- initAudioCapturer方法用于初始化AudioCapturer对象。
- start方法用于采集音频。
- stop方法比较简单,只是单纯执行audioCapturer的stop方法。

- release方法用于释放资源。

16.4 实战：短音播放

本节将演示如何进行短音播放。短音播放主要负责管理音频资源的加载与播放、tone音的生成与播放以及系统音播放。

16.4.1 接口说明

短音播放开放能力分为音频资源、tone音和系统音3部分，均定义在SoundPlayer类中。

1. 音频资源的加载与播放类 SoundPlayer 的主要接口

音频资源的加载与播放类SoundPlayer的主要接口有：

- SoundPlayer(int taskType)：构造函数，仅用于音频资源。
- createSound(String path)：从指定的路径加载音频数据生成短音资源。
- createSound(Context context, int resourceId)：根据应用程序上下文和音频资源ID加载音频数据生成短音资源。
- createSound(AssetFD assetFD)：从指定的AssetFD实例加载音频数据生成短音资源。
- createSound(java.io.FileDescriptor fd, long offset, long length)：根据文件描述符从文件加载音频数据生成音频资源。
- createSound(java.lang.String path, AudioRendererInfo rendererInfo)：根据指定路径和播放信息加载音频数据生成短音资源。
- setOnCreateCompleteListener(SoundPlayer.OnCreateCompleteListener listener)：设置声音创建完成的回调。
- setOnCreateCompleteListener(SoundPlayer.OnCreateCompleteListener listener, boolean isDiscarded)：设置声音创建完成的回调，并根据指定的isDiscarded标志位确定是否丢弃队列中的原始回调通知消息。
- deleteSound(int soundID)：删除短音，同时释放短音所占的资源。
- pause(int taskID)：根据播放任务ID暂停对应的短音播放。
- play(int soundID)：使用默认参数播放短音。
- play(int soundID, SoundPlayerParameters parameters)：使用指定参数播放短音。
- resume(int taskID)：恢复短音播放任务。
- setLoop(int taskID, int loopNum)：设置短音播放任务的循环次数。
- setPlaySpeedRate(int taskID, float speedRate)：设置短音播放任务的播放速度。
- setPriority(int taskID, int priority)：设置短音播放任务的优先级。
- setVolume(int taskID, AudioVolumes audioVolumes)：设置短音播放任务的播放音量。
- setVolume(int taskID, float volume)：设置短音播放任务的所有音频声道的播放音量。

- stop(int taskID)：停止短音播放任务。
- pauseAll()：暂停所有正在播放的任务。
- resumeAll()：恢复所有已暂停的播放任务。

2. tone 音的生成与播放 API 接口功能介绍

tone音的生成与播放API接口功能介绍：

- SoundPlayer()：构造函数，仅用于tone音。
- createSound(ToneDescriptor.ToneType type, int durationMs)：创建具有音调频率描述和持续时间（毫秒）的tone音。
- createSound(AudioStreamInfo.StreamType streamType, float volume)：根据音量和音频流类型创建tone音。
- play(ToneDescriptor.ToneType toneType, int durationMs)：播放指定时长和tone音类型的tone音。
- pause()：暂停tone音播放。
- play()：播放创建好的tone音。
- release()：释放tone音资源。

3. 系统音的播放 API 接口功能介绍

系统音的播放API接口功能介绍：

- SoundPlayer(String packageName)：构造函数，仅用于系统音。
- playSound(SoundType type)：播放系统音。
- playSound(SoundType type, float volume)：指定音量播放系统音。

16.4.2 创建应用

为了演示短音播放的功能，创建一个名为SoundPlayer的应用。
在应用的界面上，通过点击按钮来触发音频的操作。
在rawfile目录下放置一个test.wav音频文件，以备测试。

16.4.3 修改 ability_main.xml

修改ability_main.xml内容如下：

```
<?xml version="1.0" encoding="utf-8"?>
<DirectionalLayout
    xmlns:ohos="http://schemas.huawei.com/res/ohos"
    ohos:height="match_parent"
    ohos:width="match_parent"
    ohos:alignment="center"
    ohos:orientation="vertical">

    <Button
```

```xml
        ohos:id="$+id:button_play"
        ohos:height="match_content"
        ohos:width="match_content"
        ohos:background_element="#F76543"
        ohos:margin="10vp"
        ohos:padding="10vp"
        ohos:text="Play"
        ohos:text_size="40vp"
        />

    <Button
        ohos:id="$+id:button_pause"
        ohos:height="match_content"
        ohos:width="match_content"
        ohos:background_element="#F76543"
        ohos:margin="10vp"
        ohos:padding="10vp"
        ohos:text="Pause"
        ohos:text_size="40vp"
        />

    <Button
        ohos:id="$+id:button_stop"
        ohos:height="match_content"
        ohos:width="match_content"
        ohos:background_element="#F76543"
        ohos:margin="10vp"
        ohos:padding="10vp"
        ohos:text="Stop"
        ohos:text_size="40vp"
        />

</DirectionalLayout>
```

界面预览效果如图16-3所示。

上述代码设置了Play、Pause、Stop按钮，以备设置点击事件，以触发音频播放相关的操作。

16.4.4 修改 MainAbilitySlice

修改MainAbilitySlice内容如下：

```
package com.waylau.hmos.soundplayer.slice;

import com.waylau.hmos.soundplayer.ResourceTable;
import ohos.aafwk.ability.AbilitySlice;
```

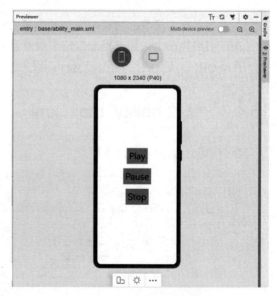

图 16-3　界面预览效果

```java
import ohos.aafwk.content.Intent;
import ohos.agp.components.Button;
import ohos.global.resource.NotExistException;
import ohos.global.resource.RawFileEntry;
import ohos.global.resource.Resource;
import ohos.hiviewdfx.HiLog;
import ohos.hiviewdfx.HiLogLabel;
import ohos.media.audio.AudioManager;
import ohos.media.audio.AudioRenderer;
import ohos.media.audio.SoundPlayer;

import java.io.IOException;


public class MainAbilitySlice extends AbilitySlice {
    private static final String TAG = MainAbilitySlice.class.getSimpleName();
    private static final HiLogLabel LABEL_LOG =
            new HiLogLabel(HiLog.LOG_APP, 0x00001, TAG);

    // 短音播放类
    private SoundPlayer soundPlayer;

    private int soundId;
    private SoundPlayer.SoundPlayerParameters parameters;

    @Override
    public void onStart(Intent intent) {
        super.onStart(intent);
        super.setUIContent(ResourceTable.Layout_ability_main);

        // 初始化SoundPlayer
        try {
            initSoundPlayer();
        } catch (IOException e) {
            e.printStackTrace();
        }

        // 为按钮设置点击事件回调
        Button buttonPlay =
                (Button) findComponentById(ResourceTable.Id_button_play);
        buttonPlay.setClickedListener(listener -> play());

        Button buttonPause =
                (Button) findComponentById(ResourceTable.Id_button_pause);
        buttonPause.setClickedListener(listener -> soundPlayer.pause());

        Button buttonStop =
                (Button) findComponentById(ResourceTable.Id_button_stop);
        buttonStop.setClickedListener(listener -> soundPlayer.stop(soundId));
    }
```

```java
private void play() {
    HiLog.info(LABEL_LOG, "before play");

    // 短音播放
    soundPlayer.play(soundId, parameters);

    HiLog.info(LABEL_LOG, "end play, soundId:%{public}s", soundId);
}


@Override
public void onStop() {
    super.onStop();

    // 关闭、释放资源
    release();
}

private void initSoundPlayer() throws IOException {
    HiLog.info(LABEL_LOG, "before initSoundPlayer");

    // 实例化SoundPlayer对象
    soundPlayer =
        new SoundPlayer(AudioManager.AudioVolumeType.STREAM_MUSIC.getValue());

    RawFileEntry rawFileEntry =
        this.getResourceManager().getRawFileEntry("resources/rawfile/test.wav");

    // 指定音频资源加载并创建短音
    soundId = soundPlayer.createSound(rawFileEntry.openRawFileDescriptor());

    // 指定音量、循环次数和播放速度
    parameters = new SoundPlayer.SoundPlayerParameters();
    parameters.setVolumes(new SoundPlayer.AudioVolumes());
    parameters.setLoop(10);
    parameters.setSpeed(1.0f);

    HiLog.info(LABEL_LOG, "end initSoundPlayer");
}

private void release() {
    HiLog.info(LABEL_LOG, "release");

    if (soundPlayer != null) {
        // 关闭、释放资源
        soundPlayer.release();
```

```
            soundPlayer = null;
        }
    }

    @Override
    public void onActive() {
        super.onActive();
    }

    @Override
    public void onForeground(Intent intent) {
        super.onForeground(intent);
    }
}
```

上述代码中：

- 在Button上设置了点击事件。
- initSoundPlayer方法用于初始化SoundPlayer对象及参数，同时读取音频文件生成了soundId。
- play方法用于播放指定soundId的音频文件。
- pause和stop方法比较简单，只是单纯执行SoundPlayer的pause和stop方法。
- release方法用于释放资源。

16.4.5 运行

运行应用后，点击Play按钮，控制台输出内容如下：

```
    02-10 17:24:31.201 3195-3195/com.waylau.hmos.soundplayer I
00001/MainAbilitySlice: before initSoundPlayer
    02-10 17:24:31.213 3195-3195/com.waylau.hmos.soundplayer I
00001/MainAbilitySlice: end initSoundPlayer
    02-10 17:24:46.267 3195-3195/com.waylau.hmos.soundplayer I
00001/MainAbilitySlice: before play
    02-10 17:25:17.515 3195-3195/com.waylau.hmos.soundplayer I
00001/MainAbilitySlice: end play, soundId:1
```

第 17 章

媒体会话管理

AVSession是一套媒体播放控制框架,对媒体服务和界面进行解耦,并提供规范的通信接口,使应用可以自由、高效地在不同的媒体之间完成切换。

17.1 媒体会话管理概述

AVSession框架有4个主要的类,控制着整个框架的核心。图17-1简单地说明4个核心媒体框架控制类的关系。

图 17-1 界面预览效果

17.1.1 AVSession 框架主要的类

AVSession框架4个主要的类分别是AVBrowser、AVController、AVBrowserService和AVSession。

1. AVBrowser

AVBrowser（媒体浏览器）通常在客户端创建，成功连接媒体服务后，通过媒体控制器 AVBrowser 向服务端发送播放控制指令。

其主要流程为，调用 connect 方法向 AVBrowserService 发起连接请求，连接成功后在回调方法 AVConnectionCallback.onConnected 中发起订阅数据请求，并在回调方法 AVSubscriptionCallback.onAVElementListLoaded 中保存请求的媒体播放数据。

2. AVController

AVController（媒体控制器）在客户端 AVBrowser 连接服务成功后的回调方法 AVConnectionCallback.onConnected 中创建，用于向 Service 发送播放控制指令，并通过实现 AVControllerCallback 回调来响应服务端媒体的状态变化，例如曲目信息变更、播放状态变更等，从而完成 UI 刷新。

3. AVBrowserService

AVBrowserService（媒体浏览器服务）通常在服务端通过媒体会话 AVSession 与媒体浏览器建立连接，并通过实现 Player 进行媒体播放。其中有两个重要的方法：

- onGetRoot：处理从媒体浏览器 AVBrowser 发来的连接请求，通过返回一个有效的 AVBrowserRoot 对象表示连接成功。
- onLoadAVElementList：处理从媒体浏览器 AVBrowser 发来的数据订阅请求，通过 AVBrowserResult.sendAVElementList(List) 方法返回媒体播放数据。

4. AVSession

AVSession（媒体会话）通常在 AVBrowserService 的 onStart 中创建，通过 setAVToken 方法设置到 AVBrowserService 中，并通过实现 AVSessionCallback 回调来接收和处理媒体控制器 AVController 发送的播放控制指令，如播放、暂停、跳转至上一曲、跳转至下一曲等。

除了上述4个类外，AVSession 框架还有 AVElement。

5. AVElement

媒体元素，用于将播放列表从 AVBrowserService 传递给 AVBrowser。

17.1.2 约束与限制

AVSession 框架包含以下使用约束与限制：

- 在使用完 AVSession 类后，需要及时进行资源释放。
- 调用 AVBrowser 的 subscribeByParentMediaId(String, AVSubscriptionCallback) 之前，需要先执行 unsubscribeByParentMediaId(String)，防止重复订阅。
- 使用 AVBrowserService 的方法 onLoadAVElementList(String, AVBrowserResult) 的 result 返回数据前，执行 detachForRetrieveAsync()。
- 播放器类需要使用 ohos.media.player.Player，否则无法正常接收按键事件。

17.2 接口说明

本节介绍AVSession框架核心类的主要接口。

17.2.1 AVBrowser 的主要接口

AVBrowser的主要接口有：

- AVBrowser(Context context, ElementName name, AVConnectionCallback callback, PacMap options)：构造AVBrowser实例，用于浏览AVBrowserService提供的媒体数据。
- connect()：连接AVBrowserService。
- disconnect()：与AVBrowserService断开连接。
- isConnected()：判断当前是否已经与AVBrowserService连接。
- getElementName()：获取AVBrowserService的ohos.bundle.ElementName实例。
- getRootMediaId()：获取默认媒体ID。
- getOptions()：获取AVBrowserService提供的附加数据。
- getAVToken()：获取媒体会话的令牌。
- getAVElement(String mediaId, AVElementCallback callback)：输入媒体的ID，查询对应的ohos.media.common.sessioncore.AVElement信息，查询结果会通过callback返回。
- subscribeByParentMediaId(String parentMediaId, AVSubscriptionCallback callback)：查询指定媒体ID包含的所有媒体元素信息，并订阅它的媒体信息更新通知。
- subscribeByParentMediaId(String parentMediaId, PacMap options, AVSubscriptionCallback callback)：基于特定于服务的参数来查询指定媒体ID中的媒体元素的信息，并订阅它的媒体信息更新通知。
- unsubscribeByParentMediaId(String parentMediaId)：取消订阅对应媒体ID的信息更新通知。
- unsubscribeByParentMediaId(String parentMediaId, AVSubscriptionCallback callback)：取消订阅与指定callback相关的媒体ID的信息更新通知。

17.2.2 AVBrowserService 的主要接口

AVBrowserService的主要接口有：

- onGetRoot(String callerPackageName, int clientUid, PacMap options)：回调方法，用于返回应用程序的媒体内容的根信息，在AVBrowser.connect()后进行回调。
- onLoadAVElementList(String parentMediaId, AVBrowserResult result)：回调方法，用于返回应用程序的媒体内容的结果信息AVBrowserResult，其中包含子节点的AVElement列表，在AVBrowser的方法subscribeByParentMediaId或notifyAVElementListUpdated执行后进行回调。

- onLoadAVElement(String mediaId, AVBrowserResult result)：回调方法，用于获取特定的媒体项目AVElement的结果信息，在AVBrowser.getAVElement方法执行后进行回调。
- getAVToken()：获取AVBrowser与AVBrowserService之间的会话令牌。
- setAVToken(AVToken token)：设置AVBrowser与AVBrowserService之间的会话令牌。
- getBrowserOptions()：获取AVBrowser在连接AVBrowserService时设置的服务参数选项。
- getCallerUserInfo()：获取当前发送请求的调用者信息。
- notifyAVElementListUpdated(String parentMediaId)：通知所有已连接的AVBrowser当前父节点的子节点已经发生改变。
- notifyAVElementListUpdated(String parentId, PacMap options)：通知所有已连接的AVBrowser当前父节点的子节点已经发生改变，可设置服务参数。

17.2.3　AVController 的主要接口

AVController的主要接口有：

- AVController(Context context, AVToken avToken)：构造AVController实例，用于应用程序与AVSession进行交互以控制媒体播放。
- setControllerForAbility(Ability ability, AVController controller)：将媒体控制器注册到ability以接收按键事件。
- setAVControllerCallback(AVControllerCallback callback)：注册一个回调以接收来自AVSession的变更，例如元数据和播放状态变更。
- releaseAVControllerCallback(AVControllerCallback callback)：释放与AVSession之间的回调实例。
- getAVQueueElement()：获取播放队列。
- getAVQueueTitle()：获取播放队列的标题。
- getAVPlaybackState()：获取播放状态。
- dispatchAVKeyEvent(KeyEvent keyEvent)：应用分发媒体按键事件给会话以控制播放。
- sendCustomCommand(String command, PacMap pacMap, GeneralReceiver receiverCb)：应用向AVSession发送自定义命令，参考ohos.media.common.sessioncore.AVSessionCallback.onCommand。
- getAVSessionAbility()：获取启动用户界面的IntentAgent。
- getAVToken()：获取应用连接到会话的令牌。此令牌用于创建媒体播放控制器。
- adjustAVPlaybackVolume(int direction, int flags)：调节播放音量。
- setAVPlaybackVolume(int value, int flags)：设置播放音量，要求支持绝对音量控制。
- getOptions()：获取与此控制器连接的AVSession的附加数据。
- getFlags()：获取AVSession的附加标识，标记在AVSession中的定义。
- getAVMetadata()：获取媒体资源的元数据ohos.media.common.AVMetadata。
- getAVPlaybackInfo()：获取播放信息。
- getSessionOwnerPackageName()：获得AVSession实例的应用程序的包名称。
- getAVSessionInfo()：获取会话的附加数据。

- getPlayControls()：获取一个PlayControls实例，将用于控制播放，比如控制媒体播放、停止、下一首等。

17.2.4　AVSession 的主要接口

AVSession的主要接口有：

- AVSession(Context context, String tag)：构造AVSession实例，用于控制媒体播放。
- AVSession(Context context, String tag, PacMap sessionInfo)：构造带有附加会话信息的AVSession实例，用于控制媒体播放。
- setAVSessionCallback(AVSessionCallback callback)：设置回调函数来控制播放器，控制逻辑由应用实现。如果callback为null，则取消控制。
- setAVSessionAbility(IntentAgent ia)：给AVSession设置一个IntentAgent，用来启动用户界面。
- setAVButtonReceiver(IntentAgent ia)：为媒体按键接收器设置一个IntentAgent，以便应用结束后，可以通过媒体按键重新拉起应用。
- enableAVSessionActive(boolean active)：设置是否激活媒体会话。当会话准备接收命令时，将输入参数设置为true。如果会话停止接收命令，则设置为false。
- isAVSessionActive()：查询会话是否激活。
- sendAVSessionEvent(String event, PacMap options)：向所有订阅此会话的控制器发送事件。
- release()：释放资源，应用播放完之后需要调用。
- getAVToken()：获取应用连接到会话的令牌。此令牌用于创建媒体播放控制器。
- getAVController()：获取会话构造时创建的控制器，方便应用使用。
- setAVPlaybackState(AVPlaybackState state)：设置当前播放状态。
- setAVMetadata(AVMetadata avMetadata)：设置媒体资源元数据ohos.media.common.AVMetadata。
- setAVQueue(List<AVQueueElement> queue)：设置播放队列。
- setAVQueueTitle(CharSequence queueTitle)：设置播放队列的标题，UI会显示此标题。
- setOptions(PacMap options)：设置此会话关联的附加数据。
- getCurrentControllerInfo()：获取发送当前请求的媒体控制器信息。

17.2.5　AVElement 的主要接口

AVElement的主要接口有：

- AVElement(AVDescription description, int flags)：构造AVElement实例。
- getFlags()：获取flags的值。
- isScannable()：判断媒体是否可扫描，如媒体有子节点，则可继续扫描获取子节点的内容。
- isPlayable()：检查媒体是否可播放。
- getAVDescription()：获取媒体的详细信息。
- getMediaId()：获取媒体的ID。

17.3 实战：AVSession 媒体框架客户端

接下来将使用AVSession媒体框架创建一个播放器示例。该示例分为创建客户端和创建服务端两部分。本节将演示如何进行AVSession媒体框架客户端的开发工作。

17.3.1 创建应用

为了演示AVSession媒体框架的功能，创建一个名为AVSession的应用。
在应用的界面上，通过点击按钮来触发音频的操作。
在rawfile目录下放置一个test.wav音频文件，以备测试。

17.3.2 修改 ability_main.xml

修改ability_main.xml内容如下：

```xml
<?xml version="1.0" encoding="utf-8"?>
<DirectionalLayout
    xmlns:ohos="http://schemas.huawei.com/res/ohos"
    ohos:height="match_parent"
    ohos:width="match_parent"
    ohos:alignment="center"
    ohos:orientation="vertical">

    <Button
        ohos:id="$+id:button_play"
        ohos:height="match_content"
        ohos:width="match_content"
        ohos:background_element="#F76543"
        ohos:margin="10vp"
        ohos:padding="10vp"
        ohos:text="Play"
        ohos:text_size="40vp"
        />

    <Button
        ohos:id="$+id:button_pause"
        ohos:height="match_content"
        ohos:width="match_content"
        ohos:background_element="#F76543"
        ohos:margin="10vp"
        ohos:padding="10vp"
        ohos:text="Pause"
        ohos:text_size="40vp"
```

```
        />

    <Button
        ohos:id="$+id:button_stop"
        ohos:height="match_content"
        ohos:width="match_content"
        ohos:background_element="#F76543"
        ohos:margin="10vp"
        ohos:padding="10vp"
        ohos:text="Stop"
        ohos:text_size="40vp"
        />

</DirectionalLayout>
```

界面预览效果如图17-2所示。

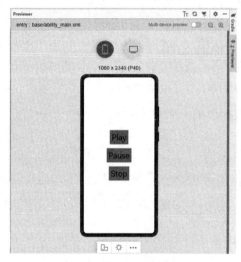

图17-2　界面预览效果

上述代码设置了Start、Pause、Stop按钮,以备设置点击事件,以触发音频播放相关的操作。

17.3.3　修改 MainAbilitySlice

修改MainAbilitySlice内容如下:

```
package com.waylau.hmos.avsession.slice;

import com.waylau.hmos.avsession.ResourceTable;
import ohos.aafwk.ability.AbilitySlice;
import ohos.aafwk.content.Intent;
import ohos.agp.components.Button;
import ohos.bundle.ElementName;
import ohos.hiviewdfx.HiLog;
import ohos.hiviewdfx.HiLogLabel;
```

```java
import ohos.media.common.AVDescription;
import ohos.media.common.AVMetadata;
import ohos.media.common.sessioncore.*;
import ohos.media.image.PixelMap;
import ohos.media.sessioncore.AVBrowser;
import ohos.media.sessioncore.AVController;

import java.util.List;

public class MainAbilitySlice extends AbilitySlice {
    private static final String TAG = MainAbilitySlice.class.getSimpleName();
    private static final HiLogLabel LABEL_LOG =
            new HiLogLabel(HiLog.LOG_APP, 0x00001, TAG);

    // 媒体浏览器
    private AVBrowser avBrowser;
    // 媒体控制器
    private AVController avController;

    private List<AVElement> list;

    @Override
    public void onStart(Intent intent) {
        super.onStart(intent);
        super.setUIContent(ResourceTable.Layout_ability_main);

        initBrowser();

        // 为按钮设置点击事件回调
        Button buttonPlay =
                (Button) findComponentById(ResourceTable.Id_button_play);
        buttonPlay.setClickedListener(listener -> play());

        Button buttonPause =
                (Button) findComponentById(ResourceTable.Id_button_pause);
        buttonPause.setClickedListener(listener -> pause());

        Button buttonStop =
                (Button) findComponentById(ResourceTable.Id_button_stop);
        buttonStop.setClickedListener(listener -> stop());
    }

    private void stop() {
        avController.getPlayControls().stop();
    }

    private void pause() {
        toPlayOrPause();
    }
```

```java
        private void play() {
            toPlayOrPause();
        }

        public void toPlayOrPause() {
            switch (avController.getAVPlaybackState().getAVPlaybackState()) {
                case AVPlaybackState.PLAYBACK_STATE_NONE: {
                    avController.getPlayControls().prepareToPlay();
                    avController.getPlayControls().play();
                    HiLog.info(LABEL_LOG, "end play");
                    break;
                }
                case AVPlaybackState.PLAYBACK_STATE_PLAYING: {
                    avController.getPlayControls().pause();
                    HiLog.info(LABEL_LOG, "end pause");
                    break;
                }
                case AVPlaybackState.PLAYBACK_STATE_PAUSED: {
                    avController.getPlayControls().play();
                    HiLog.info(LABEL_LOG, "end play");
                    break;
                }
                default: {
                    break;
                }
            }
        }

        private void initBrowser() {
            HiLog.info(LABEL_LOG, "before initBrowser");

            // 用于指向媒体浏览器服务的包路径和类名
            ElementName elementName =
                    new ElementName("", "com.waylau.hmos.avsession",
 "com.waylau.hmos.avsession.AVService");
            // connectionCallback在调用avBrowser.connect方法后进行回调
            avBrowser = new AVBrowser(this, elementName, connectionCallback, null);
            // avBrowser发送对媒体浏览器服务的连接请求
            avBrowser.connect();
            // 将媒体控制器注册到ability以接收按键事件
            AVController.setControllerForAbility(this.getAbility(),
 avController);

            HiLog.info(LABEL_LOG, "end initBrowser");
        }

        // 发起连接（avBrowser.connect）后的回调方法实现
        private AVConnectionCallback connectionCallback = new
 AVConnectionCallback() {
            @Override
```

```
                public void onConnected() {
                    // 成功连接媒体浏览器服务时回调该方法，否则回调onConnectionFailed()
                    // 重复订阅会报错，所以先解除订阅
avBrowser.unsubscribeByParentMediaId(avBrowser.getRootMediaId());
                    // 第二个参数AVSubscriptionCallback，用于处理订阅信息的回调
                    avBrowser.subscribeByParentMediaId(avBrowser.getRootMediaId(),
avSubscriptionCallback);
                    AVToken token = avBrowser.getAVToken();
                    avController = new AVController(getContext(), token); //
AVController第一个参数为当前类的context
                    // 参数AVControllerCallback，用于处理服务端播放状态及信息变化时回调
                    avController.setAVControllerCallback(avControllerCallback);

                    HiLog.info(LABEL_LOG, "end onConnected");
                }
        };

        // 发起订阅信息(avBrowser.subscribeByParentMediaId)后的回调方法实现
        private AVSubscriptionCallback avSubscriptionCallback = new
AVSubscriptionCallback() {
            @Override
            public void onAVElementListLoaded(String parentId, List<AVElement>
children) {
                // 订阅成功时回调该方法，parentID为标识，children为服务端回传的媒体列表
                super.onAVElementListLoaded(parentId, children);
                list.addAll(children);

                HiLog.info(LABEL_LOG, "end onAVElementListLoaded,
size: %{public}s", list.size());
            }
        };

        // 服务对客户端的媒体数据或播放状态变更后的回调
        private AVControllerCallback avControllerCallback = new
AVControllerCallback() {
            @Override
            public void onAVMetadataChanged(AVMetadata metadata) {
                // 当服务端调用avSession.setAVMetadata(avMetadata)时，此方法会被回调
                super.onAVMetadataChanged(metadata);
                AVDescription description = metadata.getAVDescription();
                String title = description.getTitle().toString();
                PixelMap pixelMap = description.getIcon();

                HiLog.info(LABEL_LOG, "end onAVMetadataChanged, title: %{public}s",
title);
            }

            @Override
```

```
                public void onAVPlaybackStateChanged(AVPlaybackState playbackState) {
                    // 当服务端调用avSession.setAVPlaybackState(...)时,此方法会被回调
                    super.onAVPlaybackStateChanged(playbackState);
                    long position = playbackState.getCurrentPosition();

                    HiLog.info(LABEL_LOG, "end onAVMetadataChanged,
position: %{public}s", position);
                }
            };

            @Override
            public void onActive() {
                super.onActive();
            }

            @Override
            public void onForeground(Intent intent) {
                super.onForeground(intent);
            }
        }
```

上述代码中：

- 在Button上设置了点击事件。
- initBrowser方法用于初始化AVBrowser和AVController对象。AVBrowser和AVController用于向服务端发送连接请求。
- AVConnectionCallback回调接口中的方法为可选实现，通常需要在onConnected中订阅媒体数据和创建媒体控制器AVController。
- 通常在订阅成功时，在AVSubscriptionCallback回调接口onAVElementListLoaded中保存服务端回传的媒体列表。
- AVControllerCallback回调接口中的方法均为可选方法，主要用于服务端播放状态及信息变化后对客户端的回调，客户端可在这些方法中实现UI的刷新。
- 在play、pause和stop方法中调用avController的方法向服务端发送播放控制指令。

17.4 实战：AVSession 媒体框架服务端

本节将演示如何进行AVSession媒体框架服务端的开发工作。

17.4.1 创建 Service

为了演示AVSession媒体框架的服务端功能，在AVSession的应用中创建一个名为AVService的服务。

AVService 继承自 AVBrowserService，需要实现 onGetRoot、onLoadAVElementList、 onLoadAVElementList和onLoadAVElement四个接口。

17.4.2 修改 AVService

修改AVService内容如下：

```java
package com.waylau.hmos.avsession;

import ohos.aafwk.content.Intent;
import ohos.global.resource.RawFileEntry;
import ohos.media.common.sessioncore.AVBrowserResult;
import ohos.media.common.sessioncore.AVBrowserRoot;
import ohos.media.common.sessioncore.AVPlaybackState;
import ohos.media.common.sessioncore.AVSessionCallback;
import ohos.media.player.Player;
import ohos.media.sessioncore.AVBrowserService;
import ohos.media.sessioncore.AVSession;
import ohos.hiviewdfx.HiLog;
import ohos.hiviewdfx.HiLogLabel;
import ohos.utils.PacMap;

import java.io.IOException;

public class AVService extends AVBrowserService {
    private static final String TAG = AVService.class.getSimpleName();
    private static final HiLogLabel LABEL_LOG =
            new HiLogLabel(HiLog.LOG_APP, 0x00001, TAG);

    // 根媒体ID
    private static final String AV_ROOT_ID = "av_root_id";
    // 媒体会话
    private AVSession avSession;
    // 媒体播放器
    private Player player;

    @Override
    public void onStart(Intent intent) {
        HiLog.info(LABEL_LOG, "AVService::onStart");
        super.onStart(intent);

        try {
            initPlayer();
        } catch (IOException e) {
            e.printStackTrace();
        }
    }

    @Override
    public AVBrowserRoot onGetRoot(String clientPackageName, int clientUid,
```

```java
PacMap rootHints) {
        // 响应客户端avBrowser.connect()方法。若同意连接，则返回有效的AVBrowserRoot
实例，否则返回null
        return new AVBrowserRoot(AV_ROOT_ID, null);
    }

    @Override
    public void onLoadAVElementList(String parentId, AVBrowserResult result) {
        HiLog.info(LABEL_LOG, "onLoadChildren");
        // 响应客户端avBrowser.subscribeByParentMediaId(...)方法
        // 先执行该方法detachForRetrieveAsync()
        result.detachForRetrieveAsync();
        // externalAudioItems缓存媒体文件，请开发者自行实现
        // result.sendAVElementList(externalAudioItems.getAudioItems());
    }

    @Override
    public void onLoadAVElementList(String s, AVBrowserResult avBrowserResult,
PacMap pacMap) {
        // 响应客户端avBrowser.subscribeByParentMediaId(String, PacMap,
AVSubscriptionCallback)方法
    }

    @Override
    public void onLoadAVElement(String s, AVBrowserResult avBrowserResult) {
        // 响应客户端avBrowser.getAVElement(String, AVElementCallback)方法
    }

    private void initPlayer() throws IOException {
        HiLog.info(LABEL_LOG, "before initPlayer");

        avSession = new AVSession(this, "AVService");
        setAVToken(avSession.getAVToken());
        // 设置sessioncallback，用于响应客户端的媒体控制器发起的播放控制指令
        avSession.setAVSessionCallback(avSessionCallback);
        // 设置播放状态初始状态为AVPlaybackState.PLAYBACK_STATE_NONE
        AVPlaybackState playbackState =
                new AVPlaybackState.Builder()
                        .setAVPlaybackState(AVPlaybackState.PLAYBACK_STATE_NONE, 0, 1.0f).build();
        avSession.setAVPlaybackState(playbackState);
        // 完成播放器的初始化，如果使用多个Player，也可以在执行播放时初始化
        player = new Player(this);

        RawFileEntry rawFileEntry =
this.getResourceManager().getRawFileEntry("resources/rawfile/test.wav");
        player.setSource(rawFileEntry.openRawFileDescriptor());

        HiLog.info(LABEL_LOG, "end initPlayer");
```

第17章　媒体会话管理

```java
        }

        private AVSessionCallback avSessionCallback = new AVSessionCallback() {
            @Override
            public void onPlay() {
                HiLog.info(LABEL_LOG, "before onPlay");

                super.onPlay();
                // 当客户端调用avController.getPlayControls().play()时，该方法会被回调
                // 响应播放请求，开始播放
                if (avSession.getAVController().getAVPlaybackState().getAVPlaybackState() == AVPlaybackState.PLAYBACK_STATE_PAUSED) {
                    if (player.play()) {
                        AVPlaybackState playbackState = new AVPlaybackState.Builder().setAVPlaybackState(
                                AVPlaybackState.PLAYBACK_STATE_PLAYING, player.getCurrentTime(),
                                player.getPlaybackSpeed()).build();
                        avSession.setAVPlaybackState(playbackState);
                    }
                }

                HiLog.info(LABEL_LOG, "end onPlay");
            }

            @Override
            public void onPause() {
                HiLog.info(LABEL_LOG, "before onPause");

                // 当客户端调用avController.getPlayControls().pause()时，该方法会被回调
                // 响应暂停请求，暂停播放
                super.onPause();

                HiLog.info(LABEL_LOG, "end onPause");
            }

            @Override
            public void onStop() {
                HiLog.info(LABEL_LOG, "before onStop");

                // 当客户端调用avController.getPlayControls().stop()时，该方法会被回调
                // 响应停止请求，停止播放
                super.onStop();

                HiLog.info(LABEL_LOG, "end onStop");
            }
        };
    }
```

上述代码中：

- 声明了AVSession和Player。
- AVSessionCallback响应客户端的媒体控制器发起的播放控制指令的回调实现。

17.4.3 运行

运行应用后，分别点击Play、Pause和Stop按钮，可以看到控制台输出内容如下：

```
09-12 18:51:26.486 11977-11977/com.waylau.hmos.avsession I 00001/MainAbilitySlice: end play
09-12 18:51:26.488 11977-11977/com.waylau.hmos.avsession I 00001/AVService: before onPlay
09-12 18:51:26.489 11977-11977/com.waylau.hmos.avsession I 00001/AVService: end onPlay
09-12 18:51:28.477 11977-11977/com.waylau.hmos.avsession I 00001/MainAbilitySlice: end play
09-12 18:51:28.479 11977-11977/com.waylau.hmos.avsession I 00001/AVService: before onPlay
09-12 18:51:28.479 11977-11977/com.waylau.hmos.avsession I 00001/AVService: end onPlay
09-12 18:51:29.776 11977-11977/com.waylau.hmos.avsession I 00001/AVService: before onStop
09-12 18:51:29.776 11977-11977/com.waylau.hmos.avsession I 00001/AVService: end onStop
```

第 18 章

媒体数据管理

HarmonyOS媒体数据管理模块支持多媒体数据管理相关的功能开发，包括获取媒体元数据、截取帧数据等。

18.1 媒体数据管理概述

HarmonyOS媒体数据管理模块支持多媒体数据管理相关的功能开发，常见操作有获取媒体元数据、截取帧数据等。

18.1.1 媒体数据管理基本概念

在进行应用的开发前，开发者应了解以下基本概念：

- PixelMap：PixelMap是图像解码后无压缩的位图格式，用于图像显示或者进一步的处理。这个在之前的章节中已经接触得比较多了。
- 媒体元数据：媒体元数据是用来描述多媒体数据的数据，例如媒体标题、媒体时长等数据信息。

18.1.2 约束与限制

使用媒体数据管理时，相关的约束与限制如下：

- 如果需要读取用户的存储文件，则需要声明读取用户存储的权限ohos.permission.READ_USER_STORAGE。

- 为及时释放native资源，建议在媒体数据管理AVMetadataHelper对象使用完成后，主动调用release()方法。

18.2 实战：媒体元数据的获取

本节演示如何实现媒体元数据的获取。媒体元数据是描述多媒体数据的数据，例如媒体标题、媒体时长、媒体的帧数据等。

18.2.1 接口说明

媒体元数据获取相关类AVMetadataHelper的主要接口有：

- setSource(String path)：读取指定路径的媒体文件，将其设置为媒体源。
- setSource(FileDescriptor fd)：读取指定的媒体文件描述符，设置媒体源。
- setSource(FileDescriptor fd, long offset, long length)：读取指定的媒体文件描述符，读取数据的起始位置的偏移量以及读取数据的长度，设置媒体源。
- setSource(String uri, Map<String, String> headers)：读取指定的媒体文件URI，设置媒体源。
- setSource(Context context, Uri uri)：读取指定媒体的URI和上下文，设置媒体源。
- resolveMetadata(int keyCode)：获取媒体元数据中指定keyCode对应的值。
- fetchVideoScaledPixelMapByTime(long timeUs, int option, int dstWidth, int dstHeight)：根据视频源中的时间戳、获取选项以及图像帧缩放大小获取帧数据。
- fetchVideoPixelMapByTime(long timeUs, int option)：根据视频源中的时间戳和获取选项获取帧数据。
- fetchVideoPixelMapByTime(long timeUs)：根据视频源中的时间戳获取最靠近时间戳的帧的数据。
- fetchVideoPixelMapByTime()：随机获取数据源中某一帧的数据。
- resolveImage()：获取音频源中包含的图像数据，比如专辑封面，如果有多个图像，则返回任意一个图像的数据。
- fetchVideoPixelMapByIndex(int frameIndex, PixelMapConfigs configs)：根据指定的图像像素格式选项获取视频源中指定一帧的数据。
- fetchVideoPixelMapByIndex(int frameIndex)：获取视频源中指定一帧的数据。
- fetchVideoPixelMapByIndex(int frameIndex, int numFrames, PixelMapConfigs configs)：根据指定的图像像素格式选项获取视频源中指定的连续多帧的数据。
- fetchVideoPixelMapByIndex(int frameIndex, int numFrames)：获取视频源中指定的连续多帧的数据。
- fetchImagePixelMapByIndex(int imageIndex, PixelMapConfigs configs)：根据指定的图像像素格式选项获取源图像中指定的图像。

- fetchImagePixelMapByIndex(int imageIndex)：获取源图像中指定的图像。
- fetchImagePrimaryPixelMap(PixelMapConfigs configs)：根据指定的图像像素格式选项获取源图像中默认的图像。
- fetchImagePrimaryPixelMap()：获取源图像中默认的图像。
- release()：释放读取的媒体资源。

18.2.2 创建应用

为了演示AVMetadataHelper的功能，创建一个名为AVMetadataHelper的应用。

在应用的界面上，通过点击按钮来触发AVMetadataHelper的操作。

在rawfile目录下放置一个big_buck_bunny.mp4视频文件，以备测试。

18.2.3 修改 ability_main.xml

修改ability_main.xml内容如下：

```xml
<?xml version="1.0" encoding="utf-8"?>
<DirectionalLayout
    xmlns:ohos="http://schemas.huawei.com/res/ohos"
    ohos:height="match_parent"
    ohos:width="match_parent"
    ohos:orientation="vertical">

    <Button
        ohos:id="$+id:button_get"
        ohos:height="match_content"
        ohos:width="match_parent"
        ohos:background_element="#F76543"
        ohos:layout_alignment="horizontal_center"
        ohos:margin="10vp"
        ohos:padding="10vp"
        ohos:text="Get"
        ohos:text_size="40vp"
        />

    <Image
        ohos:id="$+id:image"
        ohos:height="match_content"
        ohos:width="match_parent"/>

</DirectionalLayout>
```

界面预览效果如图18-1所示。

图 18-1　界面预览效果

上述代码中：

- 设置了Get按钮，以备设置点击事件，以触发元数据相关的操作。
- Image组件用于展示获取到的帧数据。

18.2.4　修改 MainAbilitySlice

修改MainAbilitySlice内容如下：

```
package com.waylau.hmos.avmetadatahelper.slice;

import com.waylau.hmos.avmetadatahelper.ResourceTable;
import ohos.aafwk.ability.AbilitySlice;
import ohos.aafwk.content.Intent;
import ohos.agp.components.Button;
import ohos.agp.components.Image;
import ohos.global.resource.RawFileDescriptor;
import ohos.hiviewdfx.HiLog;
import ohos.hiviewdfx.HiLogLabel;
import ohos.media.image.PixelMap;
import ohos.media.photokit.metadata.AVMetadataHelper;

import java.io.IOException;

public class MainAbilitySlice extends AbilitySlice {
    private static final String TAG = MainAbilitySlice.class.getSimpleName();
    private static final HiLogLabel LABEL_LOG =
            new HiLogLabel(HiLog.LOG_APP, 0x00001, TAG);

    private AVMetadataHelper avMetadataHelper;
```

```java
@Override
public void onStart(Intent intent) {
    super.onStart(intent);
    super.setUIContent(ResourceTable.Layout_ability_main);

    // 初始化AVMetadataHelper
    try {
        initAVMetadataHelper();
    } catch (IOException e) {
        e.printStackTrace();
    }

    // 为按钮设置点击事件回调
    Button buttonGet =
            (Button) findComponentById(ResourceTable.Id_button_get);
    buttonGet.setClickedListener(listener -> getInfo());
}

@Override
public void onStop() {
    super.onStop();

    // 关闭、释放资源
    release();
}

private void initAVMetadataHelper() throws IOException {
    HiLog.info(LABEL_LOG, "before initAVMetadataHelper");
    RawFileDescriptor rawFileDescriptor = this.getResourceManager()
            .getRawFileEntry("resources/rawfile/big_buck_bunny.mp4").openRawFileDescriptor();

    // 创建媒体数据管理AVMetadataHelper对象
    avMetadataHelper = new AVMetadataHelper();

    // 读取指定的媒体文件描述符，读取数据的起始位置的偏移量以及读取数据的长度，设置媒体源
    avMetadataHelper.setSource(rawFileDescriptor.getFileDescriptor(),
            rawFileDescriptor.getStartPosition(),
            rawFileDescriptor.getFileSize());

    HiLog.info(LABEL_LOG, "end initAVMetadataHelper, fd:%{public}s," +
                    "StartPosition:%{public}s,FileSize:%{public}s, ",
            rawFileDescriptor.getFileDescriptor(),
            rawFileDescriptor.getStartPosition(),
            rawFileDescriptor.getFileSize());
}

private void release() {
```

```java
            HiLog.info(LABEL_LOG, "release");

            if (avMetadataHelper != null) {
                // 关闭、释放资源
                avMetadataHelper.release();
                avMetadataHelper = null;
            }
        }

        private void getInfo() {
            HiLog.info(LABEL_LOG, "before getInfo");

            // 获取媒体的时长信息
            String duration = avMetadataHelper.resolveMetadata(AVMetadataHelper.AV_KEY_DURATION);

            // 获取媒体的类型
            String mimetype = avMetadataHelper.resolveMetadata(AVMetadataHelper.AV_KEY_MIMETYPE);

            // 获取媒体的高度
            String videoHeight = avMetadataHelper.resolveMetadata(AVMetadataHelper.AV_KEY_VIDEO_HEIGHT);

            // 获取媒体的宽度
            String videoWidth = avMetadataHelper.resolveMetadata(AVMetadataHelper.AV_KEY_VIDEO_WIDTH);

            HiLog.info(LABEL_LOG, "resolveMetadata duration: %{public}s, mimetype: %{public}s, videoHeight: %{public}s, videoWidth: %{public}s", duration, mimetype, videoHeight, videoWidth);

            // 随机获取帧数据
            PixelMap pixelMap = avMetadataHelper.fetchVideoPixelMapByTime();

            HiLog.info(LABEL_LOG, "end getInfo, pixelMap: %{public}s", pixelMap);

            // 展示帧数据
            showImage(pixelMap);
        }

        private void showImage(PixelMap pixelMap) {
            Image image =
                    (Image) findComponentById(ResourceTable.Id_image);
            image.setPixelMap(pixelMap);
        }

        @Override
        public void onActive() {
            super.onActive();
```

```
    }

    @Override
    public void onForeground(Intent intent) {
        super.onForeground(intent);
    }
}
```

上述代码中:

- 在Button上设置了点击事件。
- initAVMetadataHelper方法用于初始化AVMetadataHelper对象。
- AVMetadataHelper对象通过setSource方法来设置数据源。数据源是预先准备好的big_buck_bunny.mp4视频文件。
- getInfo用于获取媒体的元数据信息。其中, resolveMetadata方法是通过指定的媒体元数据的key来获取媒体元数据的。上述代码演示了获取媒体的时长信息、媒体的类型、媒体的高度、媒体的宽度等元数据。
- fetchVideoPixelMapByTime方法用于获取帧数据。如果上述代码没有指定视频源中的时间戳,就会随机获取帧数据。
- showImage方法用于将上一步获取到的帧数据以图片形式展示出来。

18.2.5 运行

运行应用后,点击界面按钮Get以触发操作的执行。此时,控制台输出内容如下:

```
    02-12 10:45:35.430 8657-8657/com.waylau.hmos.avmetadatahelper I
00001/MainAbilitySlice: before initAVMetadataHelper
    02-12 10:45:35.440 8657-8657/com.waylau.hmos.avmetadatahelper I
00001/MainAbilitySlice: end initAVMetadataHelper,
fd:java.io.FileDescriptor@e828917,StartPosition:9992,FileSize:5510872,
    02-12 10:45:38.526 8657-8657/com.waylau.hmos.avmetadatahelper I
00001/MainAbilitySlice: before getInfo
    02-12 10:45:38.528 8657-8657/com.waylau.hmos.avmetadatahelper I
00001/MainAbilitySlice: resolveMetadata duration: 60140, mimetype: video/mp4,
videoHeight: 360, videoWidth: 640
    02-12 10:45:38.637 8657-8657/com.waylau.hmos.avmetadatahelper I
00001/MainAbilitySlice: end getInfo, pixelMap: ohos.media.image.PixelMap@541e607
```

界面效果如图18-2所示。

图 18-2　界面效果

18.3　实战：媒体存储数据操作

本节演示如何实现媒体存储数据的操作。媒体存储提供了操作媒体图片、视频、音频等元数据的URI链接信息。

18.3.1　接口说明

媒体存储相关类AVStorage的主要接口有：

- appendPendingResource(Uri uri)：更新给定的URI，用于处理包含待处理标记的媒体项。
- appendRequireOriginalResource(Uri uri)：更新给定的URI，用于调用者获取原始文件内容。
- fetchVolumeName(Uri uri)：获取给定URI所属的卷名。
- fetchExternalVolumeNames(Context context)：获取所有组成External的特定卷名的列表。
- fetchMediaResource(Context context, Uri documentUri)：根据文档式的URI获取对应的媒体式的URI。
- fetchDocumentResource(Context context, Uri mediaUri)：根据媒体式的URI获取对应的文档式的URI。
- fetchVersion(Context context)：获取卷名为external_primary的不透明版本信息。
- fetchVersion(Context context, String volumeName)：获取指定卷名的不透明版本信息。
- fetchLoggerResource()：获取用于查询媒体扫描状态的URI。
- Audio.convertNameToKey(String name)：将艺术家或者专辑名称转换为可用于分组、排序和搜索的key。
- Audio.Media.fetchResource(String volumeName)：获取用于处理音频媒体信息的URI。
- Audio.Genres.fetchResource(String volumeName)：获取用于处理音频流派信息的URI。
- Audio.Genres.fetchResourceForAudioId(String volumeName, int audioId)：获取用户处理音频文件对应的流派信息的URI。
- Audio.Genres.Members.fetchResource(String volumeName, long genreId)：获取用于处理音频流派子目录的成员信息的URI。
- Audio.Playlists.fetchResource(String volumeName)：获取用于处理音频播放列表信息的URI。
- Audio.Playlists.Members.fetchResource(String volumeName, long playlistId)：获取用于处理音频播放列表子目录的成员信息的URI。
- Audio.Playlists.Members.updatePlaylistItem(DataAbilityHelper dataAbilityHelper, long playlistId, int oldLocation, int newLocation)：移动播放列表到新位置。
- Audio.Albums.fetchResource(String volumeName)：获取用于处理音频专辑信息的URI。
- Audio.Artists.fetchResource(String volumeName)：获取用于处理音频艺术家信息的URI。
- Audio.Artists.Albums.fetchResource(String volumeName, long id)：获取用于处理所有专辑出现艺术家的歌曲信息的URI。
- Audio.Downloads.fetchResource(String volumeName)：获取用于处理下载条目信息的URI。
- Audio.Files.fetchResource(String volumeName)：获取用于处理媒体文件及非媒体文件（文本、HTML、PDF等）的URI。
- Audio.Images.Media.fetchResource(String volumeName)：获取用于处理图像媒体信息的URI。
- Audio.Video.Media.fetchResource(String volumeName)：获取用于处理视频媒体信息的URI。

18.3.2 创建应用

为了演示AVStorage的功能，创建一个名为AVStorage的应用。
在应用的界面上，通过点击按钮来触发AVStorage的操作。

18.3.3 声明权限

修改配置文件，声明读取用户存储的权限如下：

```
// 声明权限
"reqPermissions": [
  {
    "name": "ohos.permission.READ_USER_STORAGE"
  }
]
```

同时，在应用启动时，显式声明用户存储权限，代码如下：

```
package com.waylau.hmos.avstorage;
```

```java
import com.waylau.hmos.avstorage.slice.MainAbilitySlice;
import ohos.aafwk.ability.Ability;
import ohos.aafwk.content.Intent;

import java.util.ArrayList;
import java.util.List;

public class MainAbility extends Ability {
    @Override
    public void onStart(Intent intent) {
        super.onStart(intent);
        super.setMainRoute(MainAbilitySlice.class.getName());

        // 显式声明需要使用的权限
        requestPermission();
    }

    // 显式声明需要使用的权限
    private void requestPermission() {
        String[] permission = {
                "ohos.permission.READ_USER_STORAGE"};
        List<String> applyPermissions = new ArrayList<>();
        for (String element : permission) {
            if (verifySelfPermission(element) != 0) {
                if (canRequestPermission(element)) {
                    applyPermissions.add(element);
                }
            }
        }
        requestPermissionsFromUser(applyPermissions.toArray(new String[0]),
0);
    }
}
```

18.3.4 修改 ability_main.xml

修改ability_main.xml内容如下：

```xml
<?xml version="1.0" encoding="utf-8"?>
<DirectionalLayout
    xmlns:ohos="http://schemas.huawei.com/res/ohos"
    ohos:height="match_parent"
    ohos:width="match_parent"
    ohos:orientation="vertical">

    <Button
        ohos:id="$+id:button_get"
        ohos:height="match_content"
        ohos:width="match_content"
```

```
        ohos:background_element="#F76543"
        ohos:margin="10vp"
        ohos:padding="10vp"
        ohos:text="Get"
        ohos:text_size="50"
        />

    <TableLayout
        ohos:id="$+id:layout_table"
        ohos:height="match_parent"
        ohos:width="match_content">

    </TableLayout>

</DirectionalLayout>
```

界面预览效果如图18-3所示。

图 18-3　界面预览效果

上述代码中：

- 设置了Get按钮，以备设置点击事件，以触发元数据相关的操作。
- TableLayout布局用于展示读取到的图片信息。

18.3.5　修改 MainAbilitySlice

修改MainAbilitySlice内容如下：

```
package com.waylau.hmos.avstorage.slice;

import com.waylau.hmos.avstorage.ResourceTable;
import ohos.aafwk.ability.AbilitySlice;
import ohos.aafwk.ability.DataAbilityHelper;
import ohos.aafwk.ability.DataAbilityRemoteException;
```

```java
import ohos.aafwk.content.Intent;
import ohos.agp.components.Button;
import ohos.agp.components.Image;
import ohos.agp.components.TableLayout;
import ohos.data.dataability.DataAbilityPredicates;
import ohos.data.resultset.ResultSet;
import ohos.hiviewdfx.HiLog;
import ohos.hiviewdfx.HiLogLabel;
import ohos.media.common.Source;
import ohos.media.image.ImageSource;
import ohos.media.image.PixelMap;
import ohos.media.photokit.metadata.AVStorage;
import ohos.media.player.Player;
import ohos.utils.net.Uri;

import java.io.FileDescriptor;
import java.io.FileNotFoundException;

public class MainAbilitySlice extends AbilitySlice {
    private static final String TAG = MainAbilitySlice.class.getSimpleName();
    private static final HiLogLabel LABEL_LOG =
            new HiLogLabel(HiLog.LOG_APP, 0x00001, TAG);

    private DataAbilityHelper helper;

    private TableLayout tableLayout;

    @Override
    public void onStart(Intent intent) {
        super.onStart(intent);
        super.setUIContent(ResourceTable.Layout_ability_main);

        helper = DataAbilityHelper.creator(this);

        // 为按钮设置点击事件回调
        Button buttonGet =
                (Button) findComponentById(ResourceTable.Id_button_get);
        buttonGet.setClickedListener(listener -> {
            try {
                getInfo();
            } catch (FileNotFoundException e) {
                e.printStackTrace();
            } catch (DataAbilityRemoteException e) {
                e.printStackTrace();
            }
        });

        tableLayout = (TableLayout) findComponentById(ResourceTable.Id_layout_table);
        tableLayout.setColumnCount(4);
```

```java
    }

    @Override
    public void onStop() {
        super.onStop();

        // 关闭、释放资源
        release();
    }

    private void getInfo() throws FileNotFoundException,
DataAbilityRemoteException {
        HiLog.info(LABEL_LOG, "before getInfo");

        // 查询条件为null，即为查询所有记录
        ResultSet result =
helper.query(AVStorage.Images.Media.EXTERNAL_DATA_ABILITY_URI,
                null, null);

        if (result == null) {
            HiLog.info(LABEL_LOG, "result is null");
            return;
        }

        while (result.goToNextRow()) {
            // 获取ID字段的值
            String mediaId = result.getString(result.
getColumnIndexForName(AVStorage.Images.Media.ID));
            HiLog.info(LABEL_LOG, "mediaId: %{public}s", mediaId);

            showImage(mediaId);
        }


        result.close();

        HiLog.info(LABEL_LOG, "end getInfo");
    }

    private void showImage(String mediaId) throws DataAbilityRemoteException,
FileNotFoundException {
        HiLog.info(LABEL_LOG, "before showImage, mediaId: %{public}s",
mediaId);

        PixelMap pixelMap = null;
        ImageSource imageSource = null;
        Image image = new Image(this);
        image.setWidth(250);
        image.setHeight(250);
```

```
        image.setMarginsLeftAndRight(10, 10);
        image.setMarginsTopAndBottom(10, 10);
        image.setScaleMode(Image.ScaleMode.CLIP_CENTER);

        Uri uri = Uri.appendEncodedPathToUri(AVStorage.Images.Media.
EXTERNAL_DATA_ABILITY_URI, mediaId);
        FileDescriptor fd = helper.openFile(uri, "r");
        try {
            imageSource = ImageSource.create(fd, null);
            pixelMap = imageSource.createPixelmap(null);
        } catch (Exception e) {
            e.printStackTrace();
        } finally {
            if (imageSource != null) {
                imageSource.release();
            }
        }

        image.setPixelMap(pixelMap);
        tableLayout.addComponent(image);

        HiLog.info(LABEL_LOG, "end play");
    }

    private void release() {
        HiLog.info(LABEL_LOG, "release");

        if (helper != null) {
            // 关闭、释放资源
            helper.release();
            helper = null;
        }
    }

    @Override
    public void onActive() {
        super.onActive();
    }

    @Override
    public void onForeground(Intent intent) {
        super.onForeground(intent);
    }
}
```

上述代码中：

- onStart方法初始化了DataAbilityHelper、TableLayout。
- 在Button上设置了点击事件。
- getInfo方法用于查询外部存储器上的图片的ID。

- showImage方法用于通过DataAbilityHelper查询对应图片ID的图片资源，并添加到TableLayout呈现出来。

18.3.6 运行

初次运行应用后，点击界面按钮Get以触发操作的执行。此时，控制台输出内容如下：

```
02-11 17:14:26.038 14589-14589/com.waylau.hmos.avstorage I
00001/MainAbilitySlice: before getInfo
02-11 17:14:26.053 14589-14589/com.waylau.hmos.avstorage I
00001/MainAbilitySlice: end getInfo
```

上述日志表明没有获取到任何图片。

进入手机的"相机"应用，连续拍摄若干张照片。相机界面的效果如图18-4所示。

图 18-4　相机界面的效果

之后，再回到AVStorage应用，点击按钮Get，此时控制台输出内容如下：

```
02-11 17:19:01.647 8039-8039/com.waylau.hmos.avstorage I
00001/MainAbilitySlice: before getInfo
02-11 17:19:01.665 8039-8039/com.waylau.hmos.avstorage I
00001/MainAbilitySlice: mediaId: 33
02-11 17:19:01.666 8039-8039/com.waylau.hmos.avstorage I
00001/MainAbilitySlice: before showImage, mediaId: 33
02-11 17:19:01.715 8039-8039/com.waylau.hmos.avstorage I
00001/MainAbilitySlice: end play
02-11 17:19:01.715 8039-8039/com.waylau.hmos.avstorage I
00001/MainAbilitySlice: mediaId: 34
02-11 17:19:01.716 8039-8039/com.waylau.hmos.avstorage I
00001/MainAbilitySlice: before showImage, mediaId: 34
02-11 17:19:01.740 8039-8039/com.waylau.hmos.avstorage I
00001/MainAbilitySlice: end play
```

```
    02-11 17:19:01.740 8039-8039/com.waylau.hmos.avstorage I
00001/MainAbilitySlice: mediaId: 35
    02-11 17:19:01.740 8039-8039/com.waylau.hmos.avstorage I
00001/MainAbilitySlice: before showImage, mediaId: 35
```

说明已经读取到了图片的ID。AVStorage应用界面的效果如图18-5所示。

图 18-5　AVStorage 应用界面的效果

18.4　实战：视频与图像缩略图获取

本节演示如何实现获取视频文件或图像文件的缩略图。

18.4.1　接口说明

视频与图像缩略图获取相关类AVThumbnailUtils的主要接口有：

- createVideoThumbnail(File file, Size size)：创建指定视频中代表性关键帧的缩略图。
- createImageThumbnail(File file, Size size)：创建指定图像的缩略图。

18.4.2　创建应用

为了演示AVThumbnailUtils的功能，创建一个名为AVThumbnailUtils的应用。

在应用的界面上，通过点击按钮来触发音频的操作。

在rawfile目录下放置一个big_buck_bunny.mp4视频文件和一个waylau_616_616.jpeg图片文件，以备测试。

18.4.3 修改 ability_main.xml

修改ability_main.xml内容如下：

```xml
<?xml version="1.0" encoding="utf-8"?>
<DirectionalLayout
    xmlns:ohos="http://schemas.huawei.com/res/ohos"
    ohos:height="match_parent"
    ohos:width="match_parent"
    ohos:orientation="vertical">

    <Button
        ohos:id="$+id:button_image"
        ohos:height="match_content"
        ohos:width="match_parent"
        ohos:background_element="#F76543"
        ohos:margin="10vp"
        ohos:padding="10vp"
        ohos:text="Get Image"
        ohos:text_size="40vp"
        />

    <Image
        ohos:id="$+id:image_image"
        ohos:height="match_content"
        ohos:width="match_parent"/>

    <Button
        ohos:id="$+id:button_video"
        ohos:height="match_content"
        ohos:width="match_parent"
        ohos:background_element="#F76543"
        ohos:margin="10vp"
        ohos:padding="10vp"
        ohos:text="Get Video"
        ohos:text_size="40vp"
        />

    <Image
        ohos:id="$+id:image_vedio"
        ohos:height="match_content"
        ohos:width="match_parent"/>

</DirectionalLayout>
```

界面预览效果如图18-6所示。

图 18-6　界面预览效果

上述代码中：

- 设置了Get Image、Get Video按钮，以备设置点击事件，以触发获取图像文件或视频文件的缩略图的操作。
- TableLayout布局用于展示获取到的缩略图。

18.4.4　修改 MainAbilitySlice

修改MainAbilitySlice内容如下：

```
public class MainAbilitySlice extends AbilitySlice {
    private static final String TAG = MainAbilitySlice.class.getSimpleName();
    private static final HiLogLabel LABEL_LOG =
            new HiLogLabel(HiLog.LOG_APP, 0x00001, TAG);

    private TableLayout tableLayout;

    @Override
    public void onStart(Intent intent) {
        super.onStart(intent);
        super.setUIContent(ResourceTable.Layout_ability_main);

        // 为按钮设置点击事件回调
        Button buttonGetImage =
                (Button) findComponentById(ResourceTable.Id_button_image);
        buttonGetImage.setClickedListener(listener -> {
            try {
                getImageInfo();
            } catch (IOException e) {
                e.printStackTrace();
            }
        });
```

```
        // 为按钮设置点击事件回调
        Button buttonGetVideo =
                (Button) findComponentById(ResourceTable.Id_button_video);
        buttonGetVideo.setClickedListener(listener -> {
            try {
                getVideoInfo();
            } catch (IOException e) {
                e.printStackTrace();
            }
        });

        tableLayout = (TableLayout)
findComponentById(ResourceTable.Id_layout_table);
        tableLayout.setColumnCount(4);
    }
    ...
}
```

上述代码中：

- onStart方法初始化了TableLayout。
- 在Button上设置了点击事件。
- getImageInfo方法用于获取图像文件的缩略图。
- getVideoInfo方法用于获取视频文件的缩略图。

1. 获取图像文件的缩略图

获取图像文件的缩略图的方法getImageInfo代码如下：

```
private void getImageInfo() throws IOException {
    HiLog.info(LABEL_LOG, "before getImageInfo");

    // 获取数据目录
    File dataDir = new File(this.getDataDir().toString());
    if (!dataDir.exists()) {
        dataDir.mkdirs();
    }

    // 构建目标文件
    File targetFile =
    new File(Paths.get(dataDir.toString(),
"waylau_616_616.jpeg").toString());

    // 获取源文件
    RawFileEntry rawFileEntry =
            this.getResourceManager().getRawFileEntry("resources/
rawfile/waylau_616_616.jpeg");
    Resource resource = rawFileEntry.openRawFile();

    // 新建目标文件
```

```
        FileOutputStream fos = new FileOutputStream(targetFile);

        byte[] buffer = new byte[4096];
        int count = 0;

        // 源文件内容写入目标文件
        while ((count = resource.read(buffer)) >= 0) {
            fos.write(buffer, 0, count);
        }

        resource.close();
        fos.close();

        Size size = new Size();
        size.height = 250;
        size.width = 250;

        PixelMap pixelMap = AVThumbnailUtils.createImageThumbnail(targetFile, size);

        // 显示图片
        Image image =
                (Image) findComponentById(ResourceTable.Id_image_image);
        image.setPixelMap(pixelMap);

        HiLog.info(LABEL_LOG, "end getImageInfo");
    }
```

上述代码中：

- 将指定的文件写入指定的位置。
- 通过AVThumbnailUtils.createImageThumbnail为指定位置的文件创建图像文件的缩略图。
- Size对象用于设置缩略图的大小。
- 将缩略图显示在TableLayout布局上。

2. 获取视频文件的缩略图

获取视频文件的缩略图的方法getImageInfo代码如下：

```
private void getVideoInfo() throws IOException {
    HiLog.info(LABEL_LOG, "before getVideoInfo");

    // 获取数据目录
    File dataDir = new File(this.getDataDir().toString());
    if (!dataDir.exists()) {
        dataDir.mkdirs();
    }

    // 构建目标文件
    File targetFile =
            new File(Paths.get(dataDir.toString(),
```

```
"big_buck_bunny.mp4").toString());

    // 获取源文件
    RawFileEntry rawFileEntry =
            this.getResourceManager().getRawFileEntry("resources/
rawfile/big_buck_bunny.mp4");
    Resource resource = rawFileEntry.openRawFile();

    // 新建目标文件
    FileOutputStream fos = new FileOutputStream(targetFile);

    byte[] buffer = new byte[4096];
    int count = 0;

    // 源文件内容写入目标文件
    while ((count = resource.read(buffer)) >= 0) {
        fos.write(buffer, 0, count);
    }

    resource.close();
    fos.close();

    Size size = new Size();
    size.height = 640;
    size.width = 320;

    PixelMap pixelMap = AVThumbnailUtils.createVideoThumbnail(targetFile,
size);

    // 显示图片
    Image image =
            (Image) findComponentById(ResourceTable.Id_image_vedio);
    image.setPixelMap(pixelMap);

    HiLog.info(LABEL_LOG, "end getVideoInfo");
}
```

上述代码与getImageInfo方法基本类似：

- 将指定的文件写入指定的位置。
- 通过AVThumbnailUtils.createVideoThumbnail为指定位置的文件创建图像文件的缩略图。
- Size对象用于设置缩略图的大小。
- 将缩略图显示在TableLayout布局上。

18.4.5 运行

运行应用后，点击界面上的Get Image按钮以触发操作的执行。此时，界面效果如图18-7所示。之后，再点击Get Video按钮，此时应用界面效果如图18-8所示。

图 18-7　界面效果

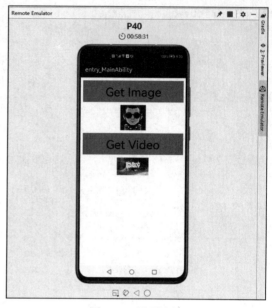

图 18-8　界面效果

第 19 章

安全管理

本章介绍HarmonyOS应用的安全管理机制。

19.1 权限的基本概念

HarmonyOS的权限包含以下基本概念：

- 应用沙盒：系统利用内核保护机制来识别和隔离应用资源，可将不同的应用隔离开，保护应用自身和系统免受恶意应用的攻击。默认情况下，应用间不能彼此交互，而且对系统的访问会受到限制。例如，如果应用A（一个单独的应用）尝试在没有权限的情况下读取应用B的数据或者调用系统的能力拨打电话，操作系统会阻止此类行为，因为应用A没有被授予相应的权限。
- 应用权限：系统通过沙盒机制管理各个应用，在默认规则下，应用只能访问有限的系统资源。但应用为了扩展功能的需要，需要访问自身沙盒之外的系统或其他应用的数据（包括用户个人数据）或能力，系统或应用也必须以明确的方式对外提供接口来共享其数据或能力。为了保证这些数据或能力不会被不当或恶意使用，就需要有一种访问控制机制来保护，这就是应用权限。应用权限是程序访问操作某种对象的许可。权限在应用层面要求明确定义且经用户授权，以便系统化地规范各类应用程序的行为准则与权限许可。
- 权限保护的对象：权限保护的对象可以分为数据和能力。
  - 数据包含个人数据（如照片、通讯录、日历、位置等）、设备数据（如设备标识、相机、麦克风等）、应用数据。
  - 能力包括设备能力（如打电话、发短信、联网等）、应用能力（如弹出悬浮框、创建快捷方式等）等。

- 权限开放范围：权限开放范围指一个权限能被哪些应用申请。按可信程度从高到低的顺序，不同权限开放范围对应的应用可分为：系统服务、系统应用、系统预置特权应用、同签名应用、系统预置普通应用、持有权限证书的后装应用、其他普通应用，开放范围依次扩大。
- 敏感权限：涉及访问个人数据（如照片、通讯录、日历、本机号码、短信等）和操作敏感能力（如相机、麦克风、拨打电话、发送短信等）的权限。
- 应用核心功能：一个应用可能提供了多种功能，其中应用为满足用户的关键需求而提供的功能称为应用的核心功能。这是一个相对宽泛的概念，主要用来辅助描述用户权限授权的预期。用户选择安装一个应用，通常是被应用的核心功能所吸引。比如导航类应用，定位导航就是这种应用的核心功能；又如媒体类应用，播放以及媒体资源管理就是核心功能，这些功能所需要的权限用户在安装时内心已经倾向于授予（否则就不会去安装）。与核心功能相对应的是辅助功能，这些功能所需要的权限需要向用户清晰说明目的、场景等信息，由用户授权。有些功能既不属于核心功能，也不是辅助功能，那么这些功能就是多余功能，这些功能所需要的权限通常被用户禁止。
- 最小必要权限：保障应用某一服务类型正常运行所需要的应用权限的最小集，一旦缺少将导致该类型服务无法实现或无法正常运行的应用权限。

19.2　权限运作机制

系统所有应用均在应用沙盒内运行。默认情况下，应用只能访问有限的系统资源。这些限制是通过DAC（Discretionary Access Control，自主访问控制）、MAC（Mandatory Access Control，强制访问控制）以及本文描述的应用权限机制等多种不同的形式实现的。因应用需要实现其某些功能而必须访问系统或其他应用的数据或操作某些器件，此时就需要系统或其他应用能提供接口，考虑到安全，需要对这些接口采用一种限制措施，这就是称为"应用权限"的安全机制。

接口的提供涉及其权限的命名和分组、对外开放的范围、被授予的应用以及用户的参与和体验。应用权限管理模块的目的就是负责管理由接口提供方（访问客体）、接口使用方（访问主体）、系统（包括云侧和端侧）和用户等共同参与的整个流程，保证受限接口在约定好的规则下被正常使用，避免接口被滥用而导致用户、应用和设备受损。

应用要使用特定的权限，就需要对权限进行声明。权限声明要求如下：

- 应用需要在config.json中使用reqPermissions属性对需要的权限逐个进行声明。
- 若使用到的第三方库也涉及权限使用，也需要统一在应用的config.json中逐个声明。
- 没有在config.json中声明的权限，应用就无法获得此权限的授权。

19.3　权限约束与限制

HarmonyOS权限在使用时应考虑约束与限制：

- 同一应用自定义的权限个数不能超过1024个。
- 同一应用申请的权限个数不能超过1024个。
- 为了避免与系统权限名冲突，应用自定义的权限名不能以ohos开头，且权限名长度不能超过256个字符。
- 自定义权限授予方式不能为user_grant。
- 自定义权限开放范围不能为restricted。

19.4 应用权限列表

HarmonyOS根据接口所涉数据的敏感程度或所涉能力的安全威胁影响定义了不同开放范围与授权方式的权限来保护数据。

19.4.1 权限分类

当前权限的开放范围分为：

- all：所有应用可用。
- signature：平台签名应用可用。
- privileged：预制特权应用可用。
- restricted：证书可控应用可用。

应用在使用对应服务的能力或数据时，需要申请对应的权限。

- 已在config.json文件中声明的非敏感权限会在应用安装时自动授予，这类权限的授权方式为系统授权（system_grant）。
- 敏感权限需要应用动态申请，通过运行时发送弹窗的方式请求用户授权，这类权限的授权方式为用户授权（user_grant）。

当应用调用服务时，服务会对应用进行权限检查，如果没有对应权限，则无法使用该服务。
接下来仅介绍HarmonyOS所有可供第三方应用申请的应用权限。HarmonyOS已定义的全量权限列表定义在ohos.security.SystemPermission类中。

19.4.2 敏感权限

敏感权限的申请需要按照动态申请流程向用户申请授权。
敏感权限包括：

- ohos.permission.LOCATION：允许应用在前台运行时获取位置信息。如果应用在后台运行时也要获取位置信息，则需要同时申请ohos.permission.LOCATION_IN_BACKGROUND权限。
- ohos.permission.LOCATION_IN_BACKGROUND：允许应用在后台运行时获取位置信息，需

要同时申请ohos.permission.LOCATION权限。
- ohos.permission.CAMERA：允许应用使用相机拍摄照片和录制视频。
- ohos.permission.MICROPHONE：允许应用使用麦克风进行录音。
- ohos.permission.READ_CALENDAR：允许应用读取日历信息。
- ohos.permission.WRITE_CALENDAR：允许应用在设备上添加、移除或修改日历活动。
- ohos.permission.ACTIVITY_MOTION：允许应用读取用户当前的运动状态。
- ohos.permission.READ_HEALTH_DATA：允许应用读取用户的健康数据。
- ohos.permission.DISTRIBUTED_DATASYNC：允许不同设备间的数据交换。
- ohos.permission.DISTRIBUTED_DATA：允许应用使用分布式数据的能力。
- ohos.permission.MEDIA_LOCATION：允许应用访问用户媒体文件中的地理位置信息。
- ohos.permission.READ_MEDIA：允许应用读取用户外部存储中的媒体文件信息。
- ohos.permission.WRITE_MEDIA：允许应用读写用户外部存储中的媒体文件信息。

19.4.3 非敏感权限

非敏感权限不涉及用户的敏感数据或危险操作，仅需在config.json中声明，应用安装后即被授权。

非敏感权限包括：

- ohos.permission.GET_NETWORK_INFO：允许应用获取数据网络信息。
- ohos.permission.GET_WIFI_INFO：允许获取WLAN信息。
- ohos.permission.USE_BLUETOOTH：允许应用查看蓝牙的配置。
- ohos.permission.DISCOVER_BLUETOOTH：允许应用配置本地蓝牙，并允许其查找远端设备且与之配对连接。
- ohos.permission.SET_NETWORK_INFO：允许应用控制数据网络。
- ohos.permission.SET_WIFI_INFO：允许配置WLAN设备。
- ohos.permission.SPREAD_STATUS_BAR：允许应用以缩略图方式呈现在状态栏。
- ohos.permission.INTERNET：允许使用网络Socket。
- ohos.permission.MODIFY_AUDIO_SETTINGS：允许应用程序修改音频设置。
- ohos.permission.RECEIVER_STARTUP_COMPLETED：允许应用接收设备启动完成广播。
- ohos.permission.RUNNING_LOCK：允许申请休眠运行锁，并执行相关操作。
- ohos.permission.ACCESS_BIOMETRIC：允许应用使用生物识别能力进行身份认证。
- ohos.permission.RCV_NFC_TRANSACTION_EVENT：允许应用接收卡模拟交易事件。
- ohos.permission.COMMONEVENT_STICKY：允许发布粘性公共事件的权限。
- ohos.permission.SYSTEM_FLOAT_WINDOW：提供显示悬浮窗的能力。
- ohos.permission.VIBRATE：允许应用程序使用马达。
- ohos.permission.USE_TRUSTCIRCLE_MANAGER：允许调用设备间认证的能力。
- ohos.permission.USE_WHOLE_SCREEN：允许通知携带一个全屏IntentAgent。
- ohos.permission.SET_WALLPAPER：允许设置静态壁纸。

- ohos.permission.SET_WALLPAPER_DIMENSION：允许设置壁纸尺寸。
- ohos.permission.REARRANGE_MISSIONS：允许调整任务栈。
- ohos.permission.CLEAN_BACKGROUND_PROCESSES：允许根据包名清理相关后台进程。
- ohos.permission.KEEP_BACKGROUND_RUNNING：允许Service Ability在后台继续运行。
- ohos.permission.GET_BUNDLE_INFO：查询其他应用的信息。
- ohos.permission.ACCELEROMETER：允许应用程序读取加速度传感器的数据。
- ohos.permission.GYROSCOPE：允许应用程序读取陀螺仪传感器的数据。
- ohos.permission.MULTIMODAL_INTERACTIVE：允许应用订阅语音或手势事件。
- ohos.permission.radio.ACCESS_FM_AM：允许用户获取收音机相关服务。
- ohos.permission.NFC_TAG：允许应用读写Tag卡片。
- ohos.permission.NFC_CARD_EMULATION：允许应用实现卡模拟功能。
- ohos.permission.DISTRIBUTED_DEVICE_STATE_CHANGE：允许获取分布式组网内设备的状态变化。
- ohos.permission.GET_DISTRIBUTED_DEVICE_INFO：允许获取分布式组网内的设备列表和设备信息。

19.4.4　受限开放的权限

受限开放的权限通常是不允许第三方应用申请的。如果有特殊场景需要使用，则需要提供相关申请材料到应用市场申请相应权限证书。如果应用未申请相应的权限证书，却试图在config.json文件中声明此类权限，则会导致应用安装失败。另外，由于此类权限涉及用户敏感数据或危险操作，当应用申请到权限证书后，还需按照动态申请权限的流程向用户申请授权。

受限开放权限包括：

- ohos.permission.READ_CONTACTS：允许应用读取联系人数据。
- ohos.permission.WRITE_CONTACTS：允许应用添加、移除和更改联系人数据。

19.5　应用权限开发流程

HarmonyOS支持开发者自定义权限来保护能力或接口，同时开发者也可申请权限来访问受权限保护的对象。

在前面几章中已经介绍了应用权限的一些使用方式，本节就来详细总结一下应用权限的开发流程。

19.5.1　权限申请

开发者需要在config.json文件中的reqPermissions字段中声明所需要的权限。示例如下：

```json
{
    "reqPermissions": [
        {
            "name": "ohos.permission.CAMERA",
            "reason": "$string:permreason_camera",
            "usedScene":
            {
                "ability": ["com.mycamera.Ability", "com.mycamera.AbilityBackground"],
                "when": "always"
            }
        },{
            ...
        }
    ]
}
```

权限申请格式采用数组格式,可支持同时申请多个权限,权限个数最多不能超过1024个。reqPermissions权限申请字段说明如表19-1所示。

表19-1 reqPermissions 权限申请字段说明

键	值 说 明	类 型	取值范围	默 认 值	规则约束
name	必需,填写需要使用的权限名称	字符串	自定义	无	未填写时,解析失败
reason	可选,当申请的权限为 user_grant 权限时此字段必填	描述申请权限的原因	字符串	显示文字长度不能超过 256 字节	空
usedScene	可选,当申请的权限为 user_grant 权限时此字段必填	描述权限使用的场景和时机。场景类型有 ability、when(调用时机)。可配置多个 ability	ability:字符串数组;when:字符串	ability: ability 的名称;when: inuse(使用时)、always(始终)	ability:空;when: inuse

如果声明使用的权限的grantMode是system_grant,则权限会在应用安装的时候被自动授予。

如果声明使用的权限的grantMode是user_grant,则必须经用户手动授权(用户在弹框中授权或进入权限设置界面授权)才可使用。用户会看到reason字段中填写的理由,来帮助用户决定是否给予授权。

注意:对于授权方式为user_grant的权限,每一次执行需要这一权限的操作时,都需要检查自身是否有该权限。当自身具有权限时,才可以继续执行,否则应用需要请求用户授予权限。示例参见动态申请权限开发步骤。

19.5.2 自定义权限

开发者需要在config.json文件中的defPermissions字段中自定义所需的权限:

```json
{
    "defPermissions": [
```

```
        {
            "name": "com.myability.permission.MYPERMISSION",
            "grantMode": "system_grant",
            "availableScope": ["signature"]
        }, {
            ...
        }
    ]
}
```

权限定义格式采用数组格式,可支持同时定义多个权限,自定义的权限个数最多不能超过1024个。

defPermissions权限定义字段说明如表19-2所示。

表19-2　defPermissions 权限定义字段说明

键	值 说 明	类 型	取值范围	默 认 值	规则约束
name	必填,权限名称。为了最大可能避免重名,采用反向域公司名+应用名+权限名组合	字符串	自定义	无	第三方应用不允许填写系统存在的权限,否则安装失败。未填写则解析失败。权限名长度不能超过256个字符
grantMode	必填,权限授予方式	字符串	user_grant(用户授权)、system_grant(系统授权)	system_grant	未填值或填写了取值范围以外的值时,自动赋予默认值
availableScope	选填,权限限制范围。不填则表示此权限对所有应用开放	字符串数组	Signature、privileged、restricted	空	填写取值范围以外的值时,权限限制范围不生效。由于第三方应用并不在restricted的范围内,很少会出现权限定义者不能访问自身定义的权限的情况,因此不允许第三方应用填写restricted
label	选填,权限的简短描述,若未填写,则使用到简短描述的地方由权限名取代	字符串	自定义	空	需要多语种适配
description	选填,权限的详细描述,若未填写,则使用到详细描述的地方由label取代	字符串	自定义	空	需要多语种适配

权限授予方式字段说明如表19-3所示。

表 19-3 权限授予方式字段说明

授予方式（grantMode）	说 明	自定义权限是否可指定该级别	取值样例
system_grant	在 config.json 里面声明，安装后系统自动授予	是	GET_NETWORK_INFO、GET_WIFI_INFO
user_grant	在 config.json 里面声明，并在使用时动态申请，用户授权后才可以使用	否，若自定义，则强制修改为 system_grant	CAMERA、MICROPHONE

权限限制范围字段说明如表19-4所示。

表 19-4 权限限制范围字段说明

权限范围（availableScope）	说 明	自定义权限是否可指定该级别	取值样例
restricted	需要开发者向华为申请后才能被使用的特殊权限	否	ANSWER_CALL、READ_CALL_LOG、RECEIVE_SMS
signature	权限定义方和使用方的签名一致。需要在 config.json 中声明后，并由权限管理模块负责签名校验一致后，才可使用	是	对应用（或 Ability）操作的系统接口上由系统定义的权限以及应用自定义的权限，如查询某 Ability，连接某 Ability
privileged	预置在系统版本中的特权应用可申请的权限	是	SET_TIME、MANAGE_USER_STORAGE

19.5.3 访问权限控制

1. Ability 的访问权限控制

在config.json中填写abilities到permissions字段，表示只有拥有该权限的应用可访问此Ability。下面的例子表明只有拥有ohos.permission.CAMERA权限的应用可以访问此Ability。

```
"abilities": [
    {
        "name": ".MainAbility",
        "description": "$string:description_main_ability",
        "icon": "$media:hiworld.png",
        "label": "HiCamera",
        "launchType": "standard",
        "orientation": "portrait",
        "visible": false,
```

```
            "permissions": [
                "ohos.permission.CAMERA"
            ],
    }
]
```

其中，permissions用以表示此Ability受哪个权限保护，只有拥有此权限的应用才可访问此Ability。目前仅支持填写一个权限名，若填写多个权限名，则仅第一个权限名有效。

2. Ability 接口的访问权限控制

在Ability实现中，若需要特定接口对调用者做访问控制，则可在服务侧的接口实现中主动通过verifyCallingPermission、verifyCallingOrSelfPermission来检查访问者是否拥有所需要的权限。示例如下：

```
    if (verifyCallingPermission("ohos.permission.CAMERA") !=
IBundleManager.PERMISSION_GRANTED) {
        // 调用者无权限，做错误处理
    }
        // 调用者权限校验通过，开始提供服务
```

19.5.4　接口说明

应用权限接口有：

- verifyPermission(String permissionName, int pid, int uid)：查询指定PID、UID的应用是否已被授予某权限。
- verifyCallingPermission(String permissionName)：查询IPC跨进程调用的调用方的进程是否已被授予某权限。
- verifySelfPermission(String permissionName)：查询自身进程是否已被授予某权限。
- verifyCallingOrSelfPermission(String permissionName)：当有远端调用时检查远端是否有权限，否则检查自身是否拥有权限。
- canRequestPermission(String permissionName)：向系统权限管理模块查询某权限是否不再弹框授权。
- requestPermissionsFromUser (String[] permissions, int requestCode)：向系统权限管理模块申请权限（接口可支持一次申请多个。若下一步操作涉及多个敏感权限，则可以这么用，其他情况建议不要这么用。因为弹框是按权限组一个一个去弹框的，耗时比较长，用到哪个权限就去申请哪个）。
- onRequestPermissionsFromUserResult (int requestCode, String[] permissions, int[] grantResults)：调用requestPermissionsFromUser后的应答接口。

19.5.5 动态申请权限的开发步骤

动态申请权限的开发步骤如下。

1. 声明所需要的权限

在config.json文件中声明所需要的权限。示例如下：

```
{
    "reqPermissions": [
      {
          "name": "ohos.permission.CAMERA",
          "reason": "$string:permreason_camera",
          "usedScene": {
              "ability": ["com.mycamera.Ability",
"com.mycamera.AbilityBackground"],
              "when": "always"}
      }, {
          ...
      }
   ]
}
```

2. 检查是否已被授予该权限

继承Ability，使用ohos.app.Context.verifySelfPermission接口查询应用是否已被授予该权限。

- 如果已被授予权限，则可结束权限申请流程。
- 如果未被授予权限，则继续执行下一步。

3. 查询是否可动态申请

使用canRequestPermission查询是否可动态申请。

- 如果不可动态申请，则说明已被用户或系统永久禁止授权，可以结束权限申请流程。
- 如果可以动态申请，则继续执行下一步。

4. 动态申请权限

使用requestPermissionFromUser动态申请权限，通过回调函数接收授予结果。样例代码如下：

```
if (verifySelfPermission("ohos.permission.CAMERA")
        != IBundleManager.PERMISSION_GRANTED) {
    // 应用未被授予权限
    if (canRequestPermission("ohos.permission.CAMERA")) {
        // 提示用户授权
        requestPermissionsFromUser(
            new String[] { "ohos.permission.CAMERA" } ,
MY_PERMISSIONS_REQUEST_CAMERA);
    } else {
        // 显示应用需要权限的理由，提示用户进入设置授权
```

```
        }
    } else {
        // 权限已被授予
    }

    @Override
    public void onRequestPermissionsFromUserResult (int requestCode, String[]
permissions, int[] grantResults) {
        switch (requestCode) {
            case MY_PERMISSIONS_REQUEST_CAMERA: {
                // 匹配requestPermissions的requestCode
                if (grantResults.length > 0
                    && grantResults[0] == IBundleManager.PERMISSION_GRANTED) {
                    // 权限被授予
                } else {
                    // 权限被拒绝
                }
                return;
            }
        }
    }
```

19.6 生物特征识别认证概述

HarmonyOS提供生物特征识别认证能力，可应用于设备解锁、支付、应用登录等身份认证场景。当前生物特征识别能力提供2D人脸识别、3D人脸识别两种人脸识别能力，设备具备哪种识别能力，取决于设备的硬件能力和技术实现。3D人脸识别技术识别率、防伪能力都优于2D人脸识别技术，但具有3D人脸识别能力（比如3D结构光、3D TOF等）的设备才可以使用3D人脸识别技术。

生物特征识别认证包含以下基本概念：

- 生物特征识别（又叫生物认证）：通过计算机与光学、声学、生物传感器和生物统计学原理等高科技手段密切结合，来进行个人身份的鉴定。
- 人脸识别：基于人的脸部特征信息进行身份识别的一种生物特征识别技术，用摄像机或摄像头采集含有人脸的图像或视频流，并自动在图像中检测和跟踪人脸，进而对检测到的人脸进行脸部识别，通常也叫作人像识别、面部识别、人脸认证。

19.7 生物特征识别运作机制

生物特征识别的运作机制说明如下：

- 人脸识别会在摄像头和TEE（Trusted Execution Environment，可信执行环境）之间建立安全通道，人脸图像信息通过安全通道传递到TEE中，由于人脸图像信息从REE（Rich Execution Environment，富执行环境）侧无法获取，从而避免了恶意软件从REE侧进行攻击。对人脸图像采集、特征提取、活体检测、特征比对等处理完全在TEE中，基于TrustZone进行安全隔离，外部的人脸框架只负责人脸的认证发起和处理认证结果等数据，不涉及人脸数据本身。

- 人脸特征数据通过TEE的安全存储区进行存储，采用高强度的密码算法对人脸特征数据进行加密和完整性保护，外部无法获取到加密人脸特征数据的密钥，保证用户的人脸特征数据不会泄露。本能力采集和存储的人脸特征数据不会在用户未授权的情况下被传出TEE，这意味着，用户未授权时，无论是系统应用还是第三方应用都无法获得人脸特征数据，也无法将人脸特征数据传送或备份到任何外部存储介质中。

19.8　生物特征识别的约束与限制

生物特征识别运作机制包含以下约束与限制：

- 当前版本提供的生物特征识别能力只包含人脸识别，且只支持本地认证，不提供认证界面。
- 要求设备上具备摄像器件，且人脸图像像素大于100×100。
- 要求设备上具有TEE安全环境，人脸特征信息高强度加密保存在TEE中。
- 对于面部特征相似的人、面部特征不断发育的儿童，人脸特征匹配率有所不同。如果对此担忧，可考虑其他认证方式。

19.9　生物特征识别的开发流程

当前生物特征识别支持2D人脸识别、3D人脸识别，可应用于设备解锁、应用登录、支付等身份认证场景。

19.9.1　接口说明

BiometricAuthentication类提供了生物认证的相关方法，包括检测认证能力、认证和取消认证等，用户可以通过人脸等生物特征信息进行认证操作。在执行认证前，需要检查设备是否支持该认证能力，具体指认证类型、安全级别和是否本地认证。如果不支持，则需要考虑使用其他认证能力。

生物特征识别开放的能力接口有：

- getInstance(Ability ability)：获取BiometricAuthentication的单例对象。
- checkAuthenticationAvailability(AuthType type, SecureLevel level, boolean isLocalAuth)：检测设备是否具有生物认证能力。
- execAuthenticationAction(AuthType type, SecureLevel level, boolean isLocalAuth,boolean

isAppAuthDialog, SystemAuthDialogInfo information)：调用者使用该方法进行生物认证。可以使用自定义的认证界面，也可以使用系统提供的认证界面。当使用系统认证界面时，调用者可以自定义提示语。该方法直到认证结束才返回认证结果。

- getAuthenticationTips()：获取生物认证过程中的提示信息。
- cancelAuthenticationAction()：取消生物认证操作。
- setSecureObjectSignature(Signature sign)：设置需要关联认证结果的Signature对象，在进行认证操作后，如果认证成功，则Signature对象被授权可以使用。设置前Signature对象需要正确初始化，且配置为认证成功才能使用。
- getSecureObjectSignature()：在认证成功后，可通过该方法获取已授权的Signature对象。如果未设置过Signature对象，则返回null。
- setSecureObjectCipher(Cipher cipher)：设置需要关联认证结果的Cipher对象，在进行认证操作后，如果认证成功，则Cipher对象被授权可以使用。设置前Cipher对象需要正确初始化，且配置为认证成功才能使用。
- getSecureObjectCipher()：在认证成功后，可通过该方法获取已授权的Cipher对象。如果未设置过Cipher对象，则返回null。
- setSecureObjectMac(Mac mac)：设置需要关联认证结果的Mac对象，在进行认证操作后，如果认证成功，则Mac对象被授权可以使用。设置前Mac对象需要正确初始化，且配置为认证成功才能使用。
- getSecureObjectMac()：在认证成功后，可通过该方法获取已授权的Mac对象。如果未设置过Mac对象，则返回null。

19.9.2　开发准备

开发前请完成以下准备工作：

- 在应用配置权限文件中，增加ohos.permission.ACCESS_BIOMETRIC的权限声明。
- 在使用生物特征识别认证能力的代码文件中，增加导入ohos.biometrics.authentication.BiometricAuthentication。

19.9.3　开发过程

开发过程大致如下。

1. 获取 BiometricAuthentication 的单例对象

获取BiometricAuthentication的单例对象，示例代码如下：

```
BiometricAuthentication biometricAuthentication =
    BiometricAuthentication.getInstance(MainAbility.mAbility);
```

2. 检测设备是否具有生物认证能力

检测设备是否具有生物认证能力：

- 2D人脸识别建议使用SECURE_LEVEL_S2。
- 3D人脸识别建议使用SECURE_LEVEL_S3。

示例代码如下：

```
int retChkAuthAvb =
    biometricAuthentication.checkAuthenticationAvailability(
        BiometricAuthentication.AuthType.AUTH_TYPE_BIOMETRIC_FACE_ONLY,
        BiometricAuthentication.SecureLevel.SECURE_LEVEL_S2, true);
```

3. 设置需要关联认证结果（可选）

设置需要关联认证结果的Signature对象、Cipher对象或Mac对象，示例代码如下：

```
// 定义一个Signature对象sign
biometricAuthentication.setSecureObjectSignature(sign);

// 定义一个Cipher对象cipher
biometricAuthentication.setSecureObjectCipher(cipher);

// 定义一个Mac对象mac
biometricAuthentication.setSecureObjectMac(mac);
```

4. 执行认证操作

在新线程中执行认证操作，避免阻塞其他操作，示例代码如下：

```
new Thread(new Runnable() {
    @Override
    public void run() {
        int retExcAuth;
        retExcAuth = biometricAuthentication.execAuthenticationAction(
            BiometricAuthentication.AuthType.AUTH_TYPE_BIOMETRIC_FACE_ONLY,
            BiometricAuthentication.SecureLevel.SECURE_LEVEL_S2, true, false, null);
    }
}).start();
```

5. 获取认证过程中的提示信息

获取认证过程中的提示信息，示例代码如下：

```
AuthenticationTips tips = biometricAuthentication.getAuthenticationTips();
```

6. 认证成功后获取结果对象（可选）

认证成功后获取已设置的Signature对象、Cipher对象或Mac对象，示例代码如下：

```
Signature sign = biometricAuthentication.getSecureObjectSignature();

Cipher cipher = biometricAuthentication.getSecureObjectCipher();

Mac mac = biometricAuthentication.getSecureObjectMac();
```

7. 认证过程中取消认证

认证过程中取消认证,示例代码如下:

```
int ret = biometricAuthentication.cancelAuthenticationAction();
```

第20章

二 维 码

HarmonyOS为应用提供丰富的AI能力,支持开箱即用。开发者可以灵活、便捷地选择AI能力,让应用变得更加智能。

本章所介绍的二维码生成功能就是HarmonyOS AI能力之一。

20.1 二维码概述

HarmonyOS支持根据开发者给定的字符串信息和二维码图片尺寸返回相应的二维码图片字节流。调用方可以通过二维码字节流生成二维码图片。

20.1.1 什么是二维码

二维码又称二维条码(2-Dimensional Bar Code),常见的二维码为QR(Quick Response,快速反应)Code,是一种近几年来移动设备上超流行的编码方式,它比传统的条形码(Bar Code)能保存更多的信息,也能表示更多的数据类型。

二维码是用某种特定的几何图形按一定规律在平面(二维方向)上分布的、黑白相间的、记录数据符号信息的图形,在代码编制上巧妙地利用构成计算机内部逻辑基础的0、1比特流的概念,使用若干个与二进制相对应的几何形体来表示文字数值信息,通过图像输入设备或光电扫描设备自动识读以实现信息自动处理。它具有条码技术的一些共性:每种码制有其特定的字符集、每个字符占有一定的宽度、具有一定的校验功能等。同时,还可以对不同行的信息自动识别及处理图形旋转变化点。

20.1.2 二维码的发展历程

国外对二维码技术的研究始于20世纪80年代末，在二维码符号表示技术研究方面已研制出多种码制，常见的有PDF417、QR Code、Code 49、Code 16K、Code One等。这些二维码的信息密度都比传统的一维码有了较大提高，如PDF417的信息密度是一维码CodeC39的20多倍。在二维码标准化研究方面，国际自动识别制造商协会（AIM）、美国标准化协会（ANSI）已完成了PDF417、QR Code、Code 49、Code 16K、Code One等码制的符号标准。国际标准技术委员会和国际电工委员会还成立了条码自动识别技术委员会（ISO/IEC/JTC1/SC31），已制定了QR Code的国际标准（ISO/IEC 18004：2000《自动识别与数据采集技术—条码符号技术规范—QR码》），起草了PDF417、Code 16K、Data Matrix、Maxi Code等二维码的ISO/IEC标准草案。在二维码设备开发研制、生产方面，美国、日本等国的设备制造商生产的识读设备、符号生成设备已广泛应用于各类二维码应用系统。二维码作为一种全新的信息存储、传递和识别技术，自诞生之日起就得到了世界上许多国家的关注。美国、德国、日本等国家不仅已将二维码技术应用于公安、外交、军事等部门对各类证件的管理，而且也将二维码应用于海关、税务等部门对各类报表和票据的管理，以及商业、交通运输等部门对商品及货物运输的管理，邮政部门对邮政包裹的管理，工业生产领域对工业生产线的自动化管理。

中国对二维码技术的研究开始于1993年。中国物品编码中心对几种常用的二维码PDF417、QRCCode、Data Matrix、Maxi Code、Code 49、Code 16K、Code One的技术规范进行了翻译和跟踪研究。随着中国市场经济的不断完善和信息技术的迅速发展，国内对二维码这一新技术的需求与日俱增。中国物品编码中心在原国家质量技术监督局和国家有关部门的大力支持下，对二维码技术的研究不断深入。在消化国外相关技术资料的基础上，制定了两个二维码的国家标准：二维码网格矩阵码（SJ/T 11349-2006）和二维码紧密矩阵码（SJ/T 11350-2006），从而大大促进了中国具有自主知识产权技术的二维码的研发。

2017年12月25日，中国人民银行以银发〔2017〕296号印发《条码支付业务规范（试行）》。该规范分总则、条码生成和受理、特约商户管理、风险管理、附则5章50条，自2018年4月1日起实施。简而言之，该规范承认了二维码的支付地位。

20.2 场景介绍

二维码的主要应用场景如下：

- 信息获取（名片、地图、WIFI密码、资料）。
- 网站跳转（跳转到微博、手机网站、网站）。
- 广告推送（用户扫码，直接浏览商家推送的视频、音频广告）。
- 手机电商（用户扫码、手机直接购物下单）。
- 防伪溯源（用户扫码即可查看生产地，同时后台可以获取最终消费地）。

- 优惠促销（用户扫码下载电子优惠券、抽奖）。
- 会员管理（用户在手机上获取电子会员信息、VIP服务）。
- 手机支付（扫描商品二维码，通过银行或第三方支付提供的手机端通道完成支付）。
- 账号登录（扫描二维码进行各个网站或软件的登录）。

20.3 接口说明

码生成提供了IBarcodeDetector()接口，常用方法的功能描述如下：

- detect(String barcodeInput, byte[] bitmapOutput, int width, int height)：根据给定的信息和二维码图片尺寸生成二维码图片字节流。
- release()：停止QR码生成服务，释放资源。

码生成具有以下约束与限制：

- 当前仅支持生成QR二维码（Quick Response Code）。由于QR二维码算法的限制，Java语言开发时字符串信息的长度不能超过2953个字符，JS语言开发时字符串信息的长度不能超过256个字符。
- Java语言开发时，生成的二维码图片的宽度不能超过1920像素，高度不能超过1680像素。由于QR二维码是通过正方形阵列承载信息的，建议二维码图片采用正方形，当二维码图片采用长方形时，会在QR二维码信息的周边区域留白。
- JS语言开发时，生成的二维码图片的宽高最小值为200px，当宽高不一致时，以二者最小值作为二维码的边长，且最终生成的二维码居中显示，支持矩形、圆形两种二维码类型（默认是矩形）。

20.4 实战：生成二维码

本节演示根据给定的字符串信息生成相应的二维码图片。

20.4.1 创建应用

为了演示生成二维码的功能，创建一个名为QuickResponseCode的应用。
在应用的界面上，通过点击按钮来触发生成二维码的操作。

20.4.2 修改 ability_main.xml

修改ability_main.xml内容如下：

```xml
<?xml version="1.0" encoding="utf-8"?>
<DirectionalLayout
    xmlns:ohos="http://schemas.huawei.com/res/ohos"
    ohos:height="match_parent"
    ohos:width="match_parent"
    ohos:orientation="vertical">

    <Button
        ohos:id="$+id:button_get"
        ohos:height="match_content"
        ohos:width="match_parent"
        ohos:background_element="#F76543"
        ohos:margin="10vp"
        ohos:padding="10vp"
        ohos:text="Get"
        ohos:text_size="40vp"
        />

    <Image
        ohos:id="$+id:image"
        ohos:height="match_content"
        ohos:width="match_parent"/>

</DirectionalLayout>
```

界面预览效果如图20-1所示。

图20-1　界面预览效果

上述代码中：

- 设置了Get按钮，以备设置点击事件，以触发元数据相关的操作。
- Image组件用于展示获取到的二维码图片。

20.4.3 修改 MainAbilitySlice

修改MainAbilitySlice内容如下:

```java
package com.waylau.hmos.quickresponsecode.slice;

import com.waylau.hmos.quickresponsecode.ResourceTable;
import ohos.aafwk.ability.AbilitySlice;
import ohos.aafwk.content.Intent;
import ohos.agp.components.Button;
import ohos.agp.components.Image;
import ohos.ai.cv.common.ConnectionCallback;
import ohos.ai.cv.common.VisionManager;
import ohos.ai.cv.qrcode.IBarcodeDetector;
import ohos.hiviewdfx.HiLog;
import ohos.hiviewdfx.HiLogLabel;
import ohos.media.image.ImageSource;
import ohos.media.image.PixelMap;
import ohos.media.image.common.PixelFormat;


public class MainAbilitySlice extends AbilitySlice {
    private static final String TAG = MainAbilitySlice.class.getSimpleName();
    private static final HiLogLabel LABEL_LOG =
            new HiLogLabel(HiLog.LOG_APP, 0x00001, TAG);

    private IBarcodeDetector barcodeDetector;

    private ConnectionCallback connectionCallback = new ConnectionCallback() {
        @Override
        public void onServiceConnect() {
            HiLog.info(LABEL_LOG, "onServiceConnect");
            getInfo();
        }

        @Override
        public void onServiceDisconnect() {
            HiLog.info(LABEL_LOG, "onServiceDisconnect");
        }
    };

    @Override
    public void onStart(Intent intent) {
        super.onStart(intent);
        super.setUIContent(ResourceTable.Layout_ability_main);

        // 为按钮设置点击事件回调
        Button buttonGet =
```

```java
            (Button) findComponentById(ResourceTable.Id_button_get);
    buttonGet.setClickedListener(listener -> {
        // 建立与能力引擎的连接
        int result = VisionManager.init(this, connectionCallback);

        HiLog.info(LABEL_LOG, "VisionManager init, result: %{public}s", result);
    });
}

@Override
public void onStop() {
    super.onStop();

    // 关闭、释放资源
    release();
}

private void release() {
    HiLog.info(LABEL_LOG, "release");

    if (barcodeDetector != null) {
        // 关闭、释放资源
        barcodeDetector.release();
        barcodeDetector = null;
    }

    // 调用VisionManager.destroy()方法断开与能力引擎的连接
    VisionManager.destroy();
}

private void getInfo() {
    HiLog.info(LABEL_LOG, "before getInfo");

    // 实例化IBarcodeDetector接口
    barcodeDetector = VisionManager.getBarcodeDetector(this);

    // 定义二维码生成图像的尺寸，并根据图像大小分配字节流数组空间
    final int SAMPLE_LENGTH = 500;
    byte[] byteArray = new byte[SAMPLE_LENGTH * SAMPLE_LENGTH * 4];

    // 根据输入的字符串信息生成相应的二维码图片字节流
    int result = barcodeDetector.detect("Welcome to waylau.com",
            byteArray, SAMPLE_LENGTH, SAMPLE_LENGTH);

    // 展示帧数据
    showImage(byteArray);

    HiLog.info(LABEL_LOG, "end getInfo, result: %{public}s", result);
}
```

```java
    private void showImage(byte[] byteArray) {
        Image image =
                (Image) findComponentById(ResourceTable.Id_image);

        // 创建图像数据源ImageSource对象
        ImageSource imageSource = ImageSource.create(byteArray, this.getSourceOptions());

        // 普通解码叠加旋转、缩放、裁剪
        PixelMap pixelMap = imageSource.createPixelmap(this.getDecodingOptions());

        Image imageDecode =
                (Image) findComponentById(ResourceTable.Id_image);

        imageDecode.setPixelMap(pixelMap);


        image.setPixelMap(pixelMap);
    }

    // 设置数据源的格式信息
    private ImageSource.SourceOptions getSourceOptions() {
        ImageSource.SourceOptions sourceOptions = new ImageSource.SourceOptions();
        sourceOptions.formatHint = "image/jpeg";

        return sourceOptions;
    }

    // 设置解码格式
    private ImageSource.DecodingOptions getDecodingOptions() {
        ImageSource.DecodingOptions decodingOpts = new ImageSource.DecodingOptions();
        decodingOpts.desiredPixelFormat= PixelFormat.ARGB_8888;

        return decodingOpts;
    }

    @Override
    public void onActive() {
        super.onActive();
    }

    @Override
    public void onForeground(Intent intent) {
        super.onForeground(intent);
    }
}
```

上述代码中：

- 定义了ConnectionCallback回调，实现连接能力引擎成功与否后的操作。
- 在Button上设置了点击事件。
- 调用VisionManager.init()方法，将此工程的context和connectionCallback作为入参，建立与能力引擎的连接。
- 与能力引擎的连接建立完成之后，会触发getInfo方法。
- 实例化IBarcodeDetector接口。
- 定义二维码生成图像的尺寸，并根据图像大小分配字节流数组空间。
- 调用IBarcodeDetector的detect()方法根据输入的字符串信息生成相应的二维码图片字节流。如果返回值为0，则表明调用成功。
- showImage方法将二维码图片字节流转成了PixelMap并最终在界面上显示出来。
- release方法用于释放资源。

20.4.4 运行

运行应用后，点击界面上的Get按钮以触发操作的执行。此时，控制台输出内容如下：

```
    02-12 23:53:35.588 20441-20441/com.waylau.hmos.quickresponsecode I
00001/MainAbilitySlice: VisionManager init, result: 0
    02-12 23:53:35.593 20441-20441/com.waylau.hmos.quickresponsecode I
00001/MainAbilitySlice: [3440b1a47f8ca3e, 3cf1be9, 1d391a2] onServiceConnect
    02-12 23:53:35.593 20441-20441/com.waylau.hmos.quickresponsecode I
00001/MainAbilitySlice: [3440b1a47f8ca3e, 3cf1be9, 1d391a2] before getInfo
    02-12 23:53:35.663 20441-20441/com.waylau.hmos.quickresponsecode I
00001/MainAbilitySlice: [3440b1a47f8ca3e, 3cf1be9, 1d391a2] end getInfo, result: 0
```

界面效果如图20-2所示。

图20-2　界面效果

第 21 章

通用文字识别

本章所介绍的通用文字识别功能也是HarmonyOS AI能力之一。

21.1 通用文字识别概述

通用文字识别的核心技术是OCR（Optical Character Recognition，光学字符识别）。OCR是一种通过拍照、扫描等光学输入方式把各种票据、卡证、表格、报刊、图书等印刷品文字转化为图像信息，再利用文字识别技术将图像信息转化为计算机等设备可以使用的字符信息的技术。

21.1.1 什么是OCR

OCR是指电子设备（例如扫描仪或数码相机）检查纸上打印的字符，通过检测暗、亮的模式确定其形状，然后用字符识别方法将形状翻译成计算机文字的过程，即针对印刷体字符，采用光学的方式将纸质文档中的文字转换成为黑白点阵的图像文件，并通过识别软件将图像中的文字转换成文本格式，供文字处理软件进一步编辑加工的技术。如何除错或利用辅助信息提高识别正确率是OCR重要的课题，ICR（Intelligent Character Recognition，智能字符识别）的名词也因此而产生。衡量一个OCR系统性能好坏的主要指标有：拒识率、误识率、识别速度、用户界面的友好性以及产品的稳定性，易用性及可行性等。

21.1.2 OCR发展简史

OCR的概念是在1929年由德国科学家Tausheck最先提出来的，后来美国科学家Handel也提出了利用技术对文字进行识别的想法。而最早对印刷体汉字识别进行研究的是IBM公司的Casey和Nagy，1966年他们发表了第一篇关于汉字识别的文章，采用模板匹配法识别了1000个印刷体汉字。

早在六七十年代，世界各国就开始有OCR的研究，而初期多以文字的识别方法研究为主，且识别的文字仅为0~9的数字。以同样拥有方块文字的日本为例，1960年左右开始研究OCR的基本识别理论，初期以数字为对象，直至1965~1970年开始有一些简单的产品，如印刷文字的邮政编码识别系统，识别邮件上的邮政编码，帮助邮局作区域分信的作业。因此，至今邮政编码一直是各国所倡导的地址书写方式。

20世纪70年代初，日本的学者开始研究汉字识别，并做了大量的工作。中国在OCR技术方面的研究工作起步较晚，在70年代才开始对数字、英文字母及符号的识别进行研究，70年代末开始进行汉字识别的研究，到1986年，我国提出863高新科技研究计划，汉字识别的研究进入一个实质性的阶段，清华大学的丁晓青教授和中科院分别开始研究，相继推出了中文OCR产品，现为中国最领先的汉字OCR技术。早期的OCR软件由于识别率及产品化等多方面因素的影响，未能达到实际要求。同时，由于硬件设备成本高，运行速度慢，也没有达到实用的程度。只有个别部门（如信息部门、新闻出版单位等）使用OCR软件。进入20世纪90年代以后，随着平台式扫描仪的广泛应用，以及我国信息自动化和办公自动化的普及，大大推动了OCR技术的进一步发展，使OCR的识别正确率、识别速度满足了广大用户的要求。

目前，随着AI技术的成熟，各大厂商相继推出了在线文字识别服务。

21.2　场景介绍

通用文字识别适用于如下场景：

- 可以进行文档翻拍、街景翻拍等图片来源的文字检测和识别，也可以集成为其他应用中，提供文字检测、识别的功能，并根据识别结果提供翻译、搜索等相关服务。
- 可以处理来自相机、图库等多种来源的图像数据，提供了自动检测文本、识别图像中的文本位置以及文本内容功能的开放接口。
- 能在一定程度上支持文本倾斜、拍摄角度倾斜、复杂光照条件以及复杂文本背景等场景的文字识别。

21.3　接口说明

通用文字识别提供了setVisionConfiguration()和detect()两个函数接口。

21.3.1 setVisionConfiguration()方法

调用ITextDetector的setVisionConfiguration()方法,通过传入的TextConfiguration选择需要调用的OCR类型。

```
void setVisionConfiguration(TextConfiguration textConfiguration);
```

其中TextConfiguration的常用设置有:

- setDetectType():OCR引擎类型定义。目前支持TextDetectType.TYPE_TEXT_DETECT_FOCUS_SHOOT(自然场景OCR)。
- setLanguage():识别语种定义。目前支持TextConfiguration.AUTO(不指定语种,会进行语种检测操作)、TextConfiguration.CHINESE(中文)、TextConfiguration.ENGLISH(英语)、TextConfiguration.SPANISH(西班牙语)、TextConfiguration.PORTUGUESE(葡萄牙语)、TextConfiguration.ITALIAN(意大利语)、TextConfiguration.GERMAN(德语)、TextConfiguration.FRENCH(法语)、TextConfiguration.RUSSIAN(俄语)、TextConfiguration.JAPANESE(日语)、TextConfiguration.KOREAN(韩语),默认值为TextConfiguration.AUTO。
- setProcessMode():进程模式定义。目前支持VisionConfiguration.MODE_IN(同进程调用)、VisionConfiguration.MODE_OUT(跨进程调用),默认值为VisionConfiguration.MODE_OUT。

21.3.2 detect()方法

调用ITextDetector的detect()方法获取识别结果。

```
int detect(VisionImage image, Text result, VisionCallback<Text> visionCallBack);
```

其中:

- image为待OCR检测识别的输入图片。
- 如果visionCallback为null,执行同步调用,结果码由方法返回,检测及识别结果由result返回。
- 如果visionCallback为有效的回调函数,则该函数为异步调用,函数返回时result中的值无效,实际识别结果由回调函数返回。回调函数的使用方法参见后续的实战示例。
- 同步模式调用成功时,该函数返回结果码0。异步模式调用请求发送成功时,该函数返回结果码700。

21.3.3 约束与限制

使用通用文字识别要注意以下约束与限制:

- 支持处理的图片格式包括JPEG、JPG、PNG。
- 通用文字识别目前支持的语言有中文、英语、日语、韩语、俄语、意大利语、西班牙语、葡萄牙语、德语以及法语（将来会增加更多语种）。

- 目前支持文档印刷体识别，不支持手写字体识别。
- 为了保证较理想的识别结果，调用通用文字识别功能时，应尽可能保证输入的图像具有合适的成像质量（建议720p以上）和高宽比例（建议2:1以下，接近手机屏幕高宽比例为宜）。当输入的图像为非建议图片尺寸时，文字识别的准确度可能会受到影响。
- 为了保证较理想的识别结果，建议文本与拍摄角度夹角在正负30度范围内。

21.4　实战：通用文字识别示例

本节演示根据给定的图片识别其中的文字。

21.4.1　创建应用

为了演示文字识别的功能，创建一个名为TextDetector的应用。

在应用的界面上，通过点击按钮来触发文字识别的操作。

在media目录下放置一个cloud_native.PNG图片文件（见图21-1），以备测试。

内 容 提 要

Cloud Native（云原生）是以云架构为优先的应用开发模式。目前，越来越多的企业已经开始大规模地“拥抱云”——在云环境下开发应用、部署应用及发布应用等。未来，越来越多的开发者也将采用 Cloud Native 来开发应用。本书是国内 Java 领域关于 Cloud Native 的著作。

本书全面讲解了基于 Cloud Native 来构建应用需要考虑的设计原则和实现方式，涵盖 REST 设计、测试、服务注册、服务发现、安全、数据管理、消息通信、批处理、任务调度、运营、容器部署、持续发布等方面的 Cloud Native 知识。同时，书中所讲解的技术方案皆为业界主流的技术，极具前瞻性。最后，本书除了讲解 Cloud Native 的理论知识，还会在每个知识点上辅以大量的代码案例，使理论可以联系实践，具备更强的可操作性。

本书主要面向对分布式系统、微服务、Cloud Native 开发感兴趣的计算机专业的学生、软件开发人员和系统架构师。

图 21-1　测试图片

图片内容节选自笔者所著的《Cloud Native分布式架构原理与实践》。

21.4.2　修改 ability_main.xml

修改ability_main.xml内容如下：

```
<?xml version="1.0" encoding="utf-8"?>
<DirectionalLayout
    xmlns:ohos="http://schemas.huawei.com/res/ohos"
    ohos:height="match_parent"
    ohos:width="match_parent"
```

```xml
    ohos:orientation="vertical">

    <Button
        ohos:id="$+id:button_get"
        ohos:height="match_content"
        ohos:width="match_parent"
        ohos:background_element="#F76543"
        ohos:margin="10vp"
        ohos:padding="10vp"
        ohos:text="Get"
        ohos:text_size="40vp"
        />

    <Text
        ohos:id="$+id:text"
        ohos:height="match_content"
        ohos:width="match_content"
        ohos:background_element="$graphic:background_ability_main"
        ohos:layout_alignment="horizontal_center"
        ohos:text="Hello World"
        ohos:text_size="30vp"
        ohos:multiple_lines="true"
        />

</DirectionalLayout>
```

界面预览效果如图21-2所示。

图 21-2　界面预览效果

上述代码中：

- 设置了Get按钮,以备设置点击事件,以触发元数据相关的操作。
- Text组件用于展示获取到的文字识别结果。

21.4.3 修改 MainAbilitySlice

修改MainAbilitySlice内容如下:

```java
package com.waylau.hmos.textdetector.slice;

import com.waylau.hmos.textdetector.ResourceTable;
import ohos.aafwk.ability.AbilitySlice;
import ohos.aafwk.content.Intent;
import ohos.agp.components.Button;
import ohos.ai.cv.common.ConnectionCallback;
import ohos.ai.cv.common.VisionImage;
import ohos.ai.cv.common.VisionManager;
import ohos.ai.cv.text.ITextDetector;
import ohos.global.resource.NotExistException;
import ohos.hiviewdfx.HiLog;
import ohos.hiviewdfx.HiLogLabel;
import ohos.media.image.ImageSource;
import ohos.media.image.PixelMap;
import ohos.media.image.common.PixelFormat;

import java.io.IOException;
import java.io.InputStream;


public class MainAbilitySlice extends AbilitySlice {
    private static final String TAG = MainAbilitySlice.class.getSimpleName();
    private static final HiLogLabel LABEL_LOG =
            new HiLogLabel(HiLog.LOG_APP, 0x00001, TAG);

    private ITextDetector textDetector;

    private ConnectionCallback connectionCallback = new ConnectionCallback() {
        @Override
        public void onServiceConnect() {
            HiLog.info(LABEL_LOG, "onServiceConnect");
            try {
                getInfo();
            } catch (IOException e) {
                e.printStackTrace();
            } catch (NotExistException e) {
                e.printStackTrace();
            }
        }
```

```java
        @Override
        public void onServiceDisconnect() {
            HiLog.info(LABEL_LOG, "onServiceDisconnect");
        }
    };

    @Override
    public void onStart(Intent intent) {
        super.onStart(intent);
        super.setUIContent(ResourceTable.Layout_ability_main);

        // 为按钮设置点击事件回调
        Button buttonGet =
                (Button) findComponentById(ResourceTable.Id_button_get);
        buttonGet.setClickedListener(listener -> {
            // 建立与能力引擎的连接
            int result = VisionManager.init(this, connectionCallback);

            HiLog.info(LABEL_LOG, "VisionManager init, result: %{public}s", result);
        });
    }

    @Override
    public void onStop() {
        super.onStop();

        // 关闭、释放资源
        release();
    }

    private void release() {
        HiLog.info(LABEL_LOG, "release");

        if (textDetector != null) {
            // 关闭、释放资源
            textDetector.release();
            textDetector = null;
        }

        // 调用VisionManager.destroy()方法断开与能力引擎的连接
        VisionManager.destroy();
    }

    private void getInfo() throws IOException, NotExistException {
        HiLog.info(LABEL_LOG, "before getInfo");

        // 实例化IBarcodeDetector接口
        textDetector = VisionManager.getTextDetector(this);
```

```java
        // 获取图片流
        InputStream drawableInputStream = getResourceManager().
getResource(ResourceTable.Media_harmonyos);

        // 创建图像数据源ImageSource对象
        ImageSource imageSource = ImageSource.create(drawableInputStream,
this.getSourceOptions());

        // 普通解码叠加旋转、缩放、裁剪
        PixelMap pixelMap = imageSource.createPixelmap(this.
getDecodingOptions());

        VisionImage image = VisionImage.fromPixelMap(pixelMap);

        // 获取文字识别结果
        ohos.ai.cv.text.Text text = new ohos.ai.cv.text.Text();
        int result = textDetector.detect(image, text, null); // 同步

        // 结果展示在Text上
        ohos.agp.components.Text textResult =
                (ohos.agp.components.Text)
findComponentById(ResourceTable.Id_text);
        textResult.setText(text.getValue());

        HiLog.info(LABEL_LOG, "end getInfo, result: %{public}s,
text: %{public}s",
                result, text.getValue());
    }

    // 设置数据源的格式信息
    private ImageSource.SourceOptions getSourceOptions() {
        ImageSource.SourceOptions sourceOptions = new
ImageSource.SourceOptions();
        sourceOptions.formatHint = "image/jpeg";

        return sourceOptions;
    }

    // 设置解码格式
    private ImageSource.DecodingOptions getDecodingOptions() {
        ImageSource.DecodingOptions decodingOpts = new
ImageSource.DecodingOptions();
        decodingOpts.desiredPixelFormat = PixelFormat.ARGB_8888;

        return decodingOpts;
    }

    @Override
    public void onActive() {
        super.onActive();
```

```java
    }

    @Override
    public void onForeground(Intent intent) {
        super.onForeground(intent);
    }
}
```

上述代码中：

- 定义了ConnectionCallback回调，实现连接能力引擎成功与否后的操作。
- 在Button上设置了点击事件。
- 调用VisionManager.init()方法，将此工程的context和connectionCallback作为入参，建立与能力引擎的连接。
- 与能力引擎的连接建立完成之后，会触发getInfo方法。
- 实例化ITextDetector接口。
- 获取图片流，并生成VisionImage对象。
- 调用ITextDetector的detect()方法，根据传入的VisionImage对象识别生成文字。
- 识别生成文字最终在界面上显示出来。
- release方法用于释放资源。

21.4.4 运行

运行应用后，点击界面上的Get按钮以触发操作的执行。此时，控制台输出内容如下：

```
    09-12 22:36:33.992 21050-21050/com.waylau.hmos.textdetector I
00001/MainAbilitySlice: VisionManager init, result: 0
    09-12 22:36:34.008 21050-21050/com.waylau.hmos.textdetector I
00001/MainAbilitySlice: [2017d11071e1d4d, 803891, 2271d5c] onServiceConnect
    09-12 22:36:34.008 21050-21050/com.waylau.hmos.textdetector I
00001/MainAbilitySlice: [2017d11071e1d4d, 803891, 2271d5c] before getInfo
    09-12 22:36:34.439 21050-21050/com.waylau.hmos.textdetector I
00001/MainAbilitySlice: [2017d11071e1d4d, 803891, 2271d5c] end getInfo, result:
0, text: 内 容 提 要
    09-12 22:36:34.439 21050-21050/com.waylau.hmos.textdetector I
00001/MainAbilitySlice: C椒nj N觚i\℃(云原生)是以云架构为优先的应用开发模式.目前,越来越
多的企业已经开始大规模地"拥抱云"——在云环境下开发应用、部署应用及发布应用等。未来,越来越多的
开发者也将采用 C椒d Nabve 来开发应用。本书是国内 hⅧ领域关于Ck飒njN砸ve 的著作。
    09-12 22:36:34.439 21050-21050/com.waylau.hmos.textdetector I
00001/MainAbilitySlice: 本书全面讲解了基于 amnj N砸ve 来构建应用需要考虑的设计原则和实
现方式,涵盖 RFgr 设计、测试、服务注册、服务发现、安全、数据管理、消息通信、批处理、任务调度、
运营、容器部署、持续发布等方面的(:knjnj N颧ve 知识。同时,书中所讲解的技术方案皆为业界主流的技
术,极具前瞻性。最后,本书除了讲解 Cbnj N砸ve 的理论知识,还会在每个知识点上辅以大量的代码案例,
使理论可以联系实践,具备更强的可操作性。
    09-12 22:36:34.439 21050-21050/com.waylau.hmos.textdetector I
00001/MainAbilitySlice: 本书主要面向对分布式系统、微服务、Cbnj N砸ve 开发感兴趣的计算
```

界面效果如图21-3所示。

图 21-3 界面效果

第22章

蓝　牙

HarmonyOS提供了对传统蓝牙和低功耗蓝牙（Bluetooth Low Energy，BLE）的支持。

22.1　蓝牙概述

蓝牙是短距离无线通信的一种方式，支持蓝牙的两个设备必须配对后才能通信。

HarmonyOS蓝牙主要分为传统蓝牙和低功耗蓝牙。传统蓝牙指的是蓝牙版本3.0以下的蓝牙，低功耗蓝牙指的是蓝牙版本4.0以上的蓝牙。

当前蓝牙的配对方式有两种：蓝牙协议2.0以下支持PIN码（Personal Identification Number，个人识别码）配对，蓝牙协议2.1以上支持简单配对。

22.1.1　传统蓝牙

HarmonyOS传统蓝牙提供的功能有：

- 传统蓝牙本机管理：打开和关闭蓝牙、设置和获取本机蓝牙名称、扫描和取消扫描周边蓝牙设备、获取本机蓝牙profile对其他设备的连接状态、获取本机蓝牙已配对的蓝牙设备列表。
- 传统蓝牙远端设备操作：查询远端蓝牙设备名称和MAC地址、设备类型和配对状态以及向远端蓝牙设备发起配对。

22.1.2　BLE

BLE设备交互时分为不同的角色：

- 中心设备和外围设备：中心设备负责扫描外围设备、发现广播。外围设备负责发送广播。
- GATT（Generic Attribute Profile，通用属性配置文件）服务端与GATT客户端：两台设备建立连接后，其中一台作为GATT服务端，另一台作为GATT客户端。

HarmonyOS低功耗蓝牙提供的功能有：

- BLE扫描和广播：根据指定状态获取外围设备，启动或停止BLE扫描、广播。
- BLE中心设备与外围设备进行数据交互：BLE外围设备和中心设备建立GATT连接后，中心设备可以查询外围设备支持的各种数据，向外围设备发起数据请求，并向其写入特征值数据。
- BLE外围设备数据管理：BLE外围设备作为服务端，可以接收来自中心设备（客户端）的GATT连接请求，应答来自中心设备的特征值内容读取和写入请求，并向中心设备提供数据。同时，外围设备还可以主动向中心设备发送数据。

22.1.3 约束与限制

使用蓝牙涉及如下约束与限制：

- 调用蓝牙的打开接口需要ohos.permission.USE_BLUETOOTH权限。
- 调用蓝牙扫描接口需要ohos.permission.LOCATION权限和ohos.permission.DISCOVER_BLUETOOTH权限。

22.2 实战：传统蓝牙本机管理

传统蓝牙本机管理主要是针对蓝牙本机的基本操作，包括打开和关闭蓝牙、设置和获取本机蓝牙名称、扫描和取消扫描周边蓝牙设备、获取本机蓝牙profile对其他设备的连接状态、获取本机蓝牙已配对的蓝牙设备列表。

本节演示如何实现打开蓝牙并扫描周边的蓝牙设备。

22.2.1 接口说明

蓝牙本机管理类BluetoothHost的主要接口有：

- getDefaultHost(Context context)：获取BluetoothHost实例，去管理本机蓝牙操作。
- enableBt()：打开本机蓝牙。
- disableBt()：关闭本机蓝牙。
- setLocalName(String name)：设置本机蓝牙名称。
- getLocalName()：获取本机蓝牙名称。
- getBtState()：获取本机蓝牙状态。
- startBtDiscovery()：发起蓝牙设备扫描。

- cancelBtDiscovery()：取消蓝牙设备扫描。
- isBtDiscovering()：检查蓝牙是否在扫描设备中。
- getProfileConnState(int profile)：获取本机蓝牙profile对其他设备的连接状态。
- getPairedDevices()：获取本机蓝牙已配对的蓝牙设备列。

22.2.2 创建应用

为了演示BluetoothHost的功能，创建一个名为BluetoothHost的应用。

在应用的界面上，通过点击按钮来触发BluetoothHost的操作。

22.2.3 声明权限

修改配置文件，声明使用蓝牙相关的权限如下：

```
// 声明权限
"reqPermissions": [
   {
   "name": "ohos.permission.USE_BLUETOOTH"
   },
   {
   "name": "ohos.permission.DISCOVER_BLUETOOTH"
   },
   {
   "name": "ohos.permission.LOCATION"
   }
]
```

同时，在应用启动时，显式声明ohos.permission.LOCATION权限。代码如下：

```
package com.waylau.hmos.bluetoothhost;

import com.waylau.hmos.bluetoothhost.slice.MainAbilitySlice;
import ohos.aafwk.ability.Ability;
import ohos.aafwk.content.Intent;

import java.util.ArrayList;
import java.util.List;

public class MainAbility extends Ability {
    @Override
    public void onStart(Intent intent) {
        super.onStart(intent);
        super.setMainRoute(MainAbilitySlice.class.getName());

        // 显式声明需要使用的权限
        requestPermission();
    }
```

```
// 显式声明需要使用的权限
private void requestPermission() {
    String[] permission = {
            "ohos.permission.LOCATION"};
    List<String> applyPermissions = new ArrayList<>();
    for (String element : permission) {
        if (verifySelfPermission(element) != 0) {
            if (canRequestPermission(element)) {
                applyPermissions.add(element);
            }
        }
    }
    requestPermissionsFromUser(applyPermissions.toArray(new String[0]), 0);
}
```

由于ohos.permission.USE_BLUETOOTH 和 ohos.permission.DISCOVER_BLUETOOTH 不是敏感权限,因此不需要在MainAbility中显式声明。

22.2.4 修改 ability_main.xml

修改ability_main.xml内容如下:

```xml
<?xml version="1.0" encoding="utf-8"?>
<DirectionalLayout
    xmlns:ohos="http://schemas.huawei.com/res/ohos"
    ohos:height="match_parent"
    ohos:width="match_parent"
    ohos:orientation="vertical">

    <Button
        ohos:id="$+id:button_get"
        ohos:height="match_content"
        ohos:width="match_parent"
        ohos:background_element="#F76543"
        ohos:margin="10vp"
        ohos:padding="10vp"
        ohos:text="Get"
        ohos:text_size="30vp"
        />

    <Text
        ohos:id="$+id:text"
        ohos:height="match_content"
        ohos:width="match_content"
        ohos:background_element="$graphic:background_ability_main"
        ohos:layout_alignment="horizontal_center"
```

```
            ohos:text="Hello World"
            ohos:text_size="20vp"
            ohos:multiple_lines="true"
            />

</DirectionalLayout>
```

界面预览效果如图22-1所示。

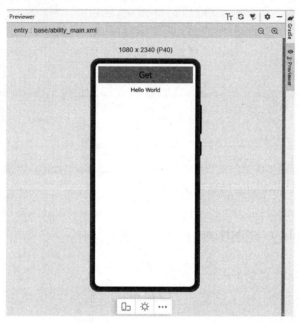

图 22-1　界面预览效果

上述代码中：

- 设置了Get按钮，以备设置点击事件，以触发蓝牙相关的操作。
- Text组件用于展示读取到的蓝牙信息。

22.2.5　修改 MainAbilitySlice

修改MainAbilitySlice内容如下：

```
package com.waylau.hmos.bluetoothhost.slice;

import com.waylau.hmos.bluetoothhost.ResourceTable;
import ohos.aafwk.ability.AbilitySlice;
import ohos.aafwk.content.Intent;
import ohos.aafwk.content.IntentParams;
import ohos.agp.components.Button;
import ohos.bluetooth.BluetoothHost;
import ohos.bluetooth.BluetoothRemoteDevice;
import ohos.event.commonevent.*;
import ohos.hiviewdfx.HiLog;
```

```java
import ohos.hiviewdfx.HiLogLabel;
import ohos.rpc.RemoteException;
import ohos.utils.zson.ZSONObject;

import java.util.Optional;


public class MainAbilitySlice extends AbilitySlice {
    private static final String TAG = MainAbilitySlice.class.getSimpleName();
    private static final HiLogLabel LABEL_LOG =
            new HiLogLabel(HiLog.LOG_APP, 0x00001, TAG);

    private BluetoothHost bluetoothHost;
    private ohos.agp.components.Text textResult;

    @Override
    public void onStart(Intent intent) {
        super.onStart(intent);
        super.setUIContent(ResourceTable.Layout_ability_main);

        // 初始化蓝牙
        initBluetooth();

        // 为按钮设置点击事件回调
        Button buttonGet =
                (Button) findComponentById(ResourceTable.Id_button_get);
        buttonGet.setClickedListener(listener -> getInfo());

        textResult =
                (ohos.agp.components.Text)
findComponentById(ResourceTable.Id_text);
    }

    private void initBluetooth() {
        HiLog.info(LABEL_LOG, "before initBluetooth");

        // 获取蓝牙本机管理对象
        bluetoothHost = BluetoothHost.getDefaultHost(this);

        // 调用打开接口
        bluetoothHost.enableBt();

        // 获取本机蓝牙名称
        Optional<String> nameOptional = bluetoothHost.getLocalName();

        // 调用获取蓝牙开关状态接口
        int state = bluetoothHost.getBtState();

        HiLog.info(LABEL_LOG, "end initBluetooth, name: %{public}s, state: %{public}s",
```

```java
                nameOptional.get(), state);
        }

        private void getInfo() {
            HiLog.info(LABEL_LOG, "before getInfo bluetooth name: %{public}s, state: %{public}s",
                    bluetoothHost.getLocalName().get(), bluetoothHost.getBtState());

            // 注册广播BluetoothRemoteDevice.EVENT_DEVICE_DISCOVERED
            MatchingSkills matchingSkills = new MatchingSkills();
            matchingSkills.addEvent(BluetoothRemoteDevice.EVENT_DEVICE_DISCOVERED); // 自定义事件
            CommonEventSubscribeInfo subscribeInfo = new CommonEventSubscribeInfo(matchingSkills);
            MyCommonEventSubscriber subscriber = new MyCommonEventSubscriber(subscribeInfo);

            try {
                CommonEventManager.subscribeCommonEvent(subscriber);
            } catch (RemoteException e) {
                HiLog.info(LABEL_LOG, "subscribeCommonEvent occur exception.");
            }

            //开始扫描
            bluetoothHost.startBtDiscovery();

            HiLog.info(LABEL_LOG, "end getInfo");
        }

        //接收系统广播
        class MyCommonEventSubscriber extends CommonEventSubscriber {
            public MyCommonEventSubscriber(CommonEventSubscribeInfo subscribeInfo) {
                super(subscribeInfo);
            }

            @Override
            public void onReceiveEvent(CommonEventData var) {

                Intent info = var.getIntent();
                if (info == null) {
                    return;
                }
                //获取系统广播的action
                String action = info.getAction();

                //判断是否扫描到设备的广播
                if (action == BluetoothRemoteDevice.EVENT_DEVICE_DISCOVERED) {
```

```
                    IntentParams myParam = info.getParams();
                    BluetoothRemoteDevice device =
                            (BluetoothRemoteDevice)
myParam.getParam(BluetoothRemoteDevice.REMOTE_DEVICE_PARAM_DEVICE);

                    // 结果展示在Text上
                    String stringResult = ZSONObject.toZSONString(device);
                    textResult.append(stringResult);

                    HiLog.info(LABEL_LOG, "end getInfo, text: %{public}s",
                            stringResult);
                }
            }
        }

        @Override
        public void onActive() {
            super.onActive();
        }

        @Override
        public void onForeground(Intent intent) {
            super.onForeground(intent);
        }
    }
}
```

上述代码中：

- initBluetooth方法用于初始化蓝牙，包括获取蓝牙本机管理对象、启动蓝牙，并获取蓝牙的名称和状态信息。
- 在Button上设置了点击事件。
- getInfo方法注册广播BluetoothRemoteDevice.EVENT_DEVICE_DISCOVERED事件，并启用蓝牙扫描。
- 接收到事件后，将扫描到的设备信息展示在界面的Text上。

22.2.6 运行

初次运行应用后，点击界面上的Get按钮以触发操作的执行。此时，控制台输出内容如下：

```
    09-12 22:49:25.968 16088-16088/com.waylau.hmos.bluetoothhost I
00001/MainAbilitySlice:  before getInfo bluetooth name: HUAWEI P40, state: 2
    09-12 22:49:25.994 16088-16088/com.waylau.hmos.bluetoothhost I
00001/MainAbilitySlice:  end getInfo
    09-12 22:49:26.193 16088-16088/com.waylau.hmos.bluetoothhost I
00001/MainAbilitySlice:  end getInfo, text:
{"aclConnected":false,"aclEncrypted":false,"bondedFromLocal":true,"deviceAddr"
:"B0:55:08:14:9D:7B","deviceAlias":{"present":true},"deviceBatteryLevel":-1,"d
eviceClass":{"present":true},"deviceName":{"present":true},"deviceType":1,"dev
```

```
iceUuids":[],"messagePermission":0,"pairState":0,"phonebookPermission":0}
    09-12 22:49:26.243 16088-16088/com.waylau.hmos.bluetoothhost I
00001/MainAbilitySlice: end getInfo, text:
{"aclConnected":false,"aclEncrypted":false,"bondedFromLocal":true,"deviceAddr"
:"20:AB:37:60:31:86","deviceAlias":{"present":true},"deviceBatteryLevel":-1,"d
eviceClass":{"present":true},"deviceName":{"present":true},"deviceType":1,"dev
iceUuids":[],"messagePermission":0,"pairState":0,"phonebookPermission":0}
    09-12 22:49:26.340 16088-16088/com.waylau.hmos.bluetoothhost I
00001/MainAbilitySlice: end getInfo, text:
{"aclConnected":false,"aclEncrypted":false,"bondedFromLocal":true,"deviceAddr"
:"50:04:B8:C1:81:F8","deviceAlias":{"present":true},"deviceBatteryLevel":-1,"d
eviceClass":{"present":true},"deviceName":{"present":true},"deviceType":1,"dev
iceUuids":[],"messagePermission":0,"pairState":0,"phonebookPermission":0}
    09-12 22:49:27.043 16088-16088/com.waylau.hmos.bluetoothhost I
00001/MainAbilitySlice: end getInfo, text:
{"aclConnected":false,"aclEncrypted":false,"bondedFromLocal":true,"deviceAddr"
:"1C:15:1F:8B:80:0E","deviceAlias":{"present":true},"deviceBatteryLevel":-1,"d
eviceClass":{"present":true},"deviceName":{"present":false},"deviceType":1,"de
viceUuids":[],"messagePermission":0,"pairState":0,"phonebookPermission":0}
    09-12 22:49:27.245 16088-16088/com.waylau.hmos.bluetoothhost I
00001/MainAbilitySlice: end getInfo, text:
{"aclConnected":false,"aclEncrypted":false,"bondedFromLocal":true,"deviceAddr"
:"50:04:B8:C1:82:CC","deviceAlias":{"present":true},"deviceBatteryLevel":-1,"d
eviceClass":{"present":true},"deviceName":{"present":false},"deviceType":1,"de
viceUuids":[],"messagePermission":0,"pairState":0,"phonebookPermission":0}
    09-12 22:49:27.426 16088-16088/com.waylau.hmos.bluetoothhost I
00001/MainAbilitySlice: end getInfo, text:
{"aclConnected":false,"aclEncrypted":false,"bondedFromLocal":true,"deviceAddr"
:"10:B1:F8:0E:B4:7D","deviceAlias":{"present":true},"deviceBatteryLevel":-1,"d
eviceClass":{"present":true},"deviceName":{"present":false},"deviceType":1,"de
viceUuids":[],"messagePermission":0,"pairState":0,"phonebookPermission":0}
    09-12 22:49:27.658 16088-16088/com.waylau.hmos.bluetoothhost I
00001/MainAbilitySlice: end getInfo, text:
{"aclConnected":false,"aclEncrypted":false,"bondedFromLocal":true,"deviceAddr"
:"0C:8F:FF:FD:49:34","deviceAlias":{"present":true},"deviceBatteryLevel":-1,"d
eviceClass":{"present":true},"deviceName":{"present":true},"deviceType":1,"dev
iceUuids":[],"messagePermission":0,"pairState":0,"phonebookPermission":0}
    09-12 22:49:27.890 16088-16088/com.waylau.hmos.bluetoothhost I
00001/MainAbilitySlice: end getInfo, text:
{"aclConnected":false,"aclEncrypted":false,"bondedFromLocal":true,"deviceAddr"
:"00:15:83:EB:61:F8","deviceAlias":{"present":true},"deviceBatteryLevel":-1,"d
eviceClass":{"present":true},"deviceName":{"present":true},"deviceType":1,"dev
iceUuids":[],"messagePermission":0,"pairState":0,"phonebookPermission":0}
    09-12 22:49:28.045 16088-16088/com.waylau.hmos.bluetoothhost I
00001/MainAbilitySlice: end getInfo, text:
{"aclConnected":false,"aclEncrypted":false,"bondedFromLocal":true,"deviceAddr"
:"0C:8F:FF:75:3F:CF","deviceAlias":{"present":true},"deviceBatteryLevel":-1,"d
eviceClass":{"present":true},"deviceName":{"present":true},"deviceType":1,"dev
iceUuids":[],"messagePermission":0,"pairState":0,"phonebookPermission":0}
    09-12 22:49:28.272 16088-16088/com.waylau.hmos.bluetoothhost I
```

```
00001/MainAbilitySlice: end getInfo, text:
{"aclConnected":false,"aclEncrypted":false,"bondedFromLocal":true,"deviceAddr"
:"C3:DB:4A:E2:A3:4B","deviceAlias":{"present":true},"deviceBatteryLevel":-1,"d
eviceClass":{"present":true},"deviceName":{"present":true},"deviceType":2,"dev
iceUuids":[],"messagePermission":0,"pairState":0,"phonebookPermission":0}
    09-12 22:49:28.366 16088-16088/com.waylau.hmos.bluetoothhost I
00001/MainAbilitySlice: end getInfo, text:
{"aclConnected":false,"aclEncrypted":false,"bondedFromLocal":true,"deviceAddr"
:"70:8A:09:15:83:9D","deviceAlias":{"present":true},"deviceBatteryLevel":-1,"d
eviceClass":{"present":true},"deviceName":{"present":true},"deviceType":2,"dev
iceUuids":[],"messagePermission":0,"pairState":0,"phonebookPermission":0}
    09-12 22:49:28.426 16088-16088/com.waylau.hmos.bluetoothhost I
00001/MainAbilitySlice: end getInfo, text:
{"aclConnected":false,"aclEncrypted":false,"bondedFromLocal":true,"deviceAddr"
:"E6:13:8B:23:B1:6E","deviceAlias":{"present":true},"deviceBatteryLevel":-1,"d
eviceClass":{"present":true},"deviceName":{"present":true},"deviceType":2,"dev
iceUuids":[],"messagePermission":0,"pairState":0,"phonebookPermission":0}
    09-12 22:49:28.489 16088-16088/com.waylau.hmos.bluetoothhost I
00001/MainAbilitySlice: end getInfo, text:
{"aclConnected":false,"aclEncrypted":false,"bondedFromLocal":true,"deviceAddr"
:"20:17:06:22:02:78","deviceAlias":{"present":true},"deviceBatteryLevel":-1,"d
eviceClass":{"present":true},"deviceName":{"present":false},"deviceType":2,"de
viceUuids":[],"messagePermission":0,"pairState":0,"phonebookPermission":0}
    09-12 22:49:28.563 16088-16088/com.waylau.hmos.bluetoothhost I
00001/MainAbilitySlice: end getInfo, text:
{"aclConnected":false,"aclEncrypted":false,"bondedFromLocal":true,"deviceAddr"
:"43:F1:F2:3B:3B:04","deviceAlias":{"present":true},"deviceBatteryLevel":-1,"d
eviceClass":{"present":true},"deviceName":{"present":false},"deviceType":2,"de
viceUuids":[],"messagePermission":0,"pairState":0,"phonebookPermission":0}
    09-12 22:49:28.640 16088-16088/com.waylau.hmos.bluetoothhost I
00001/MainAbilitySlice: end getInfo, text:
{"aclConnected":false,"aclEncrypted":false,"bondedFromLocal":true,"deviceAddr"
:"40:96:54:76:07:BB","deviceAlias":{"present":true},"deviceBatteryLevel":-1,"d
eviceClass":{"present":true},"deviceName":{"present":false},"deviceType":2,"de
viceUuids":[],"messagePermission":0,"pairState":0,"phonebookPermission":0}
    09-12 22:49:28.785 16088-16088/com.waylau.hmos.bluetoothhost I
00001/MainAbilitySlice: end getInfo, text:
{"aclConnected":false,"aclEncrypted":false,"bondedFromLocal":true,"deviceAddr"
:"30:74:96:55:F5:E3","deviceAlias":{"present":true},"deviceBatteryLevel":-1,"d
eviceClass":{"present":true},"deviceName":{"present":true},"deviceType":2,"dev
iceUuids":[],"messagePermission":0,"pairState":0,"phonebookPermission":0}
    09-12 22:49:28.889 16088-16088/com.waylau.hmos.bluetoothhost I
00001/MainAbilitySlice: end getInfo, text:
{"aclConnected":false,"aclEncrypted":false,"bondedFromLocal":true,"deviceAddr"
:"20:17:06:22:02:78","deviceAlias":{"present":true},"deviceBatteryLevel":-1,"d
eviceClass":{"present":true},"deviceName":{"present":false},"deviceType":2,"de
viceUuids":[],"messagePermission":0,"pairState":0,"phonebookPermission":0}
    09-12 22:49:28.948 16088-16088/com.waylau.hmos.bluetoothhost I
00001/MainAbilitySlice: end getInfo, text:
{"aclConnected":false,"aclEncrypted":false,"bondedFromLocal":true,"deviceAddr"
```

```
:"43:F1:F2:3B:3B:04","deviceAlias":{"present":true},"deviceBatteryLevel":-1,"d
eviceClass":{"present":true},"deviceName":{"present":false},"deviceType":2,"de
viceUuids":[],"messagePermission":0,"pairState":0,"phonebookPermission":0}
    09-12 22:49:29.004 16088-16088/com.waylau.hmos.bluetoothhost I
00001/MainAbilitySlice: end getInfo, text:
{"aclConnected":false,"aclEncrypted":false,"bondedFromLocal":true,"deviceAddr"
:"40:96:54:76:07:BB","deviceAlias":{"present":true},"deviceBatteryLevel":-1,"d
eviceClass":{"present":true},"deviceName":{"present":false},"deviceType":2,"de
viceUuids":[],"messagePermission":0,"pairState":0,"phonebookPermission":0}
```

界面效果如图22-2所示。

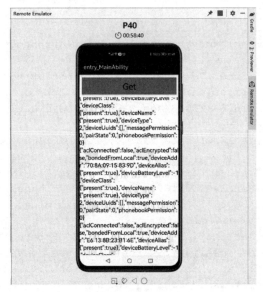

图 22-2　界面效果

22.3　实战：传统蓝牙远端设备操作

传统蓝牙远端管理操作主要是针对远端蓝牙设备的基本操作，包括获取远端蓝牙设备地址、类型、名称和配对状态，以及向远端设备发起配对。

本节演示如何实现打开蓝牙，扫描周边蓝牙设备，并向远端设备发起配对。

22.3.1　接口说明

蓝牙远端设备管理类BluetoothRemoteDevice的主要接口有：

- getDeviceAddr()：获取远端蓝牙的设备地址。
- getDeviceClass()：获取远端蓝牙的设备类型。
- getDeviceName()：获取远端蓝牙的设备名称。

- getPairState()：获取远端设备的配对状态。
- startPair()：向远端设备发起配对。

22.3.2 创建应用

为了演示BluetoothRemoteDevice的功能，创建一个名为BluetoothRemoteDevice的应用。在应用的界面上，通过点击按钮来触发BluetoothRemoteDevice的操作。

22.3.3 声明权限

修改配置文件，声明使用蓝牙相关的权限如下：

```
// 声明权限
"reqPermissions": [
   {
   "name": "ohos.permission.USE_BLUETOOTH"
   },
   {
   "name": "ohos.permission.DISCOVER_BLUETOOTH"
   },
   {
   "name": "ohos.permission.LOCATION"
   }
]
```

同时，在应用启动时，显式声明ohos.permission.LOCATION权限，代码如下：

```
package com.waylau.hmos.bluetoothremotedevice;

import com.waylau.hmos.bluetoothremotedevice.slice.MainAbilitySlice;
import ohos.aafwk.ability.Ability;
import ohos.aafwk.content.Intent;

import java.util.ArrayList;
import java.util.List;

public class MainAbility extends Ability {
    @Override
    public void onStart(Intent intent) {
        super.onStart(intent);
        super.setMainRoute(MainAbilitySlice.class.getName());

        // 显式声明需要使用的权限
        requestPermission();
    }

    // 显式声明需要使用的权限
    private void requestPermission() {
```

```
            String[] permission = {
                "ohos.permission.LOCATION"};
            List<String> applyPermissions = new ArrayList<>();
            for (String element : permission) {
                if (verifySelfPermission(element) != 0) {
                    if (canRequestPermission(element)) {
                        applyPermissions.add(element);
                    }
                }
            }
            requestPermissionsFromUser(applyPermissions.toArray(new String[0]),
0);
    }
}
```

由于ohos.permission.USE_BLUETOOTH和ohos.permission.DISCOVER_BLUETOOTH不是敏感权限，因此不需要在MainAbility中显式声明。

22.3.4 修改 ability_main.xml

修改ability_main.xml内容如下：

```
<?xml version="1.0" encoding="utf-8"?>
<DirectionalLayout
    xmlns:ohos="http://schemas.huawei.com/res/ohos"
    ohos:height="match_parent"
    ohos:width="match_parent"
    ohos:orientation="vertical">

    <Button
        ohos:id="$+id:button_get"
        ohos:height="match_content"
        ohos:width="match_parent"
        ohos:background_element="#F76543"
        ohos:layout_alignment="horizontal_center"
        ohos:margin="10vp"
        ohos:padding="10vp"
        ohos:text="Get"
        ohos:text_size="40vp"
    />

    <Button
        ohos:id="$+id:button_pair"
        ohos:height="match_content"
        ohos:width="match_parent"
        ohos:background_element="#F76543"
        ohos:layout_alignment="horizontal_center"
        ohos:margin="10vp"
        ohos:padding="10vp"
```

```
            ohos:text="Pair"
            ohos:text_size="40vp"
            />

        <Text
            ohos:id="$+id:text"
            ohos:height="match_content"
            ohos:width="match_content"
            ohos:background_element="$graphic:background_ability_main"
            ohos:layout_alignment="horizontal_center"
            ohos:multiple_lines="true"
            ohos:text="Hello World"
            ohos:text_size="40vp"
            />

</DirectionalLayout>
```

界面预览效果如图22-3所示。

图22-3　界面预览效果

上述代码中：

- 设置了Get按钮，以备设置点击事件，以触发蓝牙扫描相关的操作。
- 设置了Pair按钮，以备设置点击事件，以触发蓝牙配对的操作。
- Text组件用于展示读取到的蓝牙信息。

22.3.5　修改 MainAbilitySlice

修改MainAbilitySlice内容如下：

```
package com.waylau.hmos.bluetoothremotedevice.slice;
```

```java
import com.waylau.hmos.bluetoothremotedevice.ResourceTable;
import ohos.aafwk.ability.AbilitySlice;
import ohos.aafwk.content.Intent;
import ohos.aafwk.content.IntentParams;
import ohos.agp.components.Button;
import ohos.agp.components.Text;
import ohos.bluetooth.BluetoothHost;
import ohos.bluetooth.BluetoothRemoteDevice;
import ohos.event.commonevent.*;
import ohos.hiviewdfx.HiLog;
import ohos.hiviewdfx.HiLogLabel;
import ohos.rpc.RemoteException;

import java.util.Optional;


public class MainAbilitySlice extends AbilitySlice {
    private static final String TAG = MainAbilitySlice.class.getSimpleName();
    private static final HiLogLabel LABEL_LOG =
            new HiLogLabel(HiLog.LOG_APP, 0x00001, TAG);

    private BluetoothHost bluetoothHost;

    private String selectedDeviceAddr;

    private Text text;

    @Override
    public void onStart(Intent intent) {
        super.onStart(intent);
        super.setUIContent(ResourceTable.Layout_ability_main);

        // 初始化蓝牙
        initBluetooth();

        // 为按钮设置点击事件回调
        Button buttonGet =
                (Button) findComponentById(ResourceTable.Id_button_get);
        buttonGet.setClickedListener(listener -> getInfo());

        Button buttonPair =
                (Button) findComponentById(ResourceTable.Id_button_pair);
        buttonPair.setClickedListener(listener -> pair());

        text =
                (Text) findComponentById(ResourceTable.Id_text);
    }

    private void pair() {
        HiLog.info(LABEL_LOG, "before pair device addr: %{public}s",
```

```java
                    selectedDeviceAddr);

            // 配对
            BluetoothRemoteDevice device =
bluetoothHost.getRemoteDev(selectedDeviceAddr);
            boolean result = device.startPair();

            HiLog.info(LABEL_LOG, "end pair device addr: %{public}s,
result: %{public}s",
                    selectedDeviceAddr, result);
        }

        private void initBluetooth() {
            HiLog.info(LABEL_LOG, "before initBluetooth");

            // 获取蓝牙本机管理对象
            bluetoothHost = BluetoothHost.getDefaultHost(this);

            // 调用打开接口
            bluetoothHost.enableBt();

            // 获取本机蓝牙名称
            Optional<String> nameOptional = bluetoothHost.getLocalName();

            // 调用获取蓝牙开关状态接口
            int state = bluetoothHost.getBtState();

            HiLog.info(LABEL_LOG, "end initBluetooth, name: %{public}s,
state: %{public}s",
                    nameOptional.get(), state);
        }

        private void getInfo() {
            HiLog.info(LABEL_LOG, "before getInfo bluetooth name: %{public}s,
state: %{public}s",
                    bluetoothHost.getLocalName().get(),
bluetoothHost.getBtState());

            // 注册广播BluetoothRemoteDevice.EVENT_DEVICE_DISCOVERED
            MatchingSkills matchingSkills = new MatchingSkills();
matchingSkills.addEvent(BluetoothRemoteDevice.EVENT_DEVICE_DISCOVERED); // 自定
义事件
            CommonEventSubscribeInfo subscribeInfo = new
CommonEventSubscribeInfo(matchingSkills);
            MyCommonEventSubscriber subscriber = new
MyCommonEventSubscriber(subscribeInfo);

            try {
                CommonEventManager.subscribeCommonEvent(subscriber);
```

```java
            } catch (RemoteException e) {
                HiLog.info(LABEL_LOG, "subscribeCommonEvent occur exception.");
            }

            // 开始扫描
            bluetoothHost.startBtDiscovery();

            HiLog.info(LABEL_LOG, "end getInfo");
        }

        // 接收系统广播
        class MyCommonEventSubscriber extends CommonEventSubscriber {
            public MyCommonEventSubscriber(CommonEventSubscribeInfo subscribeInfo) {
                super(subscribeInfo);
            }

            @Override
            public void onReceiveEvent(CommonEventData var) {
                Intent info = var.getIntent();
                if (info == null) {
                    return;
                }
                //获取系统广播的action
                String action = info.getAction();

                //判断是否为扫描到设备的广播
                if (action == BluetoothRemoteDevice.EVENT_DEVICE_DISCOVERED) {
                    IntentParams myParam = info.getParams();
                    BluetoothRemoteDevice device =
                            (BluetoothRemoteDevice)
myParam.getParam(BluetoothRemoteDevice.REMOTE_DEVICE_PARAM_DEVICE);

                    // 获取远端蓝牙设备地址
                    String deviceAddr = device.getDeviceAddr();

                    // 获取远端蓝牙设备名称
                    Optional<String> deviceNameOptional = device.getDeviceName();
                    String deviceName = deviceNameOptional.orElse("");

                    // 获取远端设备配对状态
                    int pairState = device.getPairState();

                    HiLog.info(LABEL_LOG, "Remote Device, deviceAddr: %{public}s, deviceName: %{public}s, pairState: %{public}s",
                            deviceAddr, deviceName, pairState);

                    // 0为可以配对,选为待配对的对象
                    if (pairState == 0) {
                        selectedDeviceAddr = deviceAddr;
```

```
                text.setText(selectedDeviceAddr);
            }
        }
    }
}

@Override
public void onActive() {
    super.onActive();
}

@Override
public void onForeground(Intent intent) {
    super.onForeground(intent);
}
}
```

上述代码中：

- initBluetooth方法用于初始化蓝牙，包括获取蓝牙本机管理对象、启动蓝牙，并获取蓝牙的名称和状态信息。
- 在Button上设置了点击事件。
- getInfo方法注册广播BluetoothRemoteDevice.EVENT_DEVICE_DISCOVERED事件，并启用蓝牙扫描。
- 接收到事件后，将待配对的蓝牙设备信息展示在界面的Text上。
- pair方法用于针对特定的蓝牙设备进行配对。

22.3.6 运行

初次运行应用后，点击界面上的Get按钮以触发操作的执行。此时，控制台输出内容如下：

```
    09-12 22:54:32.576 17662-17662/com.waylau.hmos.bluetoothremotedevice I
00001/MainAbilitySlice: before getInfo bluetooth name: HUAWEI P40, state: 2
    09-12 22:54:32.600 17662-17662/com.waylau.hmos.bluetoothremotedevice I
00001/MainAbilitySlice: end getInfo
    09-12 22:54:32.722 17662-17662/com.waylau.hmos.bluetoothremotedevice I
00001/MainAbilitySlice: Remote Device, deviceAddr: B0:55:08:14:9D:7B,
deviceName: Honor V10, pairState: 0
    09-12 22:54:32.756 17662-17662/com.waylau.hmos.bluetoothremotedevice I
00001/MainAbilitySlice: Remote Device, deviceAddr: 20:AB:37:60:31:86,
deviceName: "Administrator"的 iPhone, pairState: 0
    09-12 22:54:32.767 17662-17662/com.waylau.hmos.bluetoothremotedevice I
00001/MainAbilitySlice: Remote Device, deviceAddr: 50:04:B8:C1:81:F8,
deviceName: HISI P10 PLUS, pairState: 0
    09-12 22:54:32.772 17662-17662/com.waylau.hmos.bluetoothremotedevice I
00001/MainAbilitySlice: Remote Device, deviceAddr: 1C:15:1F:8B:80:0E,
deviceName: , pairState: 0
    09-12 22:54:32.836 17662-17662/com.waylau.hmos.bluetoothremotedevice I
```

```
00001/MainAbilitySlice: Remote Device, deviceAddr: 50:04:B8:C1:82:CC,
deviceName: , pairState: 0
    09-12 22:54:32.863 17662-17662/com.waylau.hmos.bluetoothremotedevice I
00001/MainAbilitySlice: Remote Device, deviceAddr: 10:B1:F8:0E:B4:7D,
deviceName: , pairState: 0
    09-12 22:54:32.876 17662-17662/com.waylau.hmos.bluetoothremotedevice I
00001/MainAbilitySlice: Remote Device, deviceAddr: 0C:8F:FF:FD:49:34,
deviceName: LOVER LIAN, pairState: 0
    09-12 22:54:32.881 17662-17662/com.waylau.hmos.bluetoothremotedevice I
00001/MainAbilitySlice: Remote Device, deviceAddr: 00:15:83:EB:61:F8,
deviceName: ABC123456-0, pairState: 0
    09-12 22:54:32.897 17662-17662/com.waylau.hmos.bluetoothremotedevice I
00001/MainAbilitySlice: Remote Device, deviceAddr: 0C:8F:FF:75:3F:CF,
deviceName: HUAWEI Mate 10 Pro, pairState: 0
    09-12 22:54:32.918 17662-17662/com.waylau.hmos.bluetoothremotedevice I
00001/MainAbilitySlice: Remote Device, deviceAddr: C3:DB:4A:E2:A3:4B,
deviceName: MX Anywhere 2S, pairState: 0
    09-12 22:54:33.003 17662-17662/com.waylau.hmos.bluetoothremotedevice I
00001/MainAbilitySlice: Remote Device, deviceAddr: 70:8A:09:15:83:9D,
deviceName: honor Band 3-39d, pairState: 0
    09-12 22:54:33.009 17662-17662/com.waylau.hmos.bluetoothremotedevice I
00001/MainAbilitySlice: Remote Device, deviceAddr: E6:13:8B:23:B1:6E,
deviceName: MI Band 2, pairState: 0
    09-12 22:54:33.029 17662-17662/com.waylau.hmos.bluetoothremotedevice I
00001/MainAbilitySlice: Remote Device, deviceAddr: 20:17:06:22:02:78,
deviceName: , pairState: 0
    09-12 22:54:33.034 17662-17662/com.waylau.hmos.bluetoothremotedevice I
00001/MainAbilitySlice: Remote Device, deviceAddr: 43:F1:F2:3B:3B:04,
deviceName: , pairState: 0
    09-12 22:54:33.087 17662-17662/com.waylau.hmos.bluetoothremotedevice I
00001/MainAbilitySlice: Remote Device, deviceAddr: 40:96:54:76:07:BB,
deviceName: , pairState: 0
    09-12 22:54:33.120 17662-17662/com.waylau.hmos.bluetoothremotedevice I
00001/MainAbilitySlice: Remote Device, deviceAddr: 30:74:96:55:F5:E3,
deviceName: honor Band 3-5e3, pairState: 0
    09-12 22:54:34.688 17662-17662/com.waylau.hmos.bluetoothremotedevice I
00001/MainAbilitySlice: Remote Device, deviceAddr: 20:17:06:22:02:78,
deviceName: , pairState: 0
    09-12 22:54:34.696 17662-17662/com.waylau.hmos.bluetoothremotedevice I
00001/MainAbilitySlice: Remote Device, deviceAddr: 43:F1:F2:3B:3B:04,
deviceName: , pairState: 0
    09-12 22:54:34.701 17662-17662/com.waylau.hmos.bluetoothremotedevice I
00001/MainAbilitySlice: Remote Device, deviceAddr: 40:96:54:76:07:BB,
deviceName: , pairState: 0
```

界面效果如图22-4所示。

图 22-4　界面效果

点击界面上的Pair按钮以触发配对的操作。此时，控制台输出内容如下：

```
09-12 22:55:02.314 17662-17662/com.waylau.hmos.bluetoothremotedevice I
00001/MainAbilitySlice: before pair device addr: 40:96:54:76:07:BB
09-12 22:55:02.339 17662-17662/com.waylau.hmos.bluetoothremotedevice I
00001/MainAbilitySlice: end pair device addr: 40:96:54:76:07:BB, result: true
```

上述日志说明配对成功。

22.4　实战：BLE 扫描和广播

通过BLE扫描和广播提供的开放能力可以根据指定状态获取外围设备，启动或停止BLE扫描、广播。

本节演示如何实现启动BLE扫描、广播。

22.4.1　接口说明

1. BleCentralManager 的主要接口

BLE中心设备管理类BleCentralManager的主要接口有：

- startScan(List<BleScanFilter> filters)：进行BLE蓝牙扫描，并使用filters对结果进行过滤。
- stopScan()：停止BLE蓝牙扫描。
- getDevicesByStates(int[] states)：根据状态获取连接的外围设备。
- BleCentralManager(BleCentralManagerCallback callback)：获取中心设备管理对象。

2. BleCentralManagerCallback 的主要接口

中心设备管理回调类BleCentralManagerCallback的主要接口有：

- scanResultEvent(BleScanResult result)：扫描到BLE设备的结果回调。
- scanFailedEvent(int resultCode)：启动扫描失败的回调。

3. BleAdvertiser 和 BleAdvertiseCallback 的主要接口

BLE广播相关的BleAdvertiser类和BleAdvertiseCallback类的主要接口有：

- BleAdvertiser(Context context, BleAdvertiseCallback callback)：用于获取广播操作对象。
- startAdvertising(BleAdvertiseSettings settings, BleAdvertiseData advData, BleAdvertiseData scanResponse)：进行BLE广播，第一个参数为广播参数，第二个参数为广播数据，第三个参数是扫描和广播数据参数的响应。
- stopAdvertising()：停止BLE广播。
- startResultEvent(int result)：广播回调结果。

22.4.2 创建应用

为了演示BLE的功能，创建一个名为BleCentralManager的应用。
在应用的界面上，通过点击按钮来触发BLE的操作。

22.4.3 声明权限

修改配置文件，声明使用蓝牙相关的权限如下：

```
// 声明权限
"reqPermissions": [
    {
    "name": "ohos.permission.USE_BLUETOOTH"
    },
    {
    "name": "ohos.permission.DISCOVER_BLUETOOTH"
    },
    {
    "name": "ohos.permission.LOCATION"
    }
]
```

同时，在应用启动时，显式声明ohos.permission.LOCATION权限，代码如下：

```
package com.waylau.hmos.blecentralmanager;


import com.waylau.hmos.blecentralmanager.slice.MainAbilitySlice;
import ohos.aafwk.ability.Ability;
import ohos.aafwk.content.Intent;
```

```java
import java.util.ArrayList;
import java.util.List;

public class MainAbility extends Ability {
    @Override
    public void onStart(Intent intent) {
        super.onStart(intent);
        super.setMainRoute(MainAbilitySlice.class.getName());

        // 显式声明需要使用的权限
        requestPermission();
    }

    // 显式声明需要使用的权限
    private void requestPermission() {
        String[] permission = {
            "ohos.permission.LOCATION"};
        List<String> applyPermissions = new ArrayList<>();
        for (String element : permission) {
            if (verifySelfPermission(element) != 0) {
                if (canRequestPermission(element)) {
                    applyPermissions.add(element);
                }
            }
        }
        requestPermissionsFromUser(applyPermissions.toArray(new String[0]), 0);
    }
}
```

由于ohos.permission.USE_BLUETOOTH和ohos.permission.DISCOVER_BLUETOOTH不是敏感权限，因此不需要在MainAbility中显式声明。

22.4.4 修改 ability_main.xml

修改ability_main.xml内容如下：

```xml
<?xml version="1.0" encoding="utf-8"?>
<DirectionalLayout
    xmlns:ohos="http://schemas.huawei.com/res/ohos"
    ohos:height="match_parent"
    ohos:width="match_parent"
    ohos:orientation="vertical">

    <Button
        ohos:id="$+id:button_scan"
        ohos:height="40vp"
        ohos:width="match_parent"
```

```xml
        ohos:background_element="#F76543"
        ohos:layout_alignment="horizontal_center"
        ohos:margin="10vp"
        ohos:padding="10vp"
        ohos:text="Scan"
        ohos:text_size="50"
        />

    <Button
        ohos:id="$+id:button_advertise"
        ohos:height="40vp"
        ohos:width="match_parent"
        ohos:background_element="#F76543"
        ohos:layout_alignment="horizontal_center"
        ohos:margin="10vp"
        ohos:padding="10vp"
        ohos:text="Advertise"
        ohos:text_size="50"
        />

    <Text
        ohos:id="$+id:text"
        ohos:height="match_content"
        ohos:width="match_content"
        ohos:background_element="$graphic:background_ability_main"
        ohos:layout_alignment="horizontal_center"
        ohos:multiple_lines="true"
        ohos:text=""
        ohos:text_size="50"
        />
</DirectionalLayout>
```

界面预览效果如图22-5所示。

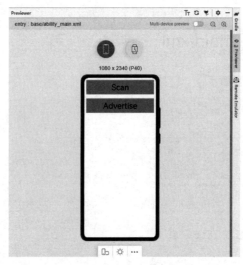

图 22-5　界面预览效果

上述代码中:

- 设置了Scan按钮,以备设置点击事件,以触发BLE扫描相关的操作。
- 设置了Advertise按钮,以备设置点击事件,以触发BLE广播的操作。
- Text组件用于展示读取到的蓝牙信息。

22.4.5 修改 MainAbilitySlice

修改MainAbilitySlice内容如下:

```java
package com.waylau.hmos.blecentralmanager.slice;

import com.waylau.hmos.blecentralmanager.ResourceTable;
import ohos.aafwk.ability.AbilitySlice;
import ohos.aafwk.content.Intent;
import ohos.agp.components.Button;
import ohos.agp.components.Text;
import ohos.bluetooth.ble.*;
import ohos.hiviewdfx.HiLog;
import ohos.hiviewdfx.HiLogLabel;
import ohos.utils.SequenceUuid;

import java.util.ArrayList;
import java.util.List;
import java.util.Optional;
import java.util.UUID;


public class MainAbilitySlice extends AbilitySlice {
    private static final String TAG = MainAbilitySlice.class.getSimpleName();
    private static final HiLogLabel LABEL_LOG =
            new HiLogLabel(HiLog.LOG_APP, 0x00001, TAG);

    private static final UUID SERVER_UUID = UUID.randomUUID();
    private static final int[] STATE_ARRAY = {
                BlePeripheralDevice.CONNECTION_PRIORITY_NORMAL,
                BlePeripheralDevice.CONNECTION_PRIORITY_HIGH,
                BlePeripheralDevice.CONNECTION_PRIORITY_LOW};

    // 获取中心设备管理对象
    private BleCentralManager centralManager;

    // 获取BLE广播对象
    private BleAdvertiser advertiser;
    // 创建BLE广播参数和数据
    private BleAdvertiseData data;
    private BleAdvertiseSettings advertiseSettings;

    private Text text;
```

```java
        @Override
        public void onStart(Intent intent) {
            super.onStart(intent);
            super.setUIContent(ResourceTable.Layout_ability_main);

            // 初始化蓝牙
            initBluetooth();

            // 为按钮设置点击事件回调
            Button buttonGet =
                    (Button) findComponentById(ResourceTable.Id_button_scan);
            buttonGet.setClickedListener(listener -> scan());

            Button buttonPair =
                    (Button) findComponentById(ResourceTable.Id_button_advertise);
            buttonPair.setClickedListener(listener -> advertise());

            text =
                    (Text) findComponentById(ResourceTable.Id_text);
        }

        private void initBluetooth() {
            HiLog.info(LABEL_LOG, "before initBluetooth");

            // 获取中心设备管理对象
            ScanCallback centralManagerCallback = new ScanCallback();
            centralManager = new BleCentralManager(this, centralManagerCallback);

            // 创建扫描过滤器，然后开始扫描
            List<BleScanFilter> filters = new ArrayList<>();
            centralManager.startScan(filters);

            // 获取BLE广播对象
            advertiser = new BleAdvertiser(this, advertiseCallback);

            // 创建BLE广播参数和数据
            data = new BleAdvertiseData.Builder()
                    .addServiceUuid(SequenceUuid.uuidFromString(SERVER_UUID.toString()))    // 添加服务的UUID
                    .addServiceData(SequenceUuid.uuidFromString(SERVER_UUID.toString()), new byte[]{0x11})    // 添加广播数据内容
                    .build();
            advertiseSettings = new BleAdvertiseSettings.Builder()
                    .setConnectable(true)  // 设置是否可连接广播
                    .setInterval(BleAdvertiseSettings.INTERVAL_SLOT_DEFAULT)// 设置广播间隔
                    .setTxPower(BleAdvertiseSettings.TX_POWER_DEFAULT)//设置广播功率
                    .build();
```

第 22 章 蓝牙 | 509

```
            HiLog.info(LABEL_LOG, "end initBluetooth");
        }

        private void scan() {
            HiLog.info(LABEL_LOG, "before scan");
            List<BlePeripheralDevice> devices =
centralManager.getDevicesByStates(STATE_ARRAY);

            for (BlePeripheralDevice device : devices) {
                Optional<String> deviceNameOptional = device.getDeviceName();
                String deviceName = deviceNameOptional.orElse("");
                String deviceAddr = device.getDeviceAddr();

                HiLog.info(LABEL_LOG, "scan deviceName: %{public}s,
deviceAddr: %{public}s",
                        deviceName, deviceAddr);

                text.append(deviceName);
                text.append(deviceAddr);
            }
            HiLog.info(LABEL_LOG, "end scan");
        }

        private void advertise() {
            HiLog.info(LABEL_LOG, "before advertise");

            // 开始广播
            advertiser.startAdvertising(advertiseSettings, data, null);

            HiLog.info(LABEL_LOG, "end advertise");
        }

        // 实现扫描回调
        private class ScanCallback implements BleCentralManagerCallback {
            List<BleScanResult> results = new ArrayList<BleScanResult>();

            @Override
            public void scanResultEvent(BleScanResult var1) {
                HiLog.info(LABEL_LOG, "scanResultEvent");

                // 对扫描结果进行处理
                results.add(var1);
            }

            @Override
            public void scanFailedEvent(int var1) {
                HiLog.info(LABEL_LOG, "Start Scan failed,Code:" + var1);
            }

            @Override
```

```java
            public void groupScanResultsEvent(List<BleScanResult> list) {
                HiLog.info(LABEL_LOG, "groupScanResultsEvent");
            }
        }

        // 实现BLE广播回调
        private BleAdvertiseCallback advertiseCallback = new
BleAdvertiseCallback() {
            @Override
            public void startResultEvent(int result) {
                if (result == BleAdvertiseCallback.RESULT_SUCC) {
                    // 开始BLE广播成功
                    HiLog.info(LABEL_LOG, "startResultEvent success");
                } else {
                    // 开始BLE广播失败
                    HiLog.error(LABEL_LOG, "startResultEvent failed");
                }
            }
        };

        @Override
        public void onActive() {
            super.onActive();
        }

        @Override
        public void onForeground(Intent intent) {
            super.onForeground(intent);
        }
    }
```

上述代码中:

- initBluetooth方法用于初始化，包括初始化中心设备管理对象、BLE广播对象、创建BLE广播参数和数据，并开始扫描。
- 在Button上设置了点击事件。
- scan启用BLE扫描。
- advertise方法启动BLE广播。

22.4.6 运行

初次运行应用后，点击界面上的Scan按钮以触发BLE扫描操作的执行。此时，控制台输出内容如下：

```
    02-13 17:37:15.186 27713-27713/? I 00001/MainAbilitySlice: before
initBluetooth
    02-13 17:37:15.195 27713-27713/? I 00001/MainAbilitySlice: end initBluetooth
    02-13 17:37:22.438 27713-27713/com.waylau.hmos.blecentralmanager I
```

```
00001/MainAbilitySlice: before scan
    02-13 17:37:22.441 27713-27713/com.waylau.hmos.blecentralmanager I
00001/MainAbilitySlice: end scan
    02-13 17:37:26.727 27713-27713/com.waylau.hmos.blecentralmanager I 00001/
```

点击界面上的Advertise按钮以触发BLE广播操作的执行。此时，控制台输出内容如下：

```
MainAbilitySlice: before advertise
    02-13 17:37:26.730 27713-27713/com.waylau.hmos.blecentralmanager E
00001/MainAbilitySlice: startResultEvent failed
    02-13 17:37:26.730 27713-27713/com.waylau.hmos.blecentralmanager I
00001/MainAbilitySlice: end advertise
```

第 23 章

WLAN

HarmonyOS WLAN（Wireless Local Area Networks，无线局域网）服务系统为用户提供WLAN基础功能、P2P功能和WLAN消息通知的相应服务，让应用可以通过WLAN和其他设备互联互通。

23.1 WLAN 概述

WLAN是通过无线电、红外光信号或者其他技术发送和接收数据的局域网，用户可以通过WLAN实现结点之间无物理连接的网络通信，常用于用户携带可移动终端的办公、公众环境中。

HarmonyOS WLAN服务系统为用户提供WLAN基础功能、P2P（Peer-to-Peer）功能和WLAN消息通知的相应服务，让应用可以通过WLAN和其他设备互联互通。

23.1.1 WLAN 简介

在WLAN发明之前，人们要想通过网络进行联络和通信，必须先用物理线缆组建一个电子运行的通路，为了提高效率和速度，后来又发明了光纤。当网络发展到一定规模后，人们又发现，这种有线网络无论组建、拆装还是在原有基础上进行重新布局和改建都非常困难，且成本和代价也非常高，于是WLAN的组网方式应运而生。

WLAN起步于1997年。当年的6月，第一个无线局域网标准IEEE802.11正式颁布实施，为无线局域网技术提供了统一的标准，但当时的传输速率只有1~2Mbps。随后，IEEE委员会又开始制定新的WLAN标准，分别取名为IEEE802.11a和IEEE802.11b。IEEE802.llb标准首先于1999年9月正式颁布，其速率为11Mbps。经过改进的IEEE802.11a标准在2001年年底才正式颁布，它的传输速率可达到54Mbps，几乎是IEEE802.llb标准的5倍。尽管如此，WLAN的应用并未真正开始，因为整个WLAN应用环境并不成熟。

WLAN的真正发展是从2003年3月Intel第一次推出带有WLAN无线网卡芯片模块的迅驰处理器开始的。尽管当时的无线网络环境还非常不成熟，最为发达的美国也不例外。但是由于Intel的捆绑销售，加上迅驰芯片的高性能、低功耗等非常明显的优点，使得许多无线网络服务商看到了商机，同时11Mbit/s的接入速率在一般的小型局域网也可以进行一些日常应用，于是各国的无线网络服务商开始在公共场所（如机场、宾馆、咖啡厅等）提供访问热点，实际上就是布置一些无线访问点（Access Point，AP），方便移动商务人士无线上网。

经过两年多的发展，基于IEEE802.11b标准的无线网络产品和应用已相当成熟，但毕竟11Mbps的接入速率还远远不能满足实际网络的应用需求。

在2003年6月，经过两年多的开发和多次改进，一种兼容原来的IEEE802.11b标准，同时也可提供54 Mbit/s接入速率的新标准——IEEE802.11g在IEEE委员会的努力下正式发布了。

目前使用最多的是802.11n（第4代）和802.11ac（第5代）标准，它们既可以工作在2.4 GHz频段，也可以工作在5GHz频段，传输速率可达600Mbps（理论值）。

23.1.2 约束与限制

HarmonyOS WLAN服务系统提供多个开发场景的指导，涉及多个API接口的调用。在调用API前，应用需要先申请对应的访问权限。

不同应用场景需要申请的权限也不同，具体涉及的权限包括：

- ohos.permission.GET_WIFI_INFO。
- ohos.permission.SET_WIFI_INFO。
- ohos.permission.LOCATION。

23.2 实战：WLAN 的基础功能

本节演示如何实现WLAN的基础功能，包括：

- 获取WLAN状态，查询WLAN是否打开。
- 发起扫描并获取扫描结果。
- 获取连接状态的详细信息，包括连接信息、IP信息等。
- 获取设备国家码。
- 获取设备是否支持指定的能力。

23.2.1 接口说明

WLAN基础功能由WifiDevice提供，其接口有：

- getInstance(Context context)：获取WLAN功能管理对象实例，通过该实例调用WLAN基本功能API。

- isWifiActive()：获取当前WLAN打开状态，需要ohos.permission.GET_WIFI_INFO权限。
- scan()：发起WLAN扫描，需要ohos.permission.SET_WIFI_INFO和 ohos.permission.LOCATION权限。
- getScanInfoList()：获取上次扫描的结果，需要ohos.permission.GET_WIFI_INFO和ohos.permission.LOCATION权限。
- isConnected()：获取当前WLAN的连接状态，需要ohos.permission.GET_WIFI_INFO权限。
- getLinkedInfo()：获取当前WLAN的连接信息，需要ohos.permission.GET_WIFI_INFO权限。
- getIpInfo()：获取当前连接的WLAN IP信息，需要ohos.permission.GET_WIFI_INFO权限。
- getSignalLevel(int rssi, int band)：通过RSSI与频段计算信号格数。
- getCountryCode()：获取设备的国家码，需要ohos.permission.LOCATION和ohos.permission.GET_WIFI_INFO权限。
- isFeatureSupported(long featureId)：获取设备是否支持指定的特性，需要ohos.permission.GET_WIFI_INFO权限。

23.2.2 创建应用

为了演示WifiDevice的功能，创建一个名为WifiDevice的应用。

在应用的界面上，通过点击按钮来触发WifiDevice的操作。

23.2.3 声明权限

修改配置文件，声明使用WLAN的权限如下：

```
// 声明权限
"reqPermissions": [
   {
   "name": "ohos.permission.SET_WIFI_INFO"
   },
   {
   "name": "ohos.permission.GET_WIFI_INFO"
   },
   {
   "name": "ohos.permission.LOCATION"
   }
]
```

同时，在应用启动时，显式声明ohos.permission.LOCATION权限，代码如下：

```
package com.waylau.hmos.wifidevice;

import com.waylau.hmos.wifidevice.slice.MainAbilitySlice;
import ohos.aafwk.ability.Ability;
import ohos.aafwk.content.Intent;


import java.util.ArrayList;
```

```java
import java.util.List;

public class MainAbility extends Ability {
    @Override
    public void onStart(Intent intent) {
        super.onStart(intent);
        super.setMainRoute(MainAbilitySlice.class.getName());

        // 显式声明需要使用的权限
        requestPermission();
    }

    // 显式声明需要使用的权限
    private void requestPermission() {
        String[] permission = {
                "ohos.permission.LOCATION"};
        List<String> applyPermissions = new ArrayList<>();
        for (String element : permission) {
            if (verifySelfPermission(element) != 0) {
                if (canRequestPermission(element)) {
                    applyPermissions.add(element);
                }
            }
        }
        requestPermissionsFromUser(applyPermissions.toArray(new String[0]), 0);
    }
}
```

由于ohos.permission.GET_WIFI_INFO和ohos.permission.SET_WIFI_INFO不是敏感权限，因此不需要在MainAbility中显式声明。

23.2.4 修改 ability_main.xml

修改ability_main.xml内容如下：

```xml
<?xml version="1.0" encoding="utf-8"?>
<DirectionalLayout
    xmlns:ohos="http://schemas.huawei.com/res/ohos"
    ohos:height="match_parent"
    ohos:width="match_parent"
    ohos:orientation="vertical">

    <Button
        ohos:id="$+id:button_get"
        ohos:height="match_content"
        ohos:width="match_parent"
        ohos:background_element="#F76543"
        ohos:layout_alignment="horizontal_center"
```

```
            ohos:margin="10vp"
            ohos:padding="10vp"
            ohos:text="Get"
            ohos:text_size="30vp"
            />

    <Text
        ohos:id="$+id:text"
        ohos:height="match_content"
        ohos:width="match_content"
        ohos:background_element="$graphic:background_ability_main"
        ohos:layout_alignment="horizontal_center"
        ohos:text=""
        ohos:text_size="20vp"
        ohos:multiple_lines="true"
        />

</DirectionalLayout>
```

界面预览效果如图23-1所示。

图 23-1　界面预览效果

上述代码中：

- 设置了Get按钮，以备设置点击事件，以触发WLAN相关的操作。
- Text组件用于展示读取到的WLAN信息。

23.2.5 修改 MainAbilitySlice

修改MainAbilitySlice内容如下：

```
package com.waylau.hmos.wifidevice.slice;
```

```java
import com.waylau.hmos.wifidevice.ResourceTable;
import ohos.aafwk.ability.AbilitySlice;
import ohos.aafwk.content.Intent;
import ohos.agp.components.Button;
import ohos.agp.components.Text;
import ohos.hiviewdfx.HiLog;
import ohos.hiviewdfx.HiLogLabel;
import ohos.wifi.*;

import java.util.List;
import java.util.Optional;


public class MainAbilitySlice extends AbilitySlice {
    private static final String TAG = MainAbilitySlice.class.getSimpleName();
    private static final HiLogLabel LABEL_LOG =
            new HiLogLabel(HiLog.LOG_APP, 0x00001, TAG);

    private WifiDevice wifiDevice;
    private ohos.agp.components.Text text;

    @Override
    public void onStart(Intent intent) {
        super.onStart(intent);
        super.setUIContent(ResourceTable.Layout_ability_main);

        // 初始化WLAN
        initWifiDevice();

        // 为按钮设置点击事件回调
        Button buttonGet =
                (Button) findComponentById(ResourceTable.Id_button_get);
        buttonGet.setClickedListener(listener -> getInfo());

        text = (Text) findComponentById(ResourceTable.Id_text);
    }

    private void initWifiDevice() {
        HiLog.info(LABEL_LOG, "before initWifiDevice");

        // 获取WLAN设备
        wifiDevice = WifiDevice.getInstance(this);

        // 调用获取WLAN开关状态接口
        // 若WLAN打开，则返回true，否则返回false
        boolean isWifiActive = wifiDevice.isWifiActive();

        HiLog.info(LABEL_LOG, "end initWifiDevice, isWifiActive: %{public}s",
                isWifiActive);
```

```java
    }

    private void getInfo() {
        HiLog.info(LABEL_LOG, "before getInfo");

        // 调用WLAN扫描接口
        boolean isScanSuccess = wifiDevice.scan(); // true
        text.append("isScanSuccess: " + isScanSuccess + "\n");

        HiLog.info(LABEL_LOG, "isScanSuccess: %{public}s", isScanSuccess);

        // 调用获取扫描结果
        List<WifiScanInfo> scanInfos = wifiDevice.getScanInfoList();

        for (WifiScanInfo scanInfo : scanInfos) {
            String ssid = scanInfo.getSsid();
            text.append("ssid: " + ssid + "\n");

            HiLog.info(LABEL_LOG, "ssid: %{public}s", ssid);
        }

        // 调用WLAN连接状态接口,确定当前设备是否连接WLAN
        boolean isConnected = wifiDevice.isConnected();
        text.append("isConnected: " + isConnected + "\n");

        HiLog.info(LABEL_LOG, "isConnected: %{public}s", isConnected);

        if (isConnected) {
            // 获取WLAN连接信息
            Optional<WifiLinkedInfo> linkedInfo = wifiDevice.getLinkedInfo();

            // 获取连接信息中的SSID
            String ssid = linkedInfo.get().getSsid();

            // 获取WLAN的IP信息
            Optional<IpInfo> ipInfo = wifiDevice.getIpInfo();

            // 获取IP信息中的IP地址与网关
            int ipAddress = ipInfo.get().getIpAddress();
            int gateway = ipInfo.get().getGateway();
            text.append("ipAddress: " + ipAddress + "; gateway:" + gateway + "\n");

            HiLog.info(LABEL_LOG, "ipAddress:%{public}s, gateway:%{public}s",
                    ipAddress, gateway);
        }

        // 获取当前设备的国家码
        String countryCode = wifiDevice.getCountryCode();
        text.append("countryCode: " + countryCode + "\n");
```

```
            HiLog.info(LABEL_LOG, "countryCode: %{public}s", countryCode);

            // 获取当前设备是否支持指定的能力
            boolean isSupport = 
wifiDevice.isFeatureSupported(WifiUtils.WIFI_FEATURE_INFRA);
            text.append("WIFI_FEATURE_INFRA: " + isSupport + "\n");
            HiLog.info(LABEL_LOG, "WIFI_FEATURE_INFRA: %{public}s", isSupport);

            isSupport = 
wifiDevice.isFeatureSupported(WifiUtils.WIFI_FEATURE_INFRA_5G);
            text.append("WIFI_FEATURE_INFRA_5G: " + isSupport + "\n");
            HiLog.info(LABEL_LOG, "WIFI_FEATURE_INFRA_5G: %{public}s",
isSupport);

            isSupport = 
wifiDevice.isFeatureSupported(WifiUtils.WIFI_FEATURE_PASSPOINT);
            text.append("WIFI_FEATURE_PASSPOINT: " + isSupport + "\n");
            HiLog.info(LABEL_LOG, "WIFI_FEATURE_PASSPOINT: %{public}s",
isSupport);

            isSupport = 
wifiDevice.isFeatureSupported(WifiUtils.WIFI_FEATURE_P2P);
            text.append("WIFI_FEATURE_P2P: " + isSupport + "\n");
            HiLog.info(LABEL_LOG, "WIFI_FEATURE_P2P: %{public}s", isSupport);

            isSupport = 
wifiDevice.isFeatureSupported(WifiUtils.WIFI_FEATURE_MOBILE_HOTSPOT);
            text.append("WIFI_FEATURE_MOBILE_HOTSPOT: " + isSupport + "\n");
            HiLog.info(LABEL_LOG, "WIFI_FEATURE_MOBILE_HOTSPOT: %{public}s",
isSupport);

            isSupport = 
wifiDevice.isFeatureSupported(WifiUtils.WIFI_FEATURE_AWARE);
            text.append("WIFI_FEATURE_AWARE: " + isSupport + "\n");
            HiLog.info(LABEL_LOG, "WIFI_FEATURE_AWARE: %{public}s", isSupport);

            isSupport = 
wifiDevice.isFeatureSupported(WifiUtils.WIFI_FEATURE_AP_STA);
            text.append("WIFI_FEATURE_AP_STA: " + isSupport + "\n");
            HiLog.info(LABEL_LOG, "WIFI_FEATURE_AP_STA: %{public}s", isSupport);

            isSupport = 
wifiDevice.isFeatureSupported(WifiUtils.WIFI_FEATURE_WPA3_SAE);
            text.append("WIFI_FEATURE_WPA3_SAE: " + isSupport + "\n");
            HiLog.info(LABEL_LOG, "WIFI_FEATURE_WPA3_SAE: %{public}s",
isSupport);

            isSupport = 
wifiDevice.isFeatureSupported(WifiUtils.WIFI_FEATURE_WPA3_SUITE_B);
```

```
            text.append("WIFI_FEATURE_WPA3_SUITE_B: " + isSupport + "\n");
            HiLog.info(LABEL_LOG, "WIFI_FEATURE_WPA3_SUITE_B: %{public}s",
isSupport);

            isSupport =
wifiDevice.isFeatureSupported(WifiUtils.WIFI_FEATURE_OWE);
            text.append("WIFI_FEATURE_OWE: " + isSupport + "\n");
            HiLog.info(LABEL_LOG, "WIFI_FEATURE_OWE: %{public}s", isSupport);

            HiLog.info(LABEL_LOG, "end getInfo");
    }


    @Override
    public void onActive() {
        super.onActive();
    }

    @Override
    public void onForeground(Intent intent) {
        super.onForeground(intent);
    }
}
```

上述代码中：

- initWifiDevice方法用于初始化WLAN设备的WifiDevice对象。
- 在Button上设置了点击事件。
- getInfo方法用于获取WLAN的信息。

23.2.6 运行

初次运行应用后，点击界面上的Get按钮以触发操作的执行。此时，控制台输出内容如下：

```
    09-12 23:28:45.366 16989-16989/com.waylau.hmos.wifidevice I
00001/MainAbilitySlice:  before initWifiDevice
    09-12 23:28:45.390 16989-16989/com.waylau.hmos.wifidevice I
00001/MainAbilitySlice:  end initWifiDevice, isWifiActive: true
    09-12 23:28:54.513 16989-16989/com.waylau.hmos.wifidevice I
00001/MainAbilitySlice:  before getInfo
    09-12 23:28:54.548 16989-16989/com.waylau.hmos.wifidevice I
00001/MainAbilitySlice:  isScanSuccess: true
    09-12 23:28:54.567 16989-16989/com.waylau.hmos.wifidevice I
00001/MainAbilitySlice:  ssid: Chinasoft
    09-12 23:28:54.569 16989-16989/com.waylau.hmos.wifidevice I
00001/MainAbilitySlice:  isConnected: true
    09-12 23:28:54.581 16989-16989/com.waylau.hmos.wifidevice I
00001/MainAbilitySlice:  ipAddress: 33882122, gateway: 17104906
    09-12 23:28:54.583 16989-16989/com.waylau.hmos.wifidevice I
```

```
00001/MainAbilitySlice: countryCode: CN
    09-12 23:28:54.584 16989-16989/com.waylau.hmos.wifidevice I
00001/MainAbilitySlice: WIFI_FEATURE_INFRA: true
    09-12 23:28:54.585 16989-16989/com.waylau.hmos.wifidevice I
00001/MainAbilitySlice: WIFI_FEATURE_INFRA_5G: false
    09-12 23:28:54.586 16989-16989/com.waylau.hmos.wifidevice I
00001/MainAbilitySlice: WIFI_FEATURE_PASSPOINT: false
    09-12 23:28:54.588 16989-16989/com.waylau.hmos.wifidevice I
00001/MainAbilitySlice: WIFI_FEATURE_P2P: true
    09-12 23:28:54.589 16989-16989/com.waylau.hmos.wifidevice I
00001/MainAbilitySlice: WIFI_FEATURE_MOBILE_HOTSPOT: true
    09-12 23:28:54.590 16989-16989/com.waylau.hmos.wifidevice I
00001/MainAbilitySlice: WIFI_FEATURE_AWARE: true
    09-12 23:28:54.591 16989-16989/com.waylau.hmos.wifidevice I
00001/MainAbilitySlice: WIFI_FEATURE_AP_STA: false
    09-12 23:28:54.592 16989-16989/com.waylau.hmos.wifidevice I
00001/MainAbilitySlice: WIFI_FEATURE_WPA3_SAE: false
    09-12 23:28:54.594 16989-16989/com.waylau.hmos.wifidevice I
00001/MainAbilitySlice: WIFI_FEATURE_WPA3_SUITE_B: false
    09-12 23:28:54.596 16989-16989/com.waylau.hmos.wifidevice I
00001/MainAbilitySlice: WIFI_FEATURE_OWE: false
    09-12 23:28:54.596 16989-16989/com.waylau.hmos.wifidevice I
00001/MainAbilitySlice: end getInfo
```

界面效果如图23-2所示。

图23-2 界面效果

23.3 实战：不信任热点配置

本节演示如何实现不信任热点的配置。应用可以添加指定的热点，其选网优先级低于已保存的热点。如果扫描后判断该热点为最合适的热点，就自动连接该热点。

应用或者其他模块可以通过接口完成以下功能：

- 设置第三方的热点配置。
- 删除第三方的热点配置。

23.3.1 接口说明

WifiDevice提供WLAN的不信任热点配置功能，其接口有：

- getInstance(Context context)：获取WLAN功能管理对象实例，通过该实例调用不信任热点配置的API。
- addUntrustedConfig(WifiDeviceConfig config)：添加不信任热点配置，选网优先级低于已保存的热点，需要ohos.permission.SET_WIFI_INFO权限。
- removeUntrustedConfig(WifiDeviceConfig config)：删除不信任的热点配置，需要ohos.permission.SET_WIFI_INFO权限。

23.3.2 创建应用

为了演示不信任热点配置功能，创建一个名为WifiDeviceUntrustedConfig的应用。
在应用的界面上，通过点击按钮来触发WifiDevice的操作。

23.3.3 声明权限

修改配置文件，声明使用WLAN的权限如下：

```
// 声明权限
"reqPermissions": [
   {
    "name": "ohos.permission.SET_WIFI_INFO"
   }
]
```

由于ohos.permission.SET_WIFI_INFO不是敏感权限，因此不需要在代码中显式声明。

23.3.4 修改 ability_main.xml

修改ability_main.xml内容如下：

```xml
<?xml version="1.0" encoding="utf-8"?>
<DirectionalLayout
    xmlns:ohos="http://schemas.huawei.com/res/ohos"
    ohos:height="match_parent"
    ohos:width="match_parent"
    ohos:orientation="vertical">

    <Button
        ohos:id="$+id:button_add"
        ohos:height="match_content"
        ohos:width="match_parent"
        ohos:background_element="#F76543"
        ohos:layout_alignment="horizontal_center"
        ohos:margin="10vp"
        ohos:padding="10vp"
        ohos:text="Add"
        ohos:text_size="30vp"
        />

    <Button
        ohos:id="$+id:button_remove"
        ohos:height="match_content"
        ohos:width="match_parent"
        ohos:background_element="#F76543"
        ohos:layout_alignment="horizontal_center"
        ohos:margin="10vp"
        ohos:padding="10vp"
        ohos:text="Remove"
        ohos:text_size="30vp"
        />

    <Text
        ohos:id="$+id:text"
        ohos:height="match_content"
        ohos:width="match_content"
        ohos:background_element="$graphic:background_ability_main"
        ohos:layout_alignment="horizontal_center"
        ohos:multiple_lines="true"
        ohos:text=""
        ohos:text_size="20vp"
        />

</DirectionalLayout>
```

界面预览效果如图23-3所示。

图 23-3　界面预览效果

上述代码中：

- 设置了Add按钮和Remove按钮，以备设置点击事件，以触发配置的添加和删除操作。
- Text组件用于展示配置操作的结果。

23.3.5　修改 MainAbilitySlice

修改MainAbilitySlice内容如下：

```java
package com.waylau.hmos.wifideviceuntrustedconfig.slice;

import com.waylau.hmos.wifideviceuntrustedconfig.ResourceTable;
import ohos.aafwk.ability.AbilitySlice;
import ohos.aafwk.content.Intent;
import ohos.agp.components.Button;
import ohos.agp.components.Text;
import ohos.hiviewdfx.HiLog;
import ohos.hiviewdfx.HiLogLabel;
import ohos.wifi.*;

public class MainAbilitySlice extends AbilitySlice {
    private static final String TAG = MainAbilitySlice.class.getSimpleName();
    private static final HiLogLabel LABEL_LOG =
            new HiLogLabel(HiLog.LOG_APP, 0x00001, TAG);

    private WifiDevice wifiDevice;
    private WifiDeviceConfig config;
    private Text text;

    @Override
```

```java
    public void onStart(Intent intent) {
        super.onStart(intent);
        super.setUIContent(ResourceTable.Layout_ability_main);

        // 初始化WLAN
        initWifiDevice();

        // 为按钮设置点击事件回调
        Button buttonAdd =
                (Button) findComponentById(ResourceTable.Id_button_add);
        buttonAdd.setClickedListener(listener -> addConfig());

        // 为按钮设置点击事件回调
        Button buttonRemove =
                (Button) findComponentById(ResourceTable.Id_button_remove);
        buttonRemove.setClickedListener(listener -> removeConfig());

        text = (Text) findComponentById(ResourceTable.Id_text);
    }

    private void initWifiDevice() {
        HiLog.info(LABEL_LOG, "before initWifiDevice");

        // 初始化热点配置
        config = new WifiDeviceConfig();
        config.setSsid("untrusted-exist");
        config.setPreSharedKey("123456789");
        config.setHiddenSsid(false);
        config.setSecurityType(WifiSecurity.PSK);

        // 获取WLAN设备
        wifiDevice = WifiDevice.getInstance(this);

        HiLog.info(LABEL_LOG, "end initWifiDevice");
    }

    private void addConfig() {
        HiLog.info(LABEL_LOG, "before addConfig");

        boolean isSuccess = wifiDevice.addUntrustedConfig(config);
        text.append("addUntrustedConfig: " + isSuccess + "\n");

        HiLog.info(LABEL_LOG, "end addConfig, isSuccess:%{public}s",
isSuccess);
    }

    private void removeConfig() {
        HiLog.info(LABEL_LOG, "before removeConfig");

        boolean isSuccess = wifiDevice.removeUntrustedConfig(config);
```

```
            text.append("removeUntrustedConfig: " + isSuccess + "\n");

            HiLog.info(LABEL_LOG, "end removeConfig, isSuccess:%{public}s",
isSuccess);
        }

        @Override
        public void onActive() {
            super.onActive();
        }

        @Override
        public void onForeground(Intent intent) {
            super.onForeground(intent);
        }
    }
```

上述代码中：

- initWifiDevice方法用于初始化WLAN设备WifiDevice对象和热点配置WifiDeviceConfig对象。
- 在Button上设置了点击事件。
- addConfig方法新增热点配置。
- removeConfig方法新增删除配置。
- Text用于在界面上展示配置结果。

23.3.6 运行

初次运行应用后，分别点击界面上的Add按钮和Remove按钮以触发操作的执行。此时，控制台输出内容如下：

```
  09-12 23:35:28.462 1341-1341/com.waylau.hmos.wifideviceuntrustedconfig I
00001/MainAbilitySlice: before initWifiDevice
  09-12 23:35:28.464 1341-1341/com.waylau.hmos.wifideviceuntrustedconfig I
00001/MainAbilitySlice: end initWifiDevice
  09-12 23:35:59.994 1341-1341/com.waylau.hmos.wifideviceuntrustedconfig I
00001/MainAbilitySlice: before addConfig
  09-12 23:36:00.058 1341-1341/com.waylau.hmos.wifideviceuntrustedconfig I
00001/MainAbilitySlice: end addConfig, isSuccess:true
  09-12 23:36:04.635 1341-1341/com.waylau.hmos.wifideviceuntrustedconfig I
00001/MainAbilitySlice: before addConfig
  09-12 23:36:04.639 1341-1341/com.waylau.hmos.wifideviceuntrustedconfig I
00001/MainAbilitySlice: end addConfig, isSuccess:false
  09-12 23:36:08.265 1341-1341/com.waylau.hmos.wifideviceuntrustedconfig I
00001/MainAbilitySlice: before removeConfig
  09-12 23:36:08.348 1341-1341/com.waylau.hmos.wifideviceuntrustedconfig I
00001/MainAbilitySlice: end removeConfig, isSuccess:true
  09-12 23:36:10.345 1341-1341/com.waylau.hmos.wifideviceuntrustedconfig I
00001/MainAbilitySlice: before removeConfig
```

```
09-12 23:36:10.354 1341-1341/com.waylau.hmos.wifideviceuntrustedconfig I
00001/MainAbilitySlice: end removeConfig, isSuccess:false
```

界面效果如图23-4所示。

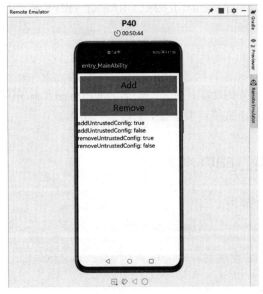

图 23-4　界面效果

23.4　实战：WLAN 消息通知

本节演示如何实现接收WLAN消息通知的功能。WLAN消息通知（Notification）是HarmonyOS内部或者与应用之间跨进程通信的机制，注册者在注册消息通知后，一旦符合条件的消息被发出，注册者即可接收到该消息并获取消息中附带的信息。

23.4.1　接口说明

WLAN消息通知的相关广播事件如下：

- WLAN状态：usual.event.wifi.POWER_STATE。
- WLAN扫描：usual.event.wifi.SCAN_FINISHED。
- WLAN RSSI变化：usual.event.wifi.RSSI_VALUE。
- WLAN连接状态：usual.event.wifi.CONN_STATE。
- Hotspot状态：usual.event.wifi.HOTSPOT_STATE。
- Hotspot连接状态：usual.event.wifi.WIFI_HS_STA_JOIN、usual.event.wifi.WIFI_HS_STA_LEAVE。
- P2P状态：usual.event.wifi.p2p.STATE_CHANGE。

- P2P连接状态：usual.event.wifi.p2p.CONN_STATE_CHANGE。
- P2P设备列表变化：usual.event.wifi.p2p.DEVICES_CHANGE。
- P2P搜索状态变化：usual.event.wifi.p2p.PEER_DISCOVERY_STATE_CHANGE。
- P2P当前设备变化：usual.event.wifi.p2p.CURRENT_DEVICE_CHANGE。

23.4.2 创建应用

为了演示WLAN消息通知的功能，创建一个名为WifiEventSubscriber的应用。

在应用的界面上，通过点击按钮来触发消息订阅的操作。

23.4.3 修改 ability_main.xml

修改ability_main.xml内容如下：

```xml
<?xml version="1.0" encoding="utf-8"?>
<DirectionalLayout
    xmlns:ohos="http://schemas.huawei.com/res/ohos"
    ohos:height="match_parent"
    ohos:width="match_parent"
    ohos:orientation="vertical">

    <Button
        ohos:id="$+id:button_subscribe"
        ohos:height="match_content"
        ohos:width="match_parent"
        ohos:background_element="#F76543"
        ohos:layout_alignment="horizontal_center"
        ohos:margin="10vp"
        ohos:padding="10vp"
        ohos:text="Subscribe"
        ohos:text_size="30vp"
        />

    <Text
        ohos:id="$+id:text"
        ohos:height="match_content"
        ohos:width="match_content"
        ohos:background_element="$graphic:background_ability_main"
        ohos:layout_alignment="horizontal_center"
        ohos:multiple_lines="true"
        ohos:text=""
        ohos:text_size="20vp"
        />

</DirectionalLayout>
```

界面预览效果如图23-5所示。

图 23-5　界面预览效果

上述代码中：

- 设置了Subscribe按钮，以备设置点击事件，以触发订阅操作。
- Text组件用于展示配置操作的结果。

23.4.4　修改 MainAbilitySlice

修改MainAbilitySlice内容如下：

```java
package com.waylau.hmos.wifieventsubscriber.slice;

import com.waylau.hmos.wifieventsubscriber.ResourceTable;
import ohos.aafwk.ability.AbilitySlice;
import ohos.aafwk.content.Intent;
import ohos.aafwk.content.IntentParams;
import ohos.agp.components.Button;
import ohos.agp.components.Text;
import ohos.event.commonevent.*;
import ohos.hiviewdfx.HiLog;
import ohos.hiviewdfx.HiLogLabel;
import ohos.rpc.RemoteException;
import ohos.wifi.*;

public class MainAbilitySlice extends AbilitySlice {
    private static final String TAG = MainAbilitySlice.class.getSimpleName();
    private static final HiLogLabel LABEL_LOG =
            new HiLogLabel(HiLog.LOG_APP, 0x00001, TAG);

    private WifiDevice wifiDevice;
    private WifiDeviceConfig config;
```

```java
        private Text text;

        @Override
        public void onStart(Intent intent) {
            super.onStart(intent);
            super.setUIContent(ResourceTable.Layout_ability_main);

            // 为按钮设置点击事件回调
            Button buttonSubscribe =
                    (Button) findComponentById(ResourceTable.Id_button_subscribe);
            buttonSubscribe.setClickedListener(listener -> subscribe());

            text = (Text) findComponentById(ResourceTable.Id_text);
        }

        private void subscribe() {
            HiLog.info(LABEL_LOG, "before subscribe");

            // 注册消息
            MatchingSkills match = new MatchingSkills();

            // 增加获取WLAN状态变化消息
            match.addEvent(WifiEvents.EVENT_ACTIVE_STATE);
            CommonEventSubscribeInfo subscribeInfo = new CommonEventSubscribeInfo(match);
            subscribeInfo.setPriority(100);
            WifiEventSubscriber subscriber = new WifiEventSubscriber(subscribeInfo);

            try {
                CommonEventManager.subscribeCommonEvent(subscriber);
            } catch (RemoteException e) {
                HiLog.warn(LABEL_LOG, "subscribe in wifi events failed!");
            }

            HiLog.info(LABEL_LOG, "end subscribe");
        }

        // 构建消息接收者/注册者
        private class WifiEventSubscriber extends CommonEventSubscriber {
            WifiEventSubscriber(CommonEventSubscribeInfo info) {
                super(info);
            }

            @Override
            public void onReceiveEvent(CommonEventData commonEventData) {
                if (WifiEvents.EVENT_ACTIVE_STATE.equals(commonEventData.getIntent().getAction())) {
                    // 获取附带参数
```

```java
                IntentParams params = commonEventData.getIntent().getParams();
                if (params == null) {
                    return;
                }

                // WLAN状态
                int wifiState = (int) params.getParam(WifiEvents.PARAM_ACTIVE_STATE);

                if (wifiState == WifiEvents.STATE_ACTIVE) {
                    // 处理WLAN被打开消息
                    HiLog.info(LABEL_LOG, "Receive WifiEvents.STATE_ACTIVE %{public}d", wifiState);
                    text.append("Receive WifiEvents.STATE_ACTIVE\n");
                } else if (wifiState == WifiEvents.STATE_INACTIVE) {
                    // 处理WLAN被关闭消息
                    HiLog.info(LABEL_LOG, "Receive WifiEvents.STATE_INACTIVE %{public}d", wifiState);
                    text.append("Receive WifiEvents.STATE_INACTIVE\n");
                } else {
                    // 处理WLAN异常状态
                    HiLog.info(LABEL_LOG, "Unknown wifi state");
                    text.append("Receive Unknown wifi state\n");
                }
            }
        }
    }

    @Override
    public void onActive() {
        super.onActive();
    }

    @Override
    public void onForeground(Intent intent) {
        super.onForeground(intent);
    }
}
```

上述代码中：

- 在Button上设置了点击事件。
- subscribe方法用于订阅事件WifiEvents.EVENT_ACTIVE_STATE。
- WLAN被打开或者被关闭时，将会接收到事件通知。

23.4.5 运行

运行应用后，点击界面上的Subscribe按钮以触发订阅事件。控制台输出内容如下：

```
    02-13 21:53:36.195 9749-9749/com.waylau.hmos.wifieventsubscriber I
00001/MainAbilitySlice: before subscribe
    02-13 21:53:36.216 9749-9749/com.waylau.hmos.wifieventsubscriber I
00001/MainAbilitySlice: end subscribe
    02-13 21:53:36.221 9749-9749/com.waylau.hmos.wifieventsubscriber I
00001/MainAbilitySlice: Receive WifiEvents.STATE_ACTIVE 1
```

此时分别执行一次关闭、开启手机WLAN功能（见图23-6），以触发事件。

图 23-6　界面效果

可以看到控制台输出内容如下：

```
    02-13 21:54:06.270 9749-9749/com.waylau.hmos.wifieventsubscriber I
00001/MainAbilitySlice: Unknown wifi state
    02-13 21:54:06.590 9749-9749/com.waylau.hmos.wifieventsubscriber I
00001/MainAbilitySlice: Receive WifiEvents.STATE_INACTIVE 0
    02-13 21:54:09.042 9749-9749/com.waylau.hmos.wifieventsubscriber I
00001/MainAbilitySlice: Unknown wifi state
    02-13 21:54:09.233 9749-9749/com.waylau.hmos.wifieventsubscriber I
00001/MainAbilitySlice: Receive WifiEvents.STATE_ACTIVE 1
```

第 24 章

网络管理

HarmonyOS提供了网络管理模块以支持多场景的网络管理。

24.1 网络管理概述

本节主要介绍HarmonyOS网络管理模块的功能，以及使用网络管理模块的相关功能时需要请求的权限。

24.1.1 功能

HarmonyOS网络管理模块主要提供以下功能：

- 数据连接管理：网卡绑定，打开URL，数据链路参数查询。
- 数据网络管理：指定数据网络传输，获取数据网络状态变更，数据网络状态查询。
- 流量统计：获取蜂窝网络、所有网卡、指定应用或指定网卡的数据流量统计值。
- HTTP缓存：有效管理HTTP缓存，减少数据流量。
- 创建本地套接字：实现本机不同进程间的通信，目前只支持流式套接字。

24.1.2 约束与限制

使用网络管理模块的相关功能时，需要请求相应的权限：

- ohos.permission.GET_NETWORK_INFO：获取网络连接信息。
- ohos.permission.SET_NETWORK_INFO：修改网络连接状态。

- ohos.permission.INTERNET：允许程序打开网络套接字，进行网络连接。

另外，请求网络的操作不应该放在主线程中，需要另外新启一个线程进行操作。

24.2 实战：使用当前网络打开一个 URL 链接

本节演示如何实现打开一个URL链接。

24.2.1 接口说明

应用使用当前网络打开一个URL链接，主要涉及NetManager和NetHandle两个类。

1. NetManager

NetManager的主要接口有：

- getInstance(Context context)：获取网络管理的实例对象。
- hasDefaultNet()：查询当前是否有默认可用的数据网络。
- getDefaultNet()：获取当前默认的数据网络句柄。
- addDefaultNetStatusCallback(NetStatusCallback callback)：获取当前默认的数据网络状态变化。
- setAppNet(NetHandle netHandle)：应用绑定该数据网络。

2. NetHandle

NetHandle的主要接口有：

- openConnection(URL url, Proxy proxy)：使用该网络打开一个URL链接。

24.2.2 创建应用

为了演示打开一个URL链接的功能，创建一个名为NetManagerHandleURL的应用。
在应用的界面上，通过点击按钮来触发打开一个URL链接的操作。

24.2.3 声明权限

修改配置文件，声明使用网络的权限如下：

```
// 声明权限
"reqPermissions": [
   {
   "name": "ohos.permission.GET_NETWORK_INFO"
   },
   {
   "name": "ohos.permission.SET_NETWORK_INFO"
   },
```

```
    {
    "name": "ohos.permission.INTERNET"
    }
]
```

由于上述权限不是敏感权限,因此不需要在代码中显式声明。

24.2.4 修改 ability_main.xml

修改ability_main.xml内容如下:

```xml
<?xml version="1.0" encoding="utf-8"?>
<DirectionalLayout
    xmlns:ohos="http://schemas.huawei.com/res/ohos"
    ohos:height="match_parent"
    ohos:width="match_parent"
    ohos:orientation="vertical">

    <Button
        ohos:id="$+id:button_open"
        ohos:height="match_content"
        ohos:width="match_parent"
        ohos:background_element="#F76543"
        ohos:layout_alignment="horizontal_center"
        ohos:margin="10vp"
        ohos:padding="10vp"
        ohos:text="Open"
        ohos:text_size="30fp"
        />

    <Image
        ohos:id="$+id:image"
        ohos:height="match_content"
        ohos:width="match_parent"/>

</DirectionalLayout>
```

界面预览效果如图24-1所示。

上述代码中:

- 设置了Open按钮,以备设置点击事件,以触发打开链接的相关操作。
- Image组件用于展示读取的图片信息。

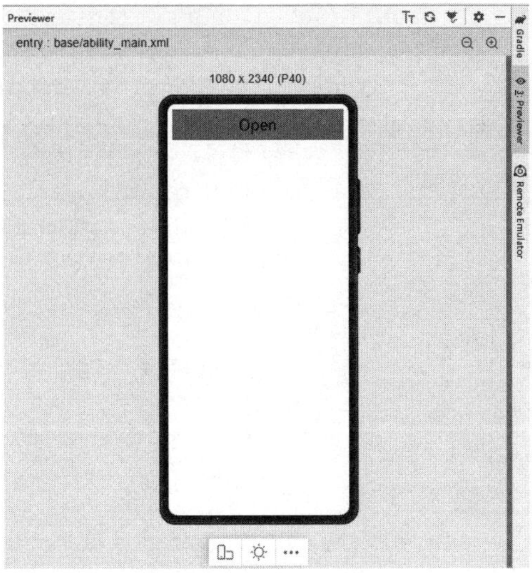

图 24-1 界面预览效果

24.2.5 修改 MainAbilitySlice

修改MainAbilitySlice内容如下:

```
package com.waylau.hmos.netmanagerhandleurl.slice;
```

```java
import com.waylau.hmos.netmanagerhandleurl.ResourceTable;
import ohos.aafwk.ability.AbilitySlice;
import ohos.aafwk.content.Intent;
import ohos.agp.components.Button;
import ohos.agp.components.Image;
import ohos.app.dispatcher.TaskDispatcher;
import ohos.app.dispatcher.task.TaskPriority;
import ohos.hiviewdfx.HiLog;
import ohos.hiviewdfx.HiLogLabel;
import ohos.media.image.ImageSource;
import ohos.media.image.PixelMap;
import ohos.media.image.common.PixelFormat;
import ohos.net.*;

import java.io.InputStream;
import java.net.*;

public class MainAbilitySlice extends AbilitySlice {
    private static final String TAG = MainAbilitySlice.class.getSimpleName();
    private static final HiLogLabel LABEL_LOG =
            new HiLogLabel(HiLog.LOG_APP, 0x00001, TAG);

    private TaskDispatcher dispatcher;
    private Image image;

    @Override
    public void onStart(Intent intent) {
        super.onStart(intent);
        super.setUIContent(ResourceTable.Layout_ability_main);

        // 为按钮设置点击事件回调
        Button buttonOpen =
                (Button) findComponentById(ResourceTable.Id_button_open);
        buttonOpen.setClickedListener(listener -> open());

        image =
                (Image) findComponentById(ResourceTable.Id_image);

        dispatcher = getGlobalTaskDispatcher(TaskPriority.DEFAULT);
    }

    private void open() {
        HiLog.info(LABEL_LOG, "before open");

        // 启动线程任务
        dispatcher.syncDispatch(() -> {
            NetManager netManager = NetManager.getInstance(getContext());

            if (!netManager.hasDefaultNet()) {
                return;
```

```
            }
            NetHandle netHandle = netManager.getDefaultNet();

            // 可以获取网络状态的变化
            netManager.addDefaultNetStatusCallback(callback);

            // 通过openConnection来获取URLConnection
            HttpURLConnection connection = null;
            try {
                String urlString =
"https://waylau.com/images/waylau_181_181.jpg";
                URL url = new URL(urlString);

                URLConnection urlConnection = netHandle.openConnection(url,
                        java.net.Proxy.NO_PROXY);
                if (urlConnection instanceof HttpURLConnection) {
                    connection = (HttpURLConnection) urlConnection;
                }
                connection.setRequestMethod("GET");
                connection.setReadTimeout(10000);
                connection.setConnectTimeout(10000);
                connection.connect();

                // 之后可进行URL的其他操作
                int code = connection.getResponseCode();
                HiLog.info(LABEL_LOG, "ResponseCode: %{public}s", code);

                if (code == HttpURLConnection.HTTP_OK) {
                    // 得到服务器返回的图片流对象在界面显示出来
                    InputStream inputStream = urlConnection.getInputStream();
                    ImageSource imageSource = ImageSource.create(inputStream,
new ImageSource.SourceOptions());
                    ImageSource.DecodingOptions decodingOptions = new
ImageSource.DecodingOptions();
                    decodingOptions.desiredPixelFormat =
PixelFormat.ARGB_8888;
                    PixelMap pixelMap =
imageSource.createPixelmap(decodingOptions);

                    image.setPixelMap(pixelMap);
                    pixelMap.release();
                }
            } catch (Exception e) {
                e.printStackTrace();
            } finally {
                connection.disconnect();
                HiLog.info(LABEL_LOG, "connection disconnect");
            }
        });

        HiLog.info(LABEL_LOG, "end open");
```

```
        }

        private NetStatusCallback callback = new NetStatusCallback() {
            public void onAvailable(NetHandle handle) {
                HiLog.info(LABEL_LOG, "onAvailable");
            }

            public void onBlockedStatusChanged(NetHandle handle, boolean blocked) {
                HiLog.info(LABEL_LOG, "onBlockedStatusChanged");
            }

            public void onLosing(NetHandle handle, long maxMsToLive) {
                HiLog.info(LABEL_LOG, "onLosing");
            }

            public void onLost(NetHandle handle) {
                HiLog.info(LABEL_LOG, "onLosing");
            }

            public void onUnavailable() {
                HiLog.info(LABEL_LOG, "onUnavailable");
            }

            public void onCapabilitiesChanged(NetHandle handle, NetCapabilities networkCapabilities) {
                HiLog.info(LABEL_LOG, "onCapabilitiesChanged");
            }

            public void onConnectionPropertiesChanged(NetHandle handle, ConnectionProperties connectionProperties) {
                HiLog.info(LABEL_LOG, "onConnectionPropertiesChanged");
            }
        };

        @Override
        public void onActive() {
            super.onActive();
        }

        @Override
        public void onForeground(Intent intent) {
            super.onForeground(intent);
        }
    }
```

上述代码中：

- onStart方法中初始化了任务分发器TaskDispatcher。
- 在Button上设置了点击事件。
- open方法用于执行打开链接的操作。

- 调用NetManager.getInstance(Context)获取网络管理的实例对象。
- 调用NetManager.getDefaultNet()获取默认的数据网络。
- 调用NetHandle.openConnection()打开一个URL。
- 通过URL链接实例访问网站。
- 得到服务器返回的图片流对象在界面显示出来。
- 网络操作避免阻塞主线程，因此放置到一个线程任务中执行。

24.2.6 运行

运行应用后，点击界面上的Open按钮以触发操作的执行。此时，控制台输出内容如下：

```
    02-14 10:29:35.982 10071-10071/com.waylau.hmos.netmanagerhandleurl I
00001/MainAbilitySlice: before open
    02-14 10:29:36.004 10071-12307/com.waylau.hmos.netmanagerhandleurl I
00001/MainAbilitySlice: onAvailable
    02-14 10:29:36.004 10071-12307/com.waylau.hmos.netmanagerhandleurl I
00001/MainAbilitySlice: onCapabilitiesChanged
    02-14 10:29:36.004 10071-12307/com.waylau.hmos.netmanagerhandleurl I
00001/MainAbilitySlice: onConnectionPropertiesChanged
    02-14 10:29:38.210 10071-12306/com.waylau.hmos.netmanagerhandleurl I
00001/MainAbilitySlice: ResponseCode: 200
    02-14 10:29:38.372 10071-12306/com.waylau.hmos.netmanagerhandleurl I
00001/MainAbilitySlice: connection disconnect
    02-14 10:29:38.372 10071-10071/com.waylau.hmos.netmanagerhandleurl I
00001/MainAbilitySlice: end open
```

界面效果如图24-2所示。

图24-2　界面效果

24.3 实战：使用当前网络进行 Socket 数据传输

本节演示如何实现Socket数据传输。

24.3.1 接口说明

应用使用当前网络进行Socket数据传输，主要涉及NetManager和NetHandle两个类。

1. NetManager

NetManager的主要接口有：

- getByName(String host)：解析主机名，获取其IP地址。
- bindSocket(Socket socket)：绑定Socket到该数据网络。

2. NetHandle

NetHandle的主要接口有：

- bindSocket(DatagramSocket socket)：绑定DatagramSocket到该数据网络。

24.3.2 创建应用

为了演示进行Socket数据传输的功能，创建一个名为NetManagerHandleSocket的应用。
在应用的界面上，通过点击按钮来触发进行Socket数据传输的操作。

24.3.3 声明权限

修改配置文件，声明使用网络的权限如下：

```
// 声明权限
"reqPermissions": [
   {
   "name": "ohos.permission.GET_NETWORK_INFO"
   },
   {
   "name": "ohos.permission.SET_NETWORK_INFO"
   },
   {
   "name": "ohos.permission.INTERNET"
   }
]
```

由于上述权限不是敏感权限，因此不需要在代码中显式声明。

24.3.4 修改 ability_main.xml

修改ability_main.xml内容如下：

```xml
<?xml version="1.0" encoding="utf-8"?>
<DirectionalLayout
    xmlns:ohos="http://schemas.huawei.com/res/ohos"
    ohos:height="match_parent"
    ohos:width="match_parent"
    ohos:orientation="vertical">

    <Button
        ohos:id="$+id:button_start"
        ohos:height="match_content"
        ohos:width="match_parent"
        ohos:background_element="#F76543"
        ohos:layout_alignment="horizontal_center"
        ohos:margin="10vp"
        ohos:padding="10vp"
        ohos:text="Start Server"
        ohos:text_size="30fp"
        />

    <Button
        ohos:id="$+id:button_open"
        ohos:height="match_content"
        ohos:width="match_parent"
        ohos:background_element="#F76543"
        ohos:layout_alignment="horizontal_center"
        ohos:margin="10vp"
        ohos:padding="10vp"
        ohos:text="Open"
        ohos:text_size="30fp"
        />

</DirectionalLayout>
```

界面预览效果如图24-3所示。

上述代码设置了Start Server、Open按钮，以备设置点击事件，以触发Socket数据传输相关的操作。

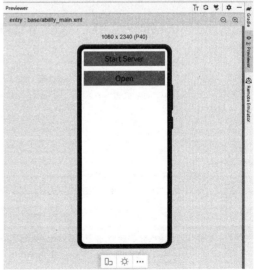

图 24-3　界面预览效果

24.3.5 修改 MainAbilitySlice

修改MainAbilitySlice内容如下：

```
package com.waylau.hmos.netmanagerhandlesocket.slice;
```

```java
import com.waylau.hmos.netmanagerhandlesocket.ResourceTable;
import ohos.aafwk.ability.AbilitySlice;
import ohos.aafwk.content.Intent;
import ohos.agp.components.Button;
import ohos.agp.components.Text;
import ohos.app.dispatcher.TaskDispatcher;
import ohos.app.dispatcher.task.TaskPriority;
import ohos.hiviewdfx.HiLog;
import ohos.hiviewdfx.HiLogLabel;
import ohos.net.*;

import java.net.*;

public class MainAbilitySlice extends AbilitySlice {
    private static final String TAG = MainAbilitySlice.class.getSimpleName();
    private static final HiLogLabel LABEL_LOG =
            new HiLogLabel(HiLog.LOG_APP, 0x00001, TAG);

    private final static String HOST = "127.0.0.1";
    private final static int PORT = 8551;

    private TaskDispatcher dispatcher;

    @Override
    public void onStart(Intent intent) {
        super.onStart(intent);
        super.setUIContent(ResourceTable.Layout_ability_main);

        // 为按钮设置点击事件回调
        Button buttonStart =
                (Button) findComponentById(ResourceTable.Id_button_start);
        buttonStart.setClickedListener(listener -> initServer());

        Button buttonOpen =
                (Button) findComponentById(ResourceTable.Id_button_open);
        buttonOpen.setClickedListener(listener -> open());

        dispatcher = getGlobalTaskDispatcher(TaskPriority.DEFAULT);
    }

    private void open() {
        HiLog.info(LABEL_LOG, "before open");

        // 启动线程任务
        dispatcher.syncDispatch(() -> {
            NetManager netManager = NetManager.getInstance(null);

            if (!netManager.hasDefaultNet()) {
                HiLog.error(LABEL_LOG, "netManager.hasDefaultNet() failed");
                return;
```

```java
            }

            NetHandle netHandle = netManager.getDefaultNet();

            // 通过Socket绑定来进行数据传输
            DatagramSocket socket = null;
            try {
                // 绑定到Socket
                InetAddress address = netHandle.getByName(HOST);
                socket = new DatagramSocket();
                netHandle.bindSocket(socket);

                // 发送数据
                String data = "Welcome to waylau.com";
                DatagramPacket request = new
DatagramPacket(data.getBytes("utf-8"), data.length(), address, PORT);
                socket.send(request);

                // 显示到界面
                HiLog.info(LABEL_LOG, "send data: " + data);
            } catch (Exception e) {
                HiLog.error(LABEL_LOG, "send IOException: " + e.toString());
            } finally {
                if (null != socket) {
                    socket.close();
                }
            }
        });

        HiLog.info(LABEL_LOG, "end open");
    }

    private void initServer() {
        HiLog.info(LABEL_LOG, "before initServer");

        // 启动线程任务
        dispatcher.asyncDispatch(() -> {
            NetManager netManager = NetManager.getInstance(null);

            if (!netManager.hasDefaultNet()) {
                HiLog.error(LABEL_LOG, "netManager.hasDefaultNet() failed");
                return;
            }

            NetHandle netHandle = netManager.getDefaultNet();

            // 通过Socket绑定来进行数据传输
            DatagramSocket socket = null;
            try {
                // 绑定到Socket
                InetAddress address = netHandle.getByName(HOST);
```

```java
            socket = new DatagramSocket(PORT, address);
            netHandle.bindSocket(socket);
            HiLog.info(LABEL_LOG, "wait for receive data");

            // 接收数据
            byte[] buffer = new byte[1024];
            DatagramPacket response = new DatagramPacket(buffer, buffer.length);
            socket.receive(response);
            int len = response.getLength();
            String data = new String(buffer, "utf-8").substring(0, len);

            // 显示到界面
            HiLog.info(LABEL_LOG, "receive data: " + data);
        } catch (Exception e) {
            HiLog.error(LABEL_LOG, "send IOException: " + e.toString());
        } finally {
            if (null != socket) {
                socket.close();
            }
        }
    });

    HiLog.info(LABEL_LOG, "end initServer");
}

private NetStatusCallback callback = new NetStatusCallback() {
    public void onAvailable(NetHandle handle) {
        HiLog.info(LABEL_LOG, "onAvailable");
    }

    public void onBlockedStatusChanged(NetHandle handle, boolean blocked) {
        HiLog.info(LABEL_LOG, "onBlockedStatusChanged");
    }

    public void onLosing(NetHandle handle, long maxMsToLive) {
        HiLog.info(LABEL_LOG, "onLosing");
    }

    public void onLost(NetHandle handle) {
        HiLog.info(LABEL_LOG, "onLosing");
    }

    public void onUnavailable() {
        HiLog.info(LABEL_LOG, "onUnavailable");
    }

    public void onCapabilitiesChanged(NetHandle handle, NetCapabilities networkCapabilities) {
        HiLog.info(LABEL_LOG, "onCapabilitiesChanged");
```

```
        }

        public void onConnectionPropertiesChanged(NetHandle handle,
ConnectionProperties connectionProperties) {
            HiLog.info(LABEL_LOG, "onConnectionPropertiesChanged");
        }
    };

    @Override
    public void onActive() {
        super.onActive();
    }

    @Override
    public void onForeground(Intent intent) {
        super.onForeground(intent);
    }
}
```

上述代码中：

- onStart方法中初始化了任务分发器TaskDispatcher。
- 在Button上设置了点击事件。
- initServer方法用于启动Socket服务端。
- open方法用于发送Socket数据。
- 网络操作避免阻塞主线程，因此Socket的操作都放置到独立的线程任务中执行。

24.3.6 运行

运行应用后，点击界面上的Start Server按钮以触发启动Socket服务端的操作执行。此时，控制台输出内容如下：

```
09-13 23:46:20.261 16498-16498/com.waylau.hmos.netmanagerhandlesocket I
00001/MainAbilitySlice: before initServer
09-13 23:46:20.261 16498-16498/com.waylau.hmos.netmanagerhandlesocket I
00001/MainAbilitySlice: end initServer
09-13 23:46:20.272 16498-5332/com.waylau.hmos.netmanagerhandlesocket I
00001/MainAbilitySlice: wait for receive data
```

点击界面上的Open按钮以触发发送Socket数据的操作执行。此时，控制台输出内容如下：

```
09-13 23:46:22.567 16498-16498/com.waylau.hmos.netmanagerhandlesocket I
00001/MainAbilitySlice: before open
09-13 23:46:22.573 16498-5332/com.waylau.hmos.netmanagerhandlesocket I
00001/MainAbilitySlice: receive data: Welcome to waylau.com
09-13 23:46:22.574 16498-5574/com.waylau.hmos.netmanagerhandlesocket I
00001/MainAbilitySlice: send data: Welcome to waylau.com
09-13 23:46:22.574 16498-16498/com.waylau.hmos.netmanagerhandlesocket I
00001/MainAbilitySlice: end open
```

24.4 实战：流量统计

应用通过调用API接口可以获取蜂窝网络、所有网卡、指定应用或指定网卡的数据流量统计值。本节演示如何实现数据流量统计。

24.4.1 接口说明

应用进行流量统计所使用的接口主要由DataFlowStatistics提供。DataFlowStatistics的主要接口有：

- getCellularRxBytes()：获取蜂窝数据网络的下行流量。
- getCellularTxBytes()：获取蜂窝数据网络的上行流量。
- getAllRxBytes()：获取所有网卡的下行流量。
- getAllTxBytes()：获取所有网卡的上行流量。
- getUidRxBytes(int uid)：获取指定UID的下行流量。
- getUidTxBytes(int uid)：获取指定UID的上行流量。
- getIfaceRxBytes(String nic)：获取指定网卡的下行流量。
- getIfaceTxBytes(String nic)：获取指定网卡的上行流量。

24.4.2 创建应用

为了演示数据流量统计的功能，创建一个名为DataFlowStatistics的应用。
在应用的界面上，通过点击按钮来触发数据流量统计的操作。

24.4.3 声明权限

修改配置文件，声明使用网络的权限如下：

```
// 声明权限
"reqPermissions": [
    {
    "name": "ohos.permission.GET_NETWORK_INFO"
    },
    {
    "name": "ohos.permission.SET_NETWORK_INFO"
    },
    {
    "name": "ohos.permission.INTERNET"
    }
]
```

由于上述权限不是敏感权限,因此不需要在代码中显式声明。

24.4.4 修改 ability_main.xml

修改ability_main.xml内容如下:

```xml
<?xml version="1.0" encoding="utf-8"?>
<DirectionalLayout
    xmlns:ohos="http://schemas.huawei.com/res/ohos"
    ohos:height="match_parent"
    ohos:width="match_parent"
    ohos:orientation="vertical">

    <Button
        ohos:id="$+id:button_open"
        ohos:height="match_content"
        ohos:width="match_parent"
        ohos:background_element="#F76543"
        ohos:layout_alignment="horizontal_center"
        ohos:margin="10vp"
        ohos:padding="10vp"
        ohos:text="Open"
        ohos:text_size="30fp"
        />

    <Image
        ohos:id="$+id:image"
        ohos:height="match_content"
        ohos:width="match_parent"/>

</DirectionalLayout>
```

界面预览效果如图24-4所示。

图 24-4　界面预览效果

上述代码中：

- 设置了Open按钮，以备设置点击事件，以触发打开链接的相关操作。
- Image组件用于展示读取的图片信息。

24.4.5 修改 MainAbilitySlice

修改MainAbilitySlice内容如下：

```
package com.waylau.hmos.dataflowstatistics.slice;

import com.waylau.hmos.dataflowstatistics.ResourceTable;
import ohos.aafwk.ability.AbilitySlice;
import ohos.aafwk.content.Intent;
import ohos.agp.components.Button;
import ohos.agp.components.Image;
import ohos.app.dispatcher.TaskDispatcher;
import ohos.app.dispatcher.task.TaskPriority;
import ohos.hiviewdfx.HiLog;
import ohos.hiviewdfx.HiLogLabel;
import ohos.media.image.ImageSource;
import ohos.media.image.PixelMap;
import ohos.media.image.common.PixelFormat;
import ohos.net.*;

import java.io.InputStream;
import java.net.*;
import java.util.concurrent.atomic.AtomicLong;

public class MainAbilitySlice extends AbilitySlice {
    private static final String TAG = MainAbilitySlice.class.getSimpleName();
    private static final HiLogLabel LABEL_LOG =
            new HiLogLabel(HiLog.LOG_APP, 0x00001, TAG);

    private TaskDispatcher dispatcher;
    private Image image;

    @Override
    public void onStart(Intent intent) {
        super.onStart(intent);
        super.setUIContent(ResourceTable.Layout_ability_main);

        // 为按钮设置点击事件回调
        Button buttonOpen =
                (Button) findComponentById(ResourceTable.Id_button_open);
        buttonOpen.setClickedListener(listener -> open());

        image =
                (Image) findComponentById(ResourceTable.Id_image);
```

```java
            dispatcher = getGlobalTaskDispatcher(TaskPriority.DEFAULT);
        }

        private void open() {
            HiLog.info(LABEL_LOG, "before open");

            // 获取所有网卡的下行流量
            AtomicLong rx = new AtomicLong(DataFlowStatistics.getAllRxBytes());

            // 获取所有网卡的上行流量
            AtomicLong tx = new AtomicLong(DataFlowStatistics.getAllTxBytes());

            // 启动线程任务
            dispatcher.syncDispatch(() -> {
                NetManager netManager = NetManager.getInstance(getContext());

                if (!netManager.hasDefaultNet()) {
                    return;
                }
                NetHandle netHandle = netManager.getDefaultNet();

                // 可以获取网络状态的变化
                netManager.addDefaultNetStatusCallback(callback);

                // 通过openConnection来获取URLConnection
                HttpURLConnection connection = null;
                try {
                    String urlString =
"https://waylau.com/images/waylau_181_181.jpg";
                    URL url = new URL(urlString);

                    URLConnection urlConnection = netHandle.openConnection(url,
                            java.net.Proxy.NO_PROXY);
                    if (urlConnection instanceof HttpURLConnection) {
                        connection = (HttpURLConnection) urlConnection;
                    }
                    connection.setRequestMethod("GET");
                    connection.setReadTimeout(10000);
                    connection.setConnectTimeout(10000);
                    connection.connect();

                    // 之后可进行URL的其他操作
                    int code = connection.getResponseCode();
                    HiLog.info(LABEL_LOG, "ResponseCode: %{public}s", code);

                    if (code == HttpURLConnection.HTTP_OK) {
                        // 得到服务器返回的图片流对象在界面显示出来
                        InputStream inputStream = urlConnection.getInputStream();
                        ImageSource imageSource = ImageSource.create(inputStream,
new ImageSource.SourceOptions());
```

```java
                    ImageSource.DecodingOptions decodingOptions = new
ImageSource.DecodingOptions();
                    decodingOptions.desiredPixelFormat =
PixelFormat.ARGB_8888;
                    PixelMap pixelMap =
imageSource.createPixelmap(decodingOptions);

                    image.setPixelMap(pixelMap);
                    pixelMap.release();
                }
            } catch (Exception e) {
                e.printStackTrace();
            } finally {
                connection.disconnect();

                // 获取所有网卡的下行流量
                rx.set(DataFlowStatistics.getAllRxBytes() - rx.get());

                // 获取所有网卡的上行流量
                tx.set(DataFlowStatistics.getAllTxBytes() - tx.get());

                HiLog.info(LABEL_LOG, "connection disconnect, rx: %{public}s, tx: %{public}s",
                        rx.get(), tx.get());
            }
        });

        HiLog.info(LABEL_LOG, "end open");
    }

    private NetStatusCallback callback = new NetStatusCallback() {
        public void onAvailable(NetHandle handle) {
            HiLog.info(LABEL_LOG, "onAvailable");
        }

        public void onBlockedStatusChanged(NetHandle handle, boolean blocked) {
            HiLog.info(LABEL_LOG, "onBlockedStatusChanged");
        }

        public void onLosing(NetHandle handle, long maxMsToLive) {
            HiLog.info(LABEL_LOG, "onLosing");
        }

        public void onLost(NetHandle handle) {
            HiLog.info(LABEL_LOG, "onLosing");
        }

        public void onUnavailable() {
            HiLog.info(LABEL_LOG, "onUnavailable");
        }
```

```
            public void onCapabilitiesChanged(NetHandle handle, NetCapabilities
networkCapabilities) {
                HiLog.info(LABEL_LOG, "onCapabilitiesChanged");
            }

            public void onConnectionPropertiesChanged(NetHandle handle,
ConnectionProperties connectionProperties) {
                HiLog.info(LABEL_LOG, "onConnectionPropertiesChanged");
            }
        };

        @Override
        public void onActive() {
            super.onActive();
        }

        @Override
        public void onForeground(Intent intent) {
            super.onForeground(intent);
        }
    }
```

上述代码与NetManagerHandleURL应用的代码一致，只是多了获取所有网卡的上行和下行流量的逻辑。

24.4.6 运行

运行应用后，点击界面上的Open按钮以触发操作的执行。此时，控制台输出内容如下：

```
02-14 15:47:34.545 9634-9634/com.waylau.hmos.dataflowstatistics I
00001/MainAbilitySlice: before open
02-14 15:47:34.600 9634-10327/com.waylau.hmos.dataflowstatistics I
00001/MainAbilitySlice: onAvailable
02-14 15:47:34.600 9634-10327/com.waylau.hmos.dataflowstatistics I
00001/MainAbilitySlice: onCapabilitiesChanged
02-14 15:47:34.600 9634-10327/com.waylau.hmos.dataflowstatistics I
00001/MainAbilitySlice: onConnectionPropertiesChanged
02-14 15:47:36.404 9634-10325/com.waylau.hmos.dataflowstatistics I
00001/MainAbilitySlice: ResponseCode: 200
02-14 15:47:36.418 9634-10325/com.waylau.hmos.dataflowstatistics I
00001/MainAbilitySlice: connection disconnect, rx: 1144951, tx: 1123654
02-14 15:47:36.419 9634-9634/com.waylau.hmos.dataflowstatistics I
00001/MainAbilitySlice: end open
```

界面效果如图24-5所示。

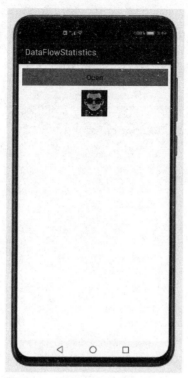

图 24-5 界面效果

第 25 章

电话服务

HarmonyOS电话服务系统提供了一系列的API,用于获取无线蜂窝网络和SIM卡相关的信息。

25.1 电话服务概述

本节主要介绍HarmonyOS电话服务系统提供的API及其约束与限制。

25.1.1 主要的API

电话服务系统主要由RadioInfoManager和SimInfoManager中的API构成。

应用可以通过调用RadioInfoManager中的API来获取当前注册网络名称、网络服务状态以及信号强度等信息。而SimInfoManager中的API主要用来获取SIM卡的相关信息。

25.1.2 约束与限制

注册获取SIM卡状态接口仅针对有SIM卡在位的场景生效,若用户拔出SIM卡,则接收不到回调事件。应用可通过调用hasSimCard接口来确定当前卡槽是否有卡在位。

25.2 实战：获取当前蜂窝网络信号信息

本节演示如何实现获取当前蜂窝网络信号信息。

应用通常需要获取用户所在蜂窝网络下的信号信息，以便获取当前驻网质量。开发者可以通过本业务获取用户指定SIM卡当前所在网络下的信号信息。

25.2.1 接口说明

RadioInfoManager类中提供了获取当前网络信号信息列表的方法。
RadioInfoManager的主要接口有：

- getInstance(Context context)：获取网络管理对象。
- getSignalInfoList(int slotId)：获取当前注册蜂窝网络信号强度信息。

25.2.2 创建应用

为了演示获取当前蜂窝网络信号信息的功能，创建一个名为RadioInfoManager的应用。
在应用的界面上，通过点击按钮来触发获取当前蜂窝网络信号信息的操作。

25.2.3 修改 ability_main.xml

修改ability_main.xml内容如下：

```xml
<?xml version="1.0" encoding="utf-8"?>
<DirectionalLayout
    xmlns:ohos="http://schemas.huawei.com/res/ohos"
    ohos:height="match_parent"
    ohos:width="match_parent"
    ohos:orientation="vertical">

    <Button
        ohos:id="$+id:button_get"
        ohos:height="match_content"
        ohos:width="match_parent"
        ohos:background_element="#F76543"
        ohos:layout_alignment="horizontal_center"
        ohos:margin="10vp"
        ohos:padding="10vp"
        ohos:text="Get"
        ohos:text_size="30fp"
        />
```

```
<Text
    ohos:id="$+id:text"
    ohos:height="match_content"
    ohos:width="match_content"
    ohos:background_element="$graphic:background_ability_main"
    ohos:layout_alignment="horizontal_center"
    ohos:text=""
    ohos:text_size="20fp"
    ohos:multiple_lines="true"
    />

</DirectionalLayout>
```

界面预览效果如图25-1所示。

图 25-1　界面预览效果

上述代码中：

- 设置了Get按钮，以备设置点击事件，以触发获取当前蜂窝网络信号信息的相关操作。
- Text组件用于展示获取到的信息结果。

25.2.4　修改 MainAbilitySlice

修改MainAbilitySlice内容如下：

```
package com.waylau.hmos.radioinfomanager.slice;

import com.waylau.hmos.radioinfomanager.ResourceTable;
import ohos.aafwk.ability.AbilitySlice;
import ohos.aafwk.content.Intent;
```

```java
import ohos.agp.components.Button;
import ohos.agp.components.Text;
import ohos.hiviewdfx.HiLog;
import ohos.hiviewdfx.HiLogLabel;
import ohos.telephony.LteSignalInformation;
import ohos.telephony.RadioInfoManager;
import ohos.telephony.SignalInformation;

import java.util.List;

public class MainAbilitySlice extends AbilitySlice {
    private static final String TAG = MainAbilitySlice.class.getSimpleName();
    private static final HiLogLabel LABEL_LOG =
            new HiLogLabel(HiLog.LOG_APP, 0x00001, TAG);

    private Text text;

    @Override
    public void onStart(Intent intent) {
        super.onStart(intent);
        super.setUIContent(ResourceTable.Layout_ability_main);

        // 为按钮设置点击事件回调
        Button buttonGet =
                (Button) findComponentById(ResourceTable.Id_button_get);
        buttonGet.setClickedListener(listener -> getInfo());

        text = (Text) findComponentById(ResourceTable.Id_text);
    }


    private void getInfo() {
        HiLog.info(LABEL_LOG, "before getInfo");

        // 获取RadioInfoManager对象
        RadioInfoManager radioInfoManager =
RadioInfoManager.getInstance(this.getContext());

        // 获取信号信息
        int slotId = 0; // 卡槽1
        List<SignalInformation> signalList =
radioInfoManager.getSignalInfoList(slotId);

        HiLog.info(LABEL_LOG, "signalList size: %{public}s",
signalList.size());

        // 检查信号信息列表大小
        if (signalList.size() == 0) {
            return;
```

```
        }

        // 依次遍历list获取当前驻网networkType对应的信号信息
        LteSignalInformation lteSignal;
        for (SignalInformation signal : signalList) {
            int signalNetworkType = signal.getNetworkType();
            int signalLevel = signal.getSignalLevel();

            HiLog.info(LABEL_LOG, "signalNetworkType: %{public}s, signalLevel: %{public}s",
                    signalNetworkType, signalLevel);
            text.append("signalNetworkType: " + signalNetworkType
                    + ", signalLevel: " + signalLevel + "\n");
        }

        HiLog.info(LABEL_LOG, "end getInfo");
    }

    @Override
    public void onActive() {
        super.onActive();
    }

    @Override
    public void onForeground(Intent intent) {
        super.onForeground(intent);
    }
}
```

上述代码中：

- 在Button上设置了点击事件。
- getInfo方法用于执行获取当前蜂窝网络信号信息的相关操作。
- 调用RadioInfoManager的getInstance接口获取RadioInfoManager实例。
- 调用getSignalInfoList(slotId)方法返回所有SignalInformation列表。
- 遍历SignalInformation列表，获取当前驻网NetworkType对应的信号信息。
- Text组件用于将结果信息展示在界面上。

25.2.5 运行

运行应用后，点击界面上的Get按钮以触发操作的执行。此时，控制台输出内容如下：

```
    02-14 16:15:34.938 23454-23454/com.waylau.hmos.radioinfomanager I
00001/MainAbilitySlice: before getInfo
    02-14 16:15:34.941 23454-23454/com.waylau.hmos.radioinfomanager I
00001/MainAbilitySlice: signalList size: 1
    02-14 16:15:34.942 23454-23454/com.waylau.hmos.radioinfomanager I
```

```
00001/MainAbilitySlice: signalNetworkType: 6, signalLevel: 5
    02-14 16:15:34.942 23454-23454/com.waylau.hmos.radioinfomanager I
00001/MainAbilitySlice: end getInfo
```

界面效果如图25-2所示。

图 25-2　界面效果

从上述结果可以看出，电话信号Level是5；NetworkType是6，是指NR网络。NR是基于OFDM的全新空口设计的全球性5G标准，也是下一代非常重要的蜂窝移动技术基础，5G技术将实现超低时延、高可靠性。

NetworkType值的含义定义在TelephonyConstants类中，源码如下：

```java
public final class TelephonyConstants {
    public static final int CALL_STATE_IDLE = 0;
    public static final int CALL_STATE_OFFHOOK = 2;
    public static final int CALL_STATE_RINGING = 1;
    public static final int CALL_STATE_UNKNOWN = -1;
    public static final int CT_NATIONAL_ROAMING_CARD = 41;
    public static final int CU_DUAL_MODE_CARD = 42;
    public static final int DEFAULT_SLOT_ID = 0;
    public static final int DUAL_MODE_CG_CARD = 40;
    public static final int DUAL_MODE_TELECOM_LTE_CARD = 43;
    public static final int DUAL_MODE_UG_CARD = 50;
    public static final int INVALID_SLOT_ID = -1;
    public static final int MAX_SIM_COUNT_DUAL = 2;
    public static final int MAX_SIM_COUNT_SINGLE = 1;
    public static final int MAX_SIM_COUNT_TRIPLE = 3;
    public static final int NETWORK_SELECTION_AUTOMATIC = 0;
    public static final int NETWORK_SELECTION_MANUAL = 1;
    public static final int NETWORK_SELECTION_UNKNOWN = -1;
    public static final int NETWORK_TYPE_CDMA = 2;
    public static final int NETWORK_TYPE_GSM = 1;
```

```java
    public static final int NETWORK_TYPE_LTE = 5;
    public static final int NETWORK_TYPE_NR = 6;
    public static final int NETWORK_TYPE_TDSCDMA = 4;
    public static final int NETWORK_TYPE_UNKNOWN = 0;
    public static final int NETWORK_TYPE_WCDMA = 3;
    public static final int NR_OPTION_NSA_AND_SA = 3;
    public static final int NR_OPTION_NSA_ONLY = 1;
    public static final int NR_OPTION_SA_ONLY = 2;
    public static final int NR_OPTION_UNKNOWN = 0;
    public static final int RADIO_TECHNOLOGY_1XRTT = 2;
    public static final int RADIO_TECHNOLOGY_EHRPD = 8;
    public static final int RADIO_TECHNOLOGY_EVDO = 7;
    public static final int RADIO_TECHNOLOGY_GSM = 1;
    public static final int RADIO_TECHNOLOGY_HSPA = 4;
    public static final int RADIO_TECHNOLOGY_HSPAP = 5;
    public static final int RADIO_TECHNOLOGY_IWLAN = 11;
    public static final int RADIO_TECHNOLOGY_LTE = 9;
    public static final int RADIO_TECHNOLOGY_LTE_CA = 10;
    public static final int RADIO_TECHNOLOGY_NR = 12;
    public static final int RADIO_TECHNOLOGY_TD_SCDMA = 6;
    public static final int RADIO_TECHNOLOGY_UNKNOWN = 0;
    public static final int RADIO_TECHNOLOGY_WCDMA = 3;
    public static final int RESULT_ERROR = -1;
    public static final int SIM_STATE_LOADED = 5;
    public static final int SIM_STATE_LOCKED = 2;
    public static final int SIM_STATE_NOT_PRESENT = 1;
    public static final int SIM_STATE_NOT_READY = 3;
    public static final int SIM_STATE_READY = 4;
    public static final int SIM_STATE_UNKNOWN = 0;
    public static final int SINGLE_MODE_RUIM_CARD = 30;
    public static final int SINGLE_MODE_SIM_CARD = 10;
    public static final int SINGLE_MODE_USIM_CARD = 20;
    public static final int SLOT_ID_0 = 0;
    public static final int SLOT_ID_1 = 1;
    public static final int SLOT_ID_2 = 2;
    public static final int UNKNOWN_CARD = -1;

    public TelephonyConstants() {
        throw new RuntimeException("Stub!");
    }
}
```

25.3 实战：观察蜂窝网络的状态变化

本节演示观察蜂窝网络的状态变化。应用可以通过观察蜂窝网络的状态变化来接收最新蜂窝网络的服务状态信息、信号信息等。

25.3.1 接口说明

应用使用当前网络观察蜂窝网络的状态变化,主要涉及RadioStateObserver和RadioInfoManager两个类。

RadioStateObserver类中提供了观察蜂窝网络状态变化的方法,为了能够实时观察蜂窝网络状态变化,应用必须包含ohos.permission.GET_NETWORK_INFO权限。

需要使用RadioInfoManager的如下接口将继承RadioStateObserver类的对象注册到系统服务。

- addObserver:添加观察。可以观察OBSERVE_MASK_NETWORK_STATE(观察蜂窝网络驻网状态信息)和OBSERVE_MASK_SIGNAL_INFO(观察蜂窝网络信号信息)。
- removeObserver:停止观察所有状态的变化。

25.3.2 创建应用

为了演示观察蜂窝网络状态变化的功能,创建一个名为RadioStateObserver的应用。
在应用的界面上,通过点击按钮来触发观察蜂窝网络状态变化的操作。

25.3.3 声明权限

修改配置文件,声明使用网络的权限如下:

```
// 声明权限
"reqPermissions": [
   {
   "name": "ohos.permission.GET_NETWORK_INFO"
   }
]
```

由于上述权限不是敏感权限,因此不需要在代码中显式声明。

25.3.4 修改 ability_main.xml

修改ability_main.xml内容如下:

```xml
<?xml version="1.0" encoding="utf-8"?>
<DirectionalLayout
    xmlns:ohos="http://schemas.huawei.com/res/ohos"
    ohos:height="match_parent"
    ohos:width="match_parent"
    ohos:orientation="vertical">

    <Button
        ohos:id="$+id:button_add"
```

```xml
    ohos:height="match_content"
    ohos:width="match_parent"
    ohos:background_element="#F76543"
    ohos:layout_alignment="horizontal_center"
    ohos:margin="10vp"
    ohos:padding="10vp"
    ohos:text="Add"
    ohos:text_size="30fp"
    />

<Button
    ohos:id="$+id:button_remove"
    ohos:height="match_content"
    ohos:width="match_parent"
    ohos:background_element="#F76543"
    ohos:layout_alignment="horizontal_center"
    ohos:margin="10vp"
    ohos:padding="10vp"
    ohos:text="Remove"
    ohos:text_size="30fp"
    />

<Text
    ohos:id="$+id:text"
    ohos:height="match_content"
    ohos:width="match_content"
    ohos:background_element="$graphic:background_ability_main"
    ohos:layout_alignment="horizontal_center"
    ohos:text=""
    ohos:text_size="20fp"
    ohos:multiple_lines="true"
    />
</DirectionalLayout>
```

界面预览效果如图25-3所示。

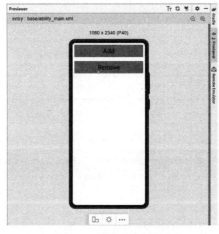

图25-3　界面预览效果

上述代码中：

- 设置了Add按钮和Remove按钮，以备设置点击事件，以触发获取当前蜂窝网络信号信息的相关操作。
- Text组件用于展示获取到的信息结果。

25.3.5 修改 MainAbilitySlice

修改MainAbilitySlice内容如下：

```java
package com.waylau.hmos.radiostateobserver.slice;

import com.waylau.hmos.radiostateobserver.ResourceTable;
import ohos.aafwk.ability.AbilitySlice;
import ohos.aafwk.content.Intent;
import ohos.agp.components.Button;
import ohos.agp.components.Text;
import ohos.eventhandler.EventRunner;
import ohos.hiviewdfx.HiLog;
import ohos.hiviewdfx.HiLogLabel;
import ohos.telephony.*;
import ohos.utils.zson.ZSONObject;

import java.util.List;


public class MainAbilitySlice extends AbilitySlice {
    private static final String TAG = MainAbilitySlice.class.getSimpleName();
    private static final HiLogLabel LABEL_LOG =
            new HiLogLabel(HiLog.LOG_APP, 0x00001, TAG);

    private Text text;

    private RadioInfoManager radioInfoManager;
    private MyRadioStateObserver observer;

    @Override
    public void onStart(Intent intent) {
        super.onStart(intent);
        super.setUIContent(ResourceTable.Layout_ability_main);

        // 为按钮设置点击事件回调
        Button buttonAdd =
                (Button) findComponentById(ResourceTable.Id_button_add);
        buttonAdd.setClickedListener(listener -> add());

        Button buttonRemove =
                (Button) findComponentById(ResourceTable.Id_button_remove);
        buttonRemove.setClickedListener(listener -> remove());
```

```java
        text = (Text) findComponentById(ResourceTable.Id_text);

        // 获取RadioInfoManager对象
        radioInfoManager = RadioInfoManager.getInstance(this.getContext());
    }

    private void remove() {
        HiLog.info(LABEL_LOG, "before remove");

        // 停止观察
        radioInfoManager.removeObserver(observer);

        text.append("remove observer\n");

        HiLog.info(LABEL_LOG, "end remove");
    }


    private void add() {
        HiLog.info(LABEL_LOG, "before add");

        // 执行回调的runner
        EventRunner runner = EventRunner.create();

        // 创建MyRadioStateObserver的对象
        int slotId = 0; // 卡槽1
        observer = new MyRadioStateObserver(slotId, runner);

        // 添加回调,以NETWORK_STATE和SIGNAL_INFO为例
        radioInfoManager.addObserver(observer, RadioStateObserver.OBSERVE_MASK_NETWORK_STATE | RadioStateObserver.OBSERVE_MASK_SIGNAL_INFO);

        text.append("add observer\n");

        HiLog.info(LABEL_LOG, "end add");
    }

    // 创建继承RadioStateObserver的类MyRadioStateObserver
    private class MyRadioStateObserver extends RadioStateObserver {
        // 构造方法,在当前线程的runner中执行回调,slotId需要传入要观察的卡槽ID(0或1)
        MyRadioStateObserver(int slotId) {
            super(slotId);
        }

        // 构造方法,在runner中执行回调
        MyRadioStateObserver(int slotId, EventRunner runner) {
            super(slotId, runner);
        }
```

```java
        // 网络注册状态变化的回调方法
        @Override
        public void onNetworkStateUpdated(NetworkState state) {
            String stateString = ZSONObject.toZSONString(state.getNsaState());
            HiLog.info(LABEL_LOG, "onNetworkStateUpdated, state: %{public}s",
                stateString);
        }

        // 信号信息变化的回调方法
        @Override
        public void onSignalInfoUpdated(List<SignalInformation> signalInfos) {
            HiLog.info(LABEL_LOG, "onSignalInfoUpdated");

            for (SignalInformation signal : signalInfos) {
                int signalNetworkType = signal.getNetworkType();
                int signalLevel = signal.getSignalLevel();

                HiLog.info(LABEL_LOG, "signalNetworkType: %{public}s, signalLevel: %{public}s",
                    signalNetworkType, signalLevel);
            }
        }
    }

    @Override
    public void onActive() {
        super.onActive();
    }

    @Override
    public void onForeground(Intent intent) {
        super.onForeground(intent);
    }
}
```

上述代码中:

- onStart方法中初始化了RadioInfoManager对象。
- 在Button上设置了点击事件。
- add方法添加事件监听用于启动Socket服务端。其中，实例化了MyRadioStateObserver的对象作为监听器。MyRadioStateObserver继承自RadioStateObserver类。
- remove方法用于移除监听器。

25.3.6 运行

运行应用后，点击界面上的Add按钮以触发监听事件的操作执行。此时，控制台输出内容如下:

```
09-13 23:59:39.692 10677-10677/com.waylau.hmos.radiostateobserver I
00001/MainAbilitySlice: before add
```

```
    09-13 23:59:39.706 10677-10677/com.waylau.hmos.radiostateobserver I
00001/MainAbilitySlice:  end add
    09-13 23:59:39.718 10677-29897/com.waylau.hmos.radiostateobserver I
00001/MainAbilitySlice:  onNetworkStateUpdated, state: 1
    09-13 23:59:39.718 10677-29897/com.waylau.hmos.radiostateobserver I
00001/MainAbilitySlice:  onSignalInfoUpdated
    09-13 23:59:39.719 10677-29897/com.waylau.hmos.radiostateobserver I
00001/MainAbilitySlice:  signalNetworkType: 6, signalLevel: 5
```

通过上述日志可以看出，当前NetworkType是6，状态是1，信号Level是5。界面效果如图25-4所示。

手动点击设置手机网络，将5G网络关闭，以模拟网络的变化。界面效果如图25-5所示。

图 25-4　界面效果　　　　　　　图 25-5　界面效果

此时，控制台输出内容如下：

```
    09-14 00:05:06.455 10677-29897/com.waylau.hmos.radiostateobserver I
00001/MainAbilitySlice:  onSignalInfoUpdated
    09-14 00:05:06.455 10677-29897/com.waylau.hmos.radiostateobserver I
00001/MainAbilitySlice:  signalNetworkType: 5, signalLevel: 5
    09-14 00:05:07.041 10677-29897/com.waylau.hmos.radiostateobserver I
00001/MainAbilitySlice:  onNetworkStateUpdated, state: 1
    09-14 00:05:07.048 10677-29897/com.waylau.hmos.radiostateobserver I
00001/MainAbilitySlice:  onSignalInfoUpdated
    09-14 00:05:07.049 10677-29897/com.waylau.hmos.radiostateobserver I
00001/MainAbilitySlice:  signalNetworkType: 5, signalLevel: 5
    09-14 00:05:07.821 10677-29897/com.waylau.hmos.radiostateobserver I
00001/MainAbilitySlice:  onNetworkStateUpdated, state: 1
```

通过上述日志可以看出，应用已经能够观察到网络的变化了。当前NetworkType是5，状态是1，信号Level是5。

从上一节的内容可以获知，NetworkType为5是LTE网络的意思。

再次将5G网络打开，以模拟网络的变化。界面效果如图25-6所示。

图 25-6 界面效果

此时，控制台输出内容如下：

```
09-14 00:05:52.807 10677-29897/com.waylau.hmos.radiostateobserver I
00001/MainAbilitySlice: onSignalInfoUpdated
    09-14 00:05:52.808 10677-29897/com.waylau.hmos.radiostateobserver I
00001/MainAbilitySlice: signalNetworkType: 6, signalLevel: 5
    09-14 00:05:53.295 10677-29897/com.waylau.hmos.radiostateobserver I
00001/MainAbilitySlice: onNetworkStateUpdated, state: 1
    09-14 00:05:53.307 10677-29897/com.waylau.hmos.radiostateobserver I
00001/MainAbilitySlice: onSignalInfoUpdated
    09-14 00:05:53.308 10677-29897/com.waylau.hmos.radiostateobserver I
00001/MainAbilitySlice: signalNetworkType: 6, signalLevel: 5
    09-14 00:05:53.815 10677-29897/com.waylau.hmos.radiostateobserver I
00001/MainAbilitySlice: onNetworkStateUpdated, state: 1
```

通过上述日志可以看出，应用又观察到网络的变化了。当前NetworkType是6，状态是1，信号Level是5。

点击界面上的Remove按钮以触发移除监听事件的操作执行。此时，控制台输出内容如下：

```
09-14 00:06:41.539 10677-10677/com.waylau.hmos.radiostateobserver I
00001/MainAbilitySlice: before remove
    09-14 00:06:41.547 10677-10677/com.waylau.hmos.radiostateobserver I
00001/MainAbilitySlice: end remove
```

界面效果如图25-7所示。

图 25-7 界面效果

此时，将5G网络关闭或者打开已经不会再监听到任何事件，证明事件监听器已经移除。

第 26 章

设备管理

HarmonyOS设备是底层硬件的一种设备抽象概念。HarmonyOS提供了一系列API针对不同的设备进行管理。

26.1 设备管理概述

HarmonyOS设备是对底层硬件的一种设备抽象概念。
HarmonyOS提供了一系列API针对不同的设备进行管理，这些设备包括：

- 传感器。传感器分为6大类：运动类传感器、环境类传感器、方向类传感器、光线类传感器、健康类传感器、其他类传感器（如霍尔传感器）。
- 控制类小器件。控制类小器件指的是设备上的LED灯和振动器。其中，LED灯主要用作指示（如充电状态）、闪烁（如三色灯）等，振动器主要用于闹钟、开关机振动、来电振动等场景。
- 位置。位置功能用于确定用户设备在哪里，系统使用位置坐标标示设备的位置，并用多种定位技术提供服务，如GNSS定位、基站定位、WLAN/蓝牙定位（基站定位、WLAN/蓝牙定位后续统称"网络定位技术"）。通过这些定位技术，无论用户设备在室内还是户外都可以准确地确定设备位置。
- 设置。应用程序可以对系统各类设置项进行查询。例如，第三方应用提前注册飞行模式设置项的回调，当用户通过系统设置修改终端的飞行模式状态时，第三方应用会检测到此设置项发生变化并进行适配。若检测到飞行模式开启，则进入离线状态；若检测到飞行模式关闭，则重新获取在线数据。

26.1.1 传感器

HarmonyOS传感器包含4个模块：Sensor API、Sensor Framework、Sensor Service、HD_IDL层。图26-1展示的是HarmonyOS传感器的架构图。

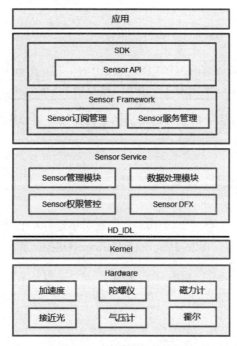

图 26-1 传感器的架构图

1. 传感器的架构

传感器的架构可以分为以下几层：

- Sensor API：提供传感器的基础API，主要包含查询传感器的列表、订阅/取消传感器的数据、执行控制命令等，简化应用开发。
- Sensor Framework：主要实现传感器的订阅管理，数据通道的创建、销毁、订阅与取消订阅，实现与SensorService的通信。
- Sensor Service：主要实现HD_IDL层数据接收、解析、分发，前后台的策略管控，对该设备Sensor的管理，Sensor权限管控等。
- HD_IDL层：对不同的FIFO、频率进行策略选择，以及对不同设备的适配。

2. 约束与限制

针对某些传感器，开发者需要请求相应的权限，才能获取相应传感器的数据。表26-1展示的是传感器权限。

表 26-1 传感器权限

传 感 器	权 限 名	敏感级别	权限描述
加速度传感器、加速度未校准传感器、线性加速度传感器	ohos.permission.ACCELEROMETER	system_grant	允许订阅 Motion 组对应的加速度传感器的数据
陀螺仪传感器、陀螺仪未校准传感器	ohos.permission.GYROSCOPE	system_grant	允许订阅 Motion 组对应的陀螺仪传感器的数据
计步器	ohos.permission.ACTIVITY_MOTION	user_grant	允许订阅运动状态
心率	ohos.permission.READ_HEALTH_DATA	user_grant	允许读取健康数据

传感器数据订阅和取消订阅接口成对调用,当不再需要订阅传感器的数据时,开发者需要调用取消订阅接口进行资源释放。

26.1.2 控制类小器件

HarmonyOS控制类小器件主要包含4个模块:控制类小器件API、控制类小器件Framework、控制类小器件Service、HD_IDL层。图26-2展示的是HarmonyOS控制类小器件的架构图。

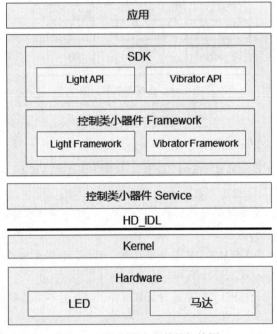

图 26-2 控制类小器件的架构图

1. 控制类小器件的架构

控制类小器件的架构可以分为以下几层:

- 控制类小器件API:提供灯和振动器基础的API,主要包含灯的列表查询、打开灯、关闭灯等接口，以及振动器的列表查询、振动器的振动器效果查询、触发/关闭振动器等接口。
- 控制类小器件Framework：主要实现灯和振动器的框架层管理，实现与控制类小器件Service的通信。
- 控制类小器件Service：实现灯和振动器的服务管理。
- HD_IDL层：对不同设备的适配。

2. 约束与限制

使用控制类小器件时需要注意以下约束与限制：
- 在调用Light API时，先通过getLightIdList接口查询设备所支持的灯的ID列表，以免调用打开接口异常。
- 在调用Vibrator API时，先通过getVibratorIdList接口查询设备所支持的振动器的ID列表，以免调用振动接口异常。
- 在使用振动器时，开发者需要配置请求振动器的权限ohos.permission.VIBRATE，才能控制振动器振动。

26.1.3 位置

移动终端设备已经深入人们日常生活的方方面面，如查看所在城市的天气、新闻轶事、出行打车、旅行导航、运动记录。这些习以为常的活动都离不开定位用户终端设备的位置。

当用户处于这些丰富的使用场景中时，系统的位置能力可以提供实时准确的位置数据。对于开发者，设计基于位置体验的服务，也可以使应用的使用体验更贴近用户。

当应用在实现基于设备位置的功能（如驾车导航、记录运动轨迹等）时，可以调用该模块的API接口完成位置信息的获取。

1. 基本概念

位置功能包含以下基本概念：
- 坐标：系统以1984年的世界大地坐标系统为参考，使用经度、纬度数据描述地球上的一个位置。
- GNSS定位：基于全球导航卫星系统，包含GPS、GLONASS、北斗、Galileo等，通过导航卫星、设备芯片提供的定位算法来确定设备的准确位置。定位过程具体使用哪些定位系统取决于用户设备的硬件能力。
- 基站定位：根据设备当前驻网基站和相邻基站的位置估算设备当前的位置。此定位方式的定位结果精度相对较低，并且需要设备可以访问蜂窝网络。
- WLAN、蓝牙定位：根据设备可搜索到的周围WLAN、蓝牙设备的位置估算设备当前的位置。此定位方式的定位结果精度依赖设备周围可见的固定WLAN、蓝牙设备的分布，密度较高时，精度也比基站定位方式更高，同时也需要设备可以访问网络。

2. 运作机制

位置功能作为系统为应用提供的一种基础服务，需要应用在所使用的业务场景向系统主动发起请求,并在业务场景结束时主动结束此请求,在此过程中系统会将实时的定位结果上报给应用。

3. 约束与限制

使用位置时需要注意以下约束与限制:

- 使用设备的位置功能需要用户进行确认并主动开启位置开关。如果位置开关没有开启,系统就不会向任何应用提供位置服务。
- 设备位置信息属于用户敏感数据,所以即使用户已经开启位置开关,应用在获取设备位置前仍需向用户申请位置访问权限。在用户确认允许后,系统才会向应用提供位置服务。

26.2 实战:传感器示例

本节演示如何实现获取方向传感器的数据。

HarmonyOS传感器提供的功能有查询传感器的列表、订阅/取消订阅传感器数据、查询传感器的最小采样时间间隔、执行控制命令。

26.2.1 接口说明

方向传感器CategoryOrientationAgent的主要接口有:

- getAllSensors():获取属于方向类别的传感器列表。
- getAllSensors(int):获取属于方向类别中特定类型的传感器列表。
- getSingleSensor(int):查询方向类别中特定类型的默认sensor(如果存在多个,则返回第一个)。
- setSensorDataCallback(ICategoryOrientationDataCallback, CategoryOrientation, long):以设定的采样间隔订阅给指定传感器的数据。
- setSensorDataCallback(ICategoryOrientationDataCallback, CategoryOrientation, long, long):以设定的采样间隔和时延订阅给指定传感器的数据。
- releaseSensorDataCallback(ICategoryOrientationDataCallback, CategoryOrientation):取消订阅指定传感器的数据。
- releaseSensorDataCallback(ICategoryOrientationDataCallback):取消订阅所有传感器的数据。

26.2.2 创建应用

为了演示获取方向传感器的数据的功能,创建一个名为CategoryOrientationAgent的应用。

在应用的界面上,通过点击按钮来触发获取方向传感器的数据的操作。

26.2.3 修改 ability_main.xml

修改ability_main.xml内容如下:

```xml
<?xml version="1.0" encoding="utf-8"?>
<DirectionalLayout
    xmlns:ohos="http://schemas.huawei.com/res/ohos"
    ohos:height="match_parent"
    ohos:width="match_parent"
    ohos:orientation="vertical">

    <Button
        ohos:id="$+id:button_add"
        ohos:height="match_content"
        ohos:width="match_parent"
        ohos:background_element="#F76543"
        ohos:layout_alignment="horizontal_center"
        ohos:margin="10vp"
        ohos:padding="10vp"
        ohos:text="Add"
        ohos:text_size="30fp"
        />

    <Button
        ohos:id="$+id:button_remove"
        ohos:height="match_content"
        ohos:width="match_parent"
        ohos:background_element="#F76543"
        ohos:layout_alignment="horizontal_center"
        ohos:margin="10vp"
        ohos:padding="10vp"
        ohos:text="Remove"
        ohos:text_size="30fp"
        />

</DirectionalLayout>
```

界面预览效果如图26-3所示。

上述代码设置了Add按钮和Remove按钮,以备设置点击事件,以触发获取方向传感器数据的相关操作。

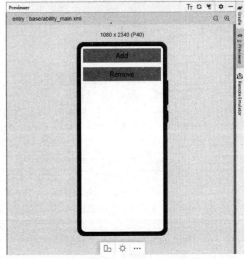

图 26-3 界面预览效果

26.2.4 修改 MainAbilitySlice

修改MainAbilitySlice内容如下:

```java
package com.waylau.hmos.categoryorientationagent.slice;

import com.waylau.hmos.categoryorientationagent.ResourceTable;
import ohos.aafwk.ability.AbilitySlice;
import ohos.aafwk.content.Intent;
import ohos.agp.components.Button;
import ohos.hiviewdfx.HiLog;
import ohos.hiviewdfx.HiLogLabel;
import ohos.sensor.agent.CategoryOrientationAgent;
import ohos.sensor.bean.CategoryOrientation;
import ohos.sensor.data.CategoryOrientationData;
import ohos.sensor.listener.ICategoryOrientationDataCallback;
import ohos.utils.zson.ZSONObject;

public class MainAbilitySlice extends AbilitySlice {
    private static final String TAG = MainAbilitySlice.class.getSimpleName();
    private static final HiLogLabel LABEL_LOG =
            new HiLogLabel(HiLog.LOG_APP, 0x00001, TAG);
    private static final int MATRIX_LENGTH = 9;
    private static final long INTERVAL = 100000000L;

    private ICategoryOrientationDataCallback orientationDataCallback;

    private CategoryOrientationAgent categoryOrientationAgent;

    private CategoryOrientation orientationSensor;

    @Override
    public void onStart(Intent intent) {
        super.onStart(intent);
        super.setUIContent(ResourceTable.Layout_ability_main);

        // 初始化对象
        initData();

        // 为按钮设置点击事件回调
        Button buttonAdd =
                (Button) findComponentById(ResourceTable.Id_button_add);
        buttonAdd.setClickedListener(listener -> add());

        Button buttonRemove =
                (Button) findComponentById(ResourceTable.Id_button_remove);
        buttonRemove.setClickedListener(listener -> remove());
    }

    private void initData() {
        orientationDataCallback = new ICategoryOrientationDataCallback() {
            @Override
            public void onSensorDataModified(CategoryOrientationData categoryOrientationData) {
```

```java
            // 对接收的categoryOrientationData传感器数据对象解析和使用
            //获取传感器的维度信息
            int dim = categoryOrientationData.getSensorDataDim();

            // 获取方向类传感器的第一维数据
            float degree = categoryOrientationData.getValues()[0];

            // 根据旋转矢量传感器的数据获得旋转矩阵
            float[] rotationMatrix = new float[MATRIX_LENGTH];
CategoryOrientationData.getDeviceRotationMatrix(rotationMatrix,
                    categoryOrientationData.values);

            // 根据计算出来的旋转矩阵获取设备的方向
            float[] rotationAngle = new float[MATRIX_LENGTH];
            float[] rotationAngleResult =
CategoryOrientationData.getDeviceOrientation(rotationMatrix, rotationAngle);

            HiLog.info(LABEL_LOG, "dim:%{public}s, degree: %{public}s, " +
                    "rotationMatrix: %{public}s, rotationAngle: %{public}s",
                    dim, degree, ZSONObject.toZSONString(rotationMatrix),
                    ZSONObject.toZSONString(rotationAngleResult));
        }

        @Override
        public void onAccuracyDataModified(CategoryOrientation categoryOrientation, int i) {
            // 使用变化的精度
            HiLog.info(LABEL_LOG, "onAccuracyDataModified");
        }

        @Override
        public void onCommandCompleted(CategoryOrientation categoryOrientation) {
            // 传感器执行命令回调
            HiLog.info(LABEL_LOG, "onCommandCompleted");
        }
    };

    categoryOrientationAgent = new CategoryOrientationAgent();
}

private void remove() {
    HiLog.info(LABEL_LOG, "before remove");

    if (orientationSensor != null) {
        // 取消订阅传感器数据
        categoryOrientationAgent.releaseSensorDataCallback(
```

```
                    orientationDataCallback, orientationSensor);
        }

        HiLog.info(LABEL_LOG, "end remove");
    }



    private void add() {
        HiLog.info(LABEL_LOG, "before add");

        // 获取传感器对象,并订阅传感器数据
        orientationSensor = categoryOrientationAgent.getSingleSensor(
                CategoryOrientation.SENSOR_TYPE_ORIENTATION);
        if (orientationSensor != null) {
            categoryOrientationAgent.setSensorDataCallback(
                    orientationDataCallback, orientationSensor, INTERVAL);
        }

        HiLog.info(LABEL_LOG, "end add");
    }

    @Override
    public void onActive() {
        super.onActive();
    }

    @Override
    public void onForeground(Intent intent) {
        super.onForeground(intent);
    }
}
```

上述代码中:

- 在Button上设置了点击事件。
- initData用于初始化ICategoryOrientationDataCallback和CategoryOrientationAgent对象。
- add方法用于订阅传感器数据。
- remove方法用于取消订阅传感器数据。

26.2.5 运行

运行应用后,点击界面上的Add按钮以触发订阅传感器数据的操作执行。此时,控制台输出内容如下:

```
09-14 00:11:42.837 17501-17501/com.waylau.hmos.categoryorientationagent I
00001/MainAbilitySlice: before add
09-14 00:11:42.878 17501-17501/com.waylau.hmos.categoryorientationagent I
00001/MainAbilitySlice: end add
09-14 00:11:43.525 17501-21999/com.waylau.hmos.categoryorientationagent I
```

```
00001/MainAbilitySlice: dim:3, degree: 304.9, rotationMatrix:
[-14.3298,-195.13599,-1676.95,-195.13599,-185942.14,1.76,-1676.95,1.76,-185927
.22], rotationAngle:
[-3.1405432,-9.465305E-6,3.1325736,0.0,0.0,0.0,0.0,0.0,0.0]
    09-14 00:11:44.115 17501-21999/com.waylau.hmos.categoryorientationagent I
00001/MainAbilitySlice: dim:3, degree: 304.77, rotationMatrix:
[-15.023399,-207.24359,-1712.8073,-207.24359,-185784.3,1.9108,-1712.8073,1.910
8,-185768.73], rotationAngle:
[-3.1404772,-1.0285039E-5,3.1323729,0.0,0.0,0.0,0.0,0.0,0.0]
    09-14 00:11:44.715 17501-21999/com.waylau.hmos.categoryorientationagent I
00001/MainAbilitySlice: dim:3, degree: 304.9, rotationMatrix:
[-14.3298,-195.13599,-1676.95,-195.13599,-185942.14,1.76,-1676.95,1.76,-185927
.22], rotationAngle:
[-3.1405432,-9.465305E-6,3.1325736,0.0,0.0,0.0,0.0,0.0,0.0]
    09-14 00:11:45.303 17501-21999/com.waylau.hmos.categoryorientationagent I
00001/MainAbilitySlice: dim:3, degree: 304.77, rotationMatrix:
[-15.023399,-207.24359,-1712.8073,-207.24359,-185784.3,1.9108,-1712.8073,1.910
8,-185768.73], rotationAngle:
[-3.1404772,-1.0285039E-5,3.1323729,0.0,0.0,0.0,0.0,0.0,0.0]
    09-14 00:11:45.891 17501-21999/com.waylau.hmos.categoryorientationagent I
00001/MainAbilitySlice: dim:3, degree: 304.9, rotationMatrix:
[-14.3298,-195.13599,-1676.95,-195.13599,-185942.14,1.76,-1676.95,1.76,-185927
.22], rotationAngle:
[-3.1405432,-9.465305E-6,3.1325736,0.0,0.0,0.0,0.0,0.0,0.0]
    09-14 00:11:46.487 17501-21999/com.waylau.hmos.categoryorientationagent I
00001/MainAbilitySlice: dim:3, degree: 304.77, rotationMatrix:
[-15.023399,-207.24359,-1712.8073,-207.24359,-185784.3,1.9108,-1712.8073,1.910
8,-185768.73], rotationAngle:
[-3.1404772,-1.0285039E-5,3.1323729,0.0,0.0,0.0,0.0,0.0,0.0]
    09-14 00:11:47.087 17501-21999/com.waylau.hmos.categoryorientationagent I
00001/MainAbilitySlice: dim:3, degree: 304.9, rotationMatrix:
[-14.3298,-195.13599,-1676.95,-195.13599,-185942.14,1.76,-1676.95,1.76,-185927
.22], rotationAngle:
[-3.1405432,-9.465305E-6,3.1325736,0.0,0.0,0.0,0.0,0.0,0.0]
    09-14 00:11:47.695 17501-21999/com.waylau.hmos.categoryorientationagent I
00001/MainAbilitySlice: dim:3, degree: 304.77, rotationMatrix:
[-15.023399,-207.24359,-1712.8073,-207.24359,-185784.3,1.9108,-1712.8073,1.910
8,-185768.73], rotationAngle:
[-3.1404772,-1.0285039E-5,3.1323729,0.0,0.0,0.0,0.0,0.0,0.0]
    09-14 00:11:48.295 17501-21999/com.waylau.hmos.categoryorientationagent I
00001/MainAbilitySlice: dim:3, degree: 304.9, rotationMatrix:
[-14.3298,-195.13599,-1676.95,-195.13599,-185942.14,1.76,-1676.95,1.76,-185927
.22], rotationAngle:
[-3.1405432,-9.465305E-6,3.1325736,0.0,0.0,0.0,0.0,0.0,0.0]
    09-14 00:11:48.883 17501-21999/com.waylau.hmos.categoryorientationagent I
00001/MainAbilitySlice: dim:3, degree: 304.77, rotationMatrix:
[-15.023399,-207.24359,-1712.8073,-207.24359,-185784.3,1.9108,-1712.8073,1.910
8,-185768.73], rotationAngle:
[-3.1404772,-1.0285039E-5,3.1323729,0.0,0.0,0.0,0.0,0.0,0.0]
    09-14 00:11:49.483 17501-21999/com.waylau.hmos.categoryorientationagent I
```

```
00001/MainAbilitySlice: dim:3, degree: 304.9, rotationMatrix:
[-14.3298,-195.13599,-1676.95,-195.13599,-185942.14,1.76,-1676.95,1.76,-185927
.22], rotationAngle:
[-3.1405432,-9.465305E-6,3.1325736,0.0,0.0,0.0,0.0,0.0,0.0]
```

从上述日志可以看到，订阅的数据以规定的频率发送过来。

点击界面上的Remove按钮以触发取消订阅传感器数据的操作执行。此时，控制台输出内容如下：

```
09-14 00:12:00.884 17501-17501/com.waylau.hmos.categoryorientationagent I
00001/MainAbilitySlice: before remove
09-14 00:12:00.918 17501-17501/com.waylau.hmos.categoryorientationagent I
00001/MainAbilitySlice: end remove
```

26.3 实战：Light 示例

本节演示如何使用Light模块。当设备需要设置不同的闪烁效果时，可以调用Light模块，例如LED灯能够设置灯颜色、灯亮和灯灭时长的闪烁效果。

26.3.1 接口说明

灯模块主要提供的功能有查询设备上灯的列表、查询某个灯设备支持的效果、打开和关闭灯设备。LightAgent类开放能力的主要接口有：

- getLightIdList()：获取硬件设备上的灯列表。
- isSupport(int)：根据指定的灯ID查询硬件设备是否有该灯。
- isEffectSupport(int, String)：查询指定的灯是否支持指定的闪烁效果。
- turnOn(int, String)：对指定的灯创建指定效果的一次性闪烁。
- turnOn(int, LightEffect)：对指定的灯创建自定义效果的一次性闪烁。
- turnOn(String)：对指定的灯创建指定效果的一次性闪烁。
- turnOn(LightEffect)：对指定的灯创建自定义效果的一次性闪烁。
- turnOff(int)：关闭指定的灯。
- turnOff()：关闭指定的灯。

26.3.2 创建应用

为了演示使用Light模块的功能，创建一个名为LightAgent的应用。

在应用的界面上，通过点击按钮来触发使用Light模块的操作。

26.3.3 修改 ability_main.xml

修改ability_main.xml内容如下：

```xml
<?xml version="1.0" encoding="utf-8"?>
<DirectionalLayout
    xmlns:ohos="http://schemas.huawei.com/res/ohos"
    ohos:height="match_parent"
    ohos:width="match_parent"
    ohos:orientation="vertical">

    <Button
        ohos:id="$+id:button_on"
        ohos:height="40vp"
        ohos:width="match_parent"
        ohos:background_element="#F76543"
        ohos:layout_alignment="horizontal_center"
        ohos:margin="10vp"
        ohos:padding="10vp"
        ohos:text="On"
        ohos:text_size="50"
        />

    <Button
        ohos:id="$+id:button_off"
        ohos:height="40vp"
        ohos:width="match_parent"
        ohos:background_element="#F76543"
        ohos:layout_alignment="horizontal_center"
        ohos:margin="10vp"
        ohos:padding="10vp"
        ohos:text="Off"
        ohos:text_size="50"
        />

</DirectionalLayout>
```

界面预览效果如图26-4所示。

上述代码设置了On按钮和Off按钮，以备设置点击事件，以触发使用Light模块的相关操作。

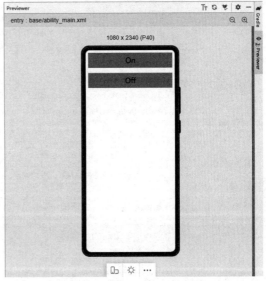

图 26-4　界面预览效果

26.3.4 修改 MainAbilitySlice

修改MainAbilitySlice内容如下：

```java
package com.waylau.hmos.lightagent.slice;

import com.waylau.hmos.lightagent.ResourceTable;
import ohos.aafwk.ability.AbilitySlice;
import ohos.aafwk.content.Intent;
import ohos.agp.components.Button;
import ohos.hiviewdfx.HiLog;
import ohos.hiviewdfx.HiLogLabel;
import ohos.light.agent.LightAgent;
import ohos.light.bean.LightBrightness;
import ohos.light.bean.LightEffect;

import java.util.List;

public class MainAbilitySlice extends AbilitySlice {
    private static final String TAG = MainAbilitySlice.class.getSimpleName();
    private static final HiLogLabel LABEL_LOG =
            new HiLogLabel(HiLog.LOG_APP, 0x00001, TAG);

    private LightAgent lightAgent;
    private List<Integer> isEffectSupportLightIdList;

    @Override
    public void onStart(Intent intent) {
        super.onStart(intent);
        super.setUIContent(ResourceTable.Layout_ability_main);

        // 初始化对象
        initData();

        // 为按钮设置点击事件回调
        Button buttonAdd =
                (Button) findComponentById(ResourceTable.Id_button_on);
        buttonAdd.setClickedListener(listener -> on());

        Button buttonRemove =
                (Button) findComponentById(ResourceTable.Id_button_off);
        buttonRemove.setClickedListener(listener -> off());
    }

    private void initData() {
        HiLog.info(LABEL_LOG, "before initData");

        lightAgent = new LightAgent();

        // 查询硬件设备上的灯列表
        isEffectSupportLightIdList = lightAgent.getLightIdList();
        if (isEffectSupportLightIdList.isEmpty()) {
            HiLog.info(LABEL_LOG, "lightIdList is empty");
        }
```

```java
        HiLog.info(LABEL_LOG, "end initData, size: %{public}s",
            isEffectSupportLightIdList.size());
    }

    private void off() {
        HiLog.info(LABEL_LOG, "before off");

        for (Integer lightId : isEffectSupportLightIdList) {
            // 关灯
            boolean turnOffResult = lightAgent.turnOff(lightId);

            HiLog.info(LABEL_LOG, "%{public}s turnOffResult : %{public}s ",
                lightId, turnOffResult);
        }

        HiLog.info(LABEL_LOG, "end off");
    }


    private void on() {
        HiLog.info(LABEL_LOG, "before on");

        for (Integer lightId : isEffectSupportLightIdList) {
            // 创建自定义效果的一次性闪烁
            LightBrightness lightBrightness =
                new LightBrightness(255, 255, 255);
            LightEffect lightEffect =
                new LightEffect(lightBrightness, 1000, 1000);

            // 开灯
            boolean turnOnResult = lightAgent.turnOn(lightId, lightEffect);

            HiLog.info(LABEL_LOG, "%{public}s turnOnResult: %{public}s ",
                lightId, turnOnResult);
        }

        HiLog.info(LABEL_LOG, "end on");
    }

    @Override
    public void onActive() {
        super.onActive();
    }

    @Override
    public void onForeground(Intent intent) {
        super.onForeground(intent);
    }
}
```

上述代码中：

- 在Button上设置了点击事件。
- initData用于初始化硬件设备上的灯列表。
- on方法用于打开灯列表上所有的灯。
- off方法用于关闭灯列表上所有的灯。
- 通过LightBrightness和LightEffect可以创建自定义效果的一次性闪烁。

26.3.5 运行

运行应用后，控制台输出内容如下：

```
09-14 00:15:04.934 11806-11806/com.waylau.hmos.lightagent I
00001/MainAbilitySlice: before initData
09-14 00:15:04.944 11806-11806/com.waylau.hmos.lightagent I
00001/MainAbilitySlice: end initData, size: 1
```

点击界面上的On按钮以触发打开灯的操作执行。此时，控制台输出内容如下：

```
09-14 00:15:15.899 11806-11806/com.waylau.hmos.lightagent I
00001/MainAbilitySlice: before on
09-14 00:15:15.922 11806-11806/com.waylau.hmos.lightagent I
00001/MainAbilitySlice: 0 turnOnResult: true
09-14 00:15:15.922 11806-11806/com.waylau.hmos.lightagent I
00001/MainAbilitySlice: end on
```

从上述日志可以看到，第0号灯已经打开了。

点击界面上的Off按钮以触发关闭灯的操作执行。此时，控制台输出内容如下：

```
09-14 00:15:29.734 11806-11806/com.waylau.hmos.lightagent I
00001/MainAbilitySlice: before off
09-14 00:15:29.767 11806-11806/com.waylau.hmos.lightagent I
00001/MainAbilitySlice: 0 turnOffResult : true
09-14 00:15:29.767 11806-11806/com.waylau.hmos.lightagent I
00001/MainAbilitySlice: end off
```

从上述日志可以看到，第0号灯已经关闭了。

26.4 实战：获取设备的位置

开发者可以调用HarmonyOS位置相关接口获取设备的实时位置或者最近的历史位置。本节演示如何获取设备的位置。

对于位置敏感的应用业务，建议获取设备的实时位置信息。如果不需要设备的实时位置信息，并且希望尽可能地节省耗电，开发者可以考虑获取最近的历史位置。

26.4.1 接口说明

获取设备的位置信息，所使用的接口有：

- Locator(Context context)：创建Locator实例对象。
- RequestParam(int scenario)：根据定位场景类型创建定位请求的RequestParam对象。
- onLocationReport(Location location)：获取定位结果。
- startLocating(RequestParam request, LocatorCallback callback)：向系统发起定位请求。
- requestOnce(RequestParam request, LocatorCallback callback)：向系统发起单次定位请求。
- stopLocating(LocatorCallback callback)：结束定位。
- getCachedLocation()：获取系统缓存的位置信息。

26.4.2 创建应用

为了演示获取设备位置的功能，创建一个名为Locator的应用。
在应用的界面上，通过点击按钮来触发获取设备位置的操作。

26.4.3 声明权限

修改配置文件，声明使用网络的权限如下：

```
// 声明权限
"reqPermissions": [
    {
    "name": "ohos.permission.LOCATION"
    }
]
```

由于上述权限是敏感权限，因此需要在代码中显式声明，代码如下：

```java
package com.waylau.hmos.locator;

import com.waylau.hmos.locator.slice.MainAbilitySlice;
import ohos.aafwk.ability.Ability;
import ohos.aafwk.content.Intent;

import java.util.ArrayList;
import java.util.List;

public class MainAbility extends Ability {
    @Override
    public void onStart(Intent intent) {
        super.onStart(intent);
        super.setMainRoute(MainAbilitySlice.class.getName());
```

```java
        // 显式声明需要使用的权限
        requestPermission();
    }

    // 显式声明需要使用的权限
    private void requestPermission() {
        String[] permission = {
                "ohos.permission.LOCATION"};
        List<String> applyPermissions = new ArrayList<>();
        for (String element : permission) {
            if (verifySelfPermission(element) != 0) {
                if (canRequestPermission(element)) {
                    applyPermissions.add(element);
                }
            }
        }
        requestPermissionsFromUser(applyPermissions.toArray(new String[0]), 0);
    }
}
```

26.4.4 修改 ability_main.xml

修改ability_main.xml内容如下：

```xml
<?xml version="1.0" encoding="utf-8"?>
<DirectionalLayout
    xmlns:ohos="http://schemas.huawei.com/res/ohos"
    ohos:height="match_parent"
    ohos:width="match_parent"
    ohos:orientation="vertical">

    <Button
        ohos:id="$+id:button_on"
        ohos:height="match_content"
        ohos:width="match_parent"
        ohos:background_element="#F76543"
        ohos:layout_alignment="horizontal_center"
        ohos:margin="10vp"
        ohos:padding="10vp"
        ohos:text="On"
        ohos:text_size="30fp"
        />

    <Button
        ohos:id="$+id:button_off"
        ohos:height="match_content"
        ohos:width="match_parent"
        ohos:background_element="#F76543"
```

```
        ohos:layout_alignment="horizontal_center"
        ohos:margin="10vp"
        ohos:padding="10vp"
        ohos:text="Off"
        ohos:text_size="30fp"
        />

</DirectionalLayout>
```

界面预览效果如图26-5所示。

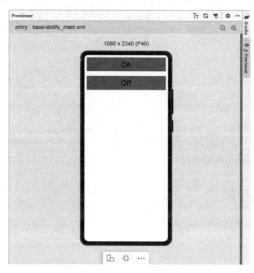

图 26-5　界面预览效果

上述代码设置了On按钮和Off按钮，以备设置点击事件，以触发使用位置的相关操作。

26.4.5　修改 MainAbilitySlice

修改MainAbilitySlice内容如下：

```
package com.waylau.hmos.locator.slice;

import com.waylau.hmos.locator.ResourceTable;
import ohos.aafwk.ability.AbilitySlice;
import ohos.aafwk.content.Intent;
import ohos.agp.components.Button;
import ohos.hiviewdfx.HiLog;
import ohos.hiviewdfx.HiLogLabel;
import ohos.location.Location;
import ohos.location.Locator;
import ohos.location.LocatorCallback;
import ohos.location.RequestParam;
import ohos.utils.zson.ZSONObject;
```

```java
public class MainAbilitySlice extends AbilitySlice {
    private static final String TAG = MainAbilitySlice.class.getSimpleName();
    private static final HiLogLabel LABEL_LOG =
            new HiLogLabel(HiLog.LOG_APP, 0x00001, TAG);

    private Locator locator;
    private MyLocatorCallback locatorCallback;

    @Override
    public void onStart(Intent intent) {
        super.onStart(intent);
        super.setUIContent(ResourceTable.Layout_ability_main);

        // 初始化对象
        initData();

        // 为按钮设置点击事件回调
        Button buttonOn =
                (Button) findComponentById(ResourceTable.Id_button_on);
        buttonOn.setClickedListener(listener -> on());

        Button buttonOff =
                (Button) findComponentById(ResourceTable.Id_button_off);
        buttonOff.setClickedListener(listener -> off());
    }

    private void initData() {
        HiLog.info(LABEL_LOG, "before initData");

        locator = new Locator(this.getContext());

        HiLog.info(LABEL_LOG, "end initData");
    }

    private void off() {
        HiLog.info(LABEL_LOG, "before off");

        locator.stopLocating(locatorCallback);

        HiLog.info(LABEL_LOG, "end off");
    }


    private void on() {
        HiLog.info(LABEL_LOG, "before on");

        locatorCallback = new MyLocatorCallback();

        // 定位精度优先策略
        RequestParam requestParam =
```

```
                new RequestParam(RequestParam.PRIORITY_ACCURACY, 0, 0);

        locator.startLocating(requestParam, locatorCallback);

        HiLog.info(LABEL_LOG, "end on");
    }

    private class MyLocatorCallback implements LocatorCallback {
        @Override
        public void onLocationReport(Location location) {
            HiLog.info(LABEL_LOG, "onLocationReport, location: %{public}s",
                    ZSONObject.toZSONString(location));
        }

        @Override
        public void onStatusChanged(int type) {
            HiLog.info(LABEL_LOG, "onStatusChanged, type: %{public}s", type);
        }

        @Override
        public void onErrorReport(int type) {
            HiLog.info(LABEL_LOG, "onErrorReport, type: %{public}s", type);
        }
    }

    @Override
    public void onActive() {
        super.onActive();
    }

    @Override
    public void onForeground(Intent intent) {
        super.onForeground(intent);
    }
}
```

上述代码中：

- 在Button上设置了点击事件。
- initData用于初始化Locator对象。
- on方法用于启动定位。
- off方法用于关闭定位。
- 实例化LocatorCallback对象，用于向系统提供位置上报的途径。系统在定位成功确定设备的实时位置结果时，会通过onLocationReport接口上报给应用。应用程序可以在onLocationReport接口的实现中完成自己的业务逻辑。

26.4.6 运行

初次运行应用后,点击界面上的On按钮以触发操作的执行。此时,控制台输出内容如下:

```
    09-14 00:18:26.630 14095-14095/com.waylau.hmos.locator I
00001/MainAbilitySlice: before initData
    09-14 00:18:26.634 14095-14095/com.waylau.hmos.locator I
00001/MainAbilitySlice: end initData
    09-14 00:18:39.554 14095-14095/com.waylau.hmos.locator I
00001/MainAbilitySlice: before on
    09-14 00:18:39.559 14095-27506/com.waylau.hmos.locator I
00001/MainAbilitySlice: onErrorReport, type: 257
    09-14 00:18:40.687 14095-27506/com.waylau.hmos.locator I
00001/MainAbilitySlice: onStatusChanged, type: 2
    09-14 00:18:40.687 14095-14095/com.waylau.hmos.locator I
00001/MainAbilitySlice: end on
```

从上述日志可以看出,并未获取到位置信息,这是因为位置信息服务并未开启。按图26-6所示的方式开启位置信息服务。

图 26-6 位置信息

再次点击界面上的On按钮以触发操作的执行。此时,控制台输出内容如下:

```
    09-14 00:20:31.449 14095-14095/com.waylau.hmos.locator I
00001/MainAbilitySlice: before on
    09-14 00:20:31.456 14095-27506/com.waylau.hmos.locator I
00001/MainAbilitySlice: onStatusChanged, type: 2
    09-14 00:20:31.456 14095-14095/com.waylau.hmos.locator I
00001/MainAbilitySlice: end on
    09-14 00:20:39.482 14095-27505/com.waylau.hmos.locator I
```

```
00001/MainAbilitySlice: onLocationReport, location:
{"accuracy":10,"altitude":51,"direction":0,"latitude":30.495864,"longitude":11
4.535703,"speed":0,"timeSinceBoot":1024134366580,"timeStamp":1489717602000}
    09-14 00:20:39.483 14095-27506/com.waylau.hmos.locator I
00001/MainAbilitySlice: onLocationReport, location:
{"accuracy":10,"altitude":51,"direction":0,"latitude":30.495864,"longitude":11
4.535703,"speed":0,"timeSinceBoot":1024134366580,"timeStamp":1489717602000}
```

从上述日志可以看出,已经能够成功获取到位置信息了。

点击界面上的Off按钮以关闭位置服务。此时,控制台输出内容如下:

```
    09-14 00:20:58.510 14095-14095/com.waylau.hmos.locator I
00001/MainAbilitySlice: before off
    09-14 00:20:58.515 14095-8091/com.waylau.hmos.locator I
00001/MainAbilitySlice: onStatusChanged, type: 3
    09-14 00:20:58.518 14095-14095/com.waylau.hmos.locator I
00001/MainAbilitySlice: end off
```

26.5 实战:(逆)地理编码转化

本节演示如何使用(逆)地理编码转化。

使用坐标描述一个位置非常准确,但是并不直观,面向用户表达并不友好。HarmonyOS向开发者提供了地理编码转化能力(将坐标转化为地理编码信息),以及逆地理编码转化能力(将地理描述转化为具体坐标)。其中地理编码由多个属性来描述位置,包括国家、行政区划、街道、门牌号、地址描述等,这样的信息更便于用户理解。

26.5.1 接口说明

GeoConvert用于坐标和地理编码信息的相互转化,所使用的接口有:

- GeoConvert():创建GeoConvert实例对象。
- GeoConvert(Locale locale):根据自定义参数创建GeoConvert实例对象。
- getAddressFromLocation(double latitude, double longitude, int maxItems):根据指定的经纬度坐标获取地理位置信息。
- getAddressFromLocationName(String description, int maxItems):根据地理位置信息获取相匹配的包含坐标数据的地址列表。
- getAddressFromLocationName(String description, double minLatitude, double minLongitude, double maxLatitude, double maxLongitude, int maxItems):根据指定的位置信息和地理区域获取相匹配的包含坐标数据的地址列表。

26.5.2 创建应用

为了演示使用（逆）地理编码转化的功能，创建一个名为GeoConvert的应用。
在应用的界面上，通过点击按钮来触发使用（逆）地理编码转化的操作。

26.5.3 修改 ability_main.xml

修改ability_main.xml内容如下：

```xml
<?xml version="1.0" encoding="utf-8"?>
<DirectionalLayout
    xmlns:ohos="http://schemas.huawei.com/res/ohos"
    ohos:height="match_parent"
    ohos:width="match_parent"
    ohos:orientation="vertical">

    <Button
        ohos:id="$+id:button_get"
        ohos:height="match_content"
        ohos:width="match_parent"
        ohos:background_element="#F76543"
        ohos:layout_alignment="horizontal_center"
        ohos:margin="10vp"
        ohos:padding="10vp"
        ohos:text="Get"
        ohos:text_size="30fp"
        />

    <Text
        ohos:id="$+id:text"
        ohos:height="match_content"
        ohos:width="match_content"
        ohos:background_element="$graphic:background_ability_main"
        ohos:layout_alignment="horizontal_center"
        ohos:multiple_lines="true"
        ohos:text=""
        ohos:text_size="20fp"
        />
</DirectionalLayout>
```

界面预览效果如图26-7所示。
上述代码设置了Get按钮，以备设置点击事件，以触发使用（逆）地理编码转化的相关操作。

图 26-7 界面预览效果

26.5.4 修改 MainAbilitySlice

修改MainAbilitySlice内容如下：

```
package com.waylau.hmos.geoconvert.slice;

import com.waylau.hmos.geoconvert.ResourceTable;
import ohos.aafwk.ability.AbilitySlice;
import ohos.aafwk.content.Intent;
import ohos.agp.components.Button;
import ohos.agp.components.Text;
import ohos.hiviewdfx.HiLog;
import ohos.hiviewdfx.HiLogLabel;
import ohos.location.*;
import ohos.utils.zson.ZSONObject;

import java.io.IOException;
import java.util.List;


public class MainAbilitySlice extends AbilitySlice {
    private static final String TAG = MainAbilitySlice.class.getSimpleName();
    private static final HiLogLabel LABEL_LOG =
            new HiLogLabel(HiLog.LOG_APP, 0x00001, TAG);


    private Text text;

    @Override
    public void onStart(Intent intent) {
```

```java
        super.onStart(intent);
        super.setUIContent(ResourceTable.Layout_ability_main);

        // 为按钮设置点击事件回调
        Button buttonOn =
                (Button) findComponentById(ResourceTable.Id_button_get);
        buttonOn.setClickedListener(listener -> {
            try {
                getInfo();
            } catch (IOException e) {
                e.printStackTrace();
            }
        });

        text = (Text) findComponentById(ResourceTable.Id_text);
    }

    private void getInfo() throws IOException {
        HiLog.info(LABEL_LOG, "before getInfo");

        // 实例化GeoConvert对象
        // 所有与(逆)地理编码转化能力相关的功能API都是通过GeoConvert提供的
        GeoConvert geoConvert = new GeoConvert();

        // 坐标转化地理位置信息
        int maxItems = 3;
        double latitude = 23.048062D;
        double longitude = 114.211902D;
        List<GeoAddress> addressList =
geoConvert.getAddressFromLocation(latitude, longitude, maxItems);

        // 显示地理位置信息
        showAddress(addressList);

        // 位置描述转化坐标
        addressList = geoConvert.getAddressFromLocationName("华为", maxItems);

        // 显示地理位置信息
        showAddress(addressList);

        HiLog.info(LABEL_LOG, "end getInfo");
    }

    private void showAddress(List<GeoAddress> addressList) {
        HiLog.info(LABEL_LOG, "showAddress");

        for (GeoAddress address : addressList) {
            String addressInfo = ZSONObject.toZSONString(address);
```

```
                    text.append(addressInfo + "\n");

                    HiLog.info(LABEL_LOG, "addressInfo:%{public}s", addressInfo);
            }
        }

        @Override
        public void onActive() {
            super.onActive();
        }

        @Override
        public void onForeground(Intent intent) {
            super.onForeground(intent);
        }
    }
```

上述代码中：

- 在Button上设置了点击事件。
- getInfo方法实例化GeoConvert对象，并执行坐标转化地理位置信息和位置描述转化坐标等操作。
- showAddress方法用于在界面上显示获取到的地理位置信息。

26.5.5 运行

运行应用后，点击界面上的Get按钮以触发（逆）地理编码转化的操作执行。此时，控制台输出内容如下：

```
    09-14 00:37:44.830 14125-14125/com.waylau.hmos.geoconvert I
00001/MainAbilitySlice:  before getInfo
    09-14 00:37:45.214 14125-14125/com.waylau.hmos.geoconvert I
00001/MainAbilitySlice:  showAddress
    09-14 00:37:45.222 14125-14125/com.waylau.hmos.geoconvert I
00001/MainAbilitySlice:  addressInfo:{"administrativeArea":"广东省
","countryCode":"CN","countryName":"中国
","descriptionsSize":0,"latitude":23.048069,"locale":"zh_CN","locality":"惠州市
","longitude":114.211912,"placeName":"潼湖镇西湖","subLocality":"惠城区"}
    09-14 00:37:45.433 14125-14125/com.waylau.hmos.geoconvert I
00001/MainAbilitySlice:  showAddress
    09-14 00:37:45.434 14125-14125/com.waylau.hmos.geoconvert I
00001/MainAbilitySlice:  addressInfo:{"administrativeArea":"广东省
","countryCode":"CN","countryName":"中国
","descriptionsSize":0,"latitude":22.547246,"locale":"zh_CN","locality":"深圳市
","longitude":114.082036,"placeName":"华为","subLocality":"福田区"}
    09-14 00:37:45.434 14125-14125/com.waylau.hmos.geoconvert I
00001/MainAbilitySlice:  end getInfo
```

界面上能够正常显示获取到的地理位置信息，如图26-8所示。

图 26-8　界面效果

第 27 章

数据管理

HarmonyOS数据管理是基于分布式软总线的能力实现应用程序数据和用户数据的分布式管理。

27.1 数据管理概述

在全场景新时代，每个人拥有的设备越来越多，单一设备的数据往往无法满足用户的诉求，数据在设备间的流转变得越来越频繁。以一组照片数据在手机、平板、智慧屏和PC之间相互浏览和编辑为例，需要考虑到照片数据在多设备间是怎么存储、怎么共享和怎么访问的。

HarmonyOS数据管理是基于分布式软总线的能力实现应用程序数据和用户数据的分布式管理。用户数据不再与单一物理设备绑定，业务逻辑与数据存储分离，跨设备的数据处理如同本地数据处理一样方便快捷，让开发者能够轻松实现全场景、多设备下的数据存储、共享和访问，为打造一致、流畅的用户体验创造了基础条件。

HarmonyOS分布式数据管理对开发者提供分布式数据库、分布式文件系统和分布式检索能力，开发者在多设备上开发应用时，对数据的操作、共享、检索可以跟使用本地数据一样处理，为开发者提供便捷、高效和安全的数据管理能力，大大降低了应用开发者实现数据分布式访问的门槛。同时，由于在系统层面实现了这样的功能，可以结合系统资源调度大大提升跨设备数据远程访问和检索性能，让更多的开发者可以快速上手实现流畅的分布式应用。

HarmonyOS分布式数据管理典型的应用场景有：

- 协同办公场景：将手机上的文档投屏到智慧屏，在智慧屏上对文档执行翻页、缩放、涂鸦等操作，文档的最新状态可以在手机上同步显示。

● 家庭出游场景：一家人出游时，妈妈用手机拍了很多照片。通过家庭照片共享，爸爸可以在自己的手机上浏览、收藏和保存这些照片，家中的爷爷奶奶也可以通过智慧屏浏览这些照片。

HarmonyOS支持多样数据管理方式，包括：

- 关系型数据库。
- 对象关系映射数据库。
- 轻量级偏好数据库。
- 分布式数据服务。
- 分布式文件服务。
- 融合搜索。
- 数据存储管理。

27.2 关系型数据库

关系型数据库（Relational Database，RDB）是一种基于关系模型来管理数据的数据库。HarmonyOS关系型数据库基于SQLite组件提供了一套完整的对本地数据库进行管理的机制，对外提供了一系列的增、删、改、查接口，也可以直接运行用户输入的SQL语句来满足复杂的场景需要。HarmonyOS提供的关系型数据库功能更加完善，查询效率更高。

27.2.1 基本概念

使用关系型数据库时，主要涉及以下概念：

- 关系型数据库：创建在关系模型基础上的数据库，以行和列的形式存储数据。
- 谓词：数据库中用来代表数据实体的性质、特征或者数据实体之间关系的词项，主要用来定义数据库的操作条件。
- 结果集：指用户查询之后的结果集合，可以对数据进行访问。结果集提供了灵活的数据访问方式，可以更方便地拿到用户想要的数据。
- SQLite数据库：一款轻型的数据库，是遵守ACID的关系型数据库管理系统。它是一个开源的项目。

27.2.2 运作机制

HarmonyOS关系型数据库对外提供通用的操作接口，底层使用SQLite作为持久化存储引擎，支持SQLite具有的所有数据库特性，包括但不限于事务、索引、视图、触发器、外键、参数化查询和预编译SQL语句。

图27-1展示了关系型数据库的运作机制架构图。

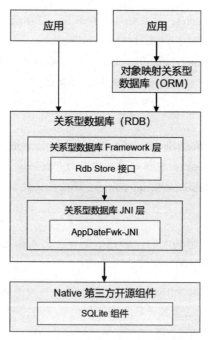

图 27-1 关系型数据库的运作机制架构图

27.2.3 默认配置

HarmonyOS关系型数据库的默认配置如下：

- 如果不指定数据库的日志模式，那么系统默认的日志模式是WAL（Write Ahead Log，预写式日志）模式。
- 如果不指定数据库的落盘模式，那么系统默认的落盘模式是FULL模式。
- HarmonyOS数据库使用的共享内存默认大小是2MB。

27.2.4 约束与限制

HarmonyOS关系型数据库的约束与限制如下：

- 数据库中连接池的最大数量是4个，用以管理用户的读写操作。
- 为了保证数据的准确性，数据库同一时间只能支持一个写操作。

27.2.5 接口说明

HarmonyOS关系型数据库的接口如下。

1. 数据库的创建和删除

关系型数据库提供了数据库创建方式以及对应的删除接口，涉及的API如下：

- StoreConfig.Builder的builder()：对数据库进行配置，包括设置数据库名、存储模式、日志模式、同步模式，是否为只读，以及对数据库加密。
- RdbOpenCallback的onCreate(RdbStore store)：数据库创建时被回调，开发者可以在该方法中初始化表结构，并添加一些应用使用到的初始化数据。
- RdbOpenCallback的onUpgrade(RdbStore store, int currentVersion, int targetVersion)：数据库升级时被回调。
- DatabaseHelper的getRdbStore(StoreConfig config, int version, RdbOpenCallback openCallback, ResultSetHook resultSetHook)：根据配置创建或打开数据库。
- DatabaseHelper的deleteRdbStore(String name)：删除指定的数据库。

2. 数据库的加密

关系型数据库提供数据库加密的能力，创建数据库时传入指定的密钥，创建加密数据库，后续打开加密数据库时，需要传入正确的密钥。

- StoreConfig.Builder的setEncryptKey(byte[] encryptKey)：为数据库配置类设置数据库加密密钥，创建或打开数据库时传入包含数据库加密密钥的配置类，即可创建或打开加密数据库。

3. 数据库的增删改查

关系型数据库提供本地数据增删改查操作的能力，相关RdbStore的API如下：

- insert(String table, ValuesBucket initialValues)：向数据库插入数据。
- update(ValuesBucket values, AbsRdbPredicates predicates)：更新数据库表中符合谓词指定条件的数据。
- delete(AbsRdbPredicates predicates)：删除数据。
- query(AbsRdbPredicates predicates, String[] columns)：查询数据。
- querySql(String sql, String[] sqlArgs)：执行原生的用于查询操作的SQL语句。

4. 数据库谓词的使用

关系型数据库提供了用于设置数据库操作条件的谓词AbsRdbPredicates，其中包括两个实现子类RdbPredicates和RawRdbPredicates：

- RdbPredicates：开发者无须编写复杂的SQL语句，仅通过调用该类中条件相关的方法，如equalTo、notEqualTo、groupBy、orderByAsc、beginsWith等，就可以自动完成SQL语句的拼接，方便用户聚焦业务操作。
- RawRdbPredicates：可满足复杂SQL语句的场景，支持开发者自己设置where条件子句和whereArgs参数，不支持equalTo等条件接口的使用。

RdbPredicates包含的API如下：

- equalTo(String field, String value)：设置谓词条件，满足filed字段与value值相等。
- notEqualTo(String field, String value)：设置谓词条件，满足filed字段与value值不相等。
- beginsWith(String field, String value)：设置谓词条件，满足field字段以value值开头。
- between(String field, int low, int high)：设置谓词条件，满足field字段在最小值low和最大值high

之间。

- orderByAsc(String field)：设置谓词条件，根据field字段升序排列。

RawRdbPredicates包含的API如下：

- setWhereClause(String whereClause)：设置where条件子句。
- setWhereArgs(List<String> whereArgs)：设置whereArgs参数，该值表示where子句中占位符的值。

5. 查询结果集的使用

关系型数据库提供了查询返回的结果集ResultSet，它指向查询结果中的一行数据，供用户对查询结果进行遍历和访问。ReusltSet的对外API如下：

- goTo(int offset)：从结果集当前位置移动指定偏移量。
- goToRow(int position)：将结果集移动到指定位置。
- goToNextRow()：将结果集向后移动一行。
- goToPreviousRow()：将结果集向前移动一行。
- isStarted()：判断结果集是否被移动过。
- isEnded()：判断结果集当前位置是否在最后一行之后。
- isAtFirstRow()：判断结果集当前位置是否在第一行。
- isAtLastRow()：判断结果集当前位置是否在最后一行。
- getRowCount()：获取当前结果集中的记录条数。
- getColumnCount()：获取结果集中的列数。
- getString(int columnIndex)：获取当前行指定索引列的值，以String类型返回。
- getBlob(int columnIndex)：获取当前行指定索引列的值，以字节数组形式返回。
- getDouble(int columnIndex)：获取当前行指定索引列的值，以Double类型返回。

6. 事务

关系型数据库提供事务机制来保证用户操作的原子性。对单条数据进行数据库操作时，无须开启事务；插入大量数据时，开启事务可以保证数据的准确性。如果中途操作出现失败，就会执行回滚操作。

RdbStore事务API如下：

- beginTransaction()：开启事务。
- markAsCommit()：设置事务的标记为成功。
- endTransaction()：结束事务。

7. 事务和结果集观察者

关系型数据库提供了事务和结果集观察者能力，当对应的事件被触发时，观察者会收到通知。

- RdbStore的beginTransactionWithObserver(TransactionObserver transactionObserver)：开启事务，并观察事务的启动、提交和回滚。
- ResultSet的registerObserver(DataObserver observer)：注册结果集的观察者。

- unregisterObserver(DataObserver observer)：注销结果集的观察者。

8. 数据库的备份和恢复

用户可以将当前数据库的数据进行保存和备份，还可以在需要的时候进行数据恢复。
RdbStore提供的数据库备份和恢复API如下：

- restore(String srcName)：数据库恢复接口，从指定的非加密数据库文件中恢复数据。
- restore(String srcName, byte[] srcEncryptKey, byte[] destEncryptKey)：数据库恢复接口，从指定的数据库文件（加密和非加密均可）中恢复数据。
- backup(String destName)：数据库备份接口，备份出的数据库文件是非加密的。
- backup(String destName, byte[] destEncryptKey)：数据库备份接口，此方法经常用在备份出的加密数据库场景。

27.2.6 开发过程

HarmonyOS关系型数据库的开发过程总结如下。

1. 创建数据库

- 配置数据库相关信息，包括数据库的名称、存储模式、是否为只读模式等。
- 初始化数据库表结构和相关数据。
- 创建数据库。

示例代码如下：

```
StoreConfig config = StoreConfig.newDefaultConfig("RdbStoreTest.db");
private static final RdbOpenCallback callback = new RdbOpenCallback() {
    @Override
    public void onCreate(RdbStore store) {
        store.executeSql("CREATE TABLE IF NOT EXISTS test "
            +"(id INTEGER PRIMARY KEY AUTOINCREMENT, name TEXT NOT NULL, "
            +"age INTEGER, salary REAL, blobType BLOB)");
    }
    @Override
    public void onUpgrade(RdbStore store, int oldVersion, int newVersion) {
    }
};


DatabaseHelper helper = new DatabaseHelper(context);
RdbStore store = helper.getRdbStore(config, 1, callback, null);
```

2. 插入数据

- 构造要插入的数据，以ValuesBucket形式存储。
- 调用关系型数据库提供的插入接口。

示例代码如下：

```
ValuesBucket values = new ValuesBucket();
values.putInteger("id", 1);
values.putString("name", "zhangsan");
values.putInteger("age", 18);
values.putDouble("salary", 100.5);
values.putByteArray("blobType", new byte[] {1, 2, 3});
long id = store.insert("test", values);
```

3. 查询数据

- 构造用于查询的谓词对象，设置查询条件。
- 指定查询返回的数据列。
- 调用查询接口查询数据。
- 调用结果集接口，遍历返回结果。

示例代码如下：

```
String[] columns = new String[] {"id", "name", "age", "salary"};
RdbPredicates rdbPredicates =
   new RdbPredicates("test").equalTo("age", 25).orderByAsc("salary");
ResultSet resultSet = store.query(rdbPredicates, columns);
resultSet.goToNextRow();
```

有关关系型数据库的详细开发示例参见5.11节。

27.3 对象关系映射数据库

HarmonyOS对象关系映射（Object Relational Mapping，ORM）数据库是一款基于SQLite的数据库框架，屏蔽了底层SQLite数据库的SQL操作，针对实体和关系提供了增删改查等一系列的面向对象接口。应用开发者不必再去编写复杂的SQL语句，以操作对象的形式来操作数据库，提升效率的同时也能聚焦于业务开发。

HarmonyOS对象关系映射数据库是建立在HarmonyOS关系型数据库的基础之上的，所以关系型数据库的一些约束与限制、默认配置可参考关系型数据库的约束与限制、默认配置。

27.3.1 基本概念

使用对象关系映射数据库时，主要涉及以下概念：

- 对象关系映射数据库的3个主要组件：
  - 数据库：被开发者用@Database 注解，且继承了 OrmDatabase 的类，对应关系型数据库。
  - 实体对象：被开发者用@Entity 注解，且继承了 OrmObject 的类，对应关系型数据库中的表。
  - 对象数据操作接口：包括数据库操作的入口 OrmContext 类和谓词接口（OrmPredicate）等。

- 谓词：数据库中用来代表数据实体的性质、特征或者数据实体之间关系的词项，主要用来定义数据库的操作条件。对象关系映射数据库将SQLite数据库中的谓词封装成了接口方法供开发者调用。开发者通过对象数据操作接口可以访问应用持久化的关系型数据。
- 对象关系映射数据库：通过将实例对象映射到关系上实现使用操作实例对象的语法来操作关系型数据库。它是在SQLite数据库的基础上提供的一个抽象层。
- SQLite数据库：一款轻型的数据库，是遵守ACID的关系型数据库管理系统。

27.3.2 运作机制

对象关系映射数据库的操作是基于关系型数据库操作接口完成的，实际上是在关系型数据库操作的基础上又实现了对象关系映射等特性。因此，对象关系映射数据库跟关系型数据库一样，都使用SQLite作为持久化引擎，底层使用的是同一套数据库连接池和数据库连接机制。

使用对象关系映射数据库的开发者需要先配置实体模型与关系映射文件。应用数据管理框架提供的类生成工具会解析这些文件，生成数据库帮助类，这样应用数据管理框架就能在运行时根据开发者的配置创建好数据库，并在存储过程中自动完成对象关系映射。开发者再通过对象数据操作接口（如OrmContext接口和谓词接口等）操作持久化数据库。

对象数据操作接口提供一组基于对象映射的数据操作接口，实现了基于SQL的关系模型数据到对象的映射，让用户不需要再和复杂的 SQL语句打交道，只需简单地操作实体对象的属性和方法。对象数据操作接口支持对象的增删改查操作，同时支持事务操作等。

图27-2展示了对象关系映射数据库的运作机制架构图。

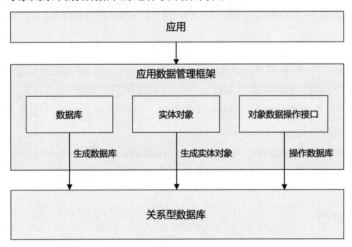

图 27-2 对象关系映射数据库的运作机制架构图

27.3.3 实体对象属性支持的类型

开发者建立实体对象类时，对象属性的类型可以在表27-1的类型中选择，不支持使用自定义类型。

表 27-1 实体对象属性支持的类型

类型名称	描 述	初 始 值
Integer	封装整型	null
int	整型	0
Long	封装长整型	null
long	长整型	0L
Double	封装双精度浮点型	null
double	双精度浮点型	0
Float	封装单精度浮点型	null
float	单精度浮点型	0
Short	封装短整型	null
short	短整型	0
String	字符串型	null
Boolean	封装布尔型	null
boolean	布尔型	0
Byte	封装字节型	null
byte	字节型	0
Character	封装字符型	null
char	字符型	' '
Date	日期类	null
Time	时间类	null
Timestamp	时间戳类	null
Calendar	日历类	null
Blob	二进制大对象	null
Clob	字符大对象	null

27.3.4 接口说明

对象关系映射数据库目前支持数据库和表的创建、对象数据的增删改查、对象数据变化回调、数据库升降级和备份等功能。

1. **数据库和表的创建**
- 创建数据库。开发者需要定义一个表示数据库的类,继承OrmDatabase,再通过@Database注解内的entities属性指定哪些数据模型类属于这个数据库。属性包括:
  - version: 数据库版本号。
  - entities: 数据库内包含的表。
- 创建数据表。开发者可通过创建一个继承了OrmObject并用@Entity注解的类,获取数据库实体对象,也就是表的对象。属性包括:
  - tableName: 表名。
  - primaryKeys: 主键名,一个表中只能有一个主键,一个主键可以由多个字段组成。

- ➢ foreignKeys：外键列表。
- ➢ indices：索引列表。

主要涉及如下注解：

- @Database：被@Database注解且继承了OrmDatabase的类对应数据库类。
- @Entity：被@Entity注解且继承了OrmObject的类对应数据表类。
- @Column：被@Column注解的变量对应数据表的字段。
- @PrimaryKey：被@PrimaryKey注解的变量对应数据表的主键。
- @ForeignKey：被@ForeignKey注解的变量对应数据表的外键。
- @Index：被@Index注解的内容对应数据表索引的属性。

2. 数据库的加密

对象关系映射数据库提供数据库加密的能力，创建数据库时传入指定的密钥，创建加密数据库，后续打开加密数据库时，需要传入正确的密钥。

OrmConfig.Builder传入密钥接口如下：

- setEncryptKey(byte[] encryptKey)：为数据库配置类设置数据库加密密钥，创建或打开数据库时传入包含数据库加密密钥的配置类，即可创建或打开加密数据库。

3. 对象数据的增删改查

通过对象数据操作接口，开发者可以对对象数据进行增删改查操作。

OrmContext对象数据操作接口如下：

- insert(T object)：添加方法。
- update(T object)：更新方法。
- query(OrmPredicates predicates)：查询方法。
- delete(T object)：删除方法。
- where(Class<T> clz)：设置谓词方法。

4. 对象数据的变化观察者设置

通过使用对象数据操作接口，开发者可以在某些数据上设置观察者，接收数据变化的通知。

OrmContext数据变化观察者接口如下：

- registerStoreObserver(String alias, OrmObjectObserver observer)：注册数据库变化回调。
- registerContextObserver(OrmContext watchedContext, OrmObjectObserver observer)：注册上下文变化回调。
- registerEntityObserver(String entityName, OrmObjectObserver observer)：注册数据库实体变化回调。
- registerObjectObserver(OrmObject ormObject, OrmObjectObserver observer)：注册对象变化回调。

5. 数据库的升降级

通过调用数据库的升降级接口，开发者可以将数据库切换到不同的版本。

OrmMigration数据库升降级接口如下：

- onMigrate(int beginVersion, int endVersion)：数据库版本升降级接口。

6. 数据库的备份与恢复

开发者可以将当前数据库的数据进行备份，在必要的时候进行数据恢复。

OrmContext数据库备份与恢复接口如下：

- backup(String destPath)：数据库备份接口。
- restore(String srcPath)：数据库恢复接口。

27.4 实战：使用对象关系映射数据库

在5.11节演示了如何通过DataAbilityHelper类来访问关系型数据库数据。本节将演示如何访问对象关系映射数据库。

创建一个名为DataAbilityHelperAccessORM的应用作为演示。

27.4.1 修改 build.gradle

修改build.gradle添加compileOptions相关的配置，否则无法识别ORM相关的注解。

完整的build.gradle文件如下：

```
apply plugin: 'com.huawei.ohos.hap'
apply plugin: 'com.huawei.ohos.decctest'

ohos {
    compileSdkVersion 5
    defaultConfig {
        compatibleSdkVersion 5
    }
    buildTypes {
        release {
            proguardOpt {
                proguardEnabled false
                rulesFiles 'proguard-rules.pro'
            }
        }
    }

    // 添加compileOptions相关的配置
    compileOptions {
        annotationEnabled true
    }
}
```

```
dependencies {
    implementation fileTree(dir: 'libs', include: ['*.jar', '*.har'])
    testImplementation 'junit:junit:4.13'
    ohosTestImplementation 'com.huawei.ohos.testkit:runner:1.0.0.200'
}
decc {
    supportType = ['html','xml']
}
```

27.4.2 新增 User

新增User类，用于表示用户实体，代码如下：

```
package com.waylau.hmos.dataabilityhelperaccessorm;

import ohos.data.orm.OrmObject;
import ohos.data.orm.annotation.Entity;
import ohos.data.orm.annotation.Index;
import ohos.data.orm.annotation.PrimaryKey;

@Entity(tableName = "user",
        indices = {@Index(value = {"userName"}, name = "name_index", unique = true)})
public class User extends OrmObject {
    // 此处将userId设为了自增的主键。注意只有在数据类型为包装类型时，自增主键才能生效
    @PrimaryKey(autoGenerate = true)
    private Integer userId;
    private String userName;
    private int age;

    public Integer getUserId() {
        return userId;
    }

    public void setUserId(Integer userId) {
        this.userId = userId;
    }

    public String getUserName() {
        return userName;
    }

    public void setUserName(String userName) {
        this.userName = userName;
    }

    public int getAge() {
        return age;
    }
```

```java
    public void setAge(int age) {
        this.age = age;
    }
}
```

上述代码构造数据表,即创建数据库实体类并配置对应的属性(如对应表的主键等)。数据表必须与其所在的数据库在同一个模块中。

上述代码定义了一个实体类User.java,对应数据库内的表名为user;indices为userName这个字段建立的索引name_index,并且索引值是唯一的。

其中,userId设为了自增的主键。注意只有在数据类型为包装类型时,自增主键才能生效。

27.4.3 新增 UserStore

新增UserStore类,用于表示数据库,代码如下:

```java
package com.waylau.hmos.dataabilityhelperaccessorm;

import ohos.data.orm.OrmDatabase;
import ohos.data.orm.annotation.Database;


@Database(entities = {User.class}, version = 1)
public abstract class UserStore extends OrmDatabase {
}
```

上述代码定义了一个数据库类UserStore.java,数据库包含User三个表,版本号为1。数据库类的getVersion方法和getHelper方法不需要实现,直接将数据库类设为虚类即可。

27.4.4 创建 DataAbility

在DevEco Studio中创建一个名为UserDataAbility的Data。UserDataAbility初始化时代码如下:

```java
package com.waylau.hmos.dataabilityhelperaccessorm;

import ohos.aafwk.ability.Ability;
import ohos.aafwk.content.Intent;
import ohos.data.resultset.ResultSet;
import ohos.data.rdb.ValuesBucket;
import ohos.data.dataability.DataAbilityPredicates;
import ohos.hiviewdfx.HiLog;
import ohos.hiviewdfx.HiLogLabel;
import ohos.utils.net.Uri;
import ohos.utils.PacMap;

import java.io.FileDescriptor;

public class UserDataAbility extends Ability {
```

```java
        private static final HiLogLabel LABEL_LOG = new HiLogLabel(3, 0xD001100,
"Demo");

    @Override
    public void onStart(Intent intent) {
        super.onStart(intent);
        HiLog.info(LABEL_LOG, "UserDataAbility onStart");
    }

    @Override
    public ResultSet query(Uri uri, String[] columns, DataAbilityPredicates
predicates) {
        return null;
    }

    @Override
    public int insert(Uri uri, ValuesBucket value) {
        HiLog.info(LABEL_LOG, "UserDataAbility insert");
        return 999;
    }

    @Override
    public int delete(Uri uri, DataAbilityPredicates predicates) {
        return 0;
    }

    @Override
    public int update(Uri uri, ValuesBucket value, DataAbilityPredicates
predicates) {
        return 0;
    }

    @Override
    public FileDescriptor openFile(Uri uri, String mode) {
        return null;
    }

    @Override
    public String[] getFileTypes(Uri uri, String mimeTypeFilter) {
        return new String[0];
    }

    @Override
    public PacMap call(String method, String arg, PacMap extras) {
        return null;
    }

    @Override
    public String getType(Uri uri) {
        return null;
```

        }
}
```

UserDataAbility自动在配置文件中添加了相应的配置,内容如下:

```
"abilities": [
    {
      "skills": [
        {
          "entities": [
            "entity.system.home"
          ],
          "actions": [
            "action.system.home"
          ]
        }
      ],
      "orientation": "unspecified",
      "name": "com.waylau.hmos.dataabilityhelperaccessorm.MainAbility",
      "icon": "$media:icon",
      "description": "$string:mainability_description",
      "label": "$string:entry_MainAbility",
      "type": "page",
      "launchType": "standard"
    },
    // 新增UserDataAbility配置
    {
      "permissions": [
"com.waylau.hmos.dataabilityhelperaccessorm.DataAbilityShellProvider.PROVIDER"
      ],
      "name": "com.waylau.hmos.dataabilityhelperaccessorm.UserDataAbility",
      "icon": "$media:icon",
      "description": "$string:userdataability_description",
      "type": "data",
      "uri":
"dataability://com.waylau.hmos.dataabilityhelperaccessorm.UserDataAbility"
    }
]
```

从上述配置可以看出:

- type: 类型设置为data。
- uri: 对外提供的访问路径,全局唯一。
- permissions: 访问该Data Ability时需要申请的访问权限。

27.4.5 初始化数据库

修改UserDataAbility代码如下:
```

```java
public class UserDataAbility extends Ability {
    private static final String TAG = UserDataAbility.class.getSimpleName();
    private static final HiLogLabel LABEL_LOG =
            new HiLogLabel(HiLog.LOG_APP, 0x00001, TAG);

    private static final String DATABASE_NAME = "RdbStoreTest.db";
    private static final String DATABASE_NAME_ALIAS = TAG;

    private DatabaseHelper manager;
    private OrmContext ormContext = null;

    @Override
    public void onStart(Intent intent) {
        super.onStart(intent);
        HiLog.info(LABEL_LOG, "UserDataAbility onStart");

        // 初始化DatabaseHelper、OrmContext
        manager = new DatabaseHelper(this);
        ormContext = manager.getOrmContext(DATABASE_NAME_ALIAS,
DATABASE_NAME, UserStore.class);
    }

    @Override
    public void onStop() {
        super.onStop();
        HiLog.info(LABEL_LOG, "UserDataAbility onStop");

        // 删除数据库
        manager.deleteRdbStore(DATABASE_NAME);
    }
    ...
```

在上述代码中：

- 在onStart方法中初始化了一个名为RdbStoreTest.db的数据库。
- 实例化了DatabaseHelper、OrmContext对象。
- 重写onStop方法，以删除数据库。

27.4.6　新增 queryAll 方法

UserDataAbility新增queryAll方法，代码如下：

```java
public List<User> queryAll() {
    OrmPredicates predicates = ormContext.where(User.class);
    List<User> users = ormContext.query(predicates);
    users.forEach(user -> {
        HiLog.info(LABEL_LOG, "query user: %{public}s", user);
    });
```

```
        return users;
    }
```

上述代码将会查询数据库中的所有用户数据。

27.4.7 新增 insert 方法

UserDataAbility新增insert方法，代码如下：

```
public int insert(User user) {
    HiLog.info(LABEL_LOG, "before insert");

    // 插入数据库
    ormContext.insert(user);
    boolean isSuccessed = ormContext.flush();

    // 获取UserId
    int userId = user.getUserId();

    HiLog.info(LABEL_LOG, "end insert: %{public}s, isSuccessed: %{public}s",
            userId, isSuccessed);
    return userId;
}
```

上述代码将在数据库中插入用户信息。

27.4.8 新增 update 方法

UserDataAbility新增update方法，代码如下：

```
public int update(User user) {
    HiLog.info(LABEL_LOG, "before update");

    ormContext.update(user);

    boolean isSuccessed = ormContext.flush();

    HiLog.info(LABEL_LOG, "end update, isSuccessed: %{public}s",
            isSuccessed);
    return 1;
}
```

上述代码将会更新指定的用户信息。

27.4.9 新增 deleteAll 方法

UserDataAbility新增deleteAll方法，代码如下：

```java
    public int deleteAll() {
        HiLog.info(LABEL_LOG, "before delete");
        OrmPredicates predicates = ormContext.where(User.class);
        List<User> users = ormContext.query(predicates);

        users.forEach(user -> {
            boolean isSuccessed = ormContext.delete(user);
            HiLog.info(LABEL_LOG, "delete user: %{public}s, isSuccessed: %{public}s",
                    user.getUserId(), isSuccessed);
        });

        boolean isSuccessed = ormContext.flush();

        HiLog.info(LABEL_LOG, "end delete, isSuccessed: %{public}s",
                isSuccessed);

        return users.size();
    }
```

上述代码将会删除所有用户信息。

27.4.10 修改 ability_main.xml

修改ability_main.xml内容如下：

```xml
<?xml version="1.0" encoding="utf-8"?>
<DirectionalLayout
    xmlns:ohos="http://schemas.huawei.com/res/ohos"
    ohos:height="match_parent"
    ohos:width="match_parent"
    ohos:orientation="vertical">

    <DirectionalLayout
        ohos:height="60vp"
        ohos:width="match_parent"
        ohos:orientation="horizontal">
        <Button
            ohos:id="$+id:button_query"
            ohos:height="40vp"
            ohos:width="0vp"
            ohos:background_element="#F76543"
            ohos:layout_alignment="horizontal_center"
            ohos:margin="10vp"
            ohos:padding="10vp"
            ohos:text="Query"
            ohos:text_size="16fp"
            ohos:weight="1"
            />
```

```xml
        <Button
            ohos:id="$+id:button_insert"
            ohos:height="40vp"
            ohos:width="0vp"
            ohos:background_element="#F76543"
            ohos:layout_alignment="horizontal_center"
            ohos:margin="10vp"
            ohos:padding="10vp"
            ohos:text="Insert"
            ohos:text_size="16fp"
            ohos:weight="1"
            />

        <Button
            ohos:id="$+id:button_update"
            ohos:height="40vp"
            ohos:width="0vp"
            ohos:background_element="#F76543"
            ohos:layout_alignment="horizontal_center"
            ohos:margin="10vp"
            ohos:padding="10vp"
            ohos:text="Update"
            ohos:text_size="16fp"
            ohos:weight="1"
            />

        <Button
            ohos:id="$+id:button_delete"
            ohos:height="40vp"
            ohos:width="0vp"
            ohos:background_element="#F76543"
            ohos:layout_alignment="horizontal_center"
            ohos:margin="10vp"
            ohos:padding="10vp"
            ohos:text="Delete"
            ohos:text_size="16fp"
            ohos:weight="1"
            />
    </DirectionalLayout>

    <Text
        ohos:id="$+id:text"
        ohos:height="match_content"
        ohos:width="match_content"
        ohos:background_element="$graphic:background_ability_main"
        ohos:multiple_lines="true"
        ohos:text=""
        ohos:text_size="20fp"
        />
</DirectionalLayout>
```

界面预览效果如图27-3所示。

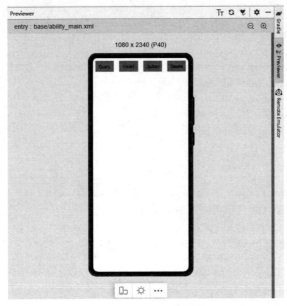

图 27-3　界面预览效果

上述代码中：

- 设置了4个按钮，用于触发操作数据库的相关操作。
- Text组件用于展示查询到的用户信息。

27.4.11　修改 MainAbilitySlice

修改MainAbilitySlice，代码如下：

```
package com.waylau.hmos.dataabilityhelperaccessorm.slice;

import com.waylau.hmos.dataabilityhelperaccessorm.ResourceTable;
import com.waylau.hmos.dataabilityhelperaccessorm.User;
import com.waylau.hmos.dataabilityhelperaccessorm.UserDataAbility;
import ohos.aafwk.ability.AbilitySlice;
import ohos.aafwk.content.Intent;
import ohos.agp.components.Button;
import ohos.agp.components.Text;
import ohos.hiviewdfx.HiLog;
import ohos.hiviewdfx.HiLogLabel;
import ohos.utils.zson.ZSONObject;

import java.util.List;
import java.util.Random;

public class MainAbilitySlice extends AbilitySlice {
    private static final String TAG = MainAbilitySlice.class.getSimpleName();
```

```java
    private static final HiLogLabel LABEL_LOG =
            new HiLogLabel(HiLog.LOG_APP, 0x00001, TAG);

    private UserDataAbility userDataAbility = new UserDataAbility();

    private Text text;

    @Override
    public void onStart(Intent intent) {
        super.onStart(intent);
        super.setUIContent(ResourceTable.Layout_ability_main);

        userDataAbility.onStart(intent);

        // 添加点击事件来触发访问数据
        Button buttonQuery = (Button) findComponentById(ResourceTable.Id_button_query);
        buttonQuery.setClickedListener(listener -> this.doQuery());

        Button buttonInsert = (Button) findComponentById(ResourceTable.Id_button_insert);
        buttonInsert.setClickedListener(listener -> this.doInsert());

        Button buttonUpdate = (Button) findComponentById(ResourceTable.Id_button_update);
        buttonUpdate.setClickedListener(listener -> this.doUpdate());

        Button buttonDelete = (Button) findComponentById(ResourceTable.Id_button_delete);
        buttonDelete.setClickedListener(listener -> this.doDelete());

        text = (Text) findComponentById(ResourceTable.Id_text);
    }

    @Override
    public void onStop() {
        super.onStop();
        userDataAbility.onStop();
    }

    private void doQuery() {
        // 查询所有用户
        List<User> users = userDataAbility.queryAll();

        // 用户的信息显示在界面上
        text.setText(ZSONObject.toZSONString(users) + "\n");
    }

    private void doInsert() {
        // 生成随机数据
```

```java
        Random random = new Random();
        int age = random.nextInt();
        String name = "n" + System.currentTimeMillis();

        // 生成用户
        User user = new User();
        user.setUserName(name);
        user.setAge(age);

        // 插入用户
        userDataAbility.insert(user);
    }

    private void doUpdate() {
        // 查询所有用户
        List<User> users = userDataAbility.queryAll();

        // 更新所有用户
        users.forEach(user -> {
            user.setAge(43);
            userDataAbility.update(user);
        });
    }

    private void doDelete() {
        // 删除所有用户
        userDataAbility.deleteAll();
    }

    @Override
    public void onActive() {
        super.onActive();
    }

    @Override
    public void onForeground(Intent intent) {
        super.onForeground(intent);
    }
}
```

在上述方法中：

- 初始化了UserDataAbility，并通过UserDataAbility来实现用户的查询、插入、更新、删除操作。
- Text组件用于展示查询到的用户信息。

27.4.12 运行

运行应用后，点击3次Insert按钮，以触发插入用户的操作，可以看到控制台输出内容如下：

```
09-14 00:32:29.836 28654-28654/com.waylau.hmos.dataabilityhelperaccessorm I
```

```
00001/UserDataAbility:  before insert
    09-14 00:32:29.842 28654-28654/com.waylau.hmos.dataabilityhelperaccessorm I
00001/UserDataAbility:  end insert: 1, isSuccessed: true
    09-14 00:32:30.055 28654-28654/com.waylau.hmos.dataabilityhelperaccessorm I
00001/UserDataAbility:  before insert
    09-14 00:32:30.061 28654-28654/com.waylau.hmos.dataabilityhelperaccessorm I
00001/UserDataAbility:  end insert: 2, isSuccessed: true
    09-14 00:32:30.351 28654-28654/com.waylau.hmos.dataabilityhelperaccessorm I
00001/UserDataAbility:  before insert
    09-14 00:32:30.358 28654-28654/com.waylau.hmos.dataabilityhelperaccessorm I
00001/UserDataAbility:  end insert: 3, isSuccessed: true
```

此时,点击"Query"按钮,以触发查询用户的操作,可以看到控制台输出内容如下:

```
    09-14 00:32:40.374 28654-28654/com.waylau.hmos.dataabilityhelperaccessorm I
00001/UserDataAbility:  query user: 
com.waylau.hmos.dataabilityhelperaccessorm.User@52c68dca
    09-14 00:32:40.374 28654-28654/com.waylau.hmos.dataabilityhelperaccessorm I
00001/UserDataAbility:  query user: 
com.waylau.hmos.dataabilityhelperaccessorm.User@52c68dea
    09-14 00:32:40.375 28654-28654/com.waylau.hmos.dataabilityhelperaccessorm I
00001/UserDataAbility:  query user: 
com.waylau.hmos.dataabilityhelperaccessorm.User@52c68e0a
```

界面效果如图27-4所示。

图27-4　界面效果

此时,点击Update按钮,以触发更新用户的操作,可以看到控制台输出内容如下:

```
    09-14 00:33:16.276 28654-28654/com.waylau.hmos.dataabilityhelperaccessorm I
00001/UserDataAbility:  query user: 
com.waylau.hmos.dataabilityhelperaccessorm.User@52c68dca
    09-14 00:33:16.276 28654-28654/com.waylau.hmos.dataabilityhelperaccessorm I
00001/UserDataAbility:  query user: 
```

```
com.waylau.hmos.dataabilityhelperaccessorm.User@52c68dea
    09-14 00:33:16.276 28654-28654/com.waylau.hmos.dataabilityhelperaccessorm I
00001/UserDataAbility:  query user:
com.waylau.hmos.dataabilityhelperaccessorm.User@52c68e0a
    09-14 00:33:16.277 28654-28654/com.waylau.hmos.dataabilityhelperaccessorm I
00001/UserDataAbility:  before update
    09-14 00:33:16.280 28654-28654/com.waylau.hmos.dataabilityhelperaccessorm I
00001/UserDataAbility:  end update, isSuccessed: true
    09-14 00:33:16.280 28654-28654/com.waylau.hmos.dataabilityhelperaccessorm I
00001/UserDataAbility:  before update
    09-14 00:33:16.284 28654-28654/com.waylau.hmos.dataabilityhelperaccessorm I
00001/UserDataAbility:  end update, isSuccessed: true
    09-14 00:33:16.284 28654-28654/com.waylau.hmos.dataabilityhelperaccessorm I
00001/UserDataAbility:  before update
    09-14 00:33:16.286 28654-28654/com.waylau.hmos.dataabilityhelperaccessorm I
00001/UserDataAbility:  end update, isSuccessed: true
```

点击Query按钮，界面效果如图27-5所示。

图27-5　界面效果

可以看到，age已经被更新为43。

此时，点击Delete按钮，以触发删除用户的操作，可以看到控制台输出内容如下：

```
    09-14 00:34:10.681 28654-28654/com.waylau.hmos.dataabilityhelperaccessorm I
00001/UserDataAbility:  before delete
    09-14 00:34:10.682 28654-28654/com.waylau.hmos.dataabilityhelperaccessorm I
00001/UserDataAbility:  delete user: 1, isSuccessed: true
    09-14 00:34:10.682 28654-28654/com.waylau.hmos.dataabilityhelperaccessorm I
00001/UserDataAbility:  delete user: 2, isSuccessed: true
    09-14 00:34:10.682 28654-28654/com.waylau.hmos.dataabilityhelperaccessorm I
00001/UserDataAbility:  delete user: 3, isSuccessed: true
    09-14 00:34:10.709 28654-28654/com.waylau.hmos.dataabilityhelperaccessorm I
```

```
00001/UserDataAbility: end delete, isSuccessed: true
```

点击Query按钮，界面效果如图27-6所示。

图27-6　界面效果

可以看到，用户信息都已经被删除了。

27.5　轻量级偏好数据库

轻量级偏好数据库主要提供轻量级Key-Value操作，支持本地应用存储少量数据，数据存储在本地文件中，同时也加载在内存中，所以访问速度更快，效率更高。轻量级偏好数据库属于非关系型数据库，不宜存储大量数据，经常用于操作键值对形式数据的场景。

27.5.1　基本概念

轻量级偏好数据库主要涉及以下概念：

- Key-Value数据库：一种以键值对存储数据的数据库，类似于Java中的Map。Key是关键字，Value是值。常见的Key-Value数据库产品有Redis、Berkley DB等。
- 非关系型数据库：区别于关系数据库，不保证遵循ACID（Atomic、Consistency、Isolation及Durability）特性，不采用关系模型来组织数据，数据之间无关系，扩展性好。除了上面几款Key-Value数据库产品外，还有Cassandra、MongoDB、HBase等。
- 偏好数据：用户经常访问和使用的数据。

有关Key-Value数据库、非关系型数据库的更多内容详见笔者所著的《分布式系统常用技术及案例分析》。

27.5.2 运作机制

HarmonyOS提供偏好型数据库的操作类，应用通过这些操作类来完成数据库操作。

- 借助DatabaseHelper的API，应用可以将指定文件的内容加载到Preferences实例，每个文件最多有一个Preferences实例，系统会通过静态容器将该实例存储在内存中，直到应用主动从内存中移除该实例或者删除该文件。
- 获取到文件对应的Preferences实例后，应用可以借助Preferences的API从Preferences实例中读取数据或者将数据写入Preferences实例，通过flush或者flushSync将Preferences实例持久化。

如图27-7所示是轻量级偏好数据库的架构图。

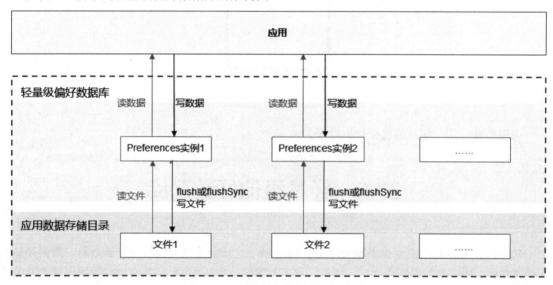

图 27-7　轻量级偏好数据库的架构图

27.5.3 约束与限制

使用偏好型数据库时，需要注意以下约束与限制：

- Key键为String类型，要求非空且大小不超过80个字符。
- 如果Value值为String类型，则可以为空，但是长度不超过8192个字符。
- 存储的数据量应该是轻量级的，建议存储的数据不超过一万条，否则会在内存方面产生较大的开销。
- 轻量级偏好数据库主要用于保存应用的一些常用配置，并不适合频繁改变数据的场景。

27.5.4　接口说明

轻量级偏好数据库向本地应用提供了操作偏好型数据库的API，支持本地应用读写少量数据及观察数据变化。数据存储形式为键值对，键的类型为字符串型，值的存储数据类型包括整型、字符串型、布尔型、浮点型、长整型、字符串型Set集合。

1. 创建数据库

通过数据库操作的辅助类可以获取要操作的Preferences实例，用于进行数据库的操作。
DatabaseHelper轻量级偏好数据库的创建接口有：

- getPreferences(String name)：获取文件对应的Preferences单实例，用于数据操作。

2. 查询数据

通过调用Get系列的方法可以查询不同类型的数据。
Preferences轻量级偏好数据库的查询接口有：

- getInt(String key, int defValue)：获取键对应的int类型的值。
- getFloat(String key, float defValue)：获取键对应的float类型的值。

3. 插入数据

通过Put系列的方法可以修改Preferences实例中的数据，通过flush或者flushSync将Preferences实例持久化。
Preferences轻量级偏好数据库的插入接口有：

- putInt(String key, int value)：设置Preferences实例中键对应的int类型的值。
- putString(String key, String value)：设置Preferences实例中键对应的String类型的值。
- flush()：将Preferences实例异步写入文件。
- flushSync()：将Preferences实例同步写入文件。

4. 观察数据变化

轻量级偏好数据库还提供了一系列的接口变化回调，用于观察数据的变化。开发者可以通过重写onChange方法来定义观察者的行为。
Preferences轻量级偏好数据库的接口变化回调有：

- registerObserver(PreferencesObserver preferencesObserver)：注册观察者，用于观察数据变化。
- unRegisterObserver(PreferencesObserver preferencesObserver)：注销观察者。
- onChange(Preferences preferences, String key)：观察者的回调方法，任意数据变化都会回调该方法。

5. 删除数据文件

DatabaseHelper轻量级偏好数据库的删除接口有：

- deletePreferences(String name)：删除文件和文件对应的Preferences单实例。

- removePreferencesFromCache(String name)：删除文件对应的Preferences单实例。

6. 移动数据库文件

DatabaseHelper轻量级偏好数据库的移动接口有：

- movePreferences(Context sourceContext, String sourceName, String targetName)：移动数据库文件。

27.6 实战：使用轻量级偏好数据库

本节将演示如何使用轻量级偏好数据库。创建一个名为Preferences的应用作为演示。

27.6.1 修改 ability_main.xml

修改ability_main.xml内容如下：

```xml
<?xml version="1.0" encoding="utf-8"?>
<DirectionalLayout
    xmlns:ohos="http://schemas.huawei.com/res/ohos"
    ohos:height="match_parent"
    ohos:width="match_parent"
    ohos:orientation="vertical">

    <DirectionalLayout
        ohos:height="60vp"
        ohos:width="match_parent"
        ohos:orientation="horizontal">
        <Button
            ohos:id="$+id:button_query"
            ohos:height="40vp"
            ohos:width="0vp"
            ohos:background_element="#F76543"
            ohos:layout_alignment="horizontal_center"
            ohos:margin="10vp"
            ohos:padding="10vp"
            ohos:text="Query"
            ohos:text_size="16fp"
            ohos:weight="1"
            />

        <Button
            ohos:id="$+id:button_insert"
            ohos:height="40vp"
            ohos:width="0vp"
            ohos:background_element="#F76543"
```

```xml
            ohos:layout_alignment="horizontal_center"
            ohos:margin="10vp"
            ohos:padding="10vp"
            ohos:text="Insert"
            ohos:text_size="16fp"
            ohos:weight="1"
            />

        <Button
            ohos:id="$+id:button_update"
            ohos:height="40vp"
            ohos:width="0vp"
            ohos:background_element="#F76543"
            ohos:layout_alignment="horizontal_center"
            ohos:margin="10vp"
            ohos:padding="10vp"
            ohos:text="Update"
            ohos:text_size="16fp"
            ohos:weight="1"
            />

        <Button
            ohos:id="$+id:button_delete"
            ohos:height="40vp"
            ohos:width="0vp"
            ohos:background_element="#F76543"
            ohos:layout_alignment="horizontal_center"
            ohos:margin="10vp"
            ohos:padding="10vp"
            ohos:text="Delete"
            ohos:text_size="16fp"
            ohos:weight="1"
            />
    </DirectionalLayout>

    <Text
        ohos:id="$+id:text"
        ohos:height="match_content"
        ohos:width="match_content"
        ohos:background_element="$graphic:background_ability_main"
        ohos:multiple_lines="true"
        ohos:text=""
        ohos:text_size="20fp"
        />
</DirectionalLayout>
```

界面预览效果如图27-8所示。

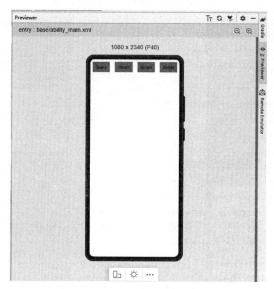

图 27-8　界面预览效果

上述代码中：

- 设置了4个按钮，用于触发操作数据库的相关操作。
- Text组件用于展示查询到的数据信息。

27.6.2　修改 MainAbilitySlice

修改MainAbilitySlice，代码如下：

```java
package com.waylau.hmos.preferences.slice;

import com.waylau.hmos.preferences.ResourceTable;
import ohos.aafwk.ability.AbilitySlice;
import ohos.aafwk.content.Intent;
import ohos.agp.components.Button;
import ohos.agp.components.Text;
import ohos.data.DatabaseHelper;
import ohos.data.preferences.Preferences;
import ohos.hiviewdfx.HiLog;
import ohos.hiviewdfx.HiLogLabel;

public class MainAbilitySlice extends AbilitySlice {
    private static final String TAG = MainAbilitySlice.class.getSimpleName();
    private static final HiLogLabel LABEL_LOG =
            new HiLogLabel(HiLog.LOG_APP, 0x00001, TAG);

    private static final String PREFERENCES_FILE = "preferences-file";
    private static final String PREFERENCES_KEY = "preferences-key";

    private DatabaseHelper databaseHelper;
```

```java
        private Preferences preferences;
        private Preferences.PreferencesObserver observer;
        private Text text;

        @Override
        public void onStart(Intent intent) {
            super.onStart(intent);
            super.setUIContent(ResourceTable.Layout_ability_main);

            // 初始化数据
            initData();

            // 添加点击事件来触发访问数据
            Button buttonQuery = (Button) findComponentById(ResourceTable.Id_button_query);
            buttonQuery.setClickedListener(listener -> this.doQuery());

            Button buttonInsert = (Button) findComponentById(ResourceTable.Id_button_insert);
            buttonInsert.setClickedListener(listener -> this.doInsert());

            Button buttonUpdate = (Button) findComponentById(ResourceTable.Id_button_update);
            buttonUpdate.setClickedListener(listener -> this.doUpdate());

            Button buttonDelete = (Button) findComponentById(ResourceTable.Id_button_delete);
            buttonDelete.setClickedListener(listener -> this.doDelete());

            text = (Text) findComponentById(ResourceTable.Id_text);
        }

        private void initData() {
            HiLog.info(LABEL_LOG, "before initData");

            databaseHelper = new DatabaseHelper(this.getContext());

            // fileName表示文件名，其取值不能为空，也不能包含路径
            // 默认存储目录可以通过context.getPreferencesDir()获取
            preferences = databaseHelper.getPreferences(PREFERENCES_FILE);

            // 观察者
            observer = new MyPreferencesObserver();

            HiLog.info(LABEL_LOG, "end initData");
        }

        @Override
        public void onStop() {
            super.onStop();
```

```java
        // 从内存中移除指定文件对应的Preferences单实例
        // 并删除指定文件及其备份文件、损坏文件
        boolean result = databaseHelper.deletePreferences(PREFERENCES_FILE);
    }

    private void doQuery() {
        HiLog.info(LABEL_LOG, "before doQuery");

        // 查询
        String result = preferences.getString(PREFERENCES_KEY, "");

        // 信息显示在界面上
        text.setText(result + "\n");

        HiLog.info(LABEL_LOG, "end doQuery, result: %{public}s", result);
    }

    private void doInsert() {
        HiLog.info(LABEL_LOG, "before doInsert");

        // 生成随机数据
        String data = "d" + System.currentTimeMillis();

        // 将数据写入Preferences实例
        preferences.putString(PREFERENCES_KEY, data);

        // 注册观察者
        preferences.registerObserver(observer);

        // 通过flush或者flushSync将Preferences实例持久化
        preferences.flush(); // 异步
        // preferences.flushSync(); // 同步

        HiLog.info(LABEL_LOG, "end doInsert, data: %{public}s", data);
    }

    private void doUpdate() {
        HiLog.info(LABEL_LOG, "before doUpdate");

        // 更新就是重新做一次插入
        doInsert();

        HiLog.info(LABEL_LOG, "end doUpdate");
    }

    private void doDelete() {
        HiLog.info(LABEL_LOG, "before doDelete");

        // 删除
```

```
            preferences.delete(PREFERENCES_KEY);

            HiLog.info(LABEL_LOG, "end doDelete");
        }

        private class MyPreferencesObserver implements
Preferences.PreferencesObserver {

            @Override
            public void onChange(Preferences preferences, String key) {
                HiLog.info(LABEL_LOG, "onChange, key: %{public}s", key);
            }
        }

        @Override
        public void onActive() {
            super.onActive();
        }

        @Override
        public void onForeground(Intent intent) {
            super.onForeground(intent);
        }
}
```

在上述方法中：

- initData方法初始化了DatabaseHelper、Preferences、PreferencesObserver对象。
- 通过Preferences对象来实现数据的查询、插入、更新、删除操作。
- onStop方法从内存中移除指定文件对应的Preferences单实例，并删除指定文件及其备份文件、损坏文件。
- Text组件用于展示查询到的数据信息。

27.6.3 运行

运行应用后，点击Query按钮，以触发查询的操作，可以看到控制台输出内容如下：

```
    09-14 00:42:37.858 15516-15516/com.waylau.hmos.preferences I
00001/MainAbilitySlice: before doQuery
    09-14 00:42:37.859 15516-15516/com.waylau.hmos.preferences I
00001/MainAbilitySlice: end doQuery, result:
```

从日志可以看出，并没有查询到任何数据，界面效果如图27-9所示。

图 27-9　界面效果

此时，点击Insert按钮，以触发插入的操作，可以看到控制台输出内容如下：

```
09-14 00:43:29.045 15516-15516/com.waylau.hmos.preferences I
00001/MainAbilitySlice:  before doInsert
09-14 00:43:29.046 15516-15516/com.waylau.hmos.preferences I
00001/MainAbilitySlice:  onChange, key: preferences-key
09-14 00:43:29.046 15516-15516/com.waylau.hmos.preferences I
00001/MainAbilitySlice:  end doInsert, data: d1631551409045
```

此时，再次点击Query按钮，就能查到刚插入的数据，效果如图27-10所示。

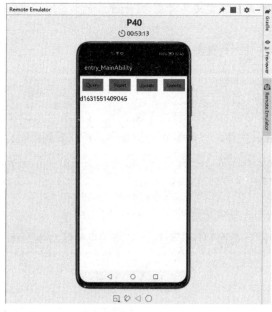

图 27-10　界面效果

点击Update按钮，以触发更新的操作，可以看到控制台输出内容如下：

```
09-14 00:44:20.546 15516-15516/com.waylau.hmos.preferences I
00001/MainAbilitySlice: before doUpdate
09-14 00:44:20.546 15516-15516/com.waylau.hmos.preferences I
00001/MainAbilitySlice: before doInsert
09-14 00:44:20.547 15516-15516/com.waylau.hmos.preferences I
00001/MainAbilitySlice: onChange, key: preferences-key
09-14 00:44:20.547 15516-15516/com.waylau.hmos.preferences I
00001/MainAbilitySlice: end doInsert, data: d1631551460546
09-14 00:44:20.547 15516-15516/com.waylau.hmos.preferences I
00001/MainAbilitySlice: end doUpdate
```

此时，点击Query按钮，就能查到刚更新的数据，效果如图27-11所示。

图27-11　界面效果

点击Delete按钮，以触发删除的操作，可以看到控制台输出内容如下：

```
09-14 00:44:52.578 15516-15516/com.waylau.hmos.preferences I
00001/MainAbilitySlice: before doDelete
09-14 00:44:52.578 15516-15516/com.waylau.hmos.preferences I
00001/MainAbilitySlice: end doDelete
```

此时，点击Query按钮，界面效果如图27-12所示。

可以看到，信息都已经被删除了。

图 27-12　界面效果

27.7　数据存储管理

数据存储管理指导开发者基于HarmonyOS进行存储设备（包含本地存储、SD卡、U盘等）的数据存储管理能力的开发，包括获取存储设备列表、获取存储设备视图等。

27.7.1　基本概念

数据存储管理涉及以下基本概念：

- 数据存储管理：数据存储管理包括获取存储设备列表、获取存储设备视图，同时也可以按照条件获取对应的存储设备视图信息。
- 设备存储视图：提供了存储设备的抽象，并且提供了访问存储设备自身信息的接口。

27.7.2　运作机制

用统一的视图结构可以表示各种存储设备，该视图结构的内部属性会因为设备的不同而不同。每个存储设备可以抽象成两部分，一部分是存储设备自身信息区域，另一部分是用来真正存放数据的区域。

图27-13展示了存储设备视图。

图 27-13　存储设备视图

27.7.3　接口说明

为了给用户展示存储设备信息，开发者可以使用数据存储管理接口获取存储设备视图信息，也可以根据用户提供的文件名获取对应存储设备的视图信息。

数据存储管理为开发者提供了下面几种功能，主要涉及DataUsage和Volume的API。

1. DataUsage

DataUsage的API有：

- getVolumes()：获取当前用户可用的设备列表视图。
- getVolume(File file)：获取存储该文件的存储设备视图。
- getVolume(Context context, Uri uri)：获取该URI对应文件所在的存储设备视图。
- getDiskMountedStatus()：获取默认存储设备的挂载状态。
- getDiskMountedStatus(File path)：获取存储该文件设备的挂载状态。
- isDiskPluggable()：默认存储设备是否为可插拔设备。
- isDiskPluggable(File path)：存储该文件的设备是否为可插拔设备。
- isDiskEmulated()：默认存储设备是否为虚拟设备。
- isDiskEmulated(File path)：存储该文件的设备是否为虚拟设备。

2. Volume

Volume的API有：

- isEmulated()：该设备是否是虚拟存储设备。
- isPluggable()：该设备是否支持插拔。
- getDescription()：获取设备描述信息。

- getState()：获取设备挂载状态。
- getVolUuid()：获取设备唯一标识符。

27.8 实战：使用数据存储管理

本节将演示如何使用数据存储管理。

创建一个名为DataUsage的应用作为演示。

27.8.1 修改 ability_main.xml

修改ability_main.xml内容如下：

```xml
<?xml version="1.0" encoding="utf-8"?>
<DirectionalLayout
    xmlns:ohos="http://schemas.huawei.com/res/ohos"
    ohos:height="match_parent"
    ohos:width="match_parent"
    ohos:orientation="vertical">

    <Button
        ohos:id="$+id:button_query"
        ohos:height="match_content"
        ohos:width="match_parent"
        ohos:background_element="#F76543"
        ohos:layout_alignment="horizontal_center"
        ohos:margin="10vp"
        ohos:padding="10vp"
        ohos:text="Query"
        ohos:text_size="30fp"
        />

    <Text
        ohos:id="$+id:text"
        ohos:height="match_content"
        ohos:width="match_content"
        ohos:background_element="$graphic:background_ability_main"
        ohos:multiple_lines="true"
        ohos:text=""
        ohos:text_size="20fp"
        />

</DirectionalLayout>
```

界面预览效果如图27-14所示。

第 27 章 数据管理 | 633

图 27-14　界面预览效果

上述代码中：

- 设置了Query按钮，以备设置点击事件，以触发查询数据的相关操作。
- Text组件用于展示查询到的数据信息。

27.8.2　修改 MainAbilitySlice

修改MainAbilitySlice，代码如下：

```
package com.waylau.hmos.datausage.slice;

import com.waylau.hmos.datausage.ResourceTable;
import ohos.aafwk.ability.AbilitySlice;
import ohos.aafwk.content.Intent;
import ohos.agp.components.Button;
import ohos.agp.components.Text;
import ohos.data.usage.DataUsage;
import ohos.data.usage.MountState;
import ohos.data.usage.Volume;
import ohos.hiviewdfx.HiLog;
import ohos.hiviewdfx.HiLogLabel;
import ohos.utils.zson.ZSONObject;

import java.util.List;
import java.util.Optional;

public class MainAbilitySlice extends AbilitySlice {
    private static final String TAG = MainAbilitySlice.class.getSimpleName();
```

```java
        private static final HiLogLabel LABEL_LOG =
            new HiLogLabel(HiLog.LOG_APP, 0x00001, TAG);

    private Text text;

    @Override
    public void onStart(Intent intent) {
        super.onStart(intent);
        super.setUIContent(ResourceTable.Layout_ability_main);

        // 添加点击事件来触发访问数据
        Button buttonQuery = (Button) findComponentById(ResourceTable.Id_button_query);
        buttonQuery.setClickedListener(listener -> this.doQuery());

        text = (Text) findComponentById(ResourceTable.Id_text);
    }

    private void doQuery() {
        HiLog.info(LABEL_LOG, "before doQuery");

        // 查询
        // 获取默认存储设备挂载状态
        MountState status = DataUsage.getDiskMountedStatus();
        String statusString = ZSONObject.toZSONString(status);
        text.append("MountState: " + statusString + "\n");

        // 默认存储设备是否为可插拔设备
        boolean isDiskPluggable = DataUsage.isDiskPluggable();
        text.append("isDiskPluggable: " + isDiskPluggable + "\n");

        // 默认存储设备是否为虚拟设备
        boolean isDiskEmulated = DataUsage.isDiskEmulated();
        text.append("isDiskEmulated: " + isDiskEmulated + "\n");

        // 获取存储设备列表
        Optional<List<Volume>> listOptional = DataUsage.getVolumes();
        if (listOptional.isPresent()) {
            text.append("Volume:\n");

            listOptional.get().forEach(volume -> {
                // 查询Volume的信息
                String volUuid = volume.getVolUuid();
                String description = volume.getDescription();
                boolean isEmulated = volume.isEmulated();
                boolean isPluggable = volume.isPluggable();

                String volumeString = ZSONObject.toZSONString(volume);
                text.append(volumeString + "\n");
```

```
                HiLog.info(LABEL_LOG, "volUuid: %{public}s, 
description: %{public}s, " +
                        "isEmulated: %{public}s, 
isPluggable: %{public}s",
                    volUuid, description, isEmulated, isPluggable);
            });
        }

        HiLog.info(LABEL_LOG, "end doQuery");
    }

    @Override
    public void onActive() {
        super.onActive();
    }

    @Override
    public void onForeground(Intent intent) {
        super.onForeground(intent);
    }
}
```

在上述方法中：

- doQuery方法通过DataUsage对象来查询默认存储设备的信息，包括挂载状态、是否为可插拔设备、是否为虚拟设备等。
- DataUsage.getVolumes可以获得所有的存储设备列表。遍历该列表，则可以获得每个Volume的信息。
- Text组件用于展示查询到的数据信息。

27.8.3 运行

运行应用后，点击Query按钮，以触发查询的操作，可以看到控制台输出内容如下：

```
    09-14 00:48:33.496 15786-15786/com.waylau.hmos.datausage I
00001/MainAbilitySlice: before doQuery
    09-14 00:48:33.509 15786-15786/com.waylau.hmos.datausage I
00001/MainAbilitySlice: volUuid: , description: 内部存储, isEmulated: true,
isPluggable: false
    09-14 00:48:33.509 15786-15786/com.waylau.hmos.datausage I
00001/MainAbilitySlice: end doQuery
```

从日志可以看出，已经查到了设备的信息。界面效果如图27-15所示。

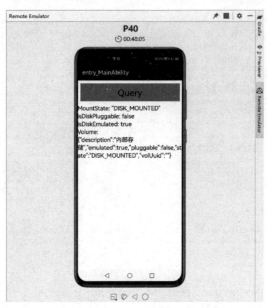

图 27-15　界面效果

第 28 章

原子化服务

HarmonyOS除了支持传统方式需要安装的应用外，还支持提供特定功能的免安装的应用，即原子化服务。

28.1 原子化服务概述

在万物互联时代，人均持有的设备数量不断攀升，设备和场景的多样性使应用开发变得更加复杂，应用入口更加丰富。在此背景下，应用提供方和用户迫切需要一种新的服务提供方式使应用开发更简单，服务（如听音乐、打车等）的获取和使用更便捷。为此，HarmonyOS提供了特定功能的免安装的应用，即原子化服务。

28.1.1 什么是原子化服务

原子化服务可以简单理解为手机端App中常见的"小程序"。

原子化服务是HarmonyOS提供的一种面向未来的服务提供方式，是有独立入口的（用户可通过点击方式直接触发）、免安装的（无须显式安装，由系统程序框架后台安装后即可使用）、可为用户提供一个或多个便捷服务的用户应用程序形态。例如，某传统方式需要安装的购物应用A，在按照原子化服务理念调整设计后，成为由"商品浏览""购物车""支付"等多个便捷服务组成的可以免安装的购物原子化服务A*。

原子化服务基于HarmonyOS API开发，支持运行在1+8+N设备上，供用户在合适的场景、合适的设备上便捷使用。原子化服务相对于传统方式需要安装的应用形态更加轻量，同时提供更丰富的入口，更精准的分发。

原子化服务由一个或多个HAP包组成，一个HAP包对应一个FA或一个PA。每个FA或PA均可独立运行，完成一个特定功能；一个或多个功能（对应FA或PA）完成一个特定的便捷服务。

表28-1展示了原子化服务与传统方式需要安装的应用对比。

表 28-1 原子化服务与传统方式的需要安装的应用对比

项 目	原子化服务	传统方式的需要安装的应用
软件包形态	App Pack（.app）	App Pack（.app）
分发平台	由原子化服务平台（Huawei Ability Gallery）管理和分发	由应用市场（AppGallery）管理和分发
安装后有无桌面icon	无桌面icon，但可手动添加到桌面，显示形式为服务卡片	有桌面icon
HAP包免安装要求	所有HAP包（包括Entry HAP和Feature HAP）均需满足免安装要求	所有HAP包（包括Entry HAP和Feature HAP）均为非免安装的

28.1.2 原子化服务特征

原子化服务具有以下特征：

1. 随处可及

原子化服务通过以下方式被发现并使用：

- 服务发现：原子化服务可在服务中心发现并使用。
- 智能推荐：原子化服务可以基于合适的场景被主动推荐给用户使用，用户可在服务中心和小艺建议中发现系统推荐的服务。

2. 服务直达

服务直达体现在以下两个方面：

- 原子化服务支持免安装使用。
- 服务卡片支持用户无须打开原子化服务便可获取服务内重要信息的展示和动态变化，如天气、关键事务备忘、热点新闻列表。

3. 跨设备

跨设备体现在以下几个方面：

- 原子化服务支持运行在1+8+N设备上，如手机、平板等设备。
- 支持跨设备分享，例如接入华为分享后，用户可分享原子化服务给好友，好友确认后即可打开分享的服务。
- 支持跨端迁移，例如手机上未完成的邮件迁移到平板继续编辑。
- 支持多端协同，例如用手机进行文档翻页和批注，配合智慧屏显示，以完成分布式办公；手机作为手柄，与智慧屏配合玩游戏。

28.1.3 原子化服务典型的使用场景

原子化服务典型的使用场景包括以下几种：

- 释放手机，让用户在更合适的设备上享受服务。打车是人们日常生活中经常使用的服务，通常人们在手机上打车需要一直停留在手机界面才能准确获取司机的状态信息，此时手机不能做其他事情。有了原子化服务的分布式能力，在手机上打车后，将司机状态实时同步到手表，无须查看手机，抬腕即可获取司机状态。
- 大小屏互动协作，打造网课新体验。在疫情期间，上网课已经成为学生们的日常，上网课的时间也越来越长。使用手机/平板上网课，屏幕较小，容易伤害眼睛，且在单一设备上无法获得良好的互动上课体验。有了原子化服务，就可以实现在智慧屏上听老师讲课，在手机/平板上互动答题，极大地提升了网课体验，并且有效保护视力。
- 快速开启服务。比如在常见的购物活动中，数字人民币原子化服务能够实现数字人民币收款、付款的功能。

28.2 服务中心

服务中心致力于为用户提供使用路径更短、体验更好的服务，向用户展示和发现便捷服务入口。便捷服务入口以服务卡片快照的形式在服务中心呈现。

28.2.1 服务中心入口

以手机为例，通过屏幕左下角或右下角向侧上方滑动进入服务中心，如图28-1所示。

28.2.2 常用服务

原子化服务在服务中心以服务卡片的形式展示，用户可将服务中心的服务卡片添加到桌面中快捷访问。"常用服务"涵盖用户常用的本地服务和云端推送的服务，为用户提供贴心便捷的服务体验，如图28-2所示。

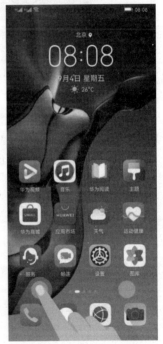

图 28-1　服务中心入口

28.2.3 我的收藏

"我的收藏"中收录用户所订阅的服务卡片，通过长按卡片可将服务添加到桌面或取消收藏，如图28-3所示。

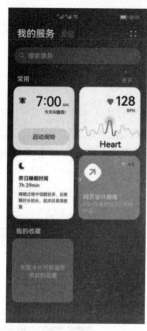
图 28-2　常用服务

图 28-3　我的收藏

28.2.4　服务发现

用户可以在"发现"版块中查找和浏览所有的服务卡片，如图28-4所示。服务以卡片（卡片由图标、名称、描述、快照组成）的形式向用户展示。轻点卡片可以选择将卡片添加到收藏或添加到桌面，随时随地查看信息获取服务。

图 28-4　服务发现

28.3 实战：原子化服务 HelloDog

本节通过创建一个名为HelloDog的原子化服务来演示原子化服务完整的开发流程。

28.3.1 初始化原子化服务

区别于传统的HarmonyOS应用，在通过DevEco Studio工程向导创建原子化服务时，Project Type应选择为Service，同时勾选Show in Service Center，如图28-5所示。

28.3.2 理解原子化服务的项目结构

如图28-6所示是初始化原子化服务时自动创建的代码结构。

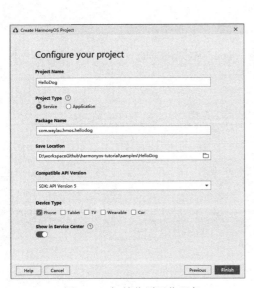

图 28-5 初始化原子化服务

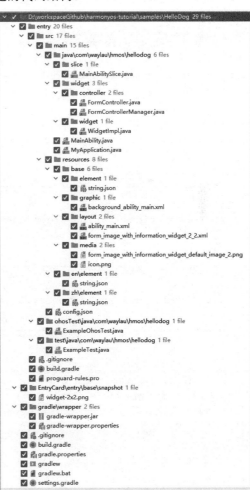

图 28-6 原子化服务代码结构

上述目录中，widget包下的代码即为服务卡片相关的代码。其中：

- controller包下的FormController是服务卡片控制器的接口。
- controller包下的FormControllerManager是服务卡片控制器的管理器。
- cwidget包下的WidgetImpl是FormController的默认实现类。

此时，会同步创建一个2×2的默认服务卡片模板form_image_with_information_widget_2_2.xml，同时还会创建该卡片对应的快照图form_image_with_information_widget_default_image_2.png。

1. FormController

FormController定义了卡片的控制器接口，代码如下：

```java
package com.waylau.hmos.hellodog.widget.controller;

import ohos.aafwk.ability.AbilitySlice;
import ohos.aafwk.ability.ProviderFormInfo;
import ohos.aafwk.content.Intent;
import ohos.app.Context;

/**
 * The api set for form controller.
 */
public abstract class FormController {
    /**
     * Context of ability
     */
    protected final Context context;

    /**
     * The name of current form service widget
     */
    protected final String formName;

    /**
     * The dimension of current form service widget
     */
    protected final int dimension;

    public FormController(Context context, String formName, Integer dimension) {
        this.context = context;
        this.formName = formName;
        this.dimension = dimension;
    }

    /**
     * Bind data for a form
     *
     * @return ProviderFormInfo
     */
```

```java
    public abstract ProviderFormInfo bindFormData();

    /**
     * Update form data
     *
     * @param formId the id of service widget to be updated
     * @param vars   the data to update for service widget, this parameter is optional
     */
    public abstract void updateFormData(long formId, Object... vars);

    /**
     * Called when receive service widget message event
     *
     * @param formId  form id
     * @param message the message context sent by service widget message event
     */
    public abstract void onTriggerFormEvent(long formId, String message);

    /**
     * Get the destination ability slice to route
     *
     * @param intent intent of current page slice
     * @return the destination ability slice name to route
     */
    public abstract Class<? extends AbilitySlice> getRoutePageSlice(Intent intent);
}
```

上述接口基本上在开发过程中无须修改，直接使用即可。

2. FormControllerManager

FormControllerManager是服务卡片控制器的管理器，代码如下：

```java
package com.waylau.hmos.hellodog.widget.controller;

import ohos.app.Context;
import ohos.data.DatabaseHelper;
import ohos.data.preferences.Preferences;
import ohos.hiviewdfx.HiLog;
import ohos.hiviewdfx.HiLogLabel;
import ohos.utils.zson.ZSONObject;

import java.lang.reflect.InvocationTargetException;
import java.util.HashMap;
import java.util.Locale;
import java.util.Map;
import java.util.List;
import java.util.ArrayList;

/**
```

```java
     * Form controller manager.
     */
    public class FormControllerManager {
        private static final HiLogLabel TAG = new HiLogLabel(HiLog.DEBUG, 0x0,
FormControllerManager.class.getName());
        private static final String PACKAGE_PATH =
"com.waylau.hmos.hellodog.widget";
        private static final String SHARED_SP_NAME = "form_info_sp.xml";
        private static final String FORM_NAME = "formName";
        private static final String DIMENSION = "dimension";
        private static FormControllerManager managerInstance = null;
        private final HashMap<Long, FormController> controllerHashMap = new
HashMap<>();

        private final Context context;

        private final Preferences preferences;

        /**
         * Constructor with context.
         *
         * @param context instance of Context.
         */
        private FormControllerManager(Context context) {
            this.context = context;
            DatabaseHelper databaseHelper = new
DatabaseHelper(this.context.getApplicationContext());
            preferences = databaseHelper.getPreferences(SHARED_SP_NAME);
        }

        /**
         * Singleton mode.
         *
         * @param context instance of Context.
         * @return FormControllerManager instance.
         */
        public static FormControllerManager getInstance(Context context) {
            if (managerInstance == null) {
                synchronized (FormControllerManager.class) {
                    if (managerInstance == null) {
                        managerInstance = new FormControllerManager(context);
                    }
                }
            }
            return managerInstance;
        }

        /**
         * Save the form id and form name.
         *
```

```java
     * @param formId    form id.
     * @param formName  form name.
     * @param dimension form dimension
     * @return FormController form controller
     */
    public FormController createFormController(long formId, String formName,
int dimension) {
        synchronized (controllerHashMap) {
            if (formId < 0 || formName.isEmpty()) {
                return null;
            }
            HiLog.info(TAG,
                    "saveFormId() formId: " + formId + ", formName: " + formName
+ ", preferences: " + preferences);
            if (preferences != null) {
                ZSONObject formObj = new ZSONObject();
                formObj.put(FORM_NAME, formName);
                formObj.put(DIMENSION, dimension);
                preferences.putString(Long.toString(formId),
ZSONObject.toZSONString(formObj));
                preferences.flushSync();
            }

            // Create controller instance.
            FormController controller = newInstance(formName, dimension,
context);

            // Cache the controller.
            if (controller != null) {
                if (!controllerHashMap.containsKey(formId)) {
                    controllerHashMap.put(formId, controller);
                }
            }

            return controller;
        }
    }

    /**
     * Get the form controller instance.
     *
     * @param formId form id.
     * @return the instance of form controller.
     */
    public FormController getController(long formId) {
        synchronized (controllerHashMap) {
            if (controllerHashMap.containsKey(formId)) {
                return controllerHashMap.get(formId);
            }
            Map<String, ?> forms = preferences.getAll();
```

```java
                String formIdString = Long.toString(formId);
                if (forms.containsKey(formIdString)) {
                    ZSONObject formObj = ZSONObject.stringToZSON((String)
forms.get(formIdString));
                    String formName = formObj.getString(FORM_NAME);
                    int dimension = formObj.getIntValue(DIMENSION);
                    FormController controller = newInstance(formName, dimension,
context);
                    controllerHashMap.put(formId, controller);
                }
                return controllerHashMap.get(formId);
            }
        }

        private FormController newInstance(String formName, int dimension, Context
context) {
            FormController ctrInstance = null;
            if (formName == null || formName.isEmpty()) {
                HiLog.error(TAG, "newInstance() get empty form name");
                return ctrInstance;
            }
            try {
                String className = PACKAGE_PATH + "." +
formName.toLowerCase(Locale.ROOT) + "."
                        + getClassNameByFormName(formName);
                Class<?> clazz = Class.forName(className);
                if (clazz != null) {
                    Object controllerInstance = clazz.getConstructor(Context.class,
String.class, Integer.class)
                            .newInstance(context, formName, dimension);
                    if (controllerInstance instanceof FormController) {
                        ctrInstance = (FormController) controllerInstance;
                    }
                }
            } catch (NoSuchMethodException | InstantiationException |
IllegalArgumentException | InvocationTargetException
                    | IllegalAccessException | ClassNotFoundException |
SecurityException exception) {
                HiLog.error(TAG, "newInstance() get exception: " +
exception.getMessage());
            }
            return ctrInstance;
        }

        /**
         * Get all form id from the share preference
         *
         * @return form id list
         */
        public List<Long> getAllFormIdFromSharePreference() {
```

```java
            List<Long> result = new ArrayList<>();
            Map<String, ?> forms = preferences.getAll();
            for (String formId : forms.keySet()) {
                result.add(Long.parseLong(formId));
            }
            return result;
        }

        /**
         * Delete a form controller
         *
         * @param formId form id
         */
        public void deleteFormController(long formId) {
            synchronized (controllerHashMap) {
                preferences.delete(Long.toString(formId));
                preferences.flushSync();
                controllerHashMap.remove(formId);
            }
        }

        private String getClassNameByFormName(String formName) {
            String[] strings = formName.split("_");
            StringBuilder result = new StringBuilder();
            for (String string : strings) {
                result.append(string);
            }
            char[] charResult = result.toString().toCharArray();
            charResult[0] = (charResult[0] >= 'a' && charResult[0] <= 'z') ? (char)
(charResult[0] - 32) : charResult[0];
            return String.copyValueOf(charResult) + "Impl";
        }
}
```

上述类基本上在开发过程中无须修改，直接使用即可。

3. WidgetImpl

WidgetImpl是FormController的默认实现类，代码如下：

```java
package com.waylau.hmos.hellodog.widget.widget;

import com.waylau.hmos.hellodog.ResourceTable;
import com.waylau.hmos.hellodog.widget.controller.FormController;

import ohos.aafwk.ability.AbilitySlice;
import ohos.aafwk.ability.ProviderFormInfo;
import ohos.aafwk.content.Intent;
import ohos.app.Context;
import ohos.hiviewdfx.HiLog;
import ohos.hiviewdfx.HiLogLabel;
```

```java
import java.util.HashMap;
import java.util.Map;

public class WidgetImpl extends FormController {
    private static final HiLogLabel TAG = new HiLogLabel(HiLog.DEBUG, 0x0, WidgetImpl.class.getName());
    private static final int DEFAULT_DIMENSION_2X2 = 2;

    private static final Map<Integer, Integer> RESOURCE_ID_MAP = new HashMap<>();

    static {
        RESOURCE_ID_MAP.put(DEFAULT_DIMENSION_2X2, ResourceTable.Layout_form_image_with_information_widget_2_2);
    }

    public WidgetImpl(Context context, String formName, Integer dimension) {
        super(context, formName, dimension);
    }

    @Override
    public ProviderFormInfo bindFormData() {
        HiLog.info(TAG, "bind form data when create form");
        return new ProviderFormInfo(RESOURCE_ID_MAP.get(dimension), context);
    }

    @Override
    public void updateFormData(long formId, Object... vars) {
        HiLog.info(TAG, "update form data timing, default 30 minutes");
    }

    @Override
    public void onTriggerFormEvent(long formId, String message) {
        HiLog.info(TAG, "handle card click event.");
    }

    @Override
    public Class<? extends AbilitySlice> getRoutePageSlice(Intent intent) {
        HiLog.info(TAG, "get the default page to route when you click card.");
        return null;
    }
}
```

上述WidgetImpl类可以根据实现开发需要进行修改，比如涉及数据的更新或者事件的监听，只需要重写上述的updateFormData或者onTriggerFormEvent方法即可。

本例力求简单，保持原有代码不做修改直接使用。

4. MainAbility

MainAbility是主页面，代码如下：

```java
package com.waylau.hmos.hellodog;

import com.waylau.hmos.hellodog.slice.MainAbilitySlice;
import ohos.aafwk.ability.Ability;
import ohos.aafwk.content.Intent;
import com.waylau.hmos.hellodog.widget.controller.FormController;
import com.waylau.hmos.hellodog.widget.controller.FormControllerManager;
import ohos.aafwk.ability.AbilitySlice;
import ohos.aafwk.ability.ProviderFormInfo;
import ohos.hiviewdfx.HiLog;
import ohos.hiviewdfx.HiLogLabel;

public class MainAbility extends Ability {
    public static final int DEFAULT_DIMENSION_2X2 = 2;
    private static final int INVALID_FORM_ID = -1;
    private static final HiLogLabel TAG = new HiLogLabel(HiLog.DEBUG, 0x0, MainAbility.class.getName());
    private String topWidgetSlice;

    @Override
    public void onStart(Intent intent) {
        super.onStart(intent);
        super.setMainRoute(MainAbilitySlice.class.getName());
        if (intentFromWidget(intent)) {
            topWidgetSlice = getRoutePageSlice(intent);
            if (topWidgetSlice != null) {
                setMainRoute(topWidgetSlice);
            }
        }
        stopAbility(intent);
    }

    @Override
    protected ProviderFormInfo onCreateForm(Intent intent) {
        HiLog.info(TAG, "onCreateForm");
        long formId = intent.getLongParam(AbilitySlice.PARAM_FORM_IDENTITY_KEY, INVALID_FORM_ID);
        String formName = intent.getStringParam(AbilitySlice.PARAM_FORM_NAME_KEY);
        int dimension = intent.getIntParam(AbilitySlice.PARAM_FORM_DIMENSION_KEY, DEFAULT_DIMENSION_2X2);
        HiLog.info(TAG, "onCreateForm: formId=" + formId + ",formName=" + formName);
        FormControllerManager formControllerManager = FormControllerManager.getInstance(this);
        FormController formController = formControllerManager.getController(formId);
        formController = (formController == null) ? formControllerManager.createFormController(formId,
```

```java
                formName, dimension) : formController;
        if (formController == null) {
            HiLog.error(TAG, "Get null controller. formId: " + formId + ", formName: " + formName);
            return null;
        }
        return formController.bindFormData();
    }

    @Override
    protected void onUpdateForm(long formId) {
        HiLog.info(TAG, "onUpdateForm");
        super.onUpdateForm(formId);
        FormControllerManager formControllerManager = FormControllerManager.getInstance(this);
        FormController formController = formControllerManager.getController(formId);
        formController.updateFormData(formId);
    }

    @Override
    protected void onDeleteForm(long formId) {
        HiLog.info(TAG, "onDeleteForm: formId=" + formId);
        super.onDeleteForm(formId);
        FormControllerManager formControllerManager = FormControllerManager.getInstance(this);
        formControllerManager.deleteFormController(formId);
    }

    @Override
    protected void onTriggerFormEvent(long formId, String message) {
        HiLog.info(TAG, "onTriggerFormEvent: " + message);
        super.onTriggerFormEvent(formId, message);
        FormControllerManager formControllerManager = FormControllerManager.getInstance(this);
        FormController formController = formControllerManager.getController(formId);
        formController.onTriggerFormEvent(formId, message);
    }

    @Override
    public void onNewIntent(Intent intent) {
        if (intentFromWidget(intent)) { // Only response to it when starting from a service widget.
            String newWidgetSlice = getRoutePageSlice(intent);
            if (topWidgetSlice == null || !topWidgetSlice.equals(newWidgetSlice)) {
                topWidgetSlice = newWidgetSlice;
                restart();
            }
```

```
        }
    }
    private boolean intentFromWidget(Intent intent) {
        long formId =
intent.getLongParam(AbilitySlice.PARAM_FORM_IDENTITY_KEY, INVALID_FORM_ID);
        return formId != INVALID_FORM_ID;
    }

    private String getRoutePageSlice(Intent intent) {
        long formId =
intent.getLongParam(AbilitySlice.PARAM_FORM_IDENTITY_KEY, INVALID_FORM_ID);
        if (formId == INVALID_FORM_ID) {
            return null;
        }
        FormControllerManager formControllerManager =
FormControllerManager.getInstance(this);
        FormController formController =
formControllerManager.getController(formId);
        if (formController == null) {
            return null;
        }
        Class<? extends AbilitySlice> clazz =
formController.getRoutePageSlice(intent);
        if (clazz == null) {
            return null;
        }
        return clazz.getName();
    }
}
```

可以看到卡片服务也是在该MainAbility类中进行管理和路由的。上述代码保持原有代码不做修改直接使用。

5. 修改 form_image_with_information_widget_2_2.xml

form_image_with_information_widget_2_2.xml是原子化服务卡片的布局，对其进行个性化的修改，代码如下：

```xml
<?xml version="1.0" encoding="utf-8"?>
<DependentLayout
    xmlns:ohos="http://schemas.huawei.com/res/ohos"
    ohos:height="match_parent"
    ohos:width="match_parent"
    ohos:background_element="#FFFFFFFF"
    ohos:remote="true">

    <Image
        ohos:height="match_parent"
        ohos:width="126vp"
        ohos:horizontal_center="true"
```

```xml
        ohos:image_src="$media:dog"
        ohos:scale_mode="zoom_start"
        ohos:top_margin="17vp"/>

    <DirectionalLayout
        ohos:height="match_content"
        ohos:width="126vp"
        ohos:align_parent_bottom="true"
        ohos:bottom_margin="12vp"
        ohos:horizontal_center="true"
        ohos:orientation="vertical">

        <Text
            ohos:height="match_content"
            ohos:width="match_parent"
            ohos:text="我的狗狗"
            ohos:text_color="#E5000000"
            ohos:text_size="16fp"
            ohos:text_weight="500"
            ohos:truncation_mode="ellipsis_at_end"/>

        <Text
            ohos:height="match_content"
            ohos:width="match_parent"
            ohos:text="狗狗撒娇！"
            ohos:text_color="#99000000"
            ohos:text_size="12fp"
            ohos:text_weight="400"
            ohos:top_margin="2vp"
            ohos:truncation_mode="ellipsis_at_end"/>
    </DirectionalLayout>
</DependentLayout>
```

上述Image组件所引用的dog.jpg图片需要事先放置于media目录下。

可以在预览器中预览原子化服务卡片，效果如图28-7所示。

图28-7　预览原子化服务

6. 修改 ability_main.xml

ability_main.xml是整个项目的主布局，修改ability_main.xml代码如下：

```xml
<?xml version="1.0" encoding="utf-8"?>
<DirectionalLayout
    xmlns:ohos="http://schemas.huawei.com/res/ohos"
    ohos:height="match_parent"
    ohos:width="match_parent"
    ohos:alignment="center"
    ohos:orientation="vertical">

    <Image
        ohos:height="match_content"
        ohos:width="match_content"
        ohos:image_src="$media:dog"
        />

    <Text
        ohos:height="match_content"
        ohos:width="match_parent"
        ohos:text="我的狗狗"
        ohos:text_size="30fp"
        />

    <Text
        ohos:height="match_content"
        ohos:width="match_parent"
        ohos:text="狗狗撒娇！"
        ohos:text_size="20fp"
        />

</DirectionalLayout>
```

可以在预览器中预览效果，如图28-8所示。

图28-8　预览项目

7. config.json

config.json是整个项目的配置，代码如下：

```json
{
  "app": {
    "bundleName": "com.waylau.hmos.hellodog",
    "vendor": "waylau",
    "version": {
      "code": 1000000,
      "name": "1.0.0"
    }
  },
  "deviceConfig": {},
  "module": {
    "package": "com.waylau.hmos.hellodog",
    "name": ".MyApplication",
    "mainAbility": "com.waylau.hmos.hellodog.MainAbility",
    "deviceType": [
      "phone"
    ],
    "distro": {
      "deliveryWithInstall": true,
      "moduleName": "entry",
      "moduleType": "entry",
      "installationFree": true
    },
    "abilities": [
      {
        "skills": [
          {
            "entities": [
              "entity.system.home"
            ],
            "actions": [
              "action.system.home"
            ]
          }
        ],
        "orientation": "unspecified",
        "name": "com.waylau.hmos.hellodog.MainAbility",
        "icon": "$media:icon",
        "description": "$string:mainability_description",
        "formsEnabled": true,
        "label": "$string:entry_MainAbility", // 服务名称
        "type": "page",
        "forms": [
          {
            "landscapeLayouts": [
              "$layout:form_image_with_information_widget_2_2"
            ],
```

```
        "isDefault": true,
        "scheduledUpdateTime": "10:30",
        "defaultDimension": "2*2",
        "name": "widget",
        "description": "This is a service widget",
        "colorMode": "auto",
        "type": "Java",
        "supportDimensions": [
          "2*2"
        ],
        "portraitLayouts": [
          "$layout:form_image_with_information_widget_2_2"
        ],
        "updateEnabled": true,
        "updateDuration": 1
      }
    ],
    "launchType": "standard"
  }
 ]
}
```

上述配置保持默认即可，无须修改，直接使用。这里最重要的是label标签。label标签是便捷服务对用户显示的名称，必须配置，且应以资源索引的方式配置，以支持多语言。不同HAP包的mainAbility的label要唯一，以免造成用户看到多个同名服务而无法区分。此外，label的命名应与服务内容强关联，能够通过显而易见的语义看出服务的关键内容。

分别对base、en.json的entry_MainAbility做了修改，修改如下：

```
{
  "string": [
    {
      "name": "entry_MainAbility",
      "value": "HelloDog"
    },
    ...
  ]
}
```

28.3.3　搜索原子化服务

安装完原子化服务后，就可以在服务中心通过服务的名称搜索到该原子化服务，如图28-9所示。可以通过长按卡片将服务添加到桌面上，或者直接点击卡片来打开原子化服务。

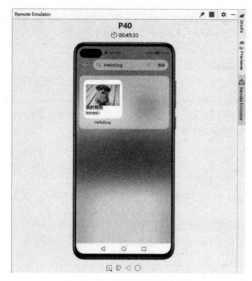

图 28-9 搜索原子化服务

28.3.4 运行原子化服务

点击卡片来打开原子化服务，就能够运行该原子化服务，效果如图28-10所示。

图 28-10 运行原子化服务

第 29 章

流 转

流转在HarmonyOS中泛指多设备分布式操作。流转能力打破了设备界限，多设备联动，使用户应用程序可分可合、可流转。流转按照体验可分为跨端迁移和多端协同。

29.1 流转概述

随着全场景多设备生活方式的不断深入，用户拥有的设备越来越多，每个设备都能在适合的场景下提供良好的体验，例如手表可以提供及时的信息查看能力，电视可以带来沉浸的观影体验。但是，每个设备也有使用场景的局限，例如在电视上输入文本相对手机来说是非常糟糕的体验。当多个设备通过分布式操作系统能够相互感知，进而整合成一个超级终端时，设备与设备之间就可以取长补短、相互帮助，为用户提供更加自然流畅的分布式体验。

因此，HarmonyOS提出了流转的概念，以提供多设备之间一致性的交互体验。流转在HarmonyOS中泛指多设备分布式操作。流转能力打破了设备界限，多设备联动，使用户应用程序可分可合、可流转，实现如邮件跨设备编辑、多设备协同健身、多屏游戏等分布式业务。流转为开发者提供了更广的使用场景和更新的产品视角，强化了产品优势，实现了体验升级。流转按照体验可分为跨端迁移和多端协同。

- 跨端迁移：一种实现用户应用程序流转的技术方案，指在A端运行的FA迁移到B端上，完成迁移后，B端FA继续任务，而A端应用退出。在用户使用设备的过程中，当使用情境发生变化时（例如从室内走到户外或者周围有更合适的设备等），之前使用的设备可能已经不适合继续当前的任务，此时用户可以选择新的设备来继续当前的任务。常见的跨端迁移场景实例：
  - ▶ 视频来电时从手机迁移到智慧屏，视频聊天体验更佳，手机视频应用退出。
  - ▶ 在手机上的阅读应用浏览文章，迁移到平板上继续查看，手机上的阅读应用退出。

● 多端协同：一种实现用户应用程序流转的技术方案，指多端上的不同FA/PA同时运行或者交替运行实现完整的业务，或者多端上的相同FA/PA同时运行实现完整的业务。多个设备作为一个整体为用户提供比单设备更加高效、沉浸的体验。例如，用户通过智慧屏的应用A拍照后，A可调用手机的应用B进行人像美颜，最终将美颜后的照片保存在智慧屏的应用A上。常见的多端协同场景实例还有：
  ➢ 手机侧应用A做游戏手柄，智慧屏侧应用B做游戏显示，为用户带来一个全新的游戏体验。
  ➢ 平板侧应用A做答题板，智慧屏侧应用B做直播，为用户带来一个全新的上网课体验。

29.2 流转架构

HarmonyOS流转提供了一组API库，可让用户应用程序更轻松、快捷地完成流转体验。HarmonyOS流转架构有如下优势：

● 统一流转管理UI，支持设备发现、选择以及任务管理。
● 支持远程服务调用等能力，可轻松设计业务。
● 支持多个应用同时进行流转。
● 支持不同形态的设备，如手机、平板、TV、手表等。

流转架构如图29-1所示。

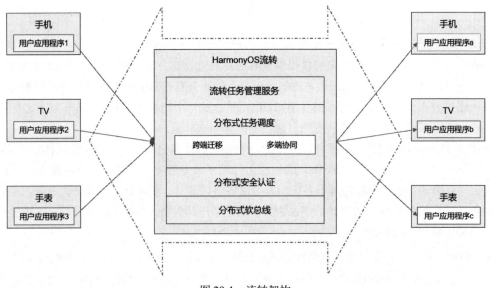

图29-1 流转架构

● 流转任务管理服务：在流转发起端，接受用户应用程序注册，提供流转入口、状态显示、退出流转等管理能力。当前仅手机、平板设备支持流转任务管理服务。
● 分布式任务调度：提供远程服务启动、远程服务连接、远程迁移等能力，并通过不同能力组合支撑用户应用程序完成跨端迁移或多端协同的业务体验。
- 分布式安全：提供E2E的加密通道，为用户应用程序提供安全的跨端传输机制，保证"正确的人，通过正确的设备，正确地使用数据"。
- 分布式软总线：使用基于手机、平板、智能穿戴、智慧屏等分布式设备的统一通信基座为设备之间的互联互通提供统一的分布式通信能力。

29.2.1 跨端迁移流程

以设备A的应用和设备B的应用进行跨端迁移为例，业务流程如图29-2所示。

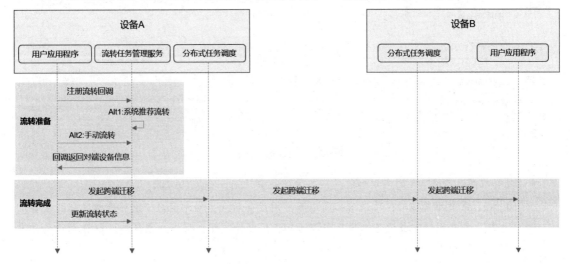

图 29-2　跨端迁移流程

- 流转准备：设备A上的应用向流转任务管理服务注册一个流转回调。
  - Alt1-系统推荐流转：系统感知周边有可用设备后，主动为用户提供可选择流转的设备信息，并在用户完成设备选择后回调通知应用开始流转，将用户选择的设备B的设备信息提供给应用。
  - Alt2-用户手动流转：系统在用户手动点击流转图标后，被动为用户提供可选择交互的设备信息，并在用户完成设备选择后回调通知应用开始流转，将用户选择的设备B的设备信息提供给应用。
- 流转完成：设备A上的应用通过调用分布式任务调度的能力（如continueAbility等）向设备B的应用发起跨端迁移。在流转中将流转状态上报到流转任务管理服务。跨端迁移后，设备A上的应用需要自行退出。

29.2.2 多端协同流程

以设备A的应用和设备B的应用进行多端协同为例，业务流程如图29-3所示。

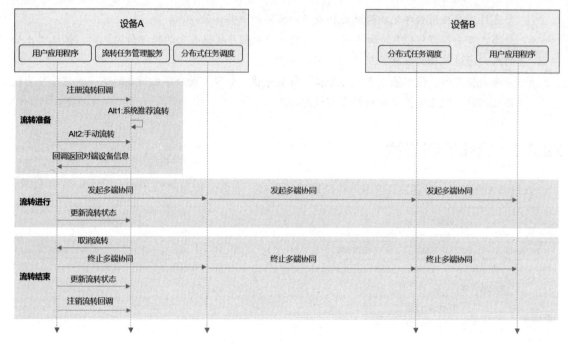

图 29-3　多端协同流程

- 流转准备：设备A上的应用向流转任务管理服务注册一个流转回调。
  - Alt1-系统推荐流转：系统感知周边有可用设备后，主动为用户提供可选择流转的设备信息，并在用户完成设备选择后回调通知应用开始流转，将用户选择的设备B的设备信息提供给应用。
  - Alt2-用户手动流转：系统在用户手动点击流转图标后，被动为用户提供可选择交互的设备信息，并在用户完成设备选择后回调通知应用开始流转，将用户选择的设备B的设备信息提供给应用。
- 流转进行：设备A上的应用通过调用分布式任务调度的能力（如startAbility、connectAbility等）向设备B的应用发起多端协同。在流转中将流转状态上报到流转任务管理服务。
- 流转结束：用户通过设备A的流转任务管理界面结束流转。用户点击结束任务后，流转任务管理服务回调通知应用取消流转。设备A上的应用通过调用分布式任务调度的能力（如stopAbility、disconnectAbility等）终止和设备B的多端协同。流转结束后，将流转状态上报到流转任务管理服务，并向流转任务管理服务注销流转回调。设备A和设备B进行多端协同后，设备A和设备C重复如上流程，可实现设备A、B、C进行多端协同，此时设备A是中心控制点。

29.2.3　兼容性设计

无论是跨端迁移还是多端协同，流转都是由两个设备上的用户应用程序共同完成的，两个设备上的用户应用程序需要做如下兼容设计。

1. 跨端应用版本兼容

建议用户程序设计版本号和最小兼容版本号。只有当设备A的用户程序版本号≥设备B的用户程序最小兼容版本号，且设备B的用户程序版本号≥设备A的用户程序最小兼容版本号时，才允许进行流转。

对于跨端迁移或多端协同场景，如果是同一个应用在A端和B端进行通信，此时流转任务管理服务提供了版本兼容性检查机制，可以将满足条件的设备筛选出来。

流转任务管理服务提供的版本兼容性检查机制要求两个设备是同一个用户应用程序。

要求用户应用程序在填写config.json时填写minCompatibleVersionCode字段（最小兼容版本号）和code字段（当前版本号）。minCompatibleVersionCode标识app pack能够兼容的最低历史版本号。该标签值为32位整型数值，非负整数。如果应用/服务未配置minCompatibleVersionCode，则系统将minCompatibleVersionCode默认填写为当前版本号。

对于多端协同，如果是不同应用之间的A、B端协同，则需要不同应用自己实现版本兼容性检查机制。

2. 跨端原子化服务安装

跨端被拉起的程序建议设计为原子化服务，即当设备A的用户应用程序向设备B的原子化服务发起多端协同时，如果设备B上没有安装对应服务，HarmonyOS会自动下载相关原子化服务，和A端的用户应用程序一起进行多端协同。如跨端被拉起的程序设计为传统方式的需要安装的应用，则当设备A的用户应用程序向设备B的原子化服务发起多端协同，如果设备B上没有安装相关应用，会提示流转失败。

29.3 跨端迁移

开发者在应用FA中通过调用流转任务管理服务、分布式任务调度的接口实现跨端迁移。

29.3.1 跨端迁移流程

跨端迁移流程如图29-4所示。

- 设备A上的应用FA向流转任务管理服务注册一个流转回调。
  - Alt1-系统推荐流转：系统感知周边有可用设备后，主动为用户提供可选择流转的设备信息，并在用户完成设备选择后回调onConnected通知应用FA开始流转，将用户选择的设备B的设备信息提供给应用FA。
  - Alt2-用户手动流转：系统在用户手动点击流转图标后，通过showDeviceList通知流转任务管理服务，被动为用户提供可选择交互的设备信息，并在用户完成设备选择后回调onConnected通知应用FA开始流转，将用户选择的设备B的设备信息提供给应用FA。
- 设备A上的应用FA通过调用分布式任务调度的能力向设备B的应用发起跨端迁移。应用FA需要自己管理流转状态，将流转状态从IDLE迁移到CONNECTING，并上报到流转任务管理服务。
  - 设备A上的FA请求迁移。
  - 系统回调设备A上的FA及其AbilitySlice栈中所有AbilitySlice实例的IAbilityContinuation.

onStartContinuation()方法，以确认当前是否可以开始迁移，onStartContinuation 方法返回 true，表示当前 FA 可以开始迁移。

如果可以开始迁移，则系统回调设备 A 上的 FA 及其 AbilitySlice 栈中所有 AbilitySlice 实例的 IAbilityContinuation.onSaveData()方法，以便保存迁移后恢复状态必需的数据。

如果保存数据成功，则系统在设备 B 上启动同一个 FA，并恢复 AbilitySlice 栈，然后回调 IAbilityContinuation.onRestoreData()方法，传递设备 A 上的 FA 保存的数据，应用可在此方法中恢复业务状态，此后设备 B 上此 FA 从 onStart()开始其生命周期回调。

- 系统回调设备 A 上的 FA 及其 AbilitySlice 栈中所有 AbilitySlice 实例的 IAbilityContinuation.onCompleteContinuation()方法，通知应用迁移成功。
- 应用将流转状态从 CONNECTING 迁移到 CONNECTED，并上报到流转任务管理服务。
- 流转任务管理服务将流转状态重新置为 IDLE，流转完成。
- 应用向流转任务管理服务注销流转回调。

● 应用自行退出。

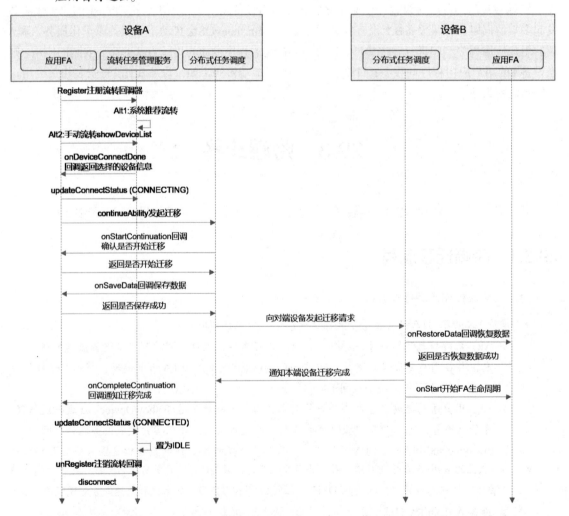

图 29-4　跨端迁移流程

29.3.2 接口说明

流转任务管理服务提供的注册、解注册、显示设备列表、上报业务状态是实现跨端迁移的前提。开发者通过跨端迁移能力可实现文档跨设备编辑、视频跨设备接续播放等功能。

1. IContinuationRegisterManager 接口

- void register(String bundleName, ExtraParams parameter, IContinuationDeviceCallback deviceCallback, RequestCallback requestCallback)：注册并连接到流转任务管理服务，并获取对应的注册token。
- void unregister(int token, RequestCallback requestCallback)：从流转任务管理服务解注册，传入注册时获取的token进行解注册。执行后，通过RequestCallback的onResult回调知道执行是否成功。
- void updateConnectStatus(int token, String deviceId, int status, RequestCallback requestCallback)：通知流转任务管理服务更新当前用户程序的连接状态，并在流转任务管理服务界面展示给用户。token、deviceId参数来自于注册流转任务管理服务的回调。status参数可以为IDLE、CONNECTING、CONNECTED、DIS_CONNECTING。如果有错误，则需要上报errorCode。执行后，通过RequestCallback的onResult回调知道执行是否成功。
- void showDeviceList(int token, ExtraParams parameter, RequestCallback requestCallback)：显示组网内可选择的设备列表信息。该接口提供手动显示设备列表的能力，parameter参数可以指定设备过滤的条件，用于手动多端协同，支持的过滤条件与register接口相同。token参数来自于注册流转任务管理服务的回调。执行后，通过RequestCallback的onResult回调知道执行是否成功。
- void disconnect()：在应用退出时，主动调用断开和流转任务管理服务的连接。

2. IContinuationDeviceCallback 接口

- void onConnected(ContinuationDeviceInfo deviceInfo)：当用户从选择设备列表选择设备时调用，返回设备ID、设备类型和设备名称供开发者使用。
- void onDisconnected(String deviceId)：当流转任务管理服务断开连接设备时调用。

3. RequestCallback 接口

- void onResult(int result)：与流转任务管理服务交互成功时调用。当作为注册流转任务管理服务的回调对象时，注册成功后给用户程序返回对应的token。

4. Ability/AbilitySlice 接口

- IContinuationRegisterManager getContinuationRegisterManager()：获取流转任务管理服务注册服务管理类，可以与流转任务管理服务进行交互，包括注册/解注册、更新设备连接状态、显示可选择设备列表等。
- void continueAbility(String deviceId)：把当前FA流转到同一个分布式网络中的另一个设备上，仅支持单向流转。

5. IAbilityContinuation 接口

- boolean onStartContinuation()：FA请求迁移后，系统首先回调此方法，开发者可以在此回调中决策当前是否可以执行迁移，比如弹框让用户确认是否开始迁移。
- boolean onSaveData(IntentParams saveData)：如果onStartContinuation()返回true，则系统回调此方法，开发者在此回调中保存必须传递到目标端设备上用于恢复FA状态的数据。
- boolean onRestoreData(IntentParams restoreData)：发起端设备上FA完成保存数据后，系统在目标端设备上回调此方法，开发者在此回调中接受用于恢复FA状态的数据。
- void onCompleteContinuation(int result)：目标端设备上恢复数据一旦完成，系统就会在发起端设备上回调FA的此方法，以便通知应用迁移流程已结束。开发者可以在此检查迁移结果是否成功，并在此处理迁移结束的动作，例如应用可以在迁移完成后终止自身生命周期。
- void onFailedContinuation(int errorCode)：迁移过程中发生异常，系统会在发起端回调FA的此方法，以便通知应用迁移流程发生的异常。并不是所有异常都会回调此方法，仅局限于该接口枚举的异常。开发者可以在此检查异常信息，并在此处理迁移异常发生后的动作，例如应用可以提醒用户此时发生的异常信息并给出处理建议。

29.3.3 约束与限制

使用跨端迁移需要注意以下约束与限制：

- 每个应用注册流转任务管理服务的Ability数量上限为5个，后续新增注册的Ability会将最开始注册的覆盖。
- 一个应用可能包含多个FA，仅需要在支持跨端迁移的FA及其所包含的AbilitySlice中调用或实现相关接口。
- 跨端迁移不支持两个设备之间分别登录不同的账号，也就是要求多个设备是同一账号。
- 跨端迁移不支持对PA的迁移，只支持对FA的迁移。
- 跨端迁移完成时，系统不会主动关闭发起迁移的FA，开发者可以根据业务需要主动调用terminateAbility或stopAbility关闭FA，并调用updateConnectStatus()更新设备的连接状态。
- 通过continueAbility进行跨端迁移的过程中，远端FA首先接收到发起端FA传输的数据，再执行启动，即onRestoreData()发生在onStart()之前。
- 跨端迁移的数据大小限制200KB以内，即onSaveData只能传递200KB以内的数据。
- 迁移传输的数据当前仅支持基础类型数据传递和系统Sequenceable对象，不支持自定义对象及文件数据传递。
- FA流转过程中，在流转未完成时再次调用continueAbility发起流转，接口将会抛出状态异常，应用需要加以限制处理。
- 跨端迁移要求HarmonyOS 2.0以上版本才能支持，注册到流转任务管理服务时jsonParams中需要增加{"harmonyVersion":"2.0.0"}过滤条件。

29.4 实战：实现跨端迁移与回迁

跨设备迁移支持将Page在同一用户的不同设备间进行迁移，以便支持用户无缝切换的诉求。以Page从设备A迁移到设备B为例，迁移动作的主要步骤如下：

- 设备A上的Page请求迁移。
- HarmonyOS处理迁移任务，并回调设备A上Page保存数据的方法，用于保存迁移必需的数据。
- HarmonyOS在设备B上启动同一个Page，并回调其恢复数据的方法。
- 请求迁移并完成后，设备A上已迁移的Page可以发起回迁，以便使用户活动重新回到此设备。

本节演示如何实现跨设备迁移与回迁。创建名为ContinueRemoteFA的项目作为演示。

29.4.1 MainAbility 实现 IAbilityContinuation 接口

MainAbility需要实现IAbilityContinuation接口，代码如下：

```java
package com.waylau.hmos.continueremotefa;

import com.waylau.hmos.continueremotefa.slice.MainAbilitySlice;
import ohos.aafwk.ability.Ability;
import ohos.aafwk.ability.IAbilityContinuation;
import ohos.aafwk.content.Intent;
import ohos.aafwk.content.IntentParams;
import ohos.bundle.IBundleManager;
import ohos.security.SystemPermission;

import java.util.ArrayList;
import java.util.List;

public class MainAbility extends Ability implements IAbilityContinuation {
    @Override
    public void onStart(Intent intent) {
        super.onStart(intent);
        super.setMainRoute(MainAbilitySlice.class.getName());
    }

    @Override
    public boolean onStartContinuation() {
        // 重写
        return true;
    }

    @Override
    public boolean onSaveData(IntentParams intentParams) {
```

```
        // 重写
        return true;
    }

    @Override
    public boolean onRestoreData(IntentParams intentParams) {
        // 重写
        return true;
    }

    @Override
    public void onCompleteContinuation(int i) {

    }
}
```

上述代码的重点是重写onStartContinuation、onSaveData、onRestoreData等方法。

29.4.2 声明权限

在项目对应的config.json中声明跨端迁移访问的权限：ohos.permission.DISTRIBUTED_DATASYNC以及获取分布式设备信息相关的权限。在config.json中的配置如下：

```
// 声明权限
"reqPermissions": [
    {
      "name": "ohos.permission.DISTRIBUTED_DATASYNC"
    },
    {
      "name": "ohos.permission.DISTRIBUTED_DEVICE_STATE_CHANGE"
    },
    {
      "name": "ohos.permission.GET_DISTRIBUTED_DEVICE_INFO"
    },
    {
      "name": "ohos.permission.GET_BUNDLE_INFO"
    }
]
```

同时修改MainAbility，显式声明ohos.permission.DISTRIBUTED_DATASYNC权限，最终MainAbility完整代码如下：

```
package com.waylau.hmos.continueremotefa;

import com.waylau.hmos.continueremotefa.slice.MainAbilitySlice;
import ohos.aafwk.ability.Ability;
import ohos.aafwk.ability.IAbilityContinuation;
import ohos.aafwk.content.Intent;
import ohos.aafwk.content.IntentParams;
```

```java
import ohos.bundle.IBundleManager;
import ohos.security.SystemPermission;

import java.util.ArrayList;
import java.util.List;

public class MainAbility extends Ability implements IAbilityContinuation {
    @Override
    public void onStart(Intent intent) {
        super.onStart(intent);
        super.setMainRoute(MainAbilitySlice.class.getName());
        requestPermission();
    }

    //获取权限
    private void requestPermission() {
        String[] permission = {
                SystemPermission.DISTRIBUTED_DATASYNC
        };
        List<String> applyPermissions = new ArrayList<>();
        for (String element : permission) {
            if (verifySelfPermission(element) != 0) {
                if (canRequestPermission(element)) {
                    applyPermissions.add(element);
                }
            }
        }
        requestPermissionsFromUser(applyPermissions.toArray(new String[0]), 0);
    }


    @Override
    public boolean onStartContinuation() {
        // 重写
        return true;
    }

    @Override
    public boolean onSaveData(IntentParams intentParams) {
        // 重写
        return true;
    }

    @Override
    public boolean onRestoreData(IntentParams intentParams) {
        // 重写
        return true;
    }
```

```
    @Override
    public void onCompleteContinuation(int i) {

    }
}
```

29.4.3 设计界面与布局

修改ability_main.xml布局文件如下:

```
<?xml version="1.0" encoding="utf-8"?>
<DirectionalLayout
    xmlns:ohos="http://schemas.huawei.com/res/ohos"
    ohos:height="match_parent"
    ohos:width="match_parent"
    ohos:alignment="center"
    ohos:orientation="vertical">

    <Button
        ohos:id="$+id:button_continue_remote_fa"
        ohos:height="match_content"
        ohos:width="match_content"
        ohos:background_element="#F76543"
        ohos:layout_alignment="horizontal_center"
        ohos:text="迁移"
        ohos:text_size="40fp"
        />

    <Button
        ohos:id="$+id:button_continue_eversibly"
        ohos:height="match_content"
        ohos:width="match_content"
        ohos:background_element="#F76543"
        ohos:layout_alignment="horizontal_center"
        ohos:text="回迁"
        ohos:text_size="40fp"
        ohos:top_margin="8vp"
        />

</DirectionalLayout>
```

最终,界面预览效果如图29-5所示。

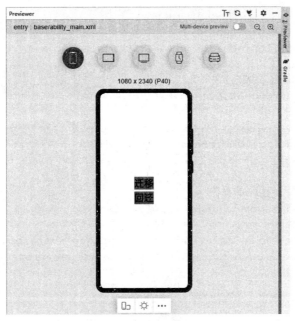

图 29-5　界面预览效果

29.4.4　修改 MainAbilitySlice

核心的逻辑都集中到了MainAbilitySlice中。MainAbilitySlice同样也要实现IAbilityContinuation接口，代码如下：

```
package com.waylau.hmos.continueremotefa.slice;

import com.waylau.hmos.continueremotefa.DeviceUtils;
import com.waylau.hmos.continueremotefa.ResourceTable;
import ohos.aafwk.ability.AbilitySlice;
import ohos.aafwk.ability.IAbilityContinuation;
import ohos.aafwk.content.Intent;
import ohos.aafwk.content.IntentParams;
import ohos.agp.components.Button;
import ohos.agp.components.TextField;
import ohos.hiviewdfx.HiLog;
import ohos.hiviewdfx.HiLogLabel;

public class MainAbilitySlice extends AbilitySlice implements
IAbilityContinuation {
    private static final String TAG = MainAbilitySlice.class.getSimpleName();
    private static final HiLogLabel LABEL_LOG =
            new HiLogLabel(HiLog.LOG_APP, 0x00001, TAG);

    private static final String MESSAGE_KEY =
"com.waylau.hmos.continueremotefa.slice.MESSAGE_KEY";

    @Override
```

```java
        public void onStart(Intent intent) {
            super.onStart(intent);
            super.setUIContent(ResourceTable.Layout_ability_main);

            // 监听跨端迁移FA的事件
            Button buttonContinueRemoteFA = (
                    Button)
findComponentById(ResourceTable.Id_button_continue_remote_fa);
            buttonContinueRemoteFA.setClickedListener(listener ->
continueRemoteFA());

            // 监听拉回迁移FA的事件
            Button buttonContinueEversibly =
                    (Button)
findComponentById(ResourceTable.Id_button_continue_eversibly);
            buttonContinueEversibly.setClickedListener(listener ->
continueEversibly());
        }

        private void continueRemoteFA() {
            HiLog.info(LABEL_LOG, "before startRemoteFA");

            String deviceId = DeviceUtils.getDeviceId();

            HiLog.info(LABEL_LOG, "get deviceId: %{public}s", deviceId);

            if (deviceId != null) {
                // 发起迁移流程
                //   continueAbility()是不可回迁的
                //   continueAbilityReversibly() 是可以回迁的
                continueAbilityReversibly(deviceId);
            }
        }

        private void continueEversibly() {
            // 发起回迁流程
            reverseContinueAbility();
        }

        @Override
        public void onActive() {
            super.onActive();
        }

        @Override
        public void onForeground(Intent intent) {
            super.onForeground(intent);
        }

        @Override
```

```java
    public boolean onStartContinuation() {
        // 重写
        return true;
    }

    @Override
    public boolean onSaveData(IntentParams intentParams) {
        // 重写
        return true;
    }

    @Override
    public boolean onRestoreData(IntentParams intentParams) {
        // 重写
        return true;
    }

    @Override
    public void onCompleteContinuation(int i) {

    }
}
```

上述代码中：

- 可以通过continueAbility或者continueAbilityReversibly方法来执行迁移。需要注意continueAbility()是不可回迁的，而continueAbilityReversibly()是可以回迁的，因此本例使用了continueAbilityReversibly方法。
- continueAbilityReversibly方法可以指定待迁移的设备的ID，该ID通过DeviceUtils工具类获取。
- 回迁调用reverseContinueAbility方法。

29.4.5 新增 DeviceUtils

DeviceUtils代码如下：

```java
package com.waylau.hmos.continueremotefa;

import ohos.distributedschedule.interwork.DeviceInfo;
import ohos.distributedschedule.interwork.DeviceManager;
import ohos.hiviewdfx.HiLog;
import ohos.hiviewdfx.HiLogLabel;

import java.util.ArrayList;
import java.util.List;

public class DeviceUtils {
    private static final String TAG = DeviceUtils.class.getSimpleName();
    private static final HiLogLabel LABEL_LOG =
```

```java
            new HiLogLabel(HiLog.LOG_APP, 0x00001, TAG);

    private DeviceUtils() {
    }

    // 获取当前组网下可迁移的设备ID列表
    public static List<String> getAvailableDeviceId() {
        List<String> deviceIds = new ArrayList<>();

        List<DeviceInfo> deviceInfoList =
                DeviceManager.getDeviceList(DeviceInfo.FLAG_GET_ALL_DEVICE);
        if (deviceInfoList == null) {
            return deviceIds;
        }

        if (deviceInfoList.size() == 0) {
            HiLog.warn(LABEL_LOG, "did not find other device");
            return deviceIds;
        }

        for (DeviceInfo deviceInfo : deviceInfoList) {
            deviceIds.add(deviceInfo.getDeviceId());
        }

        return deviceIds;
    }

    // 获取当前组网下可迁移的设备ID
    // 如果有多个，则取第一个
    public static String getDeviceId() {
        String deviceId = "";
        List<String> outerDevices = DeviceUtils.getAvailableDeviceId();

        if (outerDevices == null || outerDevices.size() == 0) {
            HiLog.warn(LABEL_LOG, "did not find other device");
        } else {
            for (String item : outerDevices) {
                HiLog.info(LABEL_LOG, "outerDevices:%{public}s", item);
            }
            deviceId = outerDevices.get(0);
        }
        HiLog.info(LABEL_LOG, "getDeviceId:%{public}s", deviceId);
        return deviceId;
    }
}
```

DeviceUtils的getDeviceId方法用于获取在线设备列表。如果有多个，则取任意一个。

29.4.6 运行

多了演示多个设备的迁移情况，需要启动Super device，同时启动两个手机模拟器。

假设左侧为设备A，作为源设备，右侧为设备B，作为目标设备。先在设备A、设备B上分别安装应用。

先启动设备A的应用，如图29-6所示。

在设备A上点击"迁移"按钮，如图29-7所示，Page已经从设备A上迁移到了设备B上，说明迁移成功了。

图 29-6　界面效果

图 29-7　界面效果

在设备A或者设备B上点击"回迁"按钮，如图29-8所示，Page从设备B上消失了，说明回迁成功了。

图 29-8　界面效果

29.5 多端协同

开发者在应用FA中通过调用流转任务管理服务、分布式任务调度的接口实现多端协同。

29.5.1 多端协同流程

多端协同流程如图29-9所示。

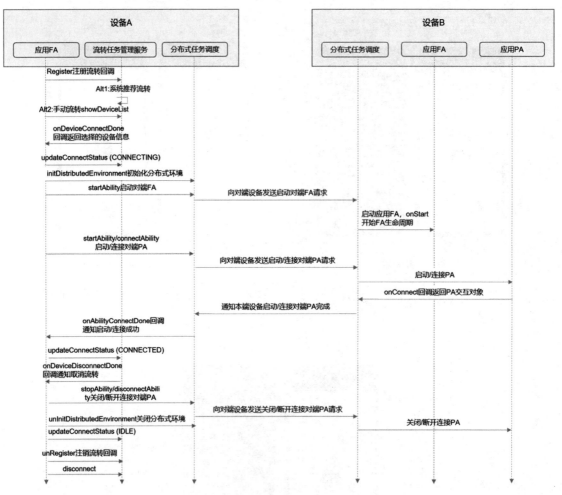

图 29-9 多端协同流程

- 设备A上的应用FA向流转任务管理服务注册一个流转回调。
  - Alt1-系统推荐流转：系统感知周边有可用设备后，主动为用户提供可选择流转的设备信息，并在用户完成设备选择后回调 onConnected 通知应用 FA 开始流转，将用户选择的设备 B 的设备信息提供给应用 FA。
  - Alt2-用户手动流转：系统在用户手动点击流转图标后，通过 showDeviceList 通知流转任务管理服务，被动为用户提供可选择交互的设备信息，并在用户完成设备选择后回调 onConnected 通知应用 FA 开始流转，将用户选择的设备 B 的设备信息提供给应用 FA。
- 设备A上的应用FA需要初始化分布式任务调度能力。
- 设备A上的应用FA通过调用分布式任务调度的能力向设备B的应用发起多端协同。应用FA需要自己管理流转状态，将流转状态从IDLE迁移到CONNECTING，并上报到流转任务管理服务。
- 开发者根据Ability模板及意图的不同，通过组合以下能力生成多端协同的业务：启动远程FA、启动远程PA、连接远程PA。这些能力都需要指定待连接设备的信息。下面以设备A（本地设备）和设备B（远端设备）为例进行场景介绍。
  - ➤ startAbility 设备 A 启动设备 B 的 FA：在设备 A 上通过本地应用提供的启动按钮，启动设备 B 上对应的 FA。例如，设备 A 控制设备 B 打开相册，只需要开发者在启动 FA 时指定打开相册的意图即可。
  - ➤ startAbility 设备 A 启动设备 B 的 PA：在设备 A 上通过本地应用提供的启动按钮，启动设备 B 上指定的 PA。例如，开发者在启动远程服务时，通过意图指定音乐播放服务，即可实现设备 A 启动设备 B 音乐播放的能力。
  - ➤ connectAbility 设备 A 连接设备 B 的 PA：在设备 A 上通过本地应用提供的连接按钮，连接设备 B 上指定的 PA。连接后，通过其他功能相关按钮实现控制对端 PA 的能力。通过连接关系，开发者可以实现跨设备的同步服务调度，实现如大型计算任务互助等价值场景。
- 设备A上的应用FA启动或连接设备B的应用FA/PA后，应用将流转状态从CONNECTING迁移到CONNECTED，并上报到流转任务管理服务。
- 用户通过设备A的流转任务管理界面结束流转。用户点击结束任务后，流转任务管理服务回调onDisconnected通知应用FA取消流转。
- 设备A上的应用通过调用分布式任务调度的能力终止和设备B的多端协同。
  - ➤ disconnectAbility 设备 A 与设备 B 的 PA 断开连接：将之前已连接的 PA 断开连接。
  - ➤ stopAbility 设备 A 关闭设备 B 的 PA：关闭设备 B 上指定的 PA。
- 应用关闭分布式任务调度能力。
- 应用将流转状态从CONNECTED迁移到IDLE，并上报到流转任务管理服务。
- 应用向流转任务管理服务注销流转回调。

29.5.2 接口说明

流转任务管理服务提供的注册、解注册、显示设备列表、上报业务状态是实现多端协同的前提。开发者使用分布式服务平台提供的连接和断开连接PA、启动远程FA、启动和关闭PA的能力可实现自定义的多端协同体验。

1. IContinuationRegisterManager 接口

- void register(String bundleName, ExtraParams parameter, IContinuationDeviceCallback deviceCallback, RequestCallback requestCallback)：注册并连接到流转任务管理服务，并获取对应的注册token。

- void unregister(int token, RequestCallback requestCallback): 从流转任务管理服务解注册，传入注册时获取的token进行解注册。执行后，通过RequestCallback的onResult回调知道执行是否成功。
- void updateConnectStatus(int token, String deviceId, int status, RequestCallback requestCallback): 通知流转任务管理服务更新当前用户程序的连接状态，并在流转任务管理服务界面展示给用户。token、deviceId参数来自于注册流转任务管理服务的回调。status参数可以为IDLE、CONNECTING、CONNECTED、DIS_CONNECTING。如果有错误，则需要上报errorCode。执行后，通过RequestCallback的onResult回调知道执行是否成功。
- void showDeviceList(int token, ExtraParams parameter, RequestCallback requestCallback): 显示组网内可选择的设备列表信息。该接口提供手动显示设备列表的能力，parameter参数可以指定设备过滤的条件，用于手动多端协同，支持的过滤条件与register接口相同。token参数来自于注册流转任务管理服务的回调。执行后，通过RequestCallback的onResult回调知道执行是否成功。
- void disconnect()：在应用退出时，主动调用断开和流转任务管理服务的连接。

2. IContinuationDeviceCallback 接口

- void onConnected(ContinuationDeviceInfo deviceInfo)：当用户从选择设备列表选择设备时调用，返回设备ID、设备类型和设备名称供开发者使用。
- void onDisconnected(String deviceId)：当流转任务管理服务断开连接设备时调用。

3. RequestCallback 接口

- void onResult(int result)：与流转任务管理服务交互成功时调用。当作为注册流转任务管理服务的回调对象时，注册成功后给用户程序返回对应的token。

4. Ability/AbilitySlice 接口

- IContinuationRegisterManager getContinuationRegisterManager()：获取流转任务管理服务注册服务管理类，可以与流转任务管理服务进行交互，包括注册/解注册、更新设备连接状态、显示可选择设备列表等。
- void continueAbility(String deviceId)：把当前FA流转到同一个分布式网络中的另一个设备上，仅支持单向流转。

5. IAbilityContinuation 接口

- void onAbilityConnectDone(ElementName element, IRemoteObject remote, int resultCode)：当连接Service服务成功时调用，可使用返回的IRemoteObject对象与连接的Service服务通信。
- void onAbilityDisconnectDone(ElementName element, int resultCode)：当已连接的Service服务被异常关闭时调用。

6. DeviceManager 接口

- void initDistributedEnvironment(String deviceId, IInitCallBack callback)：初始化分布式服务平台。其中deviceId指明设备ID，callback指明初始化分布式环境状态回调，参考IInitCallBack接口描述。

- void unInitDistributedEnvironment(String deviceId, IInitCallBack callback)：不再使用分布式服务平台。

7. IInitCallBack 接口功能介绍
- void onInitSuccess(String deviceId)：成功回调。
- void onInitFailure(String deviceId, int errorCode)：失败回调。

29.5.3 约束与限制

使用多端协同需要注意以下约束与限制：

- 每个应用注册流转任务管理服务的Ability数量上限为5个，后续新增注册的Ability会将最开始注册的覆盖。
- startAbility、connectAbility中跨设备传递的intent数据大小限制在200KB以内，不支持使用connectAbility触发远端PA的免安装。
- connectAbility中跨设备传递的remoteObject数据大小限制在200KB以内。
- 多端协同要求HarmonyOS 2.0以上版本才能支持，注册到流转任务管理服务时jsonParams中需要增加{"harmonyVersion":"2.0.0"}过滤条件。
- stopAbility不支持两个设备之间分别登录不同的账号，也就是要求多个设备是同一账号。

29.6 实战：实现多端协同

前面章节已经介绍了如何实现跨设备迁移。本节在前面章节的基础上稍作改造，以实现多端协同。创建名为ContinueRemoteFACollaboration的项目作为演示。

29.6.1 MainAbility 实现 IAbilityContinuation 接口

MainAbility需要实现IAbilityContinuation接口，代码如下：

```
package com.waylau.hmos.continueremotefacollaboration;

import com.waylau.hmos.continueremotefacollaboration.slice.MainAbilitySlice;
import ohos.aafwk.ability.Ability;
import ohos.aafwk.ability.IAbilityContinuation;
import ohos.aafwk.content.Intent;
import ohos.aafwk.content.IntentParams;
import ohos.bundle.IBundleManager;
import ohos.security.SystemPermission;

import java.util.ArrayList;
```

```java
import java.util.List;

public class MainAbility extends Ability implements IAbilityContinuation {
    @Override
    public void onStart(Intent intent) {
        super.onStart(intent);
        super.setMainRoute(MainAbilitySlice.class.getName());
    }

    @Override
    public boolean onStartContinuation() {
        // 重写
        return true;
    }

    @Override
    public boolean onSaveData(IntentParams intentParams) {
        // 重写
        return true;
    }

    @Override
    public boolean onRestoreData(IntentParams intentParams) {
        // 重写
        return true;
    }

    @Override
    public void onCompleteContinuation(int i) {

    }
}
```

上述代码的重点是重写onStartContinuation、onSaveData、onRestoreData等方法。

29.6.2 声明权限

在项目对应的config.json中声明跨端迁移访问的权限：ohos.permission.DISTRIBUTED_DATASYNC以及获取分布式设备信息的相关权限。在config.json中的配置如下：

```json
// 声明权限
"reqPermissions": [
    {
        "name": "ohos.permission.DISTRIBUTED_DATASYNC"
    },
    {
        "name": "ohos.permission.DISTRIBUTED_DEVICE_STATE_CHANGE"
    },
```

```
    {
      "name": "ohos.permission.GET_DISTRIBUTED_DEVICE_INFO"
    },
    {
      "name": "ohos.permission.GET_BUNDLE_INFO"
    }
]
```

同时修改MainAbility，显式声明ohos.permission.DISTRIBUTED_DATASYNC权限，最终MainAbility完整代码如下：

```
package com.waylau.hmos.continueremotefacollaboration;

import com.waylau.hmos.continueremotefacollaboration.slice.MainAbilitySlice;
import ohos.aafwk.ability.Ability;
import ohos.aafwk.ability.IAbilityContinuation;
import ohos.aafwk.content.Intent;
import ohos.aafwk.content.IntentParams;
import ohos.bundle.IBundleManager;
import ohos.security.SystemPermission;

import java.util.ArrayList;
import java.util.List;

public class MainAbility extends Ability implements IAbilityContinuation {
    @Override
    public void onStart(Intent intent) {
        super.onStart(intent);
        super.setMainRoute(MainAbilitySlice.class.getName());
        requestPermission();
    }

    //获取权限
    private void requestPermission() {
        String[] permission = {
                SystemPermission.DISTRIBUTED_DATASYNC
        };
        List<String> applyPermissions = new ArrayList<>();
        for (String element : permission) {
            if (verifySelfPermission(element) != 0) {
                if (canRequestPermission(element)) {
                    applyPermissions.add(element);
                }
            }
        }
        requestPermissionsFromUser(applyPermissions.toArray(new String[0]), 0);
    }
```

```java
    @Override
    public boolean onStartContinuation() {
        // 重写
        return true;
    }

    @Override
    public boolean onSaveData(IntentParams intentParams) {
        // 重写
        return true;
    }

    @Override
    public boolean onRestoreData(IntentParams intentParams) {
        // 重写
        return true;
    }

    @Override
    public void onCompleteContinuation(int i) {

    }
}
```

29.6.3　设计界面与布局

修改ability_main.xml布局文件如下：

```xml
<?xml version="1.0" encoding="utf-8"?>
<DirectionalLayout
    xmlns:ohos="http://schemas.huawei.com/res/ohos"
    ohos:height="match_parent"
    ohos:width="match_parent"
    ohos:alignment="center"
    ohos:orientation="vertical">

    <TextField
        ohos:id="$+id:message_textfield"
        ohos:height="match_content"
        ohos:width="match_parent"
        ohos:hint="输入信息"
        ohos:top_margin="100vp"
        ohos:left_margin="24vp"
        ohos:padding="5vp"
        ohos:background_element="$graphic:background_ability_main"
        ohos:text_alignment="center"
        ohos:right_margin="24vp"
        ohos:text_size="40fp"/>
```

```xml
<Button
    ohos:id="$+id:button_continue_remote_fa"
    ohos:height="match_content"
    ohos:width="match_content"
    ohos:top_margin="8vp"
    ohos:background_element="#F76543"
    ohos:layout_alignment="horizontal_center"
    ohos:text="迁移"
    ohos:text_size="40fp"
    />

</DirectionalLayout>
```

最终，界面预览效果如图29-10所示。

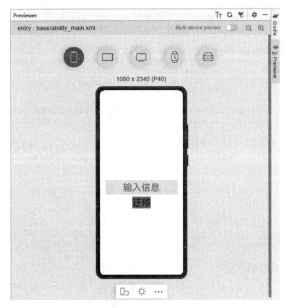

图 29-10　界面预览效果

29.6.4　修改 MainAbilitySlice

核心的逻辑都集中到了MainAbilitySlice中。MainAbilitySlice同样要实现IAbilityContinuation接口，代码如下：

```java
public class MainAbilitySlice extends AbilitySlice implements
IAbilityContinuation {
    private static final String TAG = MainAbilitySlice.class.getSimpleName();
    private static final HiLogLabel LABEL_LOG =
            new HiLogLabel(HiLog.LOG_APP, 0x00001, TAG);

    @Override
    public void onStart(Intent intent) {
```

```java
        super.onStart(intent);
        super.setUIContent(ResourceTable.Layout_ability_main);
    }

    @Override
    public void onActive() {
        super.onActive();
    }

    @Override
    public void onForeground(Intent intent) {
        super.onForeground(intent);
    }

    @Override
    public boolean onStartContinuation() {
        // 重写
        return true;
    }

    @Override
    public boolean onSaveData(IntentParams intentParams) {
        // 重写
        // 保存回迁后恢复状态必需的数据
        intentParams.setParam(MESSAGE_KEY, messageTextField.getText());
        return true;
    }

    @Override
    public boolean onRestoreData(IntentParams intentParams) {
        // 重写
        // 传递此前保存的数据
        if (intentParams.getParam(MESSAGE_KEY) instanceof String) {
            message = (String) intentParams.getParam(MESSAGE_KEY);
            isContinued = true;
        }
        return true;
    }

    @Override
    public void onCompleteContinuation(int i) {
        // 终止
        terminate();
    }
}
```

上述代码的重点是重写onStartContinuation、onSaveData、onRestoreData、onCompleteContinuation等方法。其中,onSaveData用于保存回迁后恢复状态必需的数据,onRestoreData传递此前保存的数据。onCompleteContinuation方法用于终止Page。

接着为"迁移""回迁"按钮设置点击事件，代码如下：

```java
    private static final String MESSAGE_KEY =
            "com.waylau.hmos.continueremotefacollaboration.slice.MESSAGE_KEY";

    private String message;

    private boolean isContinued;

    private TextField messageTextField;

    @Override
    public void onStart(Intent intent) {
        super.onStart(intent);
        super.setUIContent(ResourceTable.Layout_ability_main);

        // 监听跨端迁移FA的事件
        Button buttonContinueRemoteFA = (
                Button) findComponentById(ResourceTable.Id_button_continue_remote_fa);
        buttonContinueRemoteFA.setClickedListener(listener -> continueRemoteFA());

        // 设置输入框内容
        messageTextField = (TextField) findComponentById(ResourceTable.Id_message_textfield);
        if (isContinued && message != null) {
            messageTextField.setText(message);
        }
    }

    private void continueRemoteFA() {
        HiLog.info(LABEL_LOG, "before startRemoteFA");

        String deviceId = DeviceUtils.getDeviceId();

        HiLog.info(LABEL_LOG, "get deviceId: %{public}s", deviceId);

        if (deviceId != null) {
            // 发起迁移流程
            //  continueAbility()是不可回迁的
            //  continueAbilityReversibly()是可以回迁的
            continueAbility(deviceId);
        }
    }
```

29.6.5 新增 DeviceUtils

DeviceUtils代码如下：

```java
package com.waylau.hmos.continueremotefacollaboration;

import ohos.distributedschedule.interwork.DeviceInfo;
import ohos.distributedschedule.interwork.DeviceManager;
import ohos.hiviewdfx.HiLog;
import ohos.hiviewdfx.HiLogLabel;

import java.util.ArrayList;
import java.util.List;

public class DeviceUtils {
    private static final String TAG = DeviceUtils.class.getSimpleName();
    private static final HiLogLabel LABEL_LOG =
            new HiLogLabel(HiLog.LOG_APP, 0x00001, TAG);

    private DeviceUtils() {
    }

    // 获取当前组网下可迁移的设备ID列表
    public static List<String> getAvailableDeviceId() {
        List<String> deviceIds = new ArrayList<>();

        List<DeviceInfo> deviceInfoList =
                DeviceManager.getDeviceList(DeviceInfo.FLAG_GET_ALL_DEVICE);
        if (deviceInfoList == null) {
            return deviceIds;
        }

        if (deviceInfoList.size() == 0) {
            HiLog.warn(LABEL_LOG, "did not find other device");
            return deviceIds;
        }

        for (DeviceInfo deviceInfo : deviceInfoList) {
            deviceIds.add(deviceInfo.getDeviceId());
        }

        return deviceIds;
    }

    // 获取当前组网下可迁移的设备ID
    // 如果有多个，则取第一个
    public static String getDeviceId() {
        String deviceId = "";
```

```
        List<String> outerDevices = DeviceUtils.getAvailableDeviceId();

        if (outerDevices == null || outerDevices.size() == 0) {
           HiLog.warn(LABEL_LOG, "did not find other device");
        } else {
           for (String item : outerDevices) {
              HiLog.info(LABEL_LOG, "outerDevices:%{public}s", item);
           }
           deviceId = outerDevices.get(0);
        }
        HiLog.info(LABEL_LOG, "getDeviceId:%{public}s", deviceId);
        return deviceId;
     }
}
```

DeviceUtils的getDeviceId方法用于获取在线设备列表。如果有多个，则取任意一个。

29.6.6 运行

多了演示多端协同的情况，需要启动Super device，同时启动两个手机模拟器。

假设左侧为设备A，作为源设备；右侧为设备B，作为目标设备。先在设备A、设备B上分别安装应用。

先启动设备A的应用，如图29-11所示。

在设备A上的输入框中输入一些信息，比如"hello"，如图29-12所示。

图 29-11　界面效果

图 29-12　界面效果

而后点击"迁移"按钮，如图29-13所示，Page已经迁移到了设备B上，说明迁移成功了。

在设备B上可以继续对输入框进行编辑，比如输入"world"，如图29-14所示。

图 29-13　界面效果　　　　　　　　　图 29-14　界面效果

而后点击"迁移"按钮,如图29-15所示,Page已经迁移到了设备A上,同时在设备B上输入的内容也迁移到了设备A上,这样就实现了多设备之间的协同。

图 29-15　界面效果

第 30 章

综合案例：俄罗斯方块游戏

本章所演示的是一个完整的手机应用——俄罗斯方块游戏。

30.1 案例概述

本节主要介绍俄罗斯方块游戏及其规则。

30.1.1 俄罗斯方块游戏概述

俄罗斯方块（俄语为Тетрис，英语为Tetris）是一款家喻户晓的游戏，无论是在掌上游戏机，还是在计算机或者移动终端都能见到这款游戏的身影。

俄罗斯方块最初是由阿列克谢·帕基特诺夫在苏联设计和编程的益智类视频游戏。游戏发布于1984年6月6日，那时他在莫斯科，俄罗斯苏维埃联邦社会主义共和国的苏联科学学院Dorodnicyn计算中心。他的名字源于希腊数字4的前缀tetra-（所有游戏中的米诺牌都被称为tetromino，包含4个块），以及网球tennis（帕吉特诺夫喜爱的运动）。

俄罗斯方块游戏随处可见，几乎所有的视频游戏机和计算机操作系统以及图形计算器、小号手机便携式媒介播放器、掌上电脑、网络音乐播放器都可用于玩这个游戏。甚至非媒体产品，诸如示波器也可以玩它，甚至还被拿到建筑物上玩。如图30-1所示的是某大楼所实现的俄罗斯方块游戏。

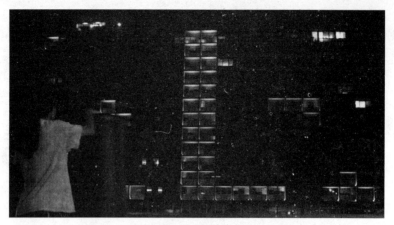

图 30-1　在大楼上所实现的俄罗斯方块游戏

30.1.2　俄罗斯方块游戏的规则

当俄罗斯方块游戏开启后，由小方格组成的不同形状的方块陆续从屏幕上方落下来，玩家通过调整方块的位置和方向使它们在屏幕底部拼出完整的一条或几条。这些完整的横条会随即消失，给新落下来的方块腾出空间，与此同时，玩家得到分数奖励。没有被消除掉的方块不断堆积起来，一旦堆到屏幕顶端，玩家便告输，游戏结束。

本章所实现的俄罗斯方块游戏，玩家可以做如下操作：

- 向左移动方块。
- 向右移动方块。
- 以90度为单位旋转方块。
- 重启游戏。

30.2　代码实现

本节演示如何在手机上实现俄罗斯方块游戏的功能。

为了演示俄罗斯方块游戏的功能，创建一个名为Tetris的应用。

30.2.1　技术重点

要实现俄罗斯方块游戏，涉及以下技术重点：

- 俄罗斯方块游戏布局。
- Canvas类的使用。
- 全屏显示。
- 任务分发器。

30.2.2 设置布局

修改ability_main.xml内容如下:

```xml
<?xml version="1.0" encoding="utf-8"?>
<DirectionalLayout
    xmlns:ohos="http://schemas.huawei.com/res/ohos"
    ohos:height="match_parent"
    ohos:width="match_parent"
    ohos:background_element="#000"
    ohos:orientation="vertical">

    <DirectionalLayout
        ohos:id="$+id:layout_tip"
        ohos:height="60vp"
        ohos:width="match_parent"
        ohos:layout_alignment="top"
        ohos:orientation="horizontal"
        ohos:margin="10vp"
        ohos:padding="10vp"
        >

        <Text
            ohos:id="$+id:text_score_label"
            ohos:height="match_parent"
            ohos:width="80vp"
            ohos:text_color="#FFFFFF"
            ohos:text_size="24fp"
            ohos:text="Score:"/>

        <Text
            ohos:id="$+id:text_score_value"
            ohos:height="match_parent"
            ohos:width="100vp"
            ohos:text_color="#FFFFFF"
            ohos:text_size="24fp"
            ohos:text="0"/>
        <Text
            ohos:id="$+id:text_gameover"
            ohos:height="match_parent"
            ohos:width="match_content"
            ohos:text_color="#FFFFFF"
            ohos:text_size="24fp"
            ohos:text="Start!"/>
    </DirectionalLayout>

    <DirectionalLayout
        ohos:id="$+id:layout_game"
```

```xml
        ohos:height="534vp"
        ohos:width="match_parent"
        ohos:orientation="vertical"
        ohos:background_element="#696969"
        >

</DirectionalLayout>

<DirectionalLayout
    ohos:id="$+id:layout_menu"
    ohos:height="60vp"
    ohos:width="match_parent"
    ohos:layout_alignment="bottom"
    ohos:orientation="horizontal"
    >

    <Button
        ohos:id="$+id:button_left"
        ohos:height="40vp"
        ohos:width="0vp"
        ohos:background_element="#87CEFA"
        ohos:layout_alignment="horizontal_center"
        ohos:margin="10vp"
        ohos:padding="10vp"
        ohos:text="←"
        ohos:text_size="30fp"
        ohos:weight="1"
        />

    <Button
        ohos:id="$+id:button_right"
        ohos:height="40vp"
        ohos:width="0vp"
        ohos:background_element="#87CEFA"
        ohos:layout_alignment="horizontal_center"
        ohos:margin="10vp"
        ohos:padding="10vp"
        ohos:text="→"
        ohos:text_size="30fp"
        ohos:weight="1"
        />

    <Button
        ohos:id="$+id:button_shift"
        ohos:height="40vp"
        ohos:width="0vp"
        ohos:background_element="#87CEFA"
        ohos:layout_alignment="horizontal_center"
        ohos:margin="10vp"
        ohos:padding="10vp"
```

```
            ohos:text="S"
            ohos:text_size="30fp"
            ohos:weight="1"
            />
        <Button
            ohos:id="$+id:button_restart"
            ohos:height="40vp"
            ohos:width="0vp"
            ohos:background_element="#87CEFA"
            ohos:layout_alignment="horizontal_center"
            ohos:margin="10vp"
            ohos:padding="10vp"
            ohos:text="R"
            ohos:text_size="30fp"
            ohos:weight="1"
            />
    </DirectionalLayout>
</DirectionalLayout>
```

界面预览效果如图30-2所示。

图 30-2　界面预览效果

上述布局大体分为3部分：

- 上方为提示区，提示当前得分和游戏状态。
- 中部为游戏画面区。
- 下方为菜单区。从左至右分别代表"左移""右移""转换""重置"4个功能按钮。

30.2.3 设置全屏

运行应用查看界面的实际效果。界面效果如图30-3所示。

图 30-3　界面效果

可以发现，界面上方entry_MainAbility字样的应用名称一栏占据了非常大的空间，极度影响美观，而且导致下方的菜单区显示不全。因此，需要去掉应用名称一栏。

修改配置文件，在module属性中增加metaData属性，修改如下：

```
"module": {
    "package": "com.waylau.hmos.tetris",
    "name": ".MyApplication",
    "mainAbility": "com.waylau.hmos.tetris.MainAbility",
    "deviceType": [
        "phone"
    ],
    "distro": {
        "deliveryWithInstall": true,
        "moduleName": "entry",
        "moduleType": "entry",
        "installationFree": false
    },
    "abilities": [
        {
            "skills": [
                {
                    "entities": [
                        "entity.system.home"
                    ],
                    "actions": [
                        "action.system.home"
                    ]
```

```
            }
        ],
        "orientation": "unspecified",
        "name": "com.waylau.hmos.tetris.MainAbility",
        "icon": "$media:icon",
        "description": "$string:mainability_description",
        "label": "$string:entry_MainAbility",
        "type": "page",
        "launchType": "standard"
      }
    ],
    // 增加metaData属性
    "metaData":{
        "customizeData":[
        {
            "name": "hwc-theme",
            "value": "androidhwext:style/Theme.Emui.NoTitleBar",
            "extra":""
        }
      ]
    }
}
```

再次运行应用，界面全屏效果如图30-4所示。

图 30-4　界面全屏效果

30.2.4　应用的主体逻辑

修改MainAbilitySlice内容，以实现俄罗斯方块游戏的主体逻辑，代码如下：

```
package com.waylau.hmos.tetris.slice;
```

```java
import com.waylau.hmos.tetris.Grid;
import com.waylau.hmos.tetris.ResourceTable;
import ohos.aafwk.ability.AbilitySlice;
import ohos.aafwk.content.Intent;
import ohos.agp.components.*;
import ohos.agp.render.Canvas;
import ohos.agp.render.Paint;
import ohos.agp.utils.Color;
import ohos.agp.utils.RectFloat;
import ohos.app.dispatcher.TaskDispatcher;
import ohos.hiviewdfx.HiLog;
import ohos.hiviewdfx.HiLogLabel;

public class MainAbilitySlice extends AbilitySlice {
    private static final String TAG = MainAbilitySlice.class.getSimpleName();
    private static final HiLogLabel LABEL_LOG =
            new HiLogLabel(HiLog.LOG_APP, 0x00001, TAG);

    private static final int LENGTH = 100;      //方格的边长
    private static final int MARGIN = 2;        //方格的间距
    private static final int GRID_NUMBER = 4;//方块所占方格的数量，固定为4

    private DirectionalLayout layoutGame;//布局
    private Text textScoreValue;
    private Text textGameover;

    private int[][] grids;         //描绘方格颜色的二维数组
    private int currentRow;        //向下移动的行数
    private int currentColumn;//向左右移动的列数，减1表示左移，加1表示右移
    private String tipValue = "Start!";
    private TaskDispatcher taskDispatcher;
    private Grid currentGrid;
    private boolean isRunning;
    private int scoreValue = 0;

    @Override
    public void onStart(Intent intent) {
        super.onStart(intent);
        super.setUIContent(ResourceTable.Layout_ability_main);

        // 初始化布局
        layoutGame =
                (DirectionalLayout) findComponentById(ResourceTable.Id_layout_game);

        // 初始化组件
        Button buttonLeft =
                (Button) findComponentById(ResourceTable.Id_button_left);
        buttonLeft.setClickedListener(listener -> goLeft(grids));
```

```
        Button buttonRight =
                (Button) findComponentById(ResourceTable.Id_button_right);
        buttonRight.setClickedListener(listener -> goRight(grids));

        Button buttonShift =
                (Button) findComponentById(ResourceTable.Id_button_shift);
        buttonShift.setClickedListener(listener -> shiftGrids(grids));

        Button buttonRestart =
                (Button) findComponentById(ResourceTable.Id_button_restart);
        buttonRestart.setClickedListener(listener -> {
            initialize();
            startGame();
        });

        textScoreValue =
                (Text) findComponentById(ResourceTable.Id_text_score_value);
        textGameover =
                (Text) findComponentById(ResourceTable.Id_text_gameover);

        // 初始化游戏
        initialize();

        // 启动游戏
        startGame();
    }
    ...
```

上述代码中:

- onStart方法初始化了DirectionalLayout、Button、Text等组件和布局。
- initialize方法用于初始化游戏的状态和数据。
- startGame方法用于启动游戏。

Grid是方格类,用于表示游戏中的方格,代码如下:

```
package com.waylau.hmos.tetris;

public class Grid {

    private int[][] currentGrids;     //当前方块的形态
    private int rowNumber;            //方块的总行数
    private int columnNumber;         //方块的总列数
    private int currentGridColor;     //当前方格的颜色
    private int columnStart;          //方块的第一个方格所在二维数组的列数

    private Grid(int[][] currentGrids, int rowNumber, int columnNumber,
            int currentGridColor, int columnStart) {
        this.currentGrids = currentGrids;
        this.rowNumber = rowNumber;
```

```
        this.columnNumber = columnNumber;
        this.currentGridColor = currentGridColor;
        this.columnStart = columnStart;
    }
    ...
```

Grid各字段含义详见注释。

30.2.5 初始化游戏

初始化游戏的方法initialize代码如下：

```
private void initialize() {
    // 初始化变量
    scoreValue = 0;
    tipValue = "Start!";
    isRunning = false;
    taskDispatcher = null;

    // 显示提示
    showTip();

    // 初始化网格
    // 15×10的二维数组
    // 数组元素都是0
    grids = new int[15][10];
    for (int row = 0; row < 15; row++) {
        for (int column = 0; column < 10; column++) {
            grids[row][column] = 0;
        }
    }

    // 创建网格数据
    createGrids(grids);

    // 画出网格
    drawGrids(grids);
}
```

上述代码中：

- 初始化了变量，显示提示。
- 初始化网格。该网格本质上是一个15×10的二维数组，数组元素值都是0。
- 同时调用createGrids和drawGrids在界面上画出网格。

showTip方法代码如下：

```
private void showTip() {
    textScoreValue.setText(scoreValue + "");
```

```
            textGameover.setText(tipValue);
    }
```

30.2.6 创建网格数据

创建网格数据的方法createGrids代码如下:

```
    private void createGrids(int[][] grids) {
        currentColumn = 0;
        currentRow = 0;

        // 当有任一行全部填满颜色方块时,消去该行
        eliminateGrids(grids);

        // 判断游戏是否完成
        // 完成时就停止定时器,并提示结束文本
        // 未完成时就生成新的颜色方块
        if (isGameOver(grids)) {
            tipValue = "Game Over!";
            isRunning = false;
            taskDispatcher = null;
            showTip();
        } else {
            //随机生成一个颜色方块
            currentGrid = Grid.generateRandomGrid();
            int[][] currentGrids = currentGrid.getCurrentGrids();
            int currentGridColor = currentGrid.getCurrentGridColor();

            //将颜色方块对应的Grids添加到二维数组中
            for (int row = 0; row < GRID_NUMBER; row++) {
                grids[currentGrids[row][0] + currentRow][currentGrids[row][1] + currentColumn] = currentGridColor;
            }
        }
    }
```

上述代码中:

- 判断是否有任一行全部填满颜色方块,如果有则执行eliminateGrids方法消去该行。
- 调用isGameOver方法进行游戏是否完成的判断:
  - ➢ 完成时就停止定时器,并提示结束文本。
  - ➢ 未完成时就生成新的颜色方块初始化网格。
- Grid.generateRandomGrid()用来生成随机的方块。

isGameOver方法代码如下:

```
    private boolean isGameOver(int[][] grids) {
        // 若新生成的颜色方块覆盖原有的颜色方块,则游戏结束
        if (currentGrid != null) {
```

```java
            int[][] currentGrids = currentGrid.getCurrentGrids();

            for (int row = 0; row < GRID_NUMBER; row++) {
                if (grids[currentGrids[row][0] + currentRow][currentGrids[row][1] + currentColumn] != 0) {
                    return true;
                }
            }
        }

        return false;
    }
```

eliminateGrids方法代码如下：

```java
private void eliminateGrids(int[][] grids) {
    boolean k;
    for (int row = 14; row >= 0; row--) {
        k = true;

        //判断是否有任一行全部填满颜色方块
        for (int column = 0; column < 10; column++) {
            if (grids[row][column] == 0) {
                k = false;
            }
        }

        //消去全部填满颜色方块的行
        if (k) {
            // 加分
            this.scoreValue++;
            this.showTip();

            // 且所有方格向下移动一格
            for (int i = row - 1; i >= 0; i--) {
                for (int j = 0; j < 10; j++) {
                    grids[i + 1][j] = grids[i][j];
                }
            }
            for (int n = 0; n < 10; n++) {
                grids[0][n] = 0;
            }
        }
    }

    drawGrids(grids);
}
```

eliminateGrids方法内部先判断是否有任一行全部填满颜色方块，有则消去全部填满颜色方块的行，同时得分scoreValue加1，并且所有的方格向下移动一格。

Grid.generateRandomGrid()方法代码如下：

```java
package com.waylau.hmos.tetris;

public class Grid {
    private static final int[][] RedGrids1 =
        {{0, 3}, {0, 4}, {1, 4}, {1, 5}};//红色方块形态1
    private static final int[][] RedGrids2 =
        {{0, 5}, {1, 5}, {1, 4}, {2, 4}};//红色方块形态2
    private static final int[][] GreenGrids1 =
        {{0, 5}, {0, 4}, {1, 4}, {1, 3}};//绿色方块形态1
    private static final int[][] GreenGrids2 =
        {{0, 4}, {1, 4}, {1, 5}, {2, 5}};//绿色方块形态2
    private static final int[][] CyanGrids1 =
        {{0, 4}, {1, 4}, {2, 4}, {3, 4}};//蓝绿色方块形态1
    private static final int[][] CyanGrids2 =
        {{0, 3}, {0, 4}, {0, 5}, {0, 6}};//蓝绿色方块形态2
    private static final int[][] MagentaGrids1 =
        {{0, 4}, {1, 3}, {1, 4}, {1, 5}};//品红色方块形态1
    private static final int[][] MagentaGrids2 =
        {{0, 4}, {1, 4}, {1, 5}, {2, 4}};//品红色方块形态2
    private static final int[][] MagentaGrids3 =
        {{0, 3}, {0, 4}, {0, 5}, {1, 4}};//品红色方块形态3
    private static final int[][] MagentaGrids4 =
        {{0, 5}, {1, 5}, {1, 4}, {2, 5}};//品红色方块形态4
    private static final int[][] BlueGrids1 =
        {{0, 3}, {1, 3}, {1, 4}, {1, 5}};//蓝色方块形态1
    private static final int[][] BlueGrids2 =
        {{0, 5}, {0, 4}, {1, 4}, {2, 4}};//蓝色方块形态2
    private static final int[][] BlueGrids3 =
        {{0, 3}, {0, 4}, {0, 5}, {1, 5}};//蓝色方块形态3
    private static final int[][] BlueGrids4 =
        {{0, 5}, {1, 5}, {2, 5}, {2, 4}};//蓝色方块形态4
    private static final int[][] WhiteGrids1 =
        {{0, 5}, {1, 5}, {1, 4}, {1, 3}};//白色方块形态1
    private static final int[][] WhiteGrids2 =
        {{0, 4}, {1, 4}, {2, 4}, {2, 5}};//白色方块形态2
    private static final int[][] WhiteGrids3 =
        {{0, 5}, {0, 4}, {0, 3}, {1, 3}};//白色方块形态3
    private static final int[][] WhiteGrids4 =
        {{0, 4}, {0, 5}, {1, 5}, {2, 5}};//白色方块形态4
    private static final int[][] YellowGrids =
        {{0, 4}, {0, 5}, {1, 5}, {1, 4}};//黄色方块形态1

    public static Grid generateRandomGrid() {
        //随机生成一个颜色方块
        double random = Math.random();
        if (random >= 0 && random < 0.2) {
            if (random >= 0 && random < 0.1)
                return createRedGrids1();
            else
                return createRedGrids2();
```

```
        } else if (random >= 0.2 && random < 0.4) {
            if (random >= 0.2 && random < 0.3)
                return createGreenGrids1();
            else
                return createGreenGrids2();
        } else if (random >= 0.4 && random < 0.45) {
            if (random >= 0.4 && random < 0.43)
                return createCyanGrids1();
            else
                return createCyanGrids2();
        } else if (random >= 0.45 && random < 0.6) {
            if (random >= 0.45 && random < 0.48)
                return createMagentaGrids1();
            else if (random >= 0.48 && random < 0.52)
                return createMagentaGrids2();
            else if (random >= 0.52 && random < 0.56)
                return createMagentaGrids3();
            else
                return createMagentaGrids4();
        } else if (random >= 0.6 && random < 0.75) {
            if (random >= 0.6 && random < 0.63)
                return createBlueGrids1();
            else if (random >= 0.63 && random < 0.67)
                return createBlueGrids2();
            else if (random >= 0.67 && random < 0.71)
                return createBlueGrids3();
            else
                return createBlueGrids4();
        } else if (random >= 0.75 && random < 0.9) {
            if (random >= 0.75 && random < 0.78)
                return createWhiteGrids1();
            else if (random >= 0.78 && random < 0.82)
                return createWhiteGrids2();
            else if (random >= 0.82 && random < 0.86)
                return createWhiteGrids3();
            else
                return createWhiteGrids4();
        } else {
            return createYellowGrids();
        }
    }

    // 以下为各种颜色、各种形状的方块
    private static Grid createRedGrids1() {
        return new Grid(RedGrids1, 2, 3, 1, 3);
    }

    private static Grid createRedGrids2() {
        return new Grid(RedGrids2, 3, 2, 1, 4);
    }
```

```java
private static Grid createGreenGrids1() {
    return new Grid(GreenGrids1, 2, 3, 2, 3);
}

private static Grid createGreenGrids2() {
    return new Grid(GreenGrids2, 3, 2, 2, 4);
}

private static Grid createCyanGrids1() {
    return new Grid(CyanGrids1, 4, 1, 3, 4);
}

private static Grid createCyanGrids2() {
    return new Grid(CyanGrids2, 1, 4, 3, 3);
}

private static Grid createMagentaGrids1() {
    return new Grid(MagentaGrids1, 2, 3, 4, 3);
}

private static Grid createMagentaGrids2() {
    return new Grid(MagentaGrids2, 3, 2, 4, 4);
}

private static Grid createMagentaGrids3() {
    return new Grid(MagentaGrids3, 2, 3, 4, 3);
}

private static Grid createMagentaGrids4() {
    return new Grid(MagentaGrids4, 3, 2, 4, 4);
}

private static Grid createBlueGrids1() {
    return new Grid(BlueGrids1, 2, 3, 5, 3);
}

private static Grid createBlueGrids2() {
    return new Grid(BlueGrids2, 3, 2, 5, 4);
}

private static Grid createBlueGrids3() {
    return new Grid(BlueGrids3, 2, 3, 5, 3);
}

private static Grid createBlueGrids4() {
    return new Grid(BlueGrids4, 3, 2, 5, 4);
}

private static Grid createWhiteGrids1() {
```

```
        return new Grid(WhiteGrids1, 2, 3, 6, 3);
    }

    private static Grid createWhiteGrids2() {
        return new Grid(WhiteGrids2, 3, 2, 6, 4);
    }

    private static Grid createWhiteGrids3() {
        return new Grid(WhiteGrids3, 2, 3, 6, 3);
    }

    private static Grid createWhiteGrids4() {
        return new Grid(WhiteGrids4, 3, 2, 6, 4);
    }

    private static Grid createYellowGrids() {
        return new Grid(YellowGrids, 2, 2, 7, 4);
    }

    ...
}
```

30.2.7 绘制网格

绘制网格的方法drawGrids代码如下:

```
private void drawGrids(int[][] grids) {
    Component.DrawTask task = new Component.DrawTask() {
        @Override
        public void onDraw(Component component, Canvas canvas) {
            Paint paint = new Paint();

            for (int row = 0; row < 15; row++) {
                for (int column = 0; column < 10; column++) {
                    // grids的值表示不同的颜色
                    int grideValue = grids[row][column];
                    Color color = null;

                    switch (grideValue) {
                        case 0:
                            color = Color.GRAY;
                            break;
                        case 1:
                            color = Color.RED;
                            break;
                        case 2:
                            color = Color.GREEN;
                            break;
                        case 3:
```

```
                    color = Color.CYAN;
                    break;
                case 4:
                    color = Color.MAGENTA;
                    break;
                case 5:
                    color = Color.BLUE;
                    break;
                case 6:
                    color = Color.WHITE;
                    break;
                case 7:
                    color = Color.YELLOW;
                    break;
                default:
                    break;
            }

            paint.setColor(color);

            RectFloat rectFloat =
                    new RectFloat(30 + column * (LENGTH + MARGIN),
                            40 + row * (LENGTH + MARGIN),
                            30 + LENGTH + column * (LENGTH + MARGIN),
                            40 + LENGTH + row * (LENGTH + MARGIN));

            //绘制方格
            canvas.drawRect(rectFloat, paint);
            }
        }
    }
};

layoutGame.addDrawTask(task);
}
```

上述代码中：

- 通过实现Component.DrawTask来自定义游戏界面组件。
- Canvas类用于绘制画面。
- 游戏界面组件最终添加到layoutGame布局上。

30.2.8 启动游戏

startGame()方法用于启动游戏，代码如下：

```
private void startGame() {
    // 设置游戏状态
    isRunning = true;
```

```
    // 派发任务
    taskDispatcher = getUITaskDispatcher();
    taskDispatcher.asyncDispatch(new GameTask());
}
```

上述代码中：

- 设置游戏状态isRunning为true。
- 通过getUITaskDispatcher获取任务分发器。
- 任务分发器调用GameTask的执行。

GameTask类代码如下：

```
private class GameTask implements Runnable {

    @Override
    public void run() {
        HiLog.info(LABEL_LOG, "before GameTask run");

        int[][] currentGrids = currentGrid.getCurrentGrids();
        int currentGridColor = currentGrid.getCurrentGridColor();
        int rowNumber = currentGrid.getRowNumber();

        //如果方块能下移则下移，否则重新随机生成新的方块
        if (couldDown(currentGrids, rowNumber)) {
            //将原来的颜色方块清除
            for (int row = 0; row < GRID_NUMBER; row++) {
                grids[currentGrids[row][0] + currentRow][currentGrids[row][1] + currentColumn] = 0;
            }

            currentRow++;

            //重新绘制颜色方块
            for (int row = 0; row < GRID_NUMBER; row++) {
                grids[currentGrids[row][0] + currentRow][currentGrids[row][1] + currentColumn] = currentGridColor;
            }
        } else {
            createGrids(grids);
        }

        drawGrids(grids);

        if (isRunning) {
            taskDispatcher.delayDispatch(this, 750);
        }

        HiLog.info(LABEL_LOG, "end GameTask run");
    }
```

上述代码中：

- 先判断方块能否下移，如果能则执行下移，否则重新随机生成新的方块。
- 通过isRunning判断游戏是否还在执行，如果是则执行分发新任务，从而实现任务的定时执行。

couldDown方法代码如下：

```java
private boolean couldDown(int[][] currentGrids, int rowNumber) {
    boolean k;

    // 如果方块向下移动到下边界，则返回false
    if (currentRow + rowNumber == 15) {
        return false;
    }

    //当下边缘方块再下一格为空时，则可以下移
    for (int row = 0; row < GRID_NUMBER; row++) {
        k = true;
        for (int i = 0; i < GRID_NUMBER; i++) {
            if (currentGrids[row][0] + 1 == currentGrids[i][0]
                    && currentGrids[row][1] == currentGrids[i][1]) {//找出非下边缘方块
                k = false;
            }
        }

        //当任一下边缘方块再下一格不为空时，返回false
        if (k) {
            if (grids[currentGrids[row][0] + currentRow + 1][currentGrids[row][1] + currentColumn] != 0)
                return false;
        }
    }

    return true;
}
```

30.2.9 左移操作

goLeft方法用于响应"左移"按钮操作，代码如下：

```java
private void goLeft(int[][] grids) {
    int currentGridColor = currentGrid.getCurrentGridColor();
    int[][] currentGrids = currentGrid.getCurrentGrids();
    int columnStart = currentGrid.getColumnStart();

    //当方块能向左移动时则左移
    if (couldLeft(currentGrids, columnStart)) {
```

```
            //将原来的颜色方块清除
            for (int row = 0; row < GRID_NUMBER; row++) {
                grids[currentGrids[row][0] + currentRow][currentGrids[row][1] + currentColumn] = 0;
            }

            currentColumn--;

            //重新绘制颜色方块
            for (int row = 0; row < GRID_NUMBER; row++) {
                grids[currentGrids[row][0] + currentRow][currentGrids[row][1] + currentColumn] = currentGridColor;
            }
        }

        drawGrids(grids);
    }
```

上述代码中：

- 先判断当前方块能否执行向左移动，如果能则执行左移。
- 左移的实际逻辑是，将原来的颜色方块清除，而后重新绘制颜色方块。

couldLeft方法代码如下：

```
    private boolean couldLeft(int[][] currentGrids, int columnStart) {
        boolean k;

        //如果方块向左移动到左边界，则返回false
        if (currentColumn + columnStart == 0) {
            return false;
        }

        //当左边缘方块再左一格为空时，则可以左移
        for (int column = 0; column < GRID_NUMBER; column++) {
            k = true;
            for (int j = 0; j < GRID_NUMBER; j++) {
                //找出非左边缘方块
                if (currentGrids[column][0] == currentGrids[j][0]
                        && currentGrids[column][1] - 1 == currentGrids[j][1]) {
                    k = false;
                }
            }

            //当任一左边缘方块再左一格不为空时，则返回false
            if (k) {
                if (grids[currentGrids[column][0] + currentRow][currentGrids[column][1] + currentColumn - 1] != 0)
                    return false;
            }
        }
```

```
        return true;
    }
```

30.2.10 右移操作

goRight方法用于响应"右移"按钮操作,代码如下:

```
private void goRight(int[][] grids) {
    int currentGridColor = currentGrid.getCurrentGridColor();
    int[][] currentGrids = currentGrid.getCurrentGrids();
    int columnNumber = currentGrid.getColumnNumber();
    int columnStart = currentGrid.getColumnStart();

    //当方块能向右移动时则右移
    if (couldRight(currentGrids, columnNumber, columnStart)) {
        //将原来的颜色方块清除
        for (int row = 0; row < GRID_NUMBER; row++) {
            grids[currentGrids[row][0] + currentRow][currentGrids[row][1] + currentColumn] = 0;
        }

        currentColumn++;

        //重新绘制颜色方块
        for (int row = 0; row < GRID_NUMBER; row++) {
            grids[currentGrids[row][0] + currentRow][currentGrids[row][1] + currentColumn] = currentGridColor;
        }
    }

    drawGrids(grids);
}
```

上述代码中:

- 先判断当方块能否执行向右移动,如果能则执行右移。
- 右移的实际逻辑是,将原来的颜色方块清除,而后重新绘制颜色方块。

couldRight方法代码如下:

```
private boolean couldRight(int[][] currentGrids, int columnNumber, int columnStart) {
    boolean k;

    //如果方块向右移动到右边界,则返回false
    if (currentColumn + columnNumber + columnStart == 10) {
        return false;
    }
```

```
        //当右边缘方块再右一格为空时，则可以右移
        for (int column = 0; column < GRID_NUMBER; column++) {
            k = true;
            for (int j = 0; j < GRID_NUMBER; j++) {
                //找出非右边缘方块
                if (currentGrids[column][0] == currentGrids[j][0]
                        && currentGrids[column][1] + 1 == currentGrids[j][1]) {
                    k = false;
                }
            }

            //当任一右边缘方块再右一格不为空时，则返回false
            if (k) {
                if (grids[currentGrids[column][0] + currentRow][currentGrids[column][1] + currentColumn + 1] != 0)
                    return false;
            }
        }

        return true;
    }
```

30.2.11 转换操作

shiftGrids方法用于响应"转换"按钮操作，代码如下：

```
    private void shiftGrids(int[][] grids) {
        int[][] currentGrids = currentGrid.getCurrentGrids();
        int columnNumber = currentGrid.getColumnNumber();
        int columnStart = currentGrid.getColumnStart();

        //将原来的颜色方块清除
        for (int row = 0; row < GRID_NUMBER; row++) {
            grids[currentGrids[row][0] + currentRow][currentGrids[row][1] + currentColumn] = 0;
        }

        if (columnNumber == 2 && currentColumn + columnStart == 0) {
            currentColumn++;
        }

        //根据Grids的颜色值调用改变方块形状的chang+Color+Grids函数
        currentGrid = Grid.shiftGrid(currentGrid);
        int currentGridColor = currentGrid.getCurrentGridColor();
        currentGrids = currentGrid.getCurrentGrids();

        //重新绘制颜色方块
        for (int row = 0; row < GRID_NUMBER; row++) {
            grids[currentGrids[row][0] + currentRow][currentGrids[row][1] +
```

```
currentColumn] = currentGridColor;
        }

        drawGrids(grids);
    }
```

上述代码中：

- 转换的实际逻辑是，将原来的颜色方块清除，而后重新绘制颜色方块。
- 新的颜色方块通过Grid.shiftGrid方法获得。

Grid.shiftGrid方法代码如下：

```
public static Grid shiftGrid(Grid grid) {
    int currentGridColor = grid.getCurrentGridColor();
    int[][] currentGrids = grid.getCurrentGrids();

    //根据Grids的颜色值调用改变方块形状的chang+Color+Grids函数
    if (currentGridColor == 1) {
        return changRedGrids(currentGrids);
    } else if (currentGridColor == 2) {
        return changeGreenGrids(currentGrids);
    } else if (currentGridColor == 3) {
        return changeCyanGrids(currentGrids);
    } else if (currentGridColor == 4) {
        return changeMagentaGrids(currentGrids);
    } else if (currentGridColor == 5) {
        return changeBlueGrids(currentGrids);
    } else if (currentGridColor == 6) {
        return changeWhiteGrids(currentGrids);
    } else {
        return null;
    }
}

private static Grid changRedGrids(int[][] currentGrids) {
    if (currentGrids == RedGrids1) {
        return createRedGrids2();
    } else if (currentGrids == RedGrids2) {
        return createRedGrids1();
    } else {
        return null;
    }
}

private static Grid changeGreenGrids(int[][] currentGrids) {
    if (currentGrids == GreenGrids1) {
        return createGreenGrids2();
    } else if (currentGrids == GreenGrids2) {
        return createGreenGrids1();
    } else {
```

```java
            return null;
        }
    }

    private static Grid changeCyanGrids(int[][] currentGrids) {
        if (currentGrids == CyanGrids1) {
            return createCyanGrids2();
        } else if (currentGrids == CyanGrids2) {
            return createCyanGrids1();
        } else {
            return null;
        }
    }

    private static Grid changeMagentaGrids(int[][] currentGrids) {
        if (currentGrids == MagentaGrids1) {
            return createMagentaGrids2();
        } else if (currentGrids == MagentaGrids2) {
            return createMagentaGrids3();
        } else if (currentGrids == MagentaGrids3) {
            return createMagentaGrids4();
        } else if (currentGrids == MagentaGrids4) {
            return createMagentaGrids1();
        } else {
            return null;
        }
    }

    private static Grid changeBlueGrids(int[][] currentGrids) {
        if (currentGrids == BlueGrids1) {
            return createBlueGrids2();
        } else if (currentGrids == BlueGrids2) {
            return createBlueGrids3();
        } else if (currentGrids == BlueGrids3) {
            return createBlueGrids4();
        } else if (currentGrids == BlueGrids4) {
            return createBlueGrids1();
        } else {
            return null;
        }
    }

    private static Grid changeWhiteGrids(int[][] currentGrids) {
        if (currentGrids == WhiteGrids1) {
            return createWhiteGrids2();
        } else if (currentGrids == WhiteGrids2) {
            return createWhiteGrids3();
        } else if (currentGrids == WhiteGrids3) {
            return createWhiteGrids4();
        } else if (currentGrids == WhiteGrids4) {
```

```
            return createWhiteGrids1();
        } else {
            return null;
        }
}
```

30.2.12　重置操作

"重置"按钮操作执行的代码如下：

```
Button buttonRestart =
            (Button) findComponentById(ResourceTable.Id_button_restart);
buttonRestart.setClickedListener(listener -> {
    initialize();
    startGame();
});
```

上述操作的本质是再执行一次initialize和startGame方法。

30.3　应用运行

初次启动应用可以直接进行游戏。游戏界面如图30-5所示。

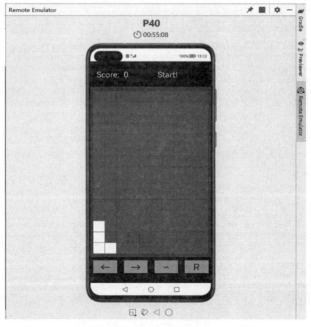

图 30-5　界面效果

当游戏结束时，则会提示Game Over!字样并统计得分，界面如图30-6所示。

图 30-6　界面效果

点击"重置"按钮可以重新开始游戏。

参考文献

[1] HarmonyOS文档[EB/OL].https://developer.harmonyos.com/cn/docs/documentation/doc-guides/harmonyos-features-0000000000011907，2021-01-01/2021-09-24.

[2] 柳伟卫. Node.js企业级应用开发实战[M]. 北京：北京大学出版社，2020.

[3] 柳伟卫. 跟老卫学HarmonyOS开发[EB/OL].https://github.com/waylau/harmonyos-tutorial，2020-12-13/2021-09-24.

[4] HarmonyOS应用开发系列课（进阶篇）[EB/OL].https://developer.huaweiuniversity.com/courses/course-v1:HuaweiX+CBGHWDCN103+Self-paced，2021-01-10/2021-01-10.

[5] 柳伟卫. 分布式系统常用技术及案例分析[M]. 北京：电子工业出版社，2017.

[6] 柳伟卫. Java核心编程[M]. 北京：清华大学出版社，2020.

[7] 柳伟卫. Netty原理解析与开发实战[M]. 北京：北京大学出版社，2020.

[8] 二维码[EB/OL].https://baike.baidu.com/item/%E4%BA%8C%E7%BB%B4%E7%A0%81，2021-02-10/2021-02-10.

[9] 光学字符识别[EB/OL].https://baike.baidu.com/item/%E5%85%89%E5%AD%A6%E5%AD%97%E7%AC%A6%E8%AF%86%E5%88%AB，2021-02-10/2021-02-10.

[10] 柳伟卫. Cloud Native 分布式架构原理与实践[M]. 北京：北京大学出版社，2019.

[11] 无线局域网[EB/OL].https://baike.baidu.com/item/%E6%97%A0%E7%BA%BF%E5%B1%80%E5%9F%9F%E7%BD%91，2021-02-10/2021-02-10.